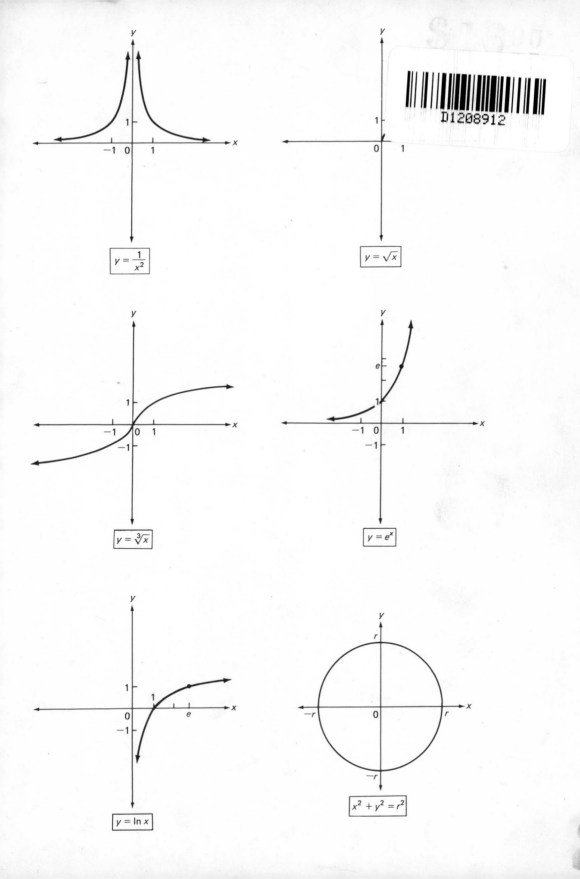

$$y = \frac{1}{x^2}$$

$$y = \sqrt{x}$$

$$y = \sqrt[3]{x}$$

$$y = e^x$$

$$y = \ln x$$

$$x^2 + y^2 = r^2$$

Algebra and Trigonometry

Algebra and

MAX A. SOBEL
Montclair State College

NORBERT LERNER
State University of New York at Cortland

Two

Fundamentals of Algebra 31

Three

Linear Functions 95

Four

Quadratic Functions 151

Trigonometry
A Pre-Calculus Approach

Prentice-Hall, Inc., Englewood Cliffs, New Jersey 07632

Library of Congress Cataloging in Publication Data

SOBEL, MAX A
   Algebra and trigonometry.

   Includes index.
   1. Algebra. 2. Trigonometry. I. Lerner,
Norbert, joint author. II. Title.
QA154.2.S59   512'.13   78–21530
ISBN 0–13–021709–3

ALGEBRA AND TRIGONOMETRY: A Pre-Calculus Approach

Max A. Sobel ╱ Norbert Lerner

Printed in the United States of America

10 9 8 7 6 5 4 3 2

Editorial/production supervision by Reynold R. Rieger
Interior design by Mark A. Binn and Reynold R. Rieger
Cover design by Mark A. Binn
Manufacturing buyer: Phil Galea
Cover illustration: Reprinted with permission from *Human Nature* Magazine,
      Copyright © 1977, Human Nature, Inc.
      Artist: Norman Adams

PRENTICE-HALL INTERNATIONAL, INC., *London*

PRENTICE-HALL OF AUSTRALIA PTY. LIMITED, *Sydney*

PRENTICE-HALL OF CANADA, LTD., *Toronto*

PRENTICE-HALL OF INDIA PRIVATE LIMITED, *New Delhi*

PRENTICE-HALL OF JAPAN, INC., *Tokyo*

PRENTICE-HALL OF SOUTHEAST ASIA PTE. LTD., *Singapore*

WHITEHALL BOOKS LIMITED, *Wellington, New Zealand*

Contents

One

Real Numbers, Equations, Inequalities   1

Five

Polynomial and Rational Functions   205

Six

Radical Functions   247

Seven

Exponential and Logarithmic Functions   281

Eight

Trigonometry   325

Nine

The Circular Functions   374

Ten

Sequences and Series   421

Tables   467

Answers to Odd-Numbered Exercises   479

Index   531

Preface

As the title suggests, *Algebra and Trigonometry: A Pre-Calculus Approach* is a multipurpose text. It can be used for the more traditional algebra and trigonometry courses, as well as for courses whose major objective is to give direct preparation for the study of calculus. Thus a major objective of this book is to provide a text that will help students make a more comfortable transition from elementary mathematics to calculus. All too often, the typical pre-calculus (PC) courses fall short in this regard; too many students are unable to adapt their knowledge of basic mathematics to calculus.

When calculus students need to spend excessive time and effort on review, they are easily frustrated and discouraged. They are seldom in a position to appreciate the beauty and strength of calculus because they are too busy with side issues.

The authors do not claim to have solved this complex problem of transition, but they do believe that some positive results can be achieved with a change in emphasis of some of the usual topics, as well as the inclusion of other material at the PC level that can be described as being *directly supportive* of topics in calculus. A few of these *supportive* items are:

Calculus Topic: The concept of a derivative.

PC Preparation: Working with difference quotients is included in a number of places. An optional section presents an informal treatment of the tangent line problem.

Calculus Topic: Using the signs of derivatives.
PC Preparation: Inequalities are considered early in the book. These concepts
are applied later to determine the signs of a variety of functions. A convenient tabular format is used throughout that
can easily be extended for working with signs of derivatives
when the student gets to calculus. Special algebraic support
material is also included.

Calculus Topic: The chain rule for derivatives.
PC Preparation: In addition to the usual work in forming composites of given
functions, special material is included that shows how to
reverse this process. For example, the student learns how to
view a *given* function, such as $f(x) = \sqrt{(2x - 1)^5}$, as the
composition of other functions. Much of the difficulty students
have later with the chain rule appears to be related to the
inability to do this type of decomposition.

The text contains a strong emphasis on graphing throughout. The objective
is to have students become familiar with basic graphs when they come to the
calculus. (For easy reference, the key graphs are reproduced on the inside covers
of the book.) The student also learns how to obtain new curves using translations,
reflections, and the like.

Improving the ability to read mathematics should be a major goal of a PC
course. To assist the student in this direction the exposition is presented in a
relaxed style that avoids strict mathematical jargon, yet maintains mathematical
accuracy. Also, in order to develop the student's self-study habits, a number of
pedagogical features have been included:

1. *Test Your Understanding:*   These are short exercise sets (in addition to
the regular section exercises) found within many sections so that the student can
test knowledge of new material just developed. Answers are given for these at the
end of each chapter.

2. *Caution Items:*   Where appropriate, students are alerted to the typical
kinds of errors that they should learn to avoid.

3. *Review Exercises:*   Each chapter has a set of review exercises that are keyed
directly to the illustrative examples developed in that chapter. Therefore, the
worked-out solutions for these review exercises are available to the student in
the body of the chapter.

4. *Chapter Tests:*   A sample chapter test is given at the end of each chapter.
The answers are in the back of the book.

5. *Illustrative Examples and Exercises:*   The text contains approximately
300 illustrative examples with detailed solutions. In addition there are approximately 3000 exercises for the student to try, with answers to the odd-numbered
problems given at the back of the book.

The authors believe that this book offers the opportunity to conduct a PC course that can result in a more effective preparation for the study of calculus than is ordinarily achieved by such courses. As an additional aid, an Instructor's Manual is available that includes answers to the even-numbered problems, alternate forms for each chapter test, chapter commentaries, and suggestions for course structure.

The authors hope that you will find this book teachable and enjoyable, and welcome your comments, criticisms, and suggestions.

Acknowledgments

The preparation of this material would not have been possible without the aid of many people to whom we owe a sincere note of thanks.

For their patience and cooperation in using preliminary manuscripts, we thank our students.

For their valuable detailed reviews, criticisms, and suggestions we thank the reviewers: Samuel Councilman, California State University at Long Beach; Leonard Deaton, California State University at Los Angeles; Mark Hale, University of Florida; James Hall, University of Wisconsin at Madison; and Ann Megaw, University of Texas at Austin.

For their opinions and overall contributions, we thank our colleagues at the State University of New York at Cortland: Paul Depue, Paul Mosbo, and especially George Feissner.

For their encouragement, support, and assistance we thank the staff at Prentice-Hall, especially Harry Gaines and Reynold Rieger.

And we thank Karin Lerner for her excellent typing of numerous versions of the manuscript as well as the accompanying Instructor's Manual.

MAX A. SOBEL
NORBERT LERNER

Suggestions for the Student

You probably are taking this course to prepare you for some future course in mathematics, such as calculus. The study of calculus is not easy. One of the major problems that students encounter involves a lack of skill with topics that should have been mastered in advance. It is this necessary preparation that we shall deal with throughout this course.

There will probably be many topics covered in this book that you have seen before in one form or another. Be careful! Don't allow yourself to become lazy because of such familiarity. All too often students will relax their efforts because of such assumed familiarity, and fail to recognize that the development of skill in these prerequisite topics is the major objective of preparatory courses such as this. They then come to the calculus or other advanced courses no better prepared than if they had never taken a pre-calculus course in the first place.

How does one succeed in a mathematics course? Unfortunately, there is no universal prescription guaranteed to work. However, the experience that the authors have had with many thousands of students throughout their teaching careers provides some guidelines that seem to help. We suggest that you make an effort to follow these suggestions that have been useful to other students who have taken similar courses in the past:

1. Read the text! We recognize that mathematics is not always easy to read, but we have made every effort to make this book as readable as possible. Don't look upon the textbook merely as a source of exercises. Stay up to date and read each section thoroughly before attempting to complete any assignment of exercises.

2. Try to complete the illustrative examples that appear within each section before studying the solutions provided. Mathematics is not "a spectator sport"; study the book with paper and pencil at hand.

3. Attempt each "Test Your Understanding" exercise, and check your results with the answers given at the back of each chapter. Re-read the section if you have difficulty with these exercises.

4. Try to complete as many of the exercises as possible at the end of each chapter. Complete the odd-numbered ones first, and check your answers with those given at the back of the book. At times your answer may be in a different form than that given in the book; if so, try to show that the two results are equivalent. Don't worry if you miss an occasional problem. And if you happen to find an incorrect answer at times, please don't hesitate to write and let us know.

5. Prior to a test you should make use of the review exercises that appear at the end of each chapter. These are collections of representative examples from within each chapter. You can check your results by referring back to the designated section from which they are taken where you will find the completely worked-out solution for each one.

6. Each chapter ends with a sample test that will help tell you whether or not you have mastered the work of that chapter. Answers to these are available at the back of the book. You may also use these tests as a preview of a particular chapter if you feel that you are familiar with the work contained therein. If you are able to complete the test with a score of 90%, you can feel comfortable about that particular segment of material.

We hope that you will have a profitable semester studying from this book. Good luck!

MAX A. SOBEL
NORBERT LERNER

Algebra and Trigonometry

One

Real Numbers, Equations, Inequalities

You first learned to count with a collection of numbers called the set of **natural numbers**, or **counting numbers**.

$$1, 2, 3, 4, 5, \ldots$$

It is always possible to add or to multiply two counting numbers and obtain an answer that is also a counting number. However, subtraction and division are not always possible if we are restricted to this collection of numbers. For example, $3 - 5$ and $3 \div 5$ do not have answers that are counting numbers.

In order to make subtraction always possible, we extend the set of counting numbers to obtain the set of **integers**.

$$\ldots, -3, -2, -1, 0, 1, 2, 3, \ldots$$

Here we assume that the *positive integers* are the same as the counting numbers; that is, $+3 = 3$. Using the set of integers, subtraction is now always possible. For example, $3 - 5 = -2$, an integer.

We can provide a geometric interpretation of the set of integers by associating them with points on a **number line**. This is done by selecting a point on a line, the **origin**, and marking it 0. Then select another point to the right of 0 and mark it 1. Using the distance between these two points as the **unit length**, mark off equally spaced segments. The positive integers are to the right of 0, and the *negative integers* are to the left of 0. Every integer is the coordinate of some point on the number line.

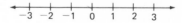

To make division always possible we extend the set of integers to obtain the set of **rational numbers**.

> A **rational number** is one that can be written in the form $\dfrac{a}{b}$, where a and b are integers, $b \neq 0$.

Now division, except for division by 0, is always possible. Thus, $3 \div 5 = \frac{3}{5}$, a rational number. Note that every integer is a rational number because every integer n can be expressed in the form $\dfrac{n}{1}$.

With the set of rational numbers we can always add, subtract, multiply, or divide (except by 0) any two rational numbers and always obtain an answer that is also a rational number. The set of rational numbers can also be associated with points on a number line. As an example, let us locate the "thirds" in this way. Divide each interval between successive integers into three equal parts. This procedure will locate the points corresponding to the rational numbers of the form $\frac{n}{3}$, where n is any integer.

Note that every rational number is the **coordinate** (name) of some point on the number line. However, not every point on the number line can be named by a rational number. For example, here is a construction that can be used to locate a point on the number line whose coordinate is $\sqrt{2}$, the *square root* of 2.

Begin by constructing a perpendicular segment of unit length at 1 on the number line, as shown in the figure. By the Pythagorean theorem the hypotenuse c of the resulting right triangle is of length $\sqrt{2}$. Use the origin, 0, as center and $c = \sqrt{2}$ as radius and mark off an arc of length $\sqrt{2}$ on the number line as shown.

$$c = \sqrt{1^2 + 1^2}$$
$$= \sqrt{2}$$

Thus we have located a point on the number line with coordinate $\sqrt{2}$. However, $\sqrt{2}$ is *not* a rational number; it cannot be written as the quotient of two integers. We call $\sqrt{2}$ an **irrational number**. Here are some other examples of irrational numbers:

$$\sqrt{5} \qquad \sqrt{18} \qquad \pi \qquad \sqrt[3]{2} \qquad \sqrt[3]{9}$$

When we combine the set of rational numbers and the set of irrational numbers, we finally have the collection of **real numbers** which will suffice for most of our work in this text.

Once we have developed the set of real numbers, we also have the **real number line**, which we refer to simply as the **number line**. Every point on the number line may be named by a real number, and each real number is the coordinate of a point on the number line. We say that there is a *one-to-one correspondence* between the set of real numbers and the set of points on the number line. Here is a number line with some of its points named by real numbers.

Example 1   Graph on a number line: $-2, -\frac{3}{2}, -1, -\frac{1}{2}, 0, \frac{1}{2}, 1$.

Solution

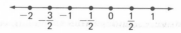

Example 2   Classify each number into as many different sets as possible: **(a)** 5; **(b)** $\frac{2}{3}$;   **(c)** $\sqrt{7}$.

Solution
(a) 5 is a counting number, an integer, a rational number, and a real number.
(b) $\frac{2}{3}$ is a rational number and a real number.
(c) $\sqrt{7}$ is an irrational number and a real number.

EXERCISES 1.1

For Exercises 1 through 10 classify each statement as true or false.

1. The number zero is a real number.
2. Every integer is a rational number.
3. Every rational number is an integer.
4. The difference between any two integers is an integer.
5. The quotient of any two integers is an integer.
6. The number $\sqrt{9}$ is an irrational number.
7. Every counting number is an integer.
8. Every irrational number is a real number.
9. The quotient of two nonzero rational numbers is a rational number.
10. Every real number is a rational or an irrational number.

For Exercises 11 through 20 check the boxes in the table to indicate the set to which each number belongs.

	Counting Numbers	Integers	Rational Numbers	Irrational Numbers	Real Numbers	
Example -15		✓	✓		✓	
11. -1		✓	✓		✓	ok
12. 72	✓	✓	✓		✓	
13. $\sqrt{5}$				✓	✓	ok
14. $-\frac{3}{4}$			✓		✓	
15. $\frac{16}{2}$ 8	✓	✓	✓		✓	ok
16. $\sqrt[3]{49}$	✓	✓	✓		✓	
17. 0.01 $\frac{1}{100}$			✓		✓	ok
18. 1000	✓	✓	✓		✓	
19. $\sqrt{12}$				✓	✓	ok
20. $\sqrt{\frac{9}{16}}$ $\frac{3}{4}$			✓		✓	

Graph each set on a number line.

21. $-3, -2, -1, 0, 1, 2$
22. $0, 1, 3, \frac{7}{2}, 4$
23. $-1, -\frac{2}{3}, -\frac{1}{3}, 0, \frac{1}{3}, \frac{2}{3}, 1, \frac{4}{3}$
24. $-3, -1, 2, \frac{5}{2}, 3$
25. $-\frac{5}{4}, -1, -\frac{1}{4}, 0, \frac{3}{4}, \frac{7}{4}$

*26. Draw a number line and locate a point with coordinate $\sqrt{2}$, as shown in this section. At this point construct a perpendicular segment of length 1. Now construct a right triangle with hypotenuse of length $\sqrt{3}$ and use this construction to locate a point on the number line with coordinate $\sqrt{3}$.

*Throughout the text, an asterisk will be used to indicate that an exercise is more difficult than usual or involves some unusual aspect.

*27. Locate a point on the number line with coordinate $\sqrt{5}$.

*28. Locate a point on the number line with coordinate $-\sqrt{2}$.

*29. Take a circular object (like a tin can) and let the diameter be 1 unit. Use this unit on a number line. Mark a point on the circular object with a dot and place the circle so that the dot coincides with zero on the number line. Roll the circle to the right and mark the point where the dot coincides with the number line after one revolution. What number corresponds to this position? Why?

*30. From a mathematical point of view, it is common practice to introduce the integers after the counting numbers have been discussed and then develop the rational numbers. Historically, however, the positive rational numbers were developed before the negative integers.

   (a) Think of some everyday situations where the negative integers are not usually used but could easily be introduced.

   (b) Speculate on the historical necessity for the development of the positive rational numbers.

1.2

Basic Properties of the Real Numbers

You are undoubtedly familiar with many of the basic properties of the real numbers, though you may not know their specific names. Here is a list of some of these important properties that you may use for reference. In each case the letters, called *variables*, are used to represent real numbers.

Closure Properties

The sum and product of two real numbers is a real number. That is, for real numbers a and b

$$a + b \text{ is a real number}$$

$$a \times b \text{ is a real number}$$

Note that $a \times b$ may be written in a number of other ways, such as $a \cdot b$, $(a)(b)$, $a(b)$, $(a)b$, or just ab.

Commutative Properties

The sum and product of two real numbers is not affected by the order in which they are combined. That is, for real numbers a and b

$$a + b = b + a$$

$$a \times b = b \times a$$

Associative Properties

The sum and product of three real numbers is the same when the third is combined with the first two or when the first is combined with the last two. That is, for real numbers a, b, and c

$$(a + b) + c = a + (b + c)$$
$$(a \times b) \times c = a \times (b \times c)$$

Distributive Property

The product of a real number times the sum of two others is the same as the sum of the products of the first number times each of the others. That is, for real numbers a, b, and c

$$a \times (b + c) = (a \times b) + (a \times c)$$

This may also be written in the form

$$(b + c) \times a = (b \times a) + (c \times a)$$

Identity Properties

The sum of any real number a and zero is the given real number, a.

$$0 + a = a + 0 = a$$

We call zero the *additive identity*.
The product of any real number a and 1 is the given real number, a.

$$1 \times a = a \times 1 = a$$

We call 1 the *multiplicative identity*.

Inverse Properties

For each real number a there exists another real number, $-a$, such that the sum of a and $-a$ is zero.

$$a + (-a) = (-a) + a = 0$$

We call $-a$ the *additive inverse* or *opposite* of a.
For each real number a different from zero there exists another real number, $\dfrac{1}{a}$, such that the product of a and $\dfrac{1}{a}$ is 1.

$$a \times \frac{1}{a} = \frac{1}{a} \times a = 1 \qquad a \neq 0$$

We call $\dfrac{1}{a}$ the *multiplicative inverse* or *reciprocal* of a.

Example 1   What basic property is being illustrated by each of the following?

(a) $6 + (17 + 4) = (17 + 4) + 6$

(b) $\frac{3}{4} + (-\frac{3}{4}) = 0$

(c) $57 \times 1 = 57$

(d) $\frac{2}{3}(12 + 36) = \frac{2}{3}(12) + \frac{2}{3}(36)$

(e) $3 + \sqrt{2}$ is a real number

Solution   **(a)** Commutative, for addition;   **(b)** Inverse, for addition;   **(c)** Identity, for multiplication;   **(d)** Distributive;   **(e)** Closure, for addition.

The preceding list represents the most important properties of the set of real numbers—the **field properties**. Any set of numbers that satisfies the properties listed is known as a **field**. There are other properties, however, that are of importance to our work.

Properties of Zero

The product of a real number a and zero is zero.

$$0 \times a = a \times 0 = 0$$

This is often referred to as the *multiplication property of zero*. If the product of two (or more) real numbers is zero, then at least one of the numbers is zero. That is, for real numbers a and b:

If $ab = 0$, then $a = 0$ or $b = 0$, or both $a = 0$ and $b = 0$.

Other Inverse Properties

The opposite of the opposite of a real number a is the number a.

$$-(-a) = a$$

The opposite of a sum is the sum of the opposites. That is, for real numbers a and b

$$-(a + b) = (-a) + (-b)$$

The opposite of a product of two real numbers is the product of one number times the opposite of the other. That is, for real numbers a and b

$$-(ab) = (-a)b = a(-b)$$

Note that all properties to date have been described in terms of addition and multiplication. We now define the other two basic operations.

Subtraction

The difference of two real numbers a and b is defined as

$$a - b = a + (-b)$$

For example, $5 - 8 = 5 + (-8) = -3$. Alternatively we say

$$a - b = c \text{ if and only if } b + c = a$$

Thus $5 - 8 = -3$ because $8 + (-3) = 5$.

Division

The quotient of two real numbers a and b is defined as

$$a \div b = a \times \frac{1}{b} \qquad b \neq 0$$

For example, $8 \div 2 = 8 \times \frac{1}{2} = 4$. Alternatively we say

$$a \div b = c \text{ if and only if } b \times c = a \qquad b \neq 0$$

Thus $8 \div 2 = 4$ because $2 \times 4 = 8$.

Using this form of division we can see why division by zero is not possible. Suppose we assume that division by zero is possible. Assume, for example, that $2 \div 0 = x$, where x is some real number. Then, by the definition of division, $0 \cdot x = 2$. But $0 \cdot x = 0$, leading to the false statement $2 = 0$. This argument can be duplicated where 2 is replaced by any nonzero number. Can you explain why $0 \div 0$ must also remain undefined? (See Exercise 23.)

As the following example illustrates, the properties for addition and multiplication do not automatically carry over for subtraction and division.

Example 2   Show that the set of real numbers is *not* commutative with respect to subtraction.

Solution   One *counterexample* is all that is needed. That is, we merely need to cite one instance where the property does *not* hold. For example, $3 - 2 = 1$ whereas $2 - 3 = -1$. Thus $3 - 2 \neq 2 - 3$; the set of real numbers does not have the commutative property for subtraction.

EXERCISES 1.2

Identify the basic property for real numbers that is illustrated by each of the following.

1. $(8 \cdot 9) \cdot 4 = 4 \cdot (8 \cdot 9)$
2. $\dfrac{1}{\sqrt{2}} \cdot \sqrt{2} = 1$

3. $(27 + 1000) + (-5) = 27 + (1000 + (-5))$

4. $10(100 + 1000) = (100 + 1000)10$
5. $(-6)(7 + a) = (-6)7 + (-6)a$

6. $0 + 1 = 1$
7. $1728 + (-\pi) = (-\pi) + 1728$

8. $[(2 \cdot 3)(4)](5 \cdot 6) = [(4)(2 \cdot 3)](5 \cdot 6)$

9. $[(2 \cdot 3)(4)](5 \cdot 6) = [(2)(3 \cdot 4)](5 \cdot 6)$

10. $\frac{1}{7} \cdot 1 = \frac{1}{7}$

11. $\frac{1}{2} + (-\frac{1}{2}) = 0$
12. $3 - 7 = 3 + (-7)$

13. $0(\sqrt{2} + \sqrt{3}) = 0$
14. $\sqrt{2} + \pi$ is a real number.

15. $(3 + 9)(7) = (3)(7) + (9)(7)$

For Exercises 16 through 20 give a specific example to illustrate each statement.

16. Subtraction is not associative.

17. Division is not commutative.

18. Division is not associative.

19. Multiplication is not distributive over division. That is, for real numbers a, b, and c, $a(b \div c) \neq ab \div ac$.

20. For real numbers a, b, and c, $a + bc \neq (a + b)(a + c)$.

21. Give the basic property that justifies each step.

$$
\begin{aligned}
d[(a + b) + c] &= d(a + b) + dc & \text{(i)} \\
&= (da + db) + dc & \text{(ii)} \\
&= da + (db + dc) & \text{(iii)} \\
&= da + (bd + cd) & \text{(iv)} \\
&= (bd + cd) + da & \text{(v)} \\
&= (b + c)d + da & \text{(vi)} \\
&= da + (b + c)d & \text{(vii)}
\end{aligned}
$$

22. Give the reasons for the following proof that multiplication distributes over subtraction.

$$
\begin{aligned}
a(b - c) &= a[b + (-c)] & \text{(i)} \\
&= ab + a(-c) & \text{(ii)} \\
&= ab + [-(ac)] & \text{(iii)} \\
&= ab - ac & \text{(iv)}
\end{aligned}
$$

*23. Explain why $\frac{0}{0}$ must remain undefined. (*Hint:* If $\frac{0}{0}$ is to be some value, then it must be a unique value.)

A statement such as $2x - 3 = 7$ is said to be a **conditional equation**. It is true for some replacements of the variable x, but not true for others. For example, $2x - 3 = 7$ is a true statement for $x = 5$ and is false for $x = 7$. On the other hand, an equation such as $3(x + 2) = 3x + 6$ is called an **identity** because it is true for *all* real numbers x.

To *solve* an equation means to find the real numbers x for which the given equation is true; these are called the **solutions** or **roots** of the given equation. Let us solve the equation $2x - 3 = 7$, showing all steps in detail.

$$2x - 3 = 7 \qquad \text{Add 3 to each side of the equation.}$$
$$(2x - 3) + 3 = 7 + 3$$
$$2x = 10 \qquad \text{Now multiply each side by } \tfrac{1}{2}.$$
$$\tfrac{1}{2}(2x) = \tfrac{1}{2}(10)$$
$$x = 5$$

We can check this solution by substituting 5 for x in the original equation.

$$2(5) - 3 = 10 - 3 = 7$$

In the preceding solution we made use of these two basic *properties of equality*.

Addition property of equality:
For all real numbers a, b, c, if $a = b$, then $a + c = b + c$.

Multiplication property of equality:
For all real numbers a, b, c, if $a = b$, then $ac = bc$.

Two equations are said to be *equivalent* if they have exactly the same roots. The strength of these two properties is that they produce equivalent equations. Thus the addition property converts $2x - 3 = 7$ into the equivalent form $2x = 10$; no roots have been lost or gained.

Example 1 Solve for x: $6x + 9 = 2x + 1$.

Solution

$$6x + 9 = 2x + 1$$
$$6x + 9 + (-9) = 2x + 1 + (-9) \qquad \text{(addition property)}$$
$$6x = 2x - 8$$
$$6x + (-2x) = 2x - 8 + (-2x) \qquad \text{(addition property)}$$
$$4x = -8$$
$$\tfrac{1}{4}(4x) = \tfrac{1}{4}(-8) \qquad \text{(multiplication property)}$$
$$x = -2$$

Check: $6(-2) + 9 = -3$; $\ 2(-2) + 1 = -3$.

You should practice showing all steps in the solution of an equation. With experience, however, you will be able to complete many of the details mentally.

†*TEST YOUR UNDERSTANDING*

Solve for x.

1. $x + 3 = 9$ **2.** $x - 5 = 12$ **3.** $x - 3 = -7$

4. $2x + 5 = x + 11$ **5.** $3x - 7 = 2x + 6$ **6.** $3(x - 1) = 2x + 7$

7. $3x + 2 = 5$ **8.** $5x - 3 = 3x + 1$ **9.** $x + 3 = 13 - x$

Many verbal problems can be solved by writing and solving an appropriate equation. The best way to learn to solve problems is to solve them! Study the following examples in detail. Then try to solve as many as possible of those given in the exercises that follow.

Example 2 The length of a rectangle is 1 centimeter less than twice the width. The perimeter is 28 centimeters. Find the dimensions.

Solution It is helpful to draw a diagram. Since the length is being compared to the width, it seems appropriate to let x represent the number of centimeters in the width; to be brief, we merely say that x represents the width.

---

†Throughout the text there are sets of exercises under the heading "Test Your Understanding." These should be used to provide a quick evaluation of your comprehension of the material just covered. Answers to the Test Your Understanding exercises are given at the end of each chapter.

2x − 1

Let x represent the width.
Then 2x − 1 represents the length.

x

The perimeter is the distance around the rectangle. This gives us the necessary information to write an equation. Try to explain each step in the solution that follows.

$$x + (2x - 1) + x + (2x - 1) = 28$$
$$6x - 2 = 28$$
$$6x = 30$$
$$x = 5$$

The original problem asked for both dimensions. If the width, x, is 5 centimeters, then the length, $2x - 1$, must be 9 centimeters. As a check we note that a rectangle with these dimensions has a perimeter of 28 centimeters.

9

5   5

9

Perimeter:
5 + 9 + 5 + 9 = 28

Example 3   Find two numbers whose sum is 42 if the larger number is 3 less than twice the smaller number.

Solution   The words may appear to be frightening, but not if taken in small doses. If we let x represent the smaller number, then the larger number is 3 less than twice x, or $2x - 3$.

Let x represent the smaller number.

Then $2x - 3$ represents the larger number.

Since the sum of the two numbers is 42, we have the following solution:

$$x + (2x - 3) = 42$$
$$3x - 3 = 42$$
$$3x = 45$$
$$x = 15$$
$$2x - 3 = 2(15) - 3 = 27$$

The two numbers are 15 and 27. Check to see that this solution is correct.

EXERCISES 1.3

Solve for x and check each result.

1. $x + 5 = 12$ **2.** $x - 7 = 15$ **3.** $x + (-8) = 6$

4. $x + 8 = -3$ **5.** $x - 2 = -7$ **6.** $3x = -27$

7. $-5x = -45$ **8.** $2x + 3 = 13$ **9.** $-2x + 1 = 9$

10. $2(x + 1) = 11$ **11.** $3x + 7 = 2x - 2$ **12.** $x + 7 = 2x - 3$

13. $4x - 7 = x + 8$ **14.** $3(x - 5) = x + 1$ **15.** $3x + 5 + x = 2x - 6$

16. Three less than five times a certain number is 37. Find the number.

17. Find the dimensions of a rectangle whose perimeter is 56 inches if the length is three times the width.

18. Find two consecutive integers whose sum is 73.

19. Find two consecutive even integers whose sum is 74.

20. Find two numbers whose sum is 27 if the larger is 1 less than three times the smaller.

21. Each of the two equal sides of an isosceles triangle is 3 inches longer than the base of the triangle. The perimeter is 21 inches. Find the length of each side.

22. Find a number such that two-thirds of the number increased by 1 is 13.

23. Maria has $169 in ones, fives, and tens. She has twice as many one-dollar bills as she has five-dollar bills, and five more ten-dollar bills than five-dollar bills. How many of each type of bill does she have?

24. Carlos spent $2.30 on stamps, in denominations of 6¢, 8¢, and 10¢. He bought one-half as many 8¢ stamps as 6¢ stamps and three more 10¢ stamps than 6¢ stamps. How many of each type did he buy?

25. Two cars leave a town and travel in opposite directions. One car travels at the rate of 40 miles per hour and the other at 60 miles per hour. In how many hours will the two cars be 350 miles apart? (*Hint:* Use rate $\times$ time $=$ distance.)

26. The width of a painting is 4 inches less than the length. The frame that surrounds the painting is 2 inches wide and has an area of 240 square inches. What are the dimensions of the painting? (*Hint:* The total area minus the area of the painting alone is equal to the area of the frame.)

27. Bob is 20 years older than Ben. In 10 years, Bob will be twice as old as Ben. How old is each now? (*Hint:* If Ben is x years old now, he will be $x + 10$ years old in 10 years.)

28. Prove that the measures of the angles of a triangle cannot be represented by consecutive odd integers. (*Hint:* The sum of the measures is 180.)

29. The units' digit of a two-digit number is 3 more than the tens' digit. The number is equal to four times the sum of the digits. Find the number. (*Hint:* We can represent a two-digit number as $10a + b$.)

*30. Find the largest possible pair of consecutive integers whose sum is less than 40.

*31. The length of a side of an equilateral triangle is a whole number of centimeters. Find all possible values for the length if the perimeter of the triangle is less than 30 centimeters but greater than 12 centimeters.

*32. The length of a rectangle is 1 inch less than three times the width. If the length is increased by 6 inches and the width is increased by 5 inches, then the length will be twice the width. Find the dimensions of the rectangle.

*33. The length of a rectangle is 1 inch less than twice the width. If the length is increased by 11 inches and the width is increased by 5 inches, the length will be twice the width. What can you conclude about the data for this problem?

*34. The *subtraction property of equality* states that for all real numbers a, b, c, if $a = b$, then $a - c = b - c$. How does this follow from the addition property?

*35. The *division property of equality* states that for all real numbers a, b, c, if $a = b$, then $a \div c = b \div c$ for all numbers $c \neq 0$. How does this follow from the multiplication property?

*36. Solve the general linear equation $ax + b = c$, $a \neq 0$, for x in terms of the constants a, b, and c.

Solve.

37. $P = 2l + 2w$ for w

38. $P = 4s$ for s

39. $N = 10t + u$ for t

40. $F = \frac{9}{5}C + 32$ for C

1.4

Properties of Order

As you continue your study of mathematics you will find a great deal of attention given to *inequalities*. We begin our discussion of this topic by considering the ordering of the real numbers on the number line. Reading from left to right in the following figure we say that *a is less than b* because a is to the left of b. In symbols we write $a < b$.

We may also use the figure to note that *b is greater than a* because b is to the right of a. In symbols we write $b > a$. Two inequalities, one using the symbol $<$ and the other $>$, are said to have the *opposite sense*.

The fundamental property behind the study of inequalities is known as the **trichotomy law**. This law states that for any two real numbers a and b exactly one of the following must be true:

$$a < b \qquad a = b \qquad a > b$$

There are a number of important properties of order that are fundamental for later work. We list them in terms of the following rules.

RULE 1.   If $a < b$ and $b < c$, then $a < c$. This is known as the **transitive property of order**. Geometrically it says that if a is to the left of b, and b is to the left of c, then a must be to the left of c on a number line.

RULE 2.   If $a < b$, then $a \pm c < b \pm c$ for any real number c. For example:

If $5 < 7$, then $5 + 10 < 7 + 10$; that is, $15 < 17$.
If $5 < 7$, then $5 - 10 < 7 - 10$; that is, $-5 < -3$.

RULE 3.   If $a < b$ and $c > 0$, then $ac < bc$. For example:

If $5 < 10$, then $5(3) < 10(3)$; that is, $15 < 30$.
If $-4 < 6$, then $-4(\frac{1}{2}) < 6(\frac{1}{2})$; that is, $-2 < 3$.

RULE 4.   If $a < b$ and $c < 0$, then $ac > bc$. For example:

If $5 < 10$, then $5(-3) > 10(-3)$; that is, $-15 > -30$.
If $-4 < 6$, then $-4(-\frac{1}{2}) > 6(-\frac{1}{2})$; that is, $2 > -3$.

RULE 5.   If $a < b$ and $c < d$, then $a + c < b + d$. For example:

If $5 < 10$ and $-15 < -4$, then $5 + (-15) < 10 + (-4)$; that is, $-10 < 6$.

RULE 6.   If $0 < a < b$ and $0 < c < d$, then $ac < bd$. For example:

If $3 < 7$ and $5 < 9$, then $(3)(5) < (7)(9)$; that is, $15 < 63$.

RULE 7.   If $a < b$ and $ab > 0$, then $\dfrac{1}{a} > \dfrac{1}{b}$. For example:

If $5 < 10$, then $\frac{1}{5} > \frac{1}{10}$.
If $-3 < -2$, then $-\frac{1}{3} > -\frac{1}{2}$.

Here are the preceding rules of order listed for easy reference.

Rules of order

1. If $a < b$ and $b < c$, then $a < c$.
2. If $a < b$ then $a \pm c < b \pm c$ for any number c.
3. If $a < b$ and $c > 0$, then $ac < bc$.
4. If $a < b$ and $c < 0$, then $ac > bc$.
5. If $a < b$ and $c < d$, then $a + c < b + d$.
6. If $0 < a < b$ and $0 < c < d$, then $ac < bd$.
7. If $a < b$ and $ab > 0$, then $\dfrac{1}{a} > \dfrac{1}{b}$.

Note that each of the preceding rules has been stated in terms of the "less than" relationship. Each rule can also be stated in terms of "greater than" as well. Rule 1, for example, can be stated in this way:

If $a > b$ and $b > c$, then $a > c$.

Try to state each of the other rules by using the "greater than" relationship.

We have developed the concept of $a < b$ by saying that a is to the left of b on a number line. It is often useful to state this relationship in terms of this formal definition:

For any two real numbers a and b, $a < b$ (and $b > a$) if and only if $b - a$ is a positive number; that is, if and only if $b - a > 0$.

For example, $2 < 5$ because $5 - 2 = 3$, a positive number; that is, $5 - 2 > 0$. Similarly, $-7 < -2$ because $-2 - (-7) = 5$, a positive number.

With this definition of order, we can prove the rules discussed earlier. The next example shows how this can be done.

Example 1   Prove Rule 1: If $a < b$ and $b < c$, then $a < c$.

Solution   We note that $a < b$ and $b < c$ means that $b - a$ and $c - b$ are positive numbers by definition. Since the sum of two positive numbers is also positive, we get

$$(b - a) + (c - b) > 0$$

But

$$(b - a) + (c - b) = c - a$$

Hence $c - a$ is positive, and the definition implies that $a < c$.

Here are several additional symbols of inequality that are often used:

$a \leq b$ means a *is less than or is equal to b;* that is, $a < b$ or $a = b$.

$a \geq b$ means a *is greater than or is equal to b;* that is, $a > b$ or $a = b$.

We may use these symbols in the rules of order previously listed. Thus Rule 1 may be stated in these forms:

If $a \leq b$ and $b \leq c$, then $a \leq c$.

If $a \geq b$ and $b \geq c$, then $a \geq c$.

The two inequalities $a < b$ and $b < c$ may be written as $a < b < c$. Also, if $a \leq b$ and $b \leq c$, then $a \leq b \leq c$. A similar statement may be made for the $\geq$ relationship.

Finally, note that a slashed line through a symbol of equality or inequality gives the negation of a statement. For example:

$a \neq b$ means *a is not equal to b.*

$a \not< b$ means *a is not less than b.*

Determine the meanings of these statements: $a \not> b$; $a \not\leq b$; $a \not\geq b$.

Example 2   Translate each verbal statement into a statement of inequality.
(a) x is to the right of zero.
(b) x is between 2 and 3.
(c) x is greater than or equal to -1.
(d) x is at least 10 but less than 100.

Solution   (a) $x > 0$;   (b) $2 < x < 3$;   (c) $x \geq -1$;   (d) $10 \leq x < 100$

EXERCISES 1.4

In Exercises 1 through 8 insert the appropriate symbol, $<$, $=$, or $>$, between the given pair of real numbers.

1. -100____2

2. $\frac{1}{2}$____$\frac{1}{3}$

3. $-\frac{4}{5}$____$-\frac{2}{3}$

4. 2.619____2.621

5. 0.7____$\frac{7}{9}$

6. $-\frac{13}{14}$____$-\frac{20}{21}$

7. $\frac{1}{2}(4.02)$____2.01

8. $\frac{1}{9}$____0.111

In Exercises 9 and 10 translate each verbal statement into a statement of inequality.

9. (a) x is to the left of 1.
(b) x is to the right of $-\frac{2}{3}$.
(c) x is at least as large as 5.
(d) x is more than -10 but less than or equal to 7.

10. (a) x is between -1 and 4.
(b) x is no more than 6.
(c) x is to the right of 12.
(d) x is a positive number less than 40.

In Exercises 11 through 18 classify each statement as true or false. If it is false, give a specific example to explain your answer.

11. If $x > 1$ and $y > 2$, then $x + y > 3$.

12. If $x < 2$, then x is negative.

13. If $x < 5$ and $y < 6$, then $xy < 30$.

14. If $0 < x$, then $-x < 0$.

15. If $x < y < -2$, then $\dfrac{1}{x} > \dfrac{1}{y}$.

16. If $0 < x$, then $x < x^2$.

17. If $x \le y$ and $y < z$, then $x < z$.

18. If $x \le -5$, then $x - 2 \le -7$.

19. Complete the following proof of Rule 3: If $a < b$ and $c > 0$, then $ac < bc$.
  (a) $a < b$ implies _____, by the definition of $<$.
  (b) Since $c > 0$, the product _____ is also positive.
  (c) But $(b - a)c = $ _____.
  (d) Therefore _____ is positive.
  (e) Thus _____ $<$ _____ by the definition of $<$.

*20. Use Rule 3 to prove that if $0 < a < 1$, then $a^2 < a$.

*21. Use Rule 3 to prove that if $1 < a$, then $a < a^2$.

*22. Prove: If $a < b < 0$ and $c < d < 0$, then $ac > bd$.

1.5

**Conditional
Inequalities**

The inequality $3x - 4 > 11$ is true for some values for x, such as 10, and false for others, such as 3. Inequalities that are not true for all allowable values of the variable are called **conditional inequalities**. Solving such an inequality means finding the set of all x for which it is true. Our basic rules of Section 1.4 will be helpful in this work.

$$3x - 4 > 11$$
$$3x - 4 + 4 > 11 + 4 \qquad \text{By Rule 2 we may add 4 to both sides.}$$
$$3x > 15$$
$$\tfrac{1}{3}(3x) > \tfrac{1}{3}(15) \qquad \text{By Rule 3 we may multiply each side by } \tfrac{1}{3}$$
$$\qquad\qquad\qquad\qquad \text{or divide by 3.}$$
$$x > 5$$

The answer then consists of all real numbers x such that $x > 5$. This answer may also be displayed graphically on a number line.

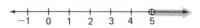

The heavily shaded arrow is used to show that all points in the indicated direction are included. The open circle indicates that 5 is *not* included in the solution. Why not? When the open circle is replaced by a solid dot we have the following graph, which represents the solution of the conditional inequality $3x - 4 \geq 11$. In this case 5 is a solution of the inequality.

Example 1   Solve for x and graph: $5(2x - 3) \leq -10$.

Solution   Multiply each side by $\frac{1}{5}$ (or divide by 5).

$$\tfrac{1}{5} \cdot 5(2x - 3) \leq \tfrac{1}{5}(-10) \qquad \text{(Rule 3)}$$
$$2x - 3 \leq -2$$

Add 3 to each side.

$$2x - 3 + 3 \leq -2 + 3 \qquad \text{(Rule 2)}$$
$$2x \leq 1$$

Multiply by $\frac{1}{2}$ (or divide by 2).

$$\tfrac{1}{2}(2x) \leq \tfrac{1}{2}(1) \qquad \text{(Rule 3)}$$
$$x \leq \tfrac{1}{2}$$

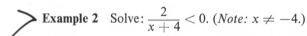

Use the solution to this example to solve the following inequality:

$$5(2x - 3) > -10$$

Example 2   Solve: $\dfrac{2}{x + 4} < 0$. (*Note:* $x \neq -4$.)

Solution   Since the numerator of the fraction is positive, the fraction will be less than zero whenever the denominator is negative. Thus

—take out of fraction $x + 4 < 0$
$$x < -4$$

Example 3   Assume $x < 2$. Is $x - 2$ positive or negative?

Solution   Algebraically we may proceed as follows:

$$x < 2$$
$$x - 2 < 2 - 2$$
$$x - 2 < 0$$

Therefore $x - 2$ is negative. Informally, we may substitute some values for x that are less than 2. Thus if $x = 1, 1 - 2 = -1$; and if $x = 0, 0 - 2 = -2$. From this we note that $x - 2$ is negative for $x < 2$.

Example 4 Assume $-2 < x < 3$. Classify as positive or negative:
(a) $x + 2$; (b) $x - 3$; (c) $x - 5$

Solution
(a) Select values for x between -2 and 3, such as $x = -1$ or $x = 2$. In each case substitution in the expression $x + 2$ shows that $x + 2 > 0$; that is, $x + 2$ is positive. Algebraically we may reason as follows. Since $-2 < x < 3$ we may add 2 throughout and get $0 < x + 2 < 5$. Thus $x + 2$ is positive. Notice that the additional information $x + 2 < 5$ does not alter the conclusion that $x + 2$ is positive.
(b) Negative.
(c) Negative.

TEST YOUR UNDERSTANDING

Solve each inequality for x. Graph the solution for Exercises 1 and 2.

1. $x - 3 < 0$ 2. $x + 5 > 0$ 3. $x + 1 < -3$
4. $x - 2 \geq \frac{1}{2}$ 5. $2x + 7 < 11$ 6. $3x + 1 \leq x - 4$

Solve for x.

7. $-2(x + 6) < 0$ 8. $-2(x + 6) \geq 0$ 9. $-\dfrac{1}{x} < 0$

10. $\dfrac{5}{3 - x} < 0$

11. Assume $x < -1$. Classify as positive or negative: (a) $x + 1$; (b) $x - 3$.
12. Assume $4 < x < 5$. Classify as positive or negative: (a) $x - 4$; (b) $x - 5$.

Let us explore in detail the solution to a more complicated statement of inequality:

$$(x - 2)(x - 5) < 0$$

It should be evident that if $x = 2$ or $x = 5$, then $(x - 2)(x - 5) = 0$. The two points, $x = 2$ and $x = 5$, divide the number line into these three regions.

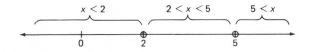

The three cases, left to right, become the three rows, top to bottom, in the following table.

	$x - 2$	$x - 5$	$(x - 2)(x - 5)$
Row 1: $x < 2$ Row 2: $2 < x < 5$ Row 3: $5 < x$	$-$ $+$ $+$	$-$ $-$ $+$	$+$ $-$ $+$

In Row 1, $x < 2$. There is a minus sign under $x - 2$ because $x < 2$ implies $x - 2 < 0$. Similarly, $x - 5 < 0$ and therefore there is a minus sign under $x - 5$. Since the product of two negative numbers is positive, there is a plus sign under $(x - 2)(x - 5)$ in Row 1.

For Row 2, $2 < x < 5$. Therefore $x - 2$ is positive and $x - 5$ is negative. Consequently $(x - 2)(x - 5)$ is negative. Similar reasoning shows that $(x - 2)(x - 5)$ is positive for $5 < x$; that is, for $x > 5$.

All values of x have now been considered, and only Row 2 produces $(x - 2)(x - 5) < 0$. Thus the answer is $2 < x < 5$. This solution may be graphed as follows:

EXERCISES 1.5

Solve for x and graph.

1. $x + 4 < 0$ **2.** $5 - x \leq 0$ **3.** $3(x - 1) < 2$

4. $5x - 10 > 0$ **5.** $-3x + 4 \geq 0$ **6.** $-3x + 4 \geq 0$

7. $2x + 7 \leq 5 - 6x$ **8.** $\frac{1}{2}x + 5 < 3$ **9.** $\frac{1}{3}(x + 4) > 2$

10. $\dfrac{2 - x}{5} \geq 0$ **11.** $\dfrac{1}{x} < 0$ **12.** $-\dfrac{2}{x + 1} > 0$

13. Assume $-2 < x$. Classify as positive or negative: **(a)** $x + 2$; **(b)** $x + 4$.

14. Assume $x < \frac{1}{2}$. Classify as positive or negative: **(a)** $x - \frac{1}{2}$; **(b)** $x - 1$.

15. Assume $-1 < x < 3$. Classify as positive or negative:
    (a) $x + 1$ **(b)** $x + 4$ **(c)** $x - 3$ **(d)** $x - 6$

16. Assume $-\frac{3}{2} < x < -\frac{1}{2}$. Classify as positive or negative:
    (a) $x + \frac{3}{2}$ **(b)** $x + 2$ **(c)** $x + \frac{1}{2}$ **(d)** $x - 1$

Solve for x.

17. $(x - 1)(x + 1) < 0$ **18.** $(x - 1)(x + 1) > 0$

19. $(x + 2)(x + 3) < 0$ **20.** $(x + 2)(x + 3) > 0$

21. $(2x - 1)(x + 5) \le 0$ **22.** $(2x - 1)(x + 5) > 0$

(*Hint:* Write $2x - 1$ as $2(x - \frac{1}{2})$.)

23. $(2x - 1)(3x + 1) > 0$ **24.** $(x + 3)(x - 3) < 0$

25. $\dfrac{x}{x + 1} > 0$ **26.** $\dfrac{x - 2}{x + 3} < 0$

27. $\dfrac{x + 5}{2 - x} < 0$ **28.** $\dfrac{1}{x} \ge 2$

*29. Here is an alternative (graphic) method for solving conditional inequalities of the type $(x - 2)(x - 5) < 0$. On the top number line, the bold (dark) half represents those x for which $x - 2$ is positive ($x > 2$), and the other half gives those x for which $x - 2$ is negative ($x < 2$). The bottom line has the same information for the factor $x - 5$.

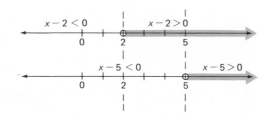

Since $(x - 2)(x - 5) < 0$ when $x - 2$ and $x - 5$ have *opposite* signs, the set of x for which this is the case are those between the vertical lines at 2 and 5. Thus the solution is $2 < x < 5$. Find the solution of $(x - 2)(x - 5) > 0$ from the preceding graph.

*30. Solve Exercise 20 by the method of Exercise 29.

1.6 Absolute Value

What do the numbers $+6$ and -6 have in common? They do not have the same sign, but each is the same *distance* from the origin; each is 6 units from 0.

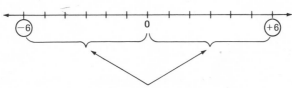

Same distance from origin

This idea can be expressed symbolically in this way:

$|+6| = 6$ read as "the absolute value of $+6$ is 6."

$|-6| = 6$ read as "the absolute value of -6 is 6."

Geometrically, for any real number x, $|x|$ is the distance (without regard to direction) that x is from the origin. Note that for a positive number, $|x| = x$; $|+6| = 6$. For a negative number, $|x| = -x$; $|-6| = -(-6) = 6$. Also, since 0 is the origin, it is natural to have $|0| = 0$.

For any real number x,

$$|x| = \begin{cases} x & \text{if } x \geq 0 \\ -x & \text{if } x < 0 \end{cases}$$

Some very useful properties of absolute value follow, presented with illustrations but without proof. Except for the usual exclusion of division by zero, the variables represent any real numbers.

PROPERTY 1.   If $|x| = a$, then $x = a$ or $-x = a$; $|0| = 0$.

This property follows immediately from the definition of absolute value.

Example 1   Solve: $|5 - x| = 7$.

Solution   Use Property 1 as follows.

$$+(5 - x) = 7 \qquad \text{or} \qquad -(5 - x) = 7$$
$$5 - x = 7 \qquad\qquad\qquad -5 + x = 7$$
$$-x = 2 \qquad\qquad\qquad\qquad x = 12$$
$$x = -2$$

These two solutions, -2 and 12, can be checked by substitution into the original equation.

Check: $|5 - (-2)| = |7| = 7$; $|5 - 12| = |-7| = 7$.

PROPERTY 2. $|x| = |-x|$.

The absolute value of a real number is the same as the absolute value of its negative. For example, $|2| = |-2| = 2$. A useful application of this property is the result that for any real numbers a and b, $|a - b| = |b - a|$. Note that $a - b$ and $b - a$ are negatives of one another; for example, $|2 - \pi| = |\pi - 2| = \pi - 2$ because $\pi - 2 > 0$.

PROPERTY 3. $-|x| \le x \le |x|$

A real number is always between its absolute value and the negative of its absolute value. For example, verify that each of the following is correct:

$$-|7| \le 7 \le |7| \qquad -|-7| \le -7 \le |-7|$$

PROPERTY 4. $|xy| = |x| \cdot |y|$

The absolute value of a product is the product of the absolute values. For example:

$$|(-3)(5)| = |-15| = 15 \quad \text{and} \quad |-3| \cdot |5| = 3 \cdot 5 = 15$$

PROPERTY 5. $\left|\dfrac{x}{y}\right| = \dfrac{|x|}{|y|}$

The absolute value of a quotient is the quotient of the absolute values. For example:

$$\left|\frac{-9}{3}\right| = \left|-\frac{9}{3}\right| = |-3| = 3 \quad \text{and} \quad \frac{|-9|}{|3|} = \frac{9}{3} = 3$$

PROPERTY 6. If $|x| < c$, then $-c < x < c$; $c > 0$.

To understand this property, study the following figure. It should be clear that if the distance from a point x to the origin is less than c, then x must lie between $-c$ and c.

$|x| < c$:

As an example, consider the inequality $|x| < 3$. This denotes the set of all numbers x whose distance from the origin is less than 3 units; that is, the set of all values for x between -3 and 3. Thus if $|x| < 3$, then $-3 < x < 3$.

$|x| < 3$:

Note that Property 6 is reversible. That is, the inequality $-c < x < c$ may be expressed in the form $|x| < c$. (Do you see why c cannot be less than zero?) For example, the inequality $-5 < x < 5$ may be written in the form $|x| < 5$.

Example 2   Solve and graph: $|x - 2| < 3$.

Solution   Let $x - 2$ play the role of x in Property 6. Consequently $|x - 2| < 3$ is equivalent to $-3 < x - 2 < 3$. We may now solve for x by solving each part separately.

$$-3 < x - 2 \qquad \text{and} \qquad x - 2 < 3$$

Thus $-1 < x$ and $x < 5$; that is, $-1 < x < 5$.

Note that the inequality $-3 < x - 2 < 3$ can also be solved by adding 2 to each member.

$$\begin{array}{r} -3 < x - 2 < 3 \\ +2 \quad +2 \quad +2 \\ \hline -1 < x + 0 < 3 \end{array} \qquad \text{Thus } -1 < x < 3.$$

PROPERTY 7. $|x + y| \le |x| + |y|$

The absolute value of a sum is less than or equal to the sum of the absolute values. Verify that each of the following examples is correct.

$$|2 + 7| \le |2| + |7| \qquad |(-2) + (-7)| \le |-2| + |-7|$$
$$|(-2) + 7| \le |-2| + |7| \qquad |2 + (-7)| \le |2| + |-7|$$

TEST YOUR UNDERSTANDING

Classify each statement as true or false.

1. $|-100| > 1$
2. $|-\frac{1}{2}| = -\frac{1}{2}$
3. $|\frac{3}{4} - \frac{5}{6}| = \frac{5}{6} - \frac{3}{4}$
4. $|-7x| = 7|x|$
5. $|(x - 2)(x + 2)| = |2 - x||x + 2|$
6. $|-3| \le -3$
7. $|4 - 9| = |4| - |9|$
8. Solve for x: $|2x - 1| = 3$.
9. Solve for x: $\dfrac{1}{|x - 2|} = 1$.
10. Solve and graph: $|x - 3| < 1$.

Finally, let us consider the inequality $|x| \geq 3$. We know that $|x| < 3$ implies $-3 < x < 3$. Therefore $|x| \geq 3$ consists of those values of x for which $x \leq -3$ or $x \geq 3$. The graph of this set may be drawn as follows.

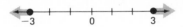

In general:

If $|x| > c$, then $x < -c$ or $x > c$, where $c \geq 0$.

If $|x| \geq c$, then $x \leq -c$ or $x \geq c$, where $c \geq 0$.

EXERCISES 1.6

Classify each statement as true or false.

1. $-|-\frac{1}{3}| = \frac{1}{3}$

2. $|-1000| < 0$

3. $|-\frac{1}{2}| = 2$

4. $|\sqrt{2} - 5| = 5 - \sqrt{2}$

5. $\left|\dfrac{x}{y}\right| = |x| \cdot \dfrac{1}{|y|}$

6. $2 \cdot |0| = 0$

7. $||x|| = |x|$

8. $|-(-1)| = -1$

9. $|a| - |b| = a - b$

Solve for x.

10. $|x| = \frac{1}{2}$

11. $|3x| = 3$

12. $|x - 1| = 3$

13. $|3x - 4| = 0$

14. $|2x - 3| = 7$

15. $|6 - 2x| = 4$

16. $\left|\dfrac{1}{x-1}\right| = 2$

17. $\dfrac{|x|}{x} = 1$

18. $\dfrac{|x|}{x} = -1$

Solve for x and graph.

19. $|2x| < 2$

20. $|x - 4| < 1$

21. $|x + 2| < 3$

22. $|2x - 1| \leq 5$

23. $|2x + 1| \leq 5$

24. $|1 - 2x| < 3$

25. $|x - 1| > 2$

26. $|x + 3| > 5$

27. $|2x + 1| \geq 3$

28. $|x - 4| \not> 1$

29. $|x - 2| \geq 0$

30. $\dfrac{1}{|x - 3|} > 0$

*31. Prove Property 3 of this section. (*Hint:* Consider the two cases $x \geq 0$ and $x < 0$.)

*32. Cite at least four different examples to confirm this inequality: $|x - y| \geq ||x| - |y||$.

*33. (a) Prove that $|xy| = |x| \cdot |y|$ for the case $x < 0$ and $y > 0$.

(b) Prove this product rule when $x < 0$ and $y < 0$.

*34. Prove $\left|\dfrac{x}{y}\right| = \dfrac{|x|}{|y|}$ for the case $x < 0$, $y > 0$.

*35. Prove that if x is a real number then $x^2 = |x^2| = |x|^2$.

REVIEW EXERCISES FOR CHAPTER 1

The solutions to the following exercises can be found within the text of Chapter 1. Try to answer each question without referring to the text.

Section 1.1

1. Classify the following numbers in terms of these categories: natural, integer, rational, irrational, real.
   (a) 5 (b) $\frac{2}{3}$ (c) $\sqrt{7}$
2. Using straightedge and compass, locate the irrational number $\sqrt{2}$ on a number line.

Section 1.2

3. What basic property of the real numbers is illustrated by each of the following?
   (a) $6 + (17 + 4) = (17 + 4) + 6$ (b) $\frac{3}{4} + (-\frac{3}{4}) = 0$
   (c) $57 \times 1 = 57$ (d) $\frac{2}{3}(12 + 36) = \frac{2}{3}(12) + \frac{2}{3}(36)$
4. If $ab = 0$, what do you know about the numbers a and b?
5. State the definition of division.
6. Show that subtraction is not commutative.

Section 1.3

7. Solve for x: $6x + 9 = 2x + 1$.
8. The length of a rectangle is 1 centimeter less than twice the width. The perimeter is 28 centimeters. Find the dimensions.
9. Find two numbers whose sum is 42 if the larger number is 3 less than twice the smaller number.

Section 1.4

10. True or false: If $a < b$ and $c < 0$, then $ac < bc$.
11. True or false: If $a < b$ and $ab > 0$, then $\dfrac{1}{a} > \dfrac{1}{b}$.
12. State the nongeometric definition of $a < b$.
13. Use the definition in Exercise 12 to prove that if $a < b$ and $b < c$, then $a < c$.
14. Translate the following into inequalities.
    (a) x is between 2 and 3.
    (b) x is at least 10 but is less than 100.
    (c) x is greater than or equal to -1.

Section 1.5

15. Solve for x and graph: $5(2x - 3) \leq -10$.
16. Solve: $(x - 2)(x - 5) < 0$.

Section 1.6

17. State the definition of absolute value.

18. Solve: $|5 - x| = 7$.

19. True or false:

    (a) $|2 - \pi| = 2 - \pi$ **(b)** $|-3| \cdot |5| = -15$

20. Solve and graph: $|x - 2| < 3$.

SAMPLE TEST FOR CHAPTER 1

1. Check the boxes in the table to indicate the set to which each number belongs.

	Counting Numbers	Integers	Rational Numbers	Irrational Numbers	Negative Numbers
$\dfrac{-6}{3}$		✓	✓		✓
0.231		✗	✓		
$\sqrt{12}$			✓ ⟶		
$\sqrt{\dfrac{1}{4}}$			✓		
1979	✓	✓	✓		

2. Explain how to locate $\sqrt{10}$ on a number line by using straightedge and compass.

3. Classify each statement as true or false.

    (a) Negative irrational numbers are not real numbers.

    (b) Every integer is a rational number.

    (c) Some irrational numbers are integers.

    (d) Zero is a rational number.

    (e) If $x < y$, then $x - 5 > y - 5$.

    (f) The absolute value of a sum equals the sum of the absolute values.

4. Match each statement in the left column with one of the properties listed at the right.

   (i) $(ab)c = (ba)c$

   (ii) $-5 - (-2) = -5 + (2)$

   (iii) $0 + 7 = 7$

   (iv) $3[(4 + 5) + 6] = 3(4 + 5) + 3(6)$

   (v) $4 \times \frac{1}{4} = 1$

   (vi) $-10 < -3$ since $-3 - (-10) > 0$

   (vii) $(2 + x) + (3 + y) = 2 + (x + (3 + y))$

   (a) Commutative for $+$

   (b) Commutative for $\times$

   (c) Associative for $+$

   (d) Associative for $\times$

   (e) Distributive

   (f) Additive identity

   (g) Multiplicative identity

   (h) Additive inverse

   (i) Multiplicative inverse

   (j) Definition of subtraction

   (k) Definition of order

   (l) Trichotomy law

5. Solve for x.

   (a) $\frac{3}{4}x - 1 = 11$ **(b)** $4x + 20 = 2x + 2$

6. Find the dimensions of a rectangle whose perimeter is 52 inches if the length is 5 inches more than twice the width.

7. Solve for x.

   (a) $|3 - 4x| = 2$ **(b)** $\dfrac{|x + 2|}{x + 2} = -1$

Solve and graph each inequality.

8. $2(5x - 1) < x$ **9.** $|2x - 1| \geq 3$

10. Solve for x: $(2x - 1)(x + 5) < 0$.

Answers to the Test Your Understanding Exercises

Page 12

1. 6   **4.** 6   **7.** 1

2. 17   **5.** 13   **8.** 2

3. -4 **6.** 10   **9.** 5

Page 21

1. $x < 3$:

2. $x > -5$:

3. $x < -4$

4. $x \geq \frac{5}{2}$

5. $x < 2$

6. $x \leq -\frac{5}{2}$

7. $x > -6$

8. $x \leq -6$

9. $x > 0$

10. $x > 3$

11. (a) Negative; (b) negative.

12. (a) Positive; (b) negative.

Page 26

1. True.   **5.** True.

2. False.   **6.** False.

3. True.   **7.** False.

4. True.   **8.** $-1; 2$.

9. 1; 3.

10. $2 < x < 4$:

Two

Fundamentals of Algebra

2.1

Integral Exponents

Much of mathematical notation can be viewed as efficient abbreviations of lengthier statements. For example:

$$4^9 = 4 \times 4 \times 4 \times 4 \times 4 \times 4 \times 4 \times 4 \times 4$$

This illustration is a specific example of the fundamental definition of a *positive integral exponent*.

Definition 1

If n is a positive integer and b is any real number, then

$$b^n = \underbrace{b \cdot b \cdot \cdots \cdot b}_{n \text{ factors}}$$

The number b is called the **base** and n is called the **exponent**.

The most common ways of referring to b^n are "b to the nth power," "b to the nth," or "the nth power of b."

Illustrations:

$$2^5 = 2 \times 2 \times 2 \times 2 \times 2 = 32$$

$$(-\tfrac{1}{10})^3 = (-\tfrac{1}{10})(-\tfrac{1}{10})(-\tfrac{1}{10}) = -\tfrac{1}{1000}$$

$$(-1)^{100} = \underbrace{(-1)(-1)(-1) \cdots (-1)}_{100 \text{ factors}} = 1$$

A number of important rules concerning positive integral exponents are easily established on the basis of the preceding definition. Here is a list of these rules in which m and n are any positive integers, a and b are any real numbers, and with the usual understanding that denominators cannot be zero.

RULE 1. $b^m b^n = b^{m+n}$

Illustrations:

$$2^3 \cdot 2^4 = 2^{3+4} = 2^7$$

$$x^3 \cdot x^4 = x^7$$

RULE 2. $\dfrac{b^m}{b^n} = \begin{cases} b^{m-n} & \text{if } m > n \\ 1 & \text{if } m = n \\ \dfrac{1}{b^{n-m}} & \text{if } m < n \end{cases}$

Illustrations:

$$(m > n) \qquad \frac{2^5}{2^2} = 2^{5-2} = 2^3$$

$$\frac{x^5}{x^2} = x^{5-2} = x^3$$

$$(m = n) \qquad \frac{5^2}{5^2} = 1$$

$$\frac{x^2}{x^2} = 1$$

$$(m < n) \qquad \frac{2^2}{2^5} = \frac{1}{2^{5-2}} = \frac{1}{2^3}$$

$$\frac{x^2}{x^5} = \frac{1}{x^{5-2}} = \frac{1}{x^3}$$

RULE 3. $(b^m)^n = b^{mn}$
Illustrations:

$$(2^3)^2 = 2^{3\cdot2} = 2^6$$
$$(x^3)^2 = x^{3\cdot2} = x^6$$

RULE 4. $(ab)^m = a^m b^m$
Illustrations:

$$(2 \cdot 3)^5 = 2^5 \cdot 3^5$$
$$(xy)^5 = x^5 y^5$$

RULE 5. $\left(\dfrac{a}{b}\right)^m = \dfrac{a^m}{b^m}$
Illustrations:

$$\left(\frac{3}{2}\right)^5 = \frac{3^5}{2^5}$$
$$\left(\frac{x}{y}\right)^5 = \frac{x^5}{y^5}$$

The proper use of these rules can simplify computations as in the following example.

Example 1 Evaluate: $12^3(\frac{1}{6})^3$.

Solution Evaluating $12^3(\frac{1}{6})^3$ without using Rule 4 can be done as follows:

$$12^3(\tfrac{1}{6})^3 = 1728(\tfrac{1}{216})$$
$$= \tfrac{1728}{216}$$
$$= 8$$

It is less work to use Rule 4.

$$12^3(\tfrac{1}{6})^3 = (12 \cdot \tfrac{1}{6})^3$$
$$= 2^3$$
$$= 8$$

Here is a proof of Rule 4. You can try proving the other rules by using similar arguments.

$$(ab)^m = (ab)(ab) \cdots (ab) \qquad \text{(by Definition 1)}$$
$$= (a \cdot a \cdot \cdots \cdot a)(b \cdot b \cdot \cdots \cdot b) \qquad \begin{cases} \text{(by repeated use of the} \\ \text{commutative and associative laws} \\ \text{for multiplication)} \end{cases}$$
$$= a^m b^m \qquad \text{(by Definition 1)}$$

Eventually, the exponential concept can be extended to include all the real numbers (not just positive integers) as exponents. And when all is said and done, the preceding rules will remain intact except for some special instances where the base numbers are negative or zero.

Example 2   Simplify: (a) $\left(-\dfrac{2}{3}\right)^5 \left(\dfrac{9}{4}\right)^5$;   (b) $\dfrac{(x^3 y)^2 y^3}{x^4 y^6}$.

Solution

(a) $\left(-\dfrac{2}{3}\right)^5 \left(\dfrac{9}{4}\right)^5 = \left(-\dfrac{2}{3} \cdot \dfrac{9}{4}\right)^5 = \left(-\dfrac{3}{2}\right)^5 = -\dfrac{243}{32}$

(b) $\dfrac{(x^3 y)^2 y^3}{x^4 y^6} = \dfrac{(x^3)^2 y^2 y^3}{x^4 y^6} = \dfrac{x^6 y^5}{x^4 y^6} = \dfrac{x^2}{y}$

TEST YOUR UNDERSTANDING

Use Definition 1 and Rules 1 through 5 to evaluate each of the following.

1. 5^3

2. $(-\tfrac{1}{2})^5$

3. $(-\tfrac{2}{3})^3 + \tfrac{8}{27}$

4. $(10^3)^2$

5. $2^3(-2)^3$

6. $(\tfrac{1}{2})^3 8^3$

7. $\dfrac{17^8}{17^9}$

8. $\dfrac{(-2)^3 + 3^2}{3^3 - 2^2}$

9. $\dfrac{(-12)^4}{4^4}$

10. $(ab^2)^3 (a^2 b)^4$

11. $\dfrac{2^2 \cdot 16^3}{(-2)^8}$

12. $\dfrac{(2x^3)^2 (3x)^2}{6x^4}$

The exponential concept may now be extended to include zero and the negative integers as exponents.

Definition 2

For any nonzero real number b we define $b^0 = 1$ and $b^{-n} = \dfrac{1}{b^n}$, where n is a positive integer.

Illustrations:

$$(-100)^0 = 1 \qquad 5^{-2} = \frac{1}{5^2} = \frac{1}{25}$$

$$(-10)^{-3} = \frac{1}{(-10)^3} \qquad (0^0 \text{ is not defined})$$

You may rightly wonder how it was decided to make these definitions. Why, for example, do we not define $5^0 = 5$ or $5^0 = 0$? In making such definitions, the major guideline is to preserve our earlier rules for exponents. That is to say, we want those rules to be correct no matter what kind of integers the letters m and n represent. So if we want them to work with $m = 0$, then Rule 1 reads

$$b^0 b^n = b^{0+n} = b^n$$

But $1 \cdot b^n = b^n$. Thus, for the sake of (future) consistency, we would *like* to have $b^0 = 1$. And once this is done, the rule for negative exponents can also be made on the basis of preserving the rules. For example, we now want

$$b^2 b^{-2} = b^{2+(-2)} = b^0 = 1$$

Dividing by b^2 produces

$$b^{-2} = \frac{1}{b^2}$$

In the following examples we shall assume that the rules for exponents apply when negative exponents are used.

Example 3   Evaluate:

(a) $\left(\dfrac{1}{7}\right)^{-2}$ **(b)** $\left(\dfrac{1}{2}\right)^3 2^{-3}$ **(c)** $\dfrac{5^{-2}}{15^{-2}}$ **(d)** $\left(\dfrac{400 - 10^4}{80^2}\right)^0$

Solution

(a) $\left(\frac{1}{7}\right)^{-2} = \frac{1}{\left(\frac{1}{7}\right)^{2}} = \frac{1}{\frac{1}{49}} = 49$ or $\left(\frac{1}{7}\right)^{-2} = (7^{-1})^{-2} = 7^{2} = 49$

(b) $\left(\frac{1}{2}\right)^{3} 2^{-3} = \left(\frac{1}{2}\right)^{3} \frac{1}{2^{3}} = \left(\frac{1}{2}\right)^{3}\left(\frac{1}{2}\right)^{3} = \left(\frac{1}{2}\right)^{6} = \frac{1}{64}$

(c) $\frac{5^{-2}}{15^{-2}} = \left(\frac{5}{15}\right)^{-2} = \left(\frac{1}{3}\right)^{-2} = (3^{-1})^{-2} = 3^{2} = 9$

(d) $\left(\frac{400 - 10^{4}}{80^{2}}\right)^{0} = 1$

As in Example 3(a), more than one correct procedure is often possible. Finding the most efficient procedure depends largely on experience.

Example 4   Simplify $\left(\frac{a^{-2}b^{3}}{a^{3}b^{-2}}\right)^{5}$ and express the answer using positive exponents only.

Solution

$$\left(\frac{a^{-2}b^{3}}{a^{3}b^{-2}}\right)^{5} = \frac{(a^{-2}b^{3})^{5}}{(a^{3}b^{-2})^{5}} \qquad \text{(Rule 5)}$$

$$= \frac{(a^{-2})^{5}(b^{3})^{5}}{(a^{3})^{5}(b^{-2})^{5}} \qquad \text{(Rule 4)}$$

$$= \frac{a^{-10}b^{15}}{a^{15}b^{-10}} \qquad \text{(Rule 3)}$$

$$= \frac{a^{-10}}{a^{15}} \cdot \frac{b^{15}}{b^{-10}} \qquad \text{(property of fractions)}$$

$$= \frac{1}{a^{25}} \cdot b^{25} \qquad \text{(Rule 2)}$$

$$= \frac{b^{25}}{a^{25}} \qquad \text{(property of fractions)}$$

With a little practice such solutions can be done more efficiently by omitting some steps. For example:

$$\left(\frac{a^{-2}b^{3}}{a^{3}b^{-2}}\right)^{5} = \frac{(a^{-2}b^{3})^{5}}{(a^{3}b^{-2})^{5}} = \frac{a^{-10}b^{15}}{a^{15}b^{-10}} = \frac{b^{25}}{a^{25}}$$

It can be shown that the rules presented in this section apply for all integral exponents. Here is a proof of Rule 4, for zero and negative integral exponents.

If $m = 0$, then $(ab)^0 = 1$. Also $a^0 b^0 = 1 \cdot 1 = 1$. Thus $(ab)^0 = 1 = a^0 b^0$.
If $m < 0$, let $m = -k$, where k is the appropriate positive integer. Thus

$$(ab)^m = (ab)^{-k} \qquad \text{(substitution)}$$

$$= \frac{1}{(ab)^k} \qquad \text{(Definition 2)}$$

$$= \frac{1}{a^k b^k} \qquad \text{(Rule 4 since } k \text{ is a positive integer)}$$

$$= \frac{1}{a^k} \cdot \frac{1}{b^k} \qquad \text{(property of fractions)}$$

$$= a^{-k} b^{-k} \qquad \text{(Definition 2)}$$

$$= a^m b^m \qquad \text{(substitution)}$$

The remaining rules can be proved in a similar manner. These proofs demonstrate that Definition 2 successfully preserves the basic Rules 1 to 5 of computing with exponents. It should be noted that in view of Definition 2 the three cases of Rule 2 can now be condensed into this single form.

RULE 2. (revised): $\dfrac{b^m}{b^n} = b^{m-n}$

Illustrations:

$$\frac{3^4}{3^2} = 3^2 \qquad \frac{3^2}{3^4} = 3^{-2} \qquad \frac{3^2}{3^2} = 3^0 = 1$$

$$\frac{x^4}{x^2} = x^2 \qquad \frac{x^2}{x^4} = x^{-2} \qquad \frac{x^2}{x^2} = x^0 = 1$$

Example 5   Simplify $\dfrac{x^3 y^4}{x^5 y}$ and express the answer using positive or negative exponents.

Solution

$$\frac{x^3 y^4}{x^5 y} = x^{3-5} y^{4-1} = x^{-2} y^3$$

Pay careful attention to details when working with exponents. Careless mistakes are often made because of misconception of the basic rules. The next page contains a list of some common errors.

CAUTION! LEARN TO AVOID MISTAKES LIKE THESE:

WRONG	RIGHT
$5^2 \cdot 5^4 = 5^8$ (Do not multiply exponents.) $5^2 \cdot 5^4 = 25^6$ (Do not multiply the base numbers.)	$5^2 \cdot 5^4 = 5^6$ (Rule 1)
$\dfrac{5^6}{5^2} = 5^3$ (Do not divide the exponents.) $\dfrac{5^6}{5^2} = 1^4$ (Do not divide the base numbers.)	$\dfrac{5^6}{5^2} = 5^4$ (Rule 2)
$(5^2)^6 = 5^8$ (Do not add the exponents.)	$(5^2)^6 = 5^{12}$ (Rule 3)
$(-2)^4 = -2^4$ (Misreading the parentheses)	$(-2)^4 = (-1)^4 2^4 = 2^4$ (Rule 4 or Definition 1)
$(-5)^0 = -1$ (Misreading Definition 2)	$(-5)^0 = 1$ (Definition 2)
$2^{-3} = -\dfrac{1}{2^3}$ (Misreading Definition 2)	$2^{-3} = \dfrac{1}{2^3}$ (Definition 2)
$\dfrac{2^3}{2^{-4}} = 2^{3-4} = 2^{-1}$ (Carelessness in subtracting exponents)	$\dfrac{2^3}{2^{-4}} = 2^{3-(-4)} = 2^7$ (Rule 2)
$5^3 + 5^3 = 5^6$ (Adding exponents does not apply because of plus sign)	$5^3 + 5^3 = (1 + 1)5^3 = 2 \cdot 5^3$ (Distributive)
$x^2 \cdot x^3 = 2x^6$ (Do not add the base numbers.)	$x^3 \cdot x^3 = x^6$ (Rule 1)
$(a + b)^{-1} = a^{-1} + b^{-1}$ (Wrong use of Definition 2)	$(a + b)^{-1} = \dfrac{1}{a + b}$ (Definition 2)

EXERCISES 2.1

Classify each statement as true or false. If it is false, correct the right side of the equality to get a true statement.

1. $(5^3)^4 = 5^7$

2. $(3^4)^5 = (5^4)^3$

3. $3^4 \cdot 3^2 = 3^8$

4. $\dfrac{10^4}{5^4} = 2^4$

5. $3^4 + 3^4 = 3^8$

6. $a^2 + a^2 = 2a^2$

7. $\dfrac{9^3}{9^3} = 9$

8. $2^5 \cdot 2^2 = 4^7$

9. $(-27)^0 = 1$

10. $\dfrac{1}{2^{-3}} = -2^3$

11. $(2^0)^3 = 2^3$

12. $[(2^3)^2]^0 = (2^3)^1$

13. $6 \cdot 6^2 \cdot 6^3 \cdot 6^4 = 6^{10}$

14. $\{[(2^1)^2]^3\}^4 = 2^{10}$

15. $(-2)^3 = -(2^3)$

16. $(2 + \pi)^{-2} = \dfrac{1}{4} + \dfrac{1}{\pi^2}$

17. $1^{-1} = 1$

18. $\dfrac{2^{-5}}{2^3} = 2^{-2}$

Evaluate.

19. 10^5

20. $\left(-\dfrac{1}{3}\right)^3$

21. $\left(\dfrac{1}{10}\right)^{-1}$

22. $\left(\dfrac{2}{3}\right)^0 + \left(\dfrac{2}{3}\right)^1 + \left(\dfrac{2}{3}\right)^2$

23. $\left(\dfrac{2}{3}\right)^{-2} + \left(\dfrac{2}{3}\right)^{-1}$

24. $\dfrac{3^2}{3^0}$

25. $\dfrac{(-5)^5}{(-5)^3}$

26. $\left[\left(\dfrac{1}{2}\right)^3\right]^2$

27. $2^4 5^2 10^{-3}$

28. $\left(\dfrac{1}{2}\right)^4 (-2)^4$

29. $(25^0)^{100}$

30. $10^{-2} + 10^{-1} + 10^0 + 10^1 + 10^2$

31. $\dfrac{8^3}{16^2}$

32. $[(-7)^2(-3)^2]^{-1}$

33. $\dfrac{(-10)^3 + 100^2}{30^2}$

34. $\left(\dfrac{1}{2}\right)^2 \left(\dfrac{1}{3}\right)^2 \left(\dfrac{1}{4}\right)^2 (12)^2$

35. $\dfrac{(-2)^3 10^3}{(-20)^3}$

36. $(-3)^5 \left(-\dfrac{1}{6}\right)^5 (-2)^5$

37. $(-0.75)^3 8^3$

38. $\dfrac{2^3 3^4 4^5}{2^2 3^3 4^4}$

39. $\dfrac{3^{-3}}{4^{-3}}$

Simplify. Express each answer using positive exponents only.

40. $(x^{-3})^2$

41. $(x^3)^{-2}$

42. $x^3 \cdot x^9$

43. $\dfrac{x^9}{x^3}$

44. $(2a)^3(3a)^2$

45. $(-2x^3y)^2(-3x^2y^2)^3$

46. $(-2a^2b^0)^4$

47. $(2x^3y^2)^0$

48. $\dfrac{(x^2y)^4}{(xy)^2}$

49. $\left(\dfrac{3a^2}{b^3}\right)^2\left(\dfrac{-2a}{3b}\right)^2$

50. $\left(\dfrac{x^3}{y^2}\right)^4\left(\dfrac{-y}{x^2}\right)^2$

51. $\dfrac{(x-2y)^6}{(x-2y)^2}$

52. $\dfrac{x^{-2}y^3}{x^3y^{-4}}$

53. $\dfrac{(x^{-2}y^2)^3}{(x^3y^{-2})^2}$

54. $\dfrac{5x^0y^{-2}}{x^{-1}y^{-2}}$

55. $\dfrac{(xy)^{-2}}{(xy)^3}$

56. $\dfrac{(a+b)^{-2}}{(a+b)^{-8}}$

*57. $\dfrac{1}{a^{-1}+b^{-2}}$

58. $x^{-2}+y^{-2}$

59. $(a^{-2}b^3)^{-1}$

60. $\left(\dfrac{x^{-2}}{y^3}\right)^{-1}$

*61. According to Definition 2, 0^0 is undefined. You may wonder why we do not allow $0^0 = 1$. The following suggests how this agreement would lead to difficulties. Suppose $0^0 = 1$. Then

$$1 = \frac{1}{1}$$

$$= \frac{1^0}{0^0}$$

$$= \left(\frac{1}{0}\right)^0$$

(a) What rule of exponents is being used in the last step?
(b) What went wrong?

2.2

**Radicals and
Rational
Exponents**

In the last section b^n was defined only for integers n. The next stage is to extend this concept to include fractional exponents—that is, to give meaning to such expressions as $4^{1/2}$ and $(-8)^{2/3}$.

Since $5^2 = 25 = (-5)^2$ we say that 5 and -5 are *square roots* of 25. The positive square root, 5, is written as $\sqrt{25}$; it is sometimes referred to as the *principal square root* of 25. The negative square root is written as $-\sqrt{25}$. It is a fundamental property of real numbers that every positive real number a has exactly one positive nth root.

For $a > 0$, there is exactly one positive number x such that $x^n = a$. We say that x is the nth root of a, written as $x = \sqrt[n]{a}$, for $n \geq 2$.

To illustrate this definition, note that $2^4 = 16$. We say that 2 is the 4th root of 16 and write $2 = \sqrt[4]{16}$. Remember, even though $(-2)^4 = 16$, we do *not* say $\sqrt[4]{16} = -2$ simply because the symbol $\sqrt[n]{a}$ is reserved for the positive root for $a > 0$.

When $a < 0$ and n is odd, there is always a negative nth root. For example, $(-2)^3 = -8$ implies that -2 is the cube root of -8. There can only be one such real root and again we use the designation $\sqrt[n]{a}$ for this principal root; that is, $\sqrt[3]{-8} = -2$.

When $a < 0$ and n is even, no real nth roots are possible. For example, there is no real number x such that $x = \sqrt{-4}$; we can never have $x^2 = -4$. The square of a number is never negative.

We can summarize these various cases as follows.

Definition 1

Let a be a real number and n a positive integer, $n \geq 2$.
   (i) If $a > 0$, then $\sqrt[n]{a}$ is the positive nth root x; that is, $x^n = a$.
  (ii) If $a < 0$ and n is odd, then $\sqrt[n]{a}$ is the negative nth root x; that is, $x^n = a$.
 (iii) If $a < 0$ and n is even, then $\sqrt[n]{a}$ is not a real number.
  (iv) $\sqrt[n]{0} = 0$.

The symbol $\sqrt[n]{a}$ is also said to be a *radical*; $\sqrt{}$ is the radical sign, n is the *index* or *root*, and a is called the *radicand*.

Here is a list of the basic laws for radicals. It is assumed here that all radicals exist according to the definition and, as usual, no denominator can be zero.

Rules for radicals

1. $(\sqrt[n]{a})^n = \sqrt[n]{a^n} = a$

2. $\sqrt[n]{a} \cdot \sqrt[n]{b} = \sqrt[n]{ab}$

3. $\dfrac{\sqrt[n]{a}}{\sqrt[n]{b}} = \sqrt[n]{\dfrac{a}{b}}$

4. $\sqrt[m]{\sqrt[n]{a}} = \sqrt[mn]{a}$

Example 1   Simplify: **(a)** $(\sqrt[4]{7})^4$ **(b)** $\sqrt{9 \cdot 25}$ **(c)** $\sqrt[4]{\frac{16}{81}}$ **(d)** $\sqrt[2]{\sqrt[3]{64}}$

Solution

(a) $(\sqrt[4]{7})^4 = 7$ by Rule 1.

(b) $\sqrt{9 \cdot 25} = \sqrt{9} \cdot \sqrt{25} = 3 \cdot 5 = 15$ by Rule 2.

(c) $\sqrt[4]{\frac{16}{81}} = \frac{\sqrt[4]{16}}{\sqrt[4]{81}} = \frac{2}{3}$ by Rule 3. (*Note:* $2^4 = 16$ and $3^4 = 81$.)

(d) $\sqrt[2]{\sqrt[3]{64}} = \sqrt[6]{64} = 2$ by Rule 4. (*Note:* $2^6 = 64$.)

In applying the rules for radicals care must be taken to avoid the type of error that results from the incorrect assumption of the existence of an nth root. For example, $\sqrt[2]{-4}$ is *not* a real number, but if this is not noticed then *false* results such as the following occur:

$$4 = \sqrt{16} = \sqrt{(-4)(-4)} = \sqrt{-4} \cdot \sqrt{-4} = (\sqrt{-4})^2 = -4$$

TEST YOUR UNDERSTANDING

Use Definition 1 and Rules 1 to 4 to evaluate each of the following.

1. $\sqrt{121}$ **2.** $\sqrt[3]{-64}$ **3.** $\sqrt[5]{32}$

4. $\sqrt{\sqrt{81}}$ **5.** $\sqrt{(25)(49)}$ **6.** $\sqrt[3]{\frac{27}{125}}$

7. $(\sqrt[4]{2})^4$ **8.** $\sqrt[3]{(-1000)(343)}$ **9.** $\sqrt[4]{\frac{256}{81}}$

10. $\sqrt{(9)(144)(225)}$ **11.** $\sqrt[3]{\frac{(-8)(125)}{27}}$ **12.** $\sqrt{\frac{144}{49}} \cdot \sqrt{\frac{196}{36}}$

We are now ready to make the promised extension of the exponential concept to include fractional exponents. Once again our guideline will be to preserve the earlier rules for integer exponents.

The first step is to consider exponents of the form $\frac{1}{n}$, where n is a positive integer. (Assume that $n \geq 2$ since the case $n = 1$ is trivial.) If our rules for exponents are to work, then $(b^{1/n})^n = b^{(1/n)(n)} = b$. So $b^{1/n}$ is the nth root of b (provided such a root exists). The suggestion is to let $b^{1/n} = \sqrt[n]{b}$, which is exactly what is done.

> **Definition 2**
>
> For a real number b and a positive integer n ($n \geq 2$)
>
> $$b^{1/n} = \sqrt[n]{b}$$
>
> provided that $\sqrt[n]{b}$ exists.

Example 2   Simplify: **(a)** $4^{1/2}$ **(b)** $(-8)^{1/3}$ **(c)** $(-1)^{1/2}$

Solution   **(a)** $4^{1/2} = \sqrt{4} = 2$;   **(b)** $(-8)^{1/3} = \sqrt[3]{-8} = -2$;
 (c) $(-1)^{1/2} = \sqrt{-1}$; this is not a real number and therefore $(-1)^{1/2}$
 has no meaning.

Squaring $8^{1/3} = \sqrt[3]{8}$ produces $(8^{1/3})^2 = (\sqrt[3]{8})^2$. But to preserve the rules
for exponents we write $(8^{1/3})^2 = 8^{2/3}$. Thus it is reasonable to let $8^{2/3} = (\sqrt[3]{8})^2$.
Likewise, $8^{2/3} = (8^2)^{1/3} = \sqrt[3]{8^2}$. These observations lead to this definition:

> **Definition 3**
>
> If $\dfrac{m}{n}$ is a rational number in lowest terms with $n > 0$, then
>
> $$b^{m/n} = (\sqrt[n]{b})^m = \sqrt[n]{b^m}$$
>
> provided that $\sqrt[n]{b}$ and $\sqrt[n]{b^m}$ exist.

Note that a rational number can always be expressed with a positive denom-
inator; for example, $\dfrac{2}{-3} = \dfrac{-2}{3}$. Note also that we may write $b^{m/n}$ as follows:
$(b^{1/n})^m = (b^m)^{1/n}$.

Example 3   Simplify: $(-64)^{2/3}$.

Solution $(-64)^{2/3} = (\sqrt[3]{-64})^2 = (-4)^2 = 16$.
We may also simplify by using Definition 3, as follows:

$$(-64)^{2/3} = \sqrt[3]{(-64)^2} = \sqrt[3]{4096} = 16$$

Obviously, the first approach is much less work. A little experience will lead you
to use the simpler method.

Definition 3 calls for the fractional exponent to be in lowest terms, so as to
avoid situations as in Example 4.

Example 4 Find what is wrong in the following "proof" that $2 = -2$.

$$-2 = \sqrt[3]{-8} = (-8)^{1/3} = (-8)^{2/6} = \sqrt[6]{(-8)^2} = \sqrt[6]{64} = 2$$

Solution Since $\frac{2}{6}$ is not in lowest terms, we are *not* permitted to write

$$(-8)^{2/6} = \sqrt[6]{(-8)^2}$$

If we restrict ourselves to the roots of only *positive* numbers, then the requirement that the exponent $\frac{m}{n}$ must be in lowest terms can be relaxed. For example:

$$2 = 8^{1/3} = 8^{2/6} = \sqrt[6]{8^2} = \sqrt[6]{64} = 2$$

Since the definition of fractional exponents was made on the basis of preserving the rules for integral exponents, it should come as no surprise that these rules apply. This principle is illustrated in the examples that follow.

Example 5 Evaluate: $27^{2/3} - 16^{-1/4}$.

Solution

$$27^{2/3} - 16^{-1/4} = (\sqrt[3]{27})^2 - \frac{1}{16^{1/4}}$$

$$= 3^2 - \frac{1}{\sqrt[4]{16}}$$

$$= 9 - \tfrac{1}{2}$$

$$= 8\tfrac{1}{2}$$

Example 6 Simplify, and express the result with positive exponents only.

$$\frac{x^{2/3}y^{-2}z^2}{x^{1/2}y^{1/2}z^{-1}}$$

Solution

$$\frac{x^{2/3}y^{-2}z^2}{x^{1/2}y^{1/2}z^{-1}} = x^{2/3-1/2}y^{-2-1/2}z^{2-(-1)}$$

$$= x^{1/6}y^{-5/2}z^3$$

$$= \frac{x^{1/6}z^3}{y^{5/2}}$$

CAUTION! LEARN TO AVOID MISTAKES LIKE THESE:

WRONG	RIGHT
$\sqrt{25} = \pm 5$	$\sqrt{25} = 5$
$16^{3/4} = (\sqrt[3]{16})^4$	$16^{3/4} = (\sqrt[4]{16})^3$ or $\sqrt[4]{16^3}$
$(-2)^{-1/3} = 2^{1/3}$	$(-2)^{-1/3} = \dfrac{1}{(-2)^{1/3}} = \dfrac{1}{\sqrt[3]{-2}}$

EXERCISES 2.2

Write with a fractional exponent.

1. $\sqrt{11}$ **2.** $\sqrt[3]{21}$ **3.** $\sqrt[4]{9}$ **4.** $\sqrt[3]{-10}$

5. $\sqrt[3]{6^2}$ **6.** $(\sqrt[3]{-7})^2$ **7.** $\left(\sqrt[5]{-\dfrac{1}{5}}\right)^3$ **8.** $\dfrac{1}{\sqrt[3]{4^2}}$

Write in radical form.

9. $3^{1/2}$ **10.** $7^{1/3}$ **11.** $(-9)^{1/3}$ **12.** $6^{2/3}$

13. $2^{-1/2}$ **14.** $7^{-3/2}$ **15.** $\left(\dfrac{3}{4}\right)^{-1/4}$ **16.** $\left(-\dfrac{1}{3}\right)^{-2/3}$

Classify each statement as true or false. If it is false, correct the right side of the equality to get a true statement.

17. $\sqrt[3]{-27} = -3$ **18.** $4^{1/2} = -2$ **19.** $(-8)^{2/3} = 4$

20. $64^{3/4} = (\sqrt[3]{64})^4$ **21.** $(-\frac{1}{8})^{-1/3} = 2$ **22.** $(\sqrt{100})^{-1} = -10$

23. $\sqrt{1.44} = 0.12$ **24.** $(0.25)^{3/2} = \frac{1}{8}$ **25.** $\sqrt{\frac{49}{9}} = \frac{7}{3}$

Evaluate.

26. $121^{1/2}$ **27.** $125^{1/3}$ **28.** $\sqrt[3]{-64}$

29. $81^{-1/2}$ **30.** $(-64)^{1/3}$ **31.** $(64)^{-2/3}$

32. $(-64)^{-2/3}$ **33.** $(\sqrt[3]{-125})^2$ **34.** $\sqrt[3]{(-125)^2}$

35. $\left(\dfrac{8}{27}\right)^{-2/3}$ **36.** $\dfrac{9^{1/2}}{\sqrt[3]{27}}$ **37.** $\dfrac{\sqrt{9}}{27^{-1/3}}$

38. $\sqrt[4]{\dfrac{16}{81}}$ **39.** $\sqrt[3]{(-125)(-1000)}$ **40.** $\sqrt{(4)(9)(49)(100)}$

41. $\left(\dfrac{1}{4}\right)^{3/2} \cdot \left(-\dfrac{1}{8}\right)^{2/3}$ **42.** $\sqrt{\dfrac{2}{3}} \cdot \sqrt{\dfrac{75}{98}}$ **43.** $\sqrt{\sqrt{625}}$

44. $\sqrt[5]{(-243)^2} \cdot (49)^{-1/2}$ **45.** $\sqrt{144 + 25}$ **46.** $\sqrt{144} + \sqrt{25}$

47. $\left(\frac{1}{8}\right)^{1/3} + \left(\frac{1}{27}\right)^{1/3}$ **48.** $\left(\frac{8}{27}\right)^{2/3} + \left(-\frac{32}{243}\right)^{2/5}$ **49.** $\sqrt[3]{\frac{-216}{8(10^3)}}$

50. $(4 \cdot 4^2 \cdot 4^4)^{1/2}$ **51.** $(\sqrt{0.01} \cdot \sqrt[3]{0.001})^2$ **52.** $(0.25)^{1/2} + (2.25)^{1/2}$

Simplify, and express all answers with positive exponents. (Assume that all letters represent positive numbers.)

53. $(8a^3b^{-9})^{2/3}$ **54.** $(27a^{-3}b^9)^{-2/3}$ **55.** $(a^{-4}b^{-8})^{3/4}$

56. $(a^{2/3}b^{1/2})(a^{1/3}b^{-1/2})$ **57.** $(a^{-1/2}b^{1/3})(a^{1/2}b^{-1/3})$ **58.** $\dfrac{a^2b^{-1/2}c^{1/3}}{a^{-3}b^0c^{-1/3}}$

59. $\left(\dfrac{64a^6}{b^{-9}}\right)^{2/3}$ **60.** $\dfrac{(49a^{-4})^{-1/2}}{(81b^6)^{-1/2}}$ **61.** $\left(\dfrac{a^{-2}b^3}{a^4b^{-3}}\right)^{-1/2}\left(\dfrac{a^4b^{-5}}{ab}\right)^{-1/3}$

Simplify, and express the answers without radicals, using only positive exponents. (Assume that n is a positive integer and all other letters represent positive numbers.)

*62. $\sqrt{\dfrac{x^n}{x^{n-2}}}$ ***63.** $\left(\dfrac{x^n}{x^{n-2}}\right)^{-1/2}$ ***64.** $\sqrt[3]{\dfrac{x^{3n+1}y^n}{x^{3n+4}y^{4n}}}$

2.3
Simplifying Radicals

Computations with radicals cause no difficulty when the radicals represent rational numbers, as in the following.

$$\sqrt{36} + \sqrt[3]{-27} = 6 + (-3) = 3$$

When the radicals involve irrational numbers, however, special methods are needed.

First let us learn how to "extract" nth roots out of the radical $\sqrt[n]{a}$. As you see in the following illustration, the property $\sqrt[n]{ab} = \sqrt[n]{a} \cdot \sqrt[n]{b}$ is fundamental in this work.

$$\sqrt{20} = \sqrt{4 \cdot 5} = \sqrt{4} \cdot \sqrt{5} = 2\sqrt{5}$$

We say that $2\sqrt{5}$ is the *simplified form* of $\sqrt{20}$. Here is another illustration.

$$\sqrt[3]{297} = \sqrt[3]{27 \cdot 11} = \sqrt[3]{27} \cdot \sqrt[3]{11} = 3\sqrt[3]{11}$$

The key to this procedure is knowing how to "break up" the radicand; that is, to find that $297 = 27 \cdot 11$. A technique that can be used here is to try to divide the radicand 297 by perfect (positive) cubes, beginning with $2^3 = 8$. Then continue until all such factors of 297 have been found. In this case there is no need to try $2^3 = 8$ since 297 is not even. Thus begin with $3^3 = 27$ and we find that $297 = 27 \cdot 11$. Since 11 has no more such factors (other than $1^3 = 1$), we stop. Here are some illustrations:

$$\sqrt{48} = \sqrt{16 \cdot 3} = \sqrt{16} \cdot \sqrt{3} = 4\sqrt{3}$$

$$\sqrt[3]{-54} = \sqrt[3]{(-27)(2)} = \sqrt[3]{-27} \cdot \sqrt[3]{2} = -3\sqrt[3]{2}$$

$$\sqrt[4]{80} = \sqrt[4]{16 \cdot 5} = \sqrt[4]{16} \cdot \sqrt[4]{5} = 2\sqrt[4]{5}$$

Let us now turn our attention to the addition of radicals. First we note that, in general, $\sqrt{a} + \sqrt{b} \neq \sqrt{a+b}$. For example, $\sqrt{4} + \sqrt{9} = 2 + 3 = 5$ but $\sqrt{4+9} = \sqrt{13} \neq 5$. However, addition and subtraction of radicals are possible under certain conditions through use of the distributive property. Consider, for example, this sum:

$$3\sqrt{5} + 4\sqrt{5}$$

Although we will usually perform the computation mentally, we can think this example through by first using x to replace $\sqrt{5}$. Study these two parallel treatments:

$$
\begin{aligned}
3\sqrt{5} + 4\sqrt{5} &= 3x + 4x \\
&= (3+4)x \\
&= 7x \\
&= 7\sqrt{5}
\end{aligned}
\qquad
\begin{aligned}
3\sqrt{5} + 4\sqrt{5} &= (3+4)\sqrt{5} \\
&= 7\sqrt{5}
\end{aligned}
$$

In order for us to be able to add or subtract radicals, they must have the same index and the same radicand. Thus the following expressions *cannot* be further simplified:

$$\sqrt{3} + \sqrt{5} \qquad \sqrt[3]{2} + \sqrt[4]{3}$$

$$\sqrt[3]{2} + \sqrt{2} \qquad \sqrt{x} + \sqrt{y}$$

Often several radicals that appear to be different can be combined if each one is simplified first, as shown in the next example.

Example 1   Simplify: $\sqrt{50} - \sqrt{18} + \sqrt{45}$.

Solution   Although each radical has the same index, they do not have the same radicand. However, each can be simplified.

$$\sqrt{50} = \sqrt{25 \cdot 2} = 5\sqrt{2}$$
$$\sqrt{18} = \sqrt{9 \cdot 2} = 3\sqrt{2}$$
$$\sqrt{45} = \sqrt{9 \cdot 5} = 3\sqrt{5}$$

Thus

$$\sqrt{50} - \sqrt{18} + \sqrt{45} = 5\sqrt{2} - 3\sqrt{2} + 3\sqrt{5}$$
$$= 2\sqrt{2} + 3\sqrt{5}$$

At times a fraction with a radical denominator can be simplified by a process known as **rationalizing the denominator**. This procedure consists of eliminating a radical from the denominator of a fraction. For example, let us consider the fraction $4/\sqrt{2}$. To rationalize the denominator, we multiply numerator and denominator by $\sqrt{2}$.

$$\frac{4}{\sqrt{2}} = \frac{4}{\sqrt{2}} \cdot 1$$
$$= \frac{4}{\sqrt{2}} \cdot \frac{\sqrt{2}}{\sqrt{2}}$$
$$= \frac{4\sqrt{2}}{2}$$
$$= 2\sqrt{2}$$

There are several reasons why we may wish to rationalize a denominator. To obtain a rational approximation for a radical, it is easier to multiply than to divide. For example, $\sqrt{2} = 1.414$ to three decimal places. It is certainly easier to multiply 1.414 by 2 than to divide 4 by 1.414.

Another reason for rationalizing a denominator is shown in the following example.

Example 2 Simplify: $\dfrac{6}{\sqrt{3}} + 2\sqrt{75} - \sqrt{3}$.

Solution Rationalize the denominator in the first term:

$$\frac{6}{\sqrt{3}} = \frac{6}{\sqrt{3}} \cdot \frac{\sqrt{3}}{\sqrt{3}} = \frac{6\sqrt{3}}{3} = 2\sqrt{3}$$

Note that $2\sqrt{75} = 2\sqrt{25 \cdot 3} = 10\sqrt{3}$ and that $\sqrt{3}$ means $1 \cdot \sqrt{3}$. Thus

$$\frac{6}{\sqrt{3}} + 2\sqrt{75} - \sqrt{3} = 2\sqrt{3} + 10\sqrt{3} - 1\sqrt{3} = 11\sqrt{3}$$

Simplify each expression, if possible.

1. $\sqrt{8} + \sqrt{32}$ **2.** $\sqrt{12} + \sqrt{48}$ **3.** $\sqrt{45} - \sqrt{20}$

4. $2\sqrt{9} - \sqrt{18}$ **5.** $\sqrt[3]{16} + \sqrt[3]{54}$ **6.** $\sqrt[3]{128} + \sqrt[3]{125}$

7. $\dfrac{8}{\sqrt{2}} + \sqrt{98}$ **8.** $\dfrac{9}{\sqrt{3}} + \sqrt{300}$ **9.** $2\sqrt{20} - \dfrac{5}{\sqrt{5}}$

10. $3\sqrt{63} - \dfrac{14}{\sqrt{7}}$

Sometimes it is necessary to simplify a radical that contains a variable, such as $\sqrt{x^2}$. In this example, the usual reaction is to claim that $\sqrt{x^2} = x$. But this is not true, since if $x = -5$ then the assumption $\sqrt{x^2} = x$ would give the false result $\sqrt{(-5)^2} = -5$. In fact, we have $\sqrt{(-5)^2} = \sqrt{25} = 5 = |-5|$, and, in general we have the following important result:

> For all real numbers n, $\sqrt{n^2} = |n|$.

This result is applied in the examples that follow.

Example 3 Simplify: $\sqrt{75x^2}$.

Solution

$$\sqrt{75x^2} = \sqrt{25 \cdot 3} \cdot \sqrt{x^2}$$
$$= 5\sqrt{3}\,|x|$$

Example 4 Simplify: $2\sqrt{8x^3} + 3x\sqrt{32x} - x\sqrt{18x}$.

Solution For this problem, $x \geq 0$. (Why?) Thus we need not make use of absolute-value notation.

$$2\sqrt{8x^3} = 2\sqrt{4 \cdot 2 \cdot x^2 \cdot x} = 4x\sqrt{2x}$$
$$3x\sqrt{32x} = 3x\sqrt{16 \cdot 2x} = 12x\sqrt{2x}$$
$$x\sqrt{18x} = x\sqrt{9 \cdot 2x} = 3x\sqrt{2x}$$

In each case the radicand is the same so that the distributive property can be used to simplify.

$$4x\sqrt{2x} + 12x\sqrt{2x} - 3x\sqrt{2x} = (4x + 12x - 3x)\sqrt{2x}$$
$$= 13x\sqrt{2x}$$

CAUTION! LEARN TO AVOID MISTAKES LIKE THESE:

WRONG	RIGHT
$\sqrt{9 + 16} = \sqrt{9} + \sqrt{16}$	$\sqrt{9 + 16} = \sqrt{25}$
$(a + b)^{1/3} = a^{1/3} + b^{1/3}$	$(a + b)^{1/3} = \sqrt[3]{a + b}$
$a^{-1/2} + b^{-1/2} = \dfrac{1}{\sqrt{a + b}}$	$a^{-1/2} + b^{-1/2} = \dfrac{1}{\sqrt{a}} + \dfrac{1}{\sqrt{b}}$
$\sqrt[3]{8} \cdot \sqrt[2]{8} = \sqrt[6]{64}$	$\sqrt[3]{8} \cdot \sqrt[2]{8} = 2 \cdot 2\sqrt{2} = 4\sqrt{2}$

EXERCISES 2.3

Simplify.

1. $\sqrt{18} + \sqrt{98}$
2. $\sqrt{48} + \sqrt{75}$
3. $2\sqrt{200} - 5\sqrt{8}$
4. $3\sqrt{45} - 2\sqrt{20}$
5. $\sqrt[3]{128} + \sqrt[3]{16}$
6. $\sqrt[3]{24} + \sqrt[3]{81}$
7. $\sqrt{50} + \sqrt{32} - \sqrt{8}$
8. $\sqrt{12} - \sqrt{3} + \sqrt{108}$
9. $\dfrac{8}{\sqrt{2}} + 2\sqrt{50}$
10. $\dfrac{12}{\sqrt{3}} - \sqrt{12}$
11. $\sqrt[3]{56x} + \sqrt[3]{7x}$
12. $\sqrt[3]{54x} + \sqrt[3]{250x}$
13. $3\sqrt{8x^2} - \sqrt{50x^2}$
14. $5\sqrt{75x^2} - 2\sqrt{12x^2}$
15. $\sqrt[4]{32} + \sqrt[4]{162}$
16. $\sqrt[5]{32} + \sqrt[5]{64}$
17. $3\sqrt{10} + 4\sqrt{90} - 5\sqrt{40}$
18. $3\sqrt{24} - \sqrt{54} + 2\sqrt{150}$
19. $\dfrac{1}{\sqrt{2}} + 3\sqrt{72} - 2\sqrt{2}$
20. $\dfrac{2}{\sqrt{3}} + 10\sqrt{3} - 2\sqrt{12}$
21. $\sqrt{24} + \sqrt{54} - \sqrt{18}$
22. $\sqrt{36} + \sqrt{28} + \sqrt{63}$
23. $\sqrt{18x} + \sqrt{50x} - \sqrt{2x}$
24. $10\sqrt{3x} - 2\sqrt{75x} + 3\sqrt{243x}$
25. $3\sqrt{9x^2} + 2\sqrt{16x^2} - \sqrt{25x^2}$
26. $\sqrt{2x^2} + 5\sqrt{32x^2} - 2\sqrt{98x^2}$
27. $\sqrt{x^2y} + \sqrt{8x^2y} + \sqrt{200x^2y}$
28. $\sqrt{72xy} + 2\sqrt{2xy} + \sqrt{128xy}$
29. $\sqrt{20a^3} + a\sqrt{5a} + \sqrt{80a^3}$
*30. $\sqrt{12ab^3} + \sqrt{27ab^3} + 2b\sqrt{ab}$

Rationalize the denominator and simplify.

31. $\dfrac{20}{\sqrt{5}}$
32. $\dfrac{24}{\sqrt{6}}$
33. $\dfrac{8x}{\sqrt{2}}$
34. $\dfrac{9y}{\sqrt{3}}$
35. $\dfrac{4}{\sqrt{2}} + \dfrac{6}{\sqrt{3}}$
36. $\dfrac{10}{\sqrt{5}} + \dfrac{8}{\sqrt{4}}$

37. $\dfrac{1}{\sqrt{18}}$

38. $\dfrac{1}{\sqrt{27}}$

39. $\dfrac{24}{\sqrt{3x^2}}$

40. $\dfrac{20}{\sqrt{5x^2}}$

41. $\dfrac{8}{\sqrt[3]{2}}$

42. $\dfrac{12}{\sqrt[3]{3}}$

43. Rationalize the denominator: $\dfrac{1}{\sqrt{3} + \sqrt{2}}$. (*Hint:* Multiply numerator and denominator by $\sqrt{3} - \sqrt{2}$.)

Follow the hint of Exercise 43 to rationalize each denominator.

44. $\dfrac{12}{\sqrt{5} - \sqrt{3}}$

45. $\dfrac{20}{3 - \sqrt{2}}$

46. $\dfrac{14}{\sqrt{2} - 3}$

*47. Use the result $\sqrt{n^2} = |n|$ and the rules for radicals to prove that $|xy| = |x| \cdot |y|$, where x and y are real numbers.

2.4
Fundamental Operations with Polynomials

The expression $5x^3 - 7x^2 + 4x - 12$ is called a **polynomial in the variable** x. Its *degree* is 3, because 3 is the largest power of the variable x. The *terms* of this polynomial are $5x^3$, $-7x^2$, $4x$, and -12. The *coefficients* are 5, -7, 4, and -12.

All the exponents of the variable of a polynomial must be nonnegative integers. Therefore $x^3 + x^{1/2}$ and $x^{-2} + 3x + 1$ are *not* polynomials because of the fractional and negative exponents.

A nonzero constant, like 7, is also classified as a polynomial of degree zero since $7 = 7x^0$. The number zero may also be referred to as a constant polynomial, but it is not assigned any degree.

A polynomial is in *standard form* if its terms are arranged so that the powers of the variable are in descending or ascending order. Here are some illustrations.

POLYNOMIAL	DEGREE	STANDARD FORM
$x^2 - 3x + 12$	2	Yes
$\frac{2}{3}x^{10} - 4x^2 + \sqrt{2}x^4$	10	No
$32 - y^5 + 2y^3$	5	No
$6 + 2x - x^2 + x^3$	3	Yes

Some of the preceding polynomials have "missing" terms. For example, $x^3 - 3x + 12$ has no x^2 term, but it is still a third-degree polynomial. The highest power of x determines the degree, and any or all lesser powers may be missing.

In general, an nth-degree polynomial in the variable x may be written in these standard forms:

$$a_n x^n + a_{n-1} x^{n-1} + \cdots + a_2 x^2 + a_1 x + a_0$$

$$a_0 + a_1 x + a_2 x^2 + \cdots + a_{n-1} x^{n-1} + a_n x^n$$

The coefficients $a_n, a_{n-1}, \ldots, a_0$ are real numbers. The *leading coefficient* is $a_n \neq 0$, and a_0 is called the *constant term*.

In a polynomial like $3x^2 - x + 4$, the variable x represents a real number. Therefore when a specific real value is substituted for x, the result will be a real number. For instance, using $x = -3$ in this polynomial gives

$$3(-3)^2 - (-3) + 4 = 34$$

In other words, polynomials represent real numbers, regardless of the specific choice of the variable. Consequently addition, subtraction, multiplication, or division of polynomials can be performed in accordance with the basic properties for real numbers.

Adding or subtracting polynomials involves the combining of *like terms* (those having the same exponent on the variable). This can be accomplished by first rearranging and regrouping the terms (associative and commutative properties) and then combining by using the distributive property.

Example 1
(a) Add: $(4x^3 - 10x^2 + 5x + 8) + (12x^2 - 9x - 1)$.
(b) Subtract: $(4t^3 - 10t^2 + 5t + 8) - (12t^2 - 9t - 1)$.

Solution
(a) $(4x^3 - 10x^2 + 5x + 8) + (12x^2 - 9x - 1)$
$$= 4x^3 + (12x^2 - 10x^2) + (5x - 9x) + (8 - 1)$$
$$= 4x^3 + (12 - 10)x^2 + (5 - 9)x + 7$$
$$= 4x^3 + 2x^2 - 4x + 7$$

(b) $(4t^3 - 10t^2 + 5t + 8) - (12t^2 - 9t - 1)$
$$= 4t^3 - 10t^2 + 5t + 8 - 12t^2 + 9t + 1$$
$$= 4t^3 - 10t^2 - 12t^2 + 5t + 9t + 8 + 1$$
$$= 4t^3 - 22t^2 + 14t + 9$$

An alternative method is to list the polynomials in column form, putting like terms in the same columns.

$$
\begin{array}{ll}
\begin{array}{l}
4x^3 - 10x^2 + 5x + 8 \\
\underline{ 12x^2 - 9x - 1} \\
4x^3 + 2x^2 - 4x + 7
\end{array} \text{ (add)}
&
\begin{array}{l}
4t^3 - 10t^2 + 5t + 8 \\
\underline{ 12t^2 - 9t - 1} \\
4t^3 - 22t^2 + 14t + 9
\end{array} \text{ (subtract)}
\end{array}
$$

The use of the distributive property is fundamental when multiplying polynomials. Perhaps the simplest situation calls for the product of a **monomial** (a polynomial having only one term) times a "lengthier" polynomial, as follows.

$$3x^2(4x^7 - 3x^4 - x^2 + 15) = 3x^2(4x^7) - 3x^2(3x^4) - 3x^2(x^2) + 3x^2(15)$$
$$= 12x^9 - 9x^6 - 3x^4 + 45x^2$$

In the first line we used an extended version of the distributive property, namely

$$a(b - c - d + e) = ab - ac - ad + ae$$

Next observe how the distributive property is used to multiply two **binomials** (polynomials having two terms).

$$
\begin{aligned}
(2x + 3)(4x + 5) &= (2x + 3)4x + (2x + 3)5 \\
&= (2x)(4x) + (3)(4x) + (2x)(5) + (3)(5) \\
&= 8x^2 + 12x + 10x + 15 \\
&= 8x^2 + 22x + 15
\end{aligned}
$$

Here is a shortcut that can be used to multiply two binomials.

$$(2x + 3)(4x + 5): \qquad (2x + 3)(4x + 5) = 8x^2 + 22x + 15$$

$8x^2$ is the product of the *first* terms in the binomials.
$10x$ and $12x$ are the products of the *outer* and *inner* terms; $10x + 12x = 22x$.
15 is the product of the *last* terms in the binomials.

Example 2   Find the product: $(5x - 9)(6x + 2)$.

Solution

$$
\begin{aligned}
(5x - 9)(6x + 2) &= 30x^2 + (10x^2 - 54x) - 18 \\
&= 30x^2 - 44x - 18
\end{aligned}
$$

In general, we may write the product $(a + b)(c + d)$ in this way:

$$(a + b)(c + d) = ac + ad + bc + bd$$

The distributive property can easily be extended to polynomials with three terms—**trinomials**. As in the next example, the column method is a convenient way to organize your work.

Example 3   Multiply $3x^3 - 8x + 4$ by $2x^2 + 5x - 1$.

Solution

$$3x^3 - 8x + 4$$
$$2x^2 + 5x - 1$$

(add) $\begin{cases} 6x^5 \qquad\quad - 16x^3 + \ 8x^2 \\ \qquad\quad 15x^4 \qquad\quad - 40x^2 + 20x \\ \qquad\qquad\quad - \ 3x^3 \qquad\quad + \ 8x - 4 \end{cases}$ $\longleftarrow (2x^2 \text{ times } 3x^3 - 8x - 14)$
$\longleftarrow (5x \text{ times } 3x^3 - 8x - 14)$
$\longleftarrow (-1 \text{ times } 3x^3 - 8x - 14)$

$$6x^5 + 15x^4 - 19x^3 - 32x^2 + 28x - 4$$

TEST YOUR UNDERSTANDING

Combine.

1. $(x^2 + 2x - 6) + (-2x + 7)$
2. $(x^2 + 2x - 6) - (-2x + 7)$
3. $(5x^4 - 4x^3 + 3x^2 - 2x + 1) + (-5x^4 + 6x^3 + 10x)$
4. $(x^2 + x + 1) + (-3x - 4) + (6x - 5) - (x^3 + x^2)$

Find the products.

5. $(-3x)(x^3 + 2x^2 - 1)$
6. $-4x(\frac{1}{8} - \frac{1}{2}x - x^2)$
7. $(2x + 3)(3x + 2)$
8. $(6x - 1)(2x + 5)$
9. $(4x + 7)(4x - 7)$
10. $(2x - 3)(2x - 3)$
11. $(x^3 + 7x^2 - 4)(x + 2)$
12. $(x^2 - 3x + 5)(4x^4 - 3x^3 + 2x^2 - x + 1)$

More than two polynomials may be involved in a product. For example, here is a product of three polynomials:

$$(x + 2)(x + 3)(x + 4) = [(x + 2)(x + 3)](x + 4)$$
$$= (x^2 + 5x + 6)(x + 4)$$
$$= x^3 + 9x^2 + 26x + 24$$

Sometimes more than one operation is involved, as demonstrated in the next example.

Example 4   Simplify by performing the indicated operations:

$$(2x - 5)(x + 3) - x^2(3 - x)$$

Solution

$$(2x - 5)(x + 3) - x^2(3 - x) = (2x^2 + x - 15) - 3x^2 + x^3$$
$$= x^3 - x^2 + x - 15$$

Division of polynomials will be considered in Chapter 5.

EXERCISES 2.4

Simplify by performing the indicated operations.

1. $(3x + 5) + (3x - 2)$
2. $5x + (1 - 2x)$
3. $(7x + 5) - (2x + 3)$
4. $(y + 2) + (2y + 1) + (3y + 3)$
5. $(-4x^3 - 6x^2 + 11x - 15) + (5x^3 - 5x - 5)$
6. $h - (h + 2)$
7. $(x^3 + 3x^2 + 3x + 1) - (x^2 + 2x + 1)$
8. $7x - (3 - x) - 2x$
9. $5y - [y - (3y + 8)]$
10. $(x + 3) + (2x - 2) - (6x + 10)$
11. $(x^2 - 3x + 4) - (5x^2 + 2x + 1)$
12. $x(3x^2 - 2x + 5)$
13. $2x^2(2x + 1 - 10x^2)$
14. $-4t(t^4 - \frac{1}{4}t^3 + 4t^2 - \frac{1}{16}t + 1)$
15. $(x + 1)(x + 1)$
16. $(2x + 1)(2x - 1)$
17. $(4x - 2)(x + 7)$
18. $(5x + 3)(4x + 6)$
19. $(5x - 3)(4x - 6)$
20. $(5x + 3)(4x - 6)$
21. $(x + \frac{1}{2})(x - \frac{1}{4})$
22. $(12x - 8)(7x + 4)$
23. $(-2x + 3)(3x + 6)$
24. $(-2x - 3)(3x + 6)$
25. $(-2x - 3)(3x - 6)$
26. $(\frac{1}{2}x + 4)(\frac{1}{2}x - 4)$
27. $(\frac{2}{3}x + 6)(\frac{2}{3}x + 6)$
28. $(7 + 3x)(9 - 4x)$
29. $(7 - 3x)(4x - 9)$
30. $(15a + 30)(15a - 30)$
31. $(ax + b)(ax - b)$
32. $(ax - b)(ax - b)$
33. $(x - 0.1)(x + 0.1)$
34. $(x + \frac{3}{4})(x + \frac{3}{4})$
35. $(\frac{1}{5}x - \frac{1}{4})(\frac{1}{5}x - \frac{1}{4})$
36. $(x - \sqrt{3})(x + \sqrt{3})$
37. $(\sqrt{x} - 10)(\sqrt{x} + 10)$
38. $(\frac{1}{10}x - \frac{1}{100})(\frac{1}{10}x - \frac{1}{100})$
39. $(\sqrt{x} + \sqrt{2})(\sqrt{x} - \sqrt{2})$
40. $(x^2 - 3)(x^2 + 3)$
41. $(\sqrt{3} - \sqrt{2})(\sqrt{3} + \sqrt{2})$
42. $(5 - 3x)(x^4 + x^3 + x^2 + x + 1)$
43. $(x^2 + x + 9)(x^2 - 3x - 4)$
44. $(x^2 + 5x - 1)(36x^2 - 30x + 25)$
45. $(2x^2 - 3)(4x^2 + 6x + 9)$
46. $(x^{1/3} - 2)(x^{2/3} + 2x^{1/3} + 4)$
47. $(x - 2)(x^2 + 2x + 4)$
48. $(x - 2)(x^3 + 2x^2 + 4x + 8)$
49. $(x - 2)(x^4 + 2x^3 + 4x^2 + 8x + 16)$

50. $(x^n - 4)(x^n + 4)$ **51.** $(x^{2n} + 1)(x^{2n} - 2)$

52. $5(x + 5)(x - 5)$ **53.** $3x(1 - x)(1 - x)$

54. $100(\frac{1}{10}x + 1)(\frac{1}{10}x + 1)$ ***55.** $x[2x - x(3x - 5) + 6]$

56. $(x + 8)(2x - 9) + (2x - 1)(3x + 2)$

57. $(a + 1)(a + 1) - (a - 1)(a - 1)$

*58. $(2x + 3)[(x + 1)(x - 1) - (2x + 3)5]$

59. $(-x^3 - x^2 - x - 1) - (3x^3 + 2x^2 + x + 1)$

60. $(x^4 - 3x^3 - x + 1) + (2x^4 - x^2 + 3x + 7) + (-5x^4 - x^3 + 2x^2 + 2)$

*61. $(x^2 - x + 1)(x^2 - x + 1)(x^2 - x + 1)$

*62. $(2x + 1)(3x - 2)(3 - 4x)$

2.5

Factoring Polynomials

Can you solve this equation: $x^3 - 3x^2 - 4x + 12 = 0$? This may not be an easy question to answer at first. The solution is not difficult, however, once you rewrite the equation in this form:

$$x^3 - 3x^2 - 4x + 12 = (x + 2)(x - 2)(x - 3) = 0$$

Recall that the product of two or more factors is zero whenever any one of the factors is zero. Therefore $(x + 2)(x - 2)(x - 3) = 0$ when $x + 2 = 0$ or when $x - 2 = 0$ or when $x - 3 = 0$. Consequently $-2, 2, 3$ are the solutions to the original equation.

The key to the preceding solution was the conversion of the polynomial $x^3 - 3x^2 - 4x + 12$ into the *factored form* $(x + 2)(x - 2)(x - 3)$. As you have just seen, it is precisely this conversion that made it possible to solve the original equation. And since solving equations is of vital importance throughout mathematics, it will be worthwhile to become familiar with some methods of factoring.

In a sense, factoring is "unmultiplying." In the last section, we multiplied polynomials like this:

$$(x + 2)(x - 2)(x - 3) = [(x + 2)(x - 2)](x - 3)$$
$$= (x^2 - 4)(x - 3)$$
$$= x^3 - 3x^2 - 4x + 12$$

Now we want to *begin* with $x^3 - 3x^2 - 4x + 12$ and factor (unmultiply) it into the form $(x + 2)(x - 2)(x - 3)$.

One of the basic methods of factoring is the reverse of multiplying by a monomial. Consider this multiplication problem:

$$6x^2(4x^7 - 3x^4 - x^2 + 15) = 24x^9 - 18x^6 - 6x^4 + 90x^2$$

As you read this equation from left to right it involves multiplication by $6x^2$. Reading from right to left it involves "factoring out" the common monomial factor $6x^2$. These are the details of this factoring process:

$$24x^9 - 18x^6 - 6x^4 + 90x^2 = 6x^2(4x^7) - 6x^2(3x^4) - 6x^2(x^2) + 6x^2(15)$$
$$= 6x^2(4x^7 - 3x^4 - x^2 + 15)$$

Note the use of the distributive property in this illustration.

Suppose, in the preceding illustration, we were to use $3x$ as the common factor:

$$24x^9 - 18x^6 - 6x^4 + 90x^2 = 3x(8x^8 - 6x^5 - 2x^3 + 30x)$$

This is a correct factorization but it is not considered the *complete factored form* because $2x$ can still be factored out of the expression in the parentheses. In general the instruction to factor calls for arriving at the complete factored form. We can extend factoring techniques to polynomials with more than one variable as in the example that follows.

Example 1   Factor:

(a) $21x^4y - 14x^5y^2 - 63x^8y^3 + 91x^{11}y^4$

(b) $8x^2(x - 1) + 4x(x - 1) + 2(x - 1)$

Solution

(a) $21x^4y - 14x^5y^2 - 63x^8y^3 + 91x^{11}y^4 = 7x^4y(3 - 2xy - 9x^4y^2 + 13x^7y^3)$

(b) $8x^2(x - 1) + 4x(x - 1) + 2(x - 1) = 2(x - 1)(4x^2 + 2x + 1)$

Note: If you have trouble seeing this, then think of $x - 1$ as a single value and use $x - 1 = a$. Then

$$8x^2(x - 1) + 4x(x - 1) + 2(x - 1) = 8x^2a + 4xa + 2a$$
$$= 2a(4x^2 + 2x + 1)$$
$$= 2(x - 1)(4x^2 + 2x + 1)$$

We will now consider several basic procedures for factoring polynomials. You should check each of the results to be given by multiplication. These forms are very useful and you need to learn to recognize and apply each one.

The difference of two squares

$$a^2 - b^2 = (a + b)(a - b)$$

Illustrations:

$$x^2 - 9 = x^2 - 3^2 = (x + 3)(x - 3)$$
$$49t^2 - 100 = \underbrace{(7t)^2 - 10^2} = (7t + 10)(7t - 10)$$

$$\begin{bmatrix} \text{This step may be omitted if you} \\ \text{recognize that } a = 7t \text{ and } b = 10. \end{bmatrix}$$

$$(x + 1)^2 - 4 = (x + 1)^2 - 2^2 = (x + 1 + 2)(x + 1 - 2)$$
$$= (x + 3)(x - 1)$$

Can $3x^2 - 75$ be factored by using the difference of two squares? It becomes possible if we first factor out the common factor 3.

$$3x^2 - 75 = 3(x^2 - 25) = 3(x + 5)(x - 5)$$

The trick is to look for a common factor first in each of the given terms. It is strongly recommended that whenever you attempt to factor, you *first look for common factors* in the given terms. This step will not only save you work in some problems, but it may well be the difference between success and failure.

The difference (sum) of two cubes

$$a^3 - b^3 = (a - b)(a^2 + ab + b^2)$$
$$a^3 + b^3 = (a + b)(a^2 - ab + b^2)$$

Illustrations:

$$8x^3 - 27 = (2x)^3 - 3^3$$
$$= (2x - 3)(4x^2 + 6x + 9)$$
$$2x^3 + 128y^3 = 2(x^3 + 64y^3)$$
$$= 2[x^3 + (4y)^3]$$
$$= 2(x + 4y)(x^2 - 4xy + 16y^2)$$

TEST YOUR UNDERSTANDING

Factor out the common monomial.

1. $3x - 9$ **2.** $-5x + 15$ **3.** $5xy + 25y^2 + 10y^5$

Factor as the difference of squares.

4. $x^2 - 36$ **5.** $4x^2 - 49$ **6.** $(a + 2)^2 - 25b^2$

Factor as the difference (sum) of two cubes.

7. $x^3 - 27$ **8.** $x^3 + 27$ **9.** $8a^3 - 125$

Factor the following by first considering common monomial factors.

10. $3x^2 - 48$ **11.** $ax^3 + ay^3$ **12.** $2hx^2 - 8h^3$

It is not possible to factor $x^2 - 5$ as the difference of two squares by using integral coefficients. There are times, however, when it is desirable to allow coefficients other than integers. Consider, for example, these factorizations:

$$x^2 - 5 = x^2 - (\sqrt{5})^2 = (x + \sqrt{5})(x - \sqrt{5})$$
$$x^3 - 5 = x^3 - (\sqrt[3]{5})^3 = (x - \sqrt[3]{5})[x^2 + \sqrt[3]{5}\,x + (\sqrt[3]{5})^2]$$

When other than polynomial factors are allowed, it becomes possible to factor $x - 8$ as the difference of two squares as well as the difference of two cubes.

$$x - 8 = (\sqrt{x})^2 - (\sqrt{8})^2 = (\sqrt{x} + \sqrt{8})(\sqrt{x} - \sqrt{8})$$
$$x - 8 = (\sqrt[3]{x})^3 - 2^3 = (\sqrt[3]{x} - 2)[(\sqrt[3]{x})^2 + 2\sqrt[3]{x} + 4]$$

Using fractional exponents this last line becomes

$$x - 8 = (x^{1/3})^3 - 2^3 = (x^{1/3} - 2)(x^{2/3} + 2x^{1/3} + 4)$$

In general, we follow this rule when factoring:

> *Unless otherwise indicated, use the same type of numerical coefficients and exponents in the factors as appear in the given unfactored form.*

TEST YOUR UNDERSTANDING

Factor the following as the difference of two squares. Irrational numbers as well as radical expressions may be used.

1. $x^2 - 2$ **2.** $7 - x^2$ **3.** $x - 9$

Factor the following as the difference (sum) of two cubes. Irrational numbers and radical expressions may be used.

4. $x^3 - 4$ **5.** $1 - h$ **6.** $x + 27$

Just as we learned how to factor the difference of two squares or cubes, we can also learn how to factor the difference of two fourth powers, two fifth powers, and so on. All these situations can be collected into the single general form that gives the factorization of the difference of two nth powers, where n is an integer greater than 1.

The difference of two nth powers

$$a^n - b^n = (a - b)(a^{n-1} + a^{n-2}b + a^{n-3}b^2 + \cdots + ab^{n-2} + b^{n-1})$$

To describe the second factor, rewrite it like this:

$$a^{n-1}b^0 + a^{n-2}b^1 + a^{n-3}b^2 + \cdots + a^1b^{n-2} + a^0b^{n-1}$$

You can see that the exponents on a begin with $n - 1$ and decrease to 0 while on b they begin with 0 and increase to $n - 1$. Note also that the sum of the exponents on a and b, for each term, is $n - 1$.

The factorization of $a^n - b^n$ is one of the most useful in mathematics; it will be needed in the calculus. Study it carefully here so that in later work you will be able to recall it with minimal effort. There is a similar form available for factoring $a^n + b^n$ when n is an odd positive integer. (This form is not so important as the one for $a^n - b^n$.) You will find some work on this point in the exercises.

Example 2   Factor: $a^5 - b^5$.

Solution   Use the general form for $a^n - b^n$ with $n = 5$.

$$a^5 - b^5 = (a - b)(a^4 + a^3b + a^2b^2 + ab^3 + b^4)$$

Example 3   Use the result of Example 2 to factor $3y^5 - 96$.

Solution

$$\begin{aligned}
3y^5 - 96 &= 3(y^5 - 32) \\
&= 3(y^5 - 2^5) \\
&= 3(y - 2)(y^4 + 2y^3 + 4y^2 + 8y + 16)
\end{aligned}$$

Example 4   Factor $x^4 - 81$ as the difference of two fourth powers and also as the difference of two squares.

Solution

$$\begin{aligned}
x^4 - 81 = x^4 - 3^4 &= (x - 3)(x^3 + 3x^2 + 9x + 27) \\
x^4 - 81 = (x^2)^2 - (9)^2 &= (x^2 + 9)(x^2 - 9) \\
&= (x^2 + 9)(x + 3)(x - 3)
\end{aligned}$$

Note: The second answer is the *complete factored form* of $x^4 - 81$ since none of its factors can be factored further (allowing only polynomials with integral coefficients). Unless otherwise directed, *always try to arrive at the complete factored form.*

The two answers in Example 4 imply

$$(x - 3)(x^3 + 3x^2 + 9x + 27) = (x^2 + 9)(x + 3)(x - 3)$$

Divide by $x - 3$ to obtain

$$x^3 + 3x^2 + 9x + 27 = (x^2 + 9)(x + 3)$$

How can this factorization of $x^3 + 3x^2 + 9x + 27$ be found directly? The answer is to *group* the terms first and then factor, as follows.

$$
\begin{aligned}
x^3 + 3x^2 + 9x + 27 &= (x^3 + 3x^2) + (9x + 27) \\
&= x^2(x + 3) + 9(x + 3) \\
&= (x^2 + 9)(x + 3)
\end{aligned}
$$

Here are alternative groupings leading to the same answer.

$$
\begin{aligned}
x^3 + 3x^2 + 9x + 27 &= (x^3 + 9x) + (3x^2 + 27) \\
&= x(x^2 + 9) + 3(x^2 + 9) \\
&= (x + 3)(x^2 + 9) \\
x^3 + 3x^2 + 9x + 27 &= (x^3 + 27) + (3x^2 + 9x) \\
&= (x + 3)(x^2 - 3x + 9) + (x + 3)(3x) \\
&= (x + 3)(x^2 - 3x + 9 + 3x) \\
&= (x + 3)(x^2 + 9)
\end{aligned}
$$

Example 5   Factor $ax^2 - 15 - 5ax + 3x$ by grouping.

Solution

$$
\begin{aligned}
ax^2 - 15 - 5ax + 3x &= (ax^2 - 5ax) + (3x - 15) \\
&= ax(x - 5) + 3(x - 5) \\
&= (ax + 3)(x - 5)
\end{aligned}
$$

Not all groupings are productive. Thus in Example 5 the grouping $(ax^2 - 15) + (3x - 5ax)$ cannot be taken any further.

EXERCISES 2.5

Factor out the common monomial.

1. $5x - 5$ **2.** $3x^2 + 12x - 6$

3. $16x^4 + 8x^3 + 4x^2 + 2x$ **4.** $4a^2b - 6$

5. $4a^2b - 6ab$ **6.** $(a + b)x^2 + (a + b)y^2$

7. $2xy + 4x^2 + 8x^4$ **8.** $-12x^3y + 9x^2y^2 - 6xy^3$

9. $10x(a - b) + 5y(a - b)$

Factor as the difference of two squares.

10. $x^2 - 9$ **11.** $81 - x^2$ **12.** $x^2 - 10,000$

13. $4x^2 - 9$ **14.** $25x^2 - 144y^2$ **15.** $a^2 - 121b^2$

Factor as the difference (sum) of two cubes.

16. $x^3 - 8$ **17.** $x^3 + 64$ **18.** $8x^3 + 1$

19. $125x^3 - 64$ **20.** $8 - 27a^3$ **21.** $8x^3 + 343y^3$

Factor as the difference of two squares, allowing irrational numbers as well as radical expressions. (All letters represent positive numbers.)

22. $a^2 - 15$ **23.** $3 - 4x^2$ **24.** $x - 1$

25. $x - 36$ **26.** $2x - 9$ **27.** $8 - 3x$

Factor as the difference (sum) of two cubes, allowing irrational numbers as well as radical expressions.

28. $x^3 - 2$ **29.** $7 + a^3$ **30.** $1 - h$

31. $27x + 1$ **32.** $27x - 64$ **33.** $3x - 4$

Factor the following by first regrouping.

34. $a^2 - 2b + 2a - ab$ **35.** $x^2 - y - x + xy$

36. $x + 1 + y + xy$ **37.** $-y - x + 1 + xy$

38. $ax + by + ay + bx$ **39.** $2 - y^2 + 2x - xy^2$

Factor completely.

40. $8a^2 - 2b^2$ **41.** $7x^3 + 7h^3$ **42.** $81x^3 - 3y^3$

43. $x^3y - xy^3$ **44.** $5 - 80x^4$ **45.** $x^4 - y^4$

46. $81x^4 - 256y^4$ **47.** $a^8 - b^8$ **48.** $40ab^3 - 5a^4$

49. $a^5 - 32$ **50.** $3x^5 - 96y^5$ **51.** $1 - h^7$

52. $(2a + b)a^2 - (2a + b)b^2$ **53.** $3(a + 1)x^3 + 24(a + 1)$

54. $x^6 + x^2y^4 - x^4y^2 - y^6$ **55.** $a^3x - b^3y + b^3x - a^3y$

56. $7a^2 - 35b + 35a - 7ab$ **57.** $x^5 - 16xy^4 - 2x^4y + 32y^5$

*58. Find these products.
  (a) $(a + b)(a^2 - ab + b^2)$
  (b) $(a + b)(a^4 - a^3b + a^2b^2 - ab^3 + b^4)$
  (c) $(a + b)(a^6 - a^5b + a^4b^2 - a^3b^3 + a^2b^4 - ab^5 + b^6)$

*59. Use the results of Exercise 58 to factor the following:
  (a) $x^5 + 32$ (b) $128x^8 + xy^7$

*60. Write the factored form of $a^n + b^n$, where n is an odd positive integer greater than 1.

2.6

Here are two more general factoring forms that can easily be verified by multiplication.

**Factoring
Trinomials**

> ### *Perfect trinomial squares*
>
> $$a^2 + 2ab + b^2 = (a + b)^2$$
> $$a^2 - 2ab + b^2 = (a - b)^2$$

Illustrations:

$$x^2 + 10x + 25 = x^2 + 2(x \cdot 5) + 5^2$$
$$= (x + 5)^2$$
$$25t^2 - 30t + 9 = (5t)^2 - 2(5t \cdot 3) + 3^2$$
$$= (5t - 3)^2$$

To help you recognize these forms observe that in $a^2 + 2ab + b^2$ the middle
term (ignoring signs) is twice the product of the square roots of the end terms.
Hence the factored form is the square of the sum (or difference) of these square
roots.

Example 1   Factor: $1 + 18b + 81b^2$.

Solution

$$1 + 18b + 81b^2 = (1 + 9b)^2$$

Example 2   Factor: $9x^3 - 42x^2 + 49x$.

Solution

$$9x^3 - 42x^2 + 49x = x(9x^2 - 42x + 49)$$
$$= x[(3x)^2 - 2(3x)(7) + 7^2]$$
$$= x(3x - 7)^2$$

The last factoring technique that will be considered deals with trinomials that are not necessarily perfect squares. Consider the product of $x - 2$ by $x + 6$:

$$(x - 2)(x + 6) = x^2 + 4x - 12$$

Our problem is to begin with $x^2 + 4x - 12$ and find the two binomial factors.

For the moment assume that the factors of $x^2 + 4x - 12$ are unknown. From our experience with multiplying binomials we can anticipate that the factors of $x^2 + 4x - 12$ will consist of two first-degree binomials. It therefore becomes reasonable to search for the factors by considering the form

$$(x + \,?)(x - \,?)$$

The question marks need to be replaced by two integers whose product is 12. The plus and minus signs are required because of the -12 in the given trinomial. Furthermore, the two integers, in conjunction with the plus and minus signs, must produce a middle term of $+4x$. Since $12 = 12 \cdot 1 = 4 \cdot 3 = 6 \cdot 2$, the possible choices for the two integers are 12 with 1, 4 with 3, and 6 with 2. To find the correct pair is now a matter of trial and error.

Try 12 with 1 in these two ways:

$$(x + 12)(x - 1) \qquad (x + 1)(x - 12)$$

Neither of these pairs produces a middle term of $4x$, so now try 4 with 3. These also fail:

$$(x + 4)(x - 3) = x^2 + x - 12$$
$$(x + 3)(x - 4) = x^2 - x - 12$$

Finally, try 6 with 2. The two trials are $(x + 2)(x - 6)$ and $(x + 6)(x - 2)$. Since

$$(x + 6)(x - 2) = x^2 + 4x - 12$$

we have found the correct factors.

With a little luck, and much more experience, you can often avoid exhausting *all* the possibilities before finding the correct factors. You will then find that such work can often be shortened significantly.

Example 3   Factor: $10x^2 - 3x - 1$.

Solution   Consider the form $(?x + 1)(?x - 1)$.   Try $10 = 10 \cdot 1$.   Both $(10x + 1)(x - 1)$ and $(x + 1)(10x - 1)$ are incorrect. Try $10 = 5 \cdot 2$. But $(2x + 1)(5x - 1)$ does not work. Why not? The correct answer is $(5x + 1)(2x - 1)$.

Example 4   Factor: $24x^2 - 35x + 4$.

Solution   Since both 24 and 4 can be broken up in more than one way, there are potentially more forms to consider. Note that

$$24 = 24 \cdot 1 = 12 \cdot 2 = 8 \cdot 3 = 6 \cdot 4$$
$$4 = 4 \cdot 1 = 2 \cdot 2$$

Consider these forms:

$$(24x - ?)(x - ?) \qquad (12x - ?)(2x - ?)$$
$$(8x - ?)(3x - ?) \qquad (6x - ?)(4x - ?)$$

It should be clear why both signs must be minus. In these forms we can try 4 with 1 and also 2 with 2. The answer is

$$(8x - 1)(3x - 4)$$

Example 5   Factor: $12x^2 + 9x + 2$.

Solution   Consider the forms

$$(12x + ?)(x + ?)$$
$$(6x + ?)(2x + ?)$$
$$(4x + ?)(3x + ?)$$

In each form try 2 with 1 both ways. None of these pairs produces a middle term of $9x$; thus we have shown that $12x^2 + 9x + 2$ is *not* factorable with integral coefficients.

TEST YOUR UNDERSTANDING

Factor the following trinomial squares.

1. $a^2 + 6a + 9$ **2.** $x^2 - 10x + 25$ **3.** $4x^2 + 12xy + 9y^2$

Factor each trinomial if possible.

4. $x^2 + 8x + 15$ **5.** $a^2 - 12a + 20$ **6.** $x^2 + 3x + 4$

7. $10x^2 - 39x + 14$ **8.** $6x^2 - 11x - 10$ **9.** $6x^2 + 6x - 5$

In some problems you may find it easier to leave out the signs in the two binomial forms as you try various cases. For example, to factor $6x^2 - 7x - 20$ consider the forms

$$(6x \quad ?)(x \quad ?) \quad\quad (3x \quad ?)(2x \quad ?)$$

Then as you try a pair, like 5 with 4, keep in mind that the signs must be opposite. As soon as you see that the difference of the two partial products is $\pm 7x$, then insert the signs to produce the desired term $-7x$. In this problem the answer is $(2x - 5)(3x + 4)$.

Example 6   Factor: $15x^2 + 7x - 8$.

Solution   Try these forms:

$$(5x \quad ?)(3x \quad ?)$$
$$(15x \quad ?)(x \quad ?)$$

In each form try 4 with 2 and 8 with 1 both ways. The factored form is $(15x - 8)(x + 1)$.

All the trinomials we have considered were of degree 2. The same methods we used can be modified to factor certain trinomials of higher degree.

Example 7   Factor each trinomial:
(a) $x^4 + x^2 - 12$ **(b)** $a^6 - 3a^3b - 18b^2$

Solution
(a) Note that $x^4 = (x^2)^2$ and let $u = x^2$. Then

$$x^4 + x^2 - 12 = u^2 + u - 12$$
$$= (u + 4)(u - 3)$$
$$= (x^2 + 4)(x^2 - 3)$$

With practice you may be able to omit the substitution step.

(b) $a^6 - 3a^3b - 18b^2 = (a^3 - 6b)(a^3 + 3b)$

CAUTION! LEARN TO AVOID MISTAKES LIKE THESE:

WRONG	RIGHT
$(x + 2)3 + (x + 2)y = (x + 2)3y$	$(x + 2)3 + (x + 2)y = (x + 2)(3 + y)$
$3x + 1 = 3(x + 3)$	$3x + 1$ is not factorable by using integers
$x^3 - y^3 = (x - y)(x^2 + y^2)$	$x^3 - y^3 = (x - y)(x^2 + xy + y^2)$
$x^3 + 8$ is not factorable	$x^3 + 8 = (x + 2)(x^2 - 2x + 4)$

EXERCISES 2.6

Factor each of the following perfect trinomial squares.

1. $x^2 + 4x + 4$

2. $x^2 - 8x + 16$

3. $a^2 - 14a + 49$

4. $r^2 - 2r + 1$

5. $1 + 2b + b^2$

6. $100 - 20x + x^2$

7. $9x^2 - 18xy + 9y^2$

8. $64a^2 + 64a + 16$

Factor each trinomial.

9. $x^2 + 5x + 6$

10. $x^2 - 7x + 10$

11. $x^2 + 20x + 51$

12. $12a^2 - 13a + 1$

13. $20a^2 - 9a + 1$

14. $4 - 5b + b^2$

15. $9x^2 + 6x + 1$

16. $x^2 - 20x + 64$

17. $25a^2 - 10a + 1$

18. $8x^2 + 14x + 3$

19. $3x^2 + 20x + 12$

20. $5x^2 + 31x + 6$

21. $14x^2 + 37x + 5$

22. $9x^2 - 18x + 5$

23. $4a^2 + 20a + 25$

24. $a^2 - 9a + 18$

25. $b^2 + 18b + 45$

26. $6x^2 + 12x + 6$

27. $8x^2 - 16x + 6$

28. $12x^2 + 92x + 15$

29. $12a^2 - 25a + 12$

30. $18t^2 - 67t + 14$

31. $12b^2 - 34b - 6$

32. $10x^2 - 7xy - 12y^2$

Factor each trinomial, if possible. When appropriate, first factor out the common monomial.

33. $3a^2 + 6a + 3$

34. $5x^2 + 25x + 20$

35. $18x^2 - 24x + 8$

36. $4ax^2 + 4ax + a$

37. $x^2 + x + 1$

38. $a^2 - 2a + 2$

39. $49r^2s - 42rs + 9s$

40. $6x^2 + 2x - 20$

41. $50a^2 - 440a - 90$

42. $6a^2 + 4a - 9$

43. $15 + 5y - 10y^2$

44. $2b^2 + 12b + 16$

45. $4a^2x^2 - 4abx^2 + b^2x^2$

46. $a^3b - 2a^2b^2 + ab^3$

47. $16x^2 - 24x - 8$

48. $25a^2 + 50ab + 25b^2$

49. $24x^2 + 39x - 18$

50. $x^3 + 3x^2 + 2x$

Factor, if possible.

*51. $x^4 - 2x^2 + 1$

*52. $a^6 - 2a^3 + 1$

*53. $3x^4 + 6x^2 + 3$

*54. $6x^4 - x^2 - 5$

*55. $a^6 - 2a^3b^3 + b^6$

*56. $12x^4 + 2x^2 - 3$

2.7
Fundamental Operations with Rational Expressions

A **rational expression** is a ratio of polynomials. Rational expressions are the "algebraic extensions" of rational numbers, so that the fundamental rules for operating with rational numbers extend to rational expressions.

Here are the basic rules for operating with *any* type of rational expression. In all cases we assume that the denominator is not zero.

1. $-\dfrac{a}{b} = \dfrac{-a}{b} = \dfrac{a}{-b} \neq \dfrac{-a}{-b}$

2. $\dfrac{ac}{bc} = \dfrac{a}{b}$

3. $\dfrac{a}{b} \cdot \dfrac{c}{d} = \dfrac{ac}{bd}$

4. $\dfrac{a}{b} \div \dfrac{c}{d} = \dfrac{a}{b} \cdot \dfrac{d}{c} = \dfrac{ad}{bc}$

5. $\dfrac{a}{d} + \dfrac{c}{d} = \dfrac{a+c}{d}$

6. $\dfrac{a}{d} - \dfrac{c}{d} = \dfrac{a-c}{d}$

7. $\dfrac{a}{b} + \dfrac{c}{d} = \dfrac{ad+bc}{bd}$

8. $\dfrac{a}{b} - \dfrac{c}{d} = \dfrac{ad-bc}{bd}$

For the remainder of this section, the replacements for a, b, c, and d will be polynomials. As you study the following illustrations, keep in mind that the work in operating with rational expressions takes place in the numerators and denominators. Since these are polynomials, we are already prepared for such work. Again, in each case, we exclude values of x for which the denominator is equal to zero.

Before illustrating these rules we note that there are a number of useful alternate forms for Rule 4. For example:

$$\frac{\dfrac{a}{b}}{\dfrac{c}{d}} = \frac{a}{b} \div \frac{c}{d} = \frac{ad}{bc}$$

$$\frac{\dfrac{a}{b}}{c} = \frac{a}{b} \div \frac{c}{1} = \frac{a}{bc}$$

$$\frac{a}{\dfrac{b}{c}} = \frac{a}{1} \div \frac{b}{c} = \frac{ac}{b}$$

RULE 1. $\quad -\dfrac{a}{b} = \dfrac{-a}{b} = \dfrac{a}{-b}$

The negative of a fraction is the same as the fraction obtained by taking the negative of the numerator or of the denominator.

Illustration:

$$-\frac{2-x}{x^2-5} = \frac{-(2-x)}{x^2-5} = \frac{2-x}{-(x^2-5)}$$

Note that since $-(2-x) = x - 2$, each of these fractions is also equal to $\dfrac{x-2}{x^2-5}$.

RULE 2. $\quad \dfrac{ac}{bc} = \dfrac{a}{b}$

Reading this formula from left to right, it shows how to reduce a fraction to *lowest* terms so that the numerator and denominator of the resulting fraction have no common polynomial factors.

Illustration:

$$\frac{x^2+5x-6}{x^2+6x} = \frac{(x-1)(x+6)}{x(x+6)}$$

$$= \frac{x-1}{x} \qquad \text{(Rule 2)}$$

Note that the numerator and denominator have been divided by the common factor $x + 6$. As you will see later, whenever we are working with rational expressions it is taken for granted that final answers have been reduced to lowest terms.

RULE 3. $\dfrac{a}{b} \cdot \dfrac{c}{d} = \dfrac{ac}{bd}$

Illustration:

$$\frac{x-1}{x+1} \cdot \frac{x^2 - x - 2}{5x} = \frac{(x-1)(x^2 - x - 2)}{(x+1)5x} \qquad \text{(Rule 3)}$$

$$= \frac{(x-1)(x+1)(x-2)}{(x+1)5x}$$

$$= \frac{(x-1)(x-2)}{5x} \qquad \text{(Rule 2)}$$

This work can be shortened.

$$\frac{x-1}{x+1} \cdot \frac{x^2 - x - 2}{5x} = \frac{x-1}{\cancel{x+1}} \cdot \frac{\cancel{(x+1)}(x-2)}{5x} = \frac{(x-1)(x-2)}{5x}$$

Note that $\dfrac{x^2 - 3x + 2}{5x}$ is an alternative form of the answer.

RULE 4. $\dfrac{a}{b} \div \dfrac{c}{d} = \dfrac{ad}{bc}$

Illustration:

$$\frac{(x+1)^2}{x^2 - 6x + 9} \div \frac{3x+3}{x-3} = \frac{(x+1)^2}{x^2 - 6x + 9} \cdot \frac{x-3}{3x+3} \qquad \text{(Rule 4)}$$

$$= \frac{(x+1)^2}{(x-3)^2} \cdot \frac{x-3}{3(x+1)}$$

$$= \frac{(x+1)^2(x-3)}{(x-3)^2 3(x+1)} \qquad \text{(Rule 3)}$$

$$= \frac{(x+1)(x+1)(x-3)}{3(x-3)(x+1)(x-3)}$$

$$= \frac{x+1}{3(x-3)} \qquad \text{(Rule 2)}$$

The preceding can be shortened as follows:

$$\frac{(x+1)^2}{x^2 - 6x + 9} \div \frac{3x+3}{x-3} = \frac{(x+1)^2}{(x-3)^2} \cdot \frac{x-3}{3(x+1)} = \frac{x+1}{3(x-3)}$$

TEST YOUR UNDERSTANDING

Simplify each expression by reducing to lowest terms.

1. $\dfrac{x^2 + 2x}{x}$

2. $\dfrac{x^2}{x^2 + 2x}$

3. $\dfrac{x^2 - 9}{x^2 - 5x + 6}$

4. $\dfrac{x^2 + 6x + 5}{x^2 - x - 2}$

5. $\dfrac{x^2 - 4}{x^4 - 16}$

6. $\dfrac{3x^2 + x - 10}{5x - 3x^2}$

Perform the indicated operation and simplify.

7. $\dfrac{2x - 4}{2} \cdot \dfrac{x + 2}{x^2 - 4}$

8. $\dfrac{x - 2}{3(x + 1)} \div \dfrac{x^2 + 2x}{x + 2}$

9. $\dfrac{a}{b} \div \dfrac{a^2 - ab}{ab + b^2}$

10. $\dfrac{x + y}{x - y} \cdot \dfrac{x^2 - 2xy + y^2}{x^2 - y^2}$

Rules 5 and 6 are very much alike. We will illustrate only one of them.

RULE 5. $\dfrac{a}{d} + \dfrac{c}{d} = \dfrac{a + c}{d}$

Illustration:

$$\frac{6x^2}{2x^2 - x - 10} + \frac{-15x}{2x^2 - x - 10} = \frac{6x^2 + (-15x)}{2x^2 - x - 10} \quad \text{(Rule 5)}$$

$$= \frac{6x^2 - 15x}{2x^2 - x - 10}$$

$$= \frac{3x(2x - 5)}{(x + 2)(2x - 5)}$$

$$= \frac{3x}{x + 2} \quad \text{(Rule 2)}$$

Rules 7 and 8 are very much alike. We will illustrate Rule 7.

RULE 7. $\dfrac{a}{b} + \dfrac{c}{d} = \dfrac{ad + bc}{bd}$

Illustration:

$$\frac{3}{x^2 + x} + \frac{2}{x^2 - 1} = \frac{3(x^2 - 1) + 2(x^2 + x)}{(x^2 + x)(x^2 - 1)} \quad \text{(Rule 7)}$$

$$= \frac{5x^2 + 2x - 3}{(x^2 + x)(x^2 - 1)}$$

$$= \frac{(5x - 3)(x + 1)}{x(x + 1)(x^2 - 1)}$$

$$= \frac{5x - 3}{x(x^2 - 1)} \quad \text{(Rule 2)}$$

Here is an alternative method for the preceding example. It makes use of the **least common denominator (LCD)** of the two fractions.

$$\frac{3}{x^2+x} + \frac{2}{x^2-1} = \frac{3}{x(x+1)} + \frac{2}{(x+1)(x-1)}$$

$$= \frac{3(x-1)}{x(x+1)(x-1)} = \frac{2x}{(x+1)(x-1)x} \qquad \text{(Rule 2)}$$

$$= \frac{3(x-1)+2x}{x(x+1)(x-1)} \qquad \text{(Rule 5)}$$

$$= \frac{5x-3}{x(x^2-1)}$$

Finding the LCD is comparable to finding the **least common multiple (LCM)** for integers. The LCM of 24, 30, and 45 is 360. It is found by first writing

$$24 = 2^3 \cdot 3 \qquad 30 = 2 \cdot 3 \cdot 5 \qquad 45 = 3^2 \cdot 5$$

Then take each factor the *maximum* number of times it appears as a factor in any of the given numbers: $2^3 \cdot 3^2 \cdot 5 = 360$.

To find the LCM for a set of polynomials, take *each* factor that appears in the factored forms the *maximum* number of times it appears, in any of the factored forms.

Example 1   Find the LCM for each set of polynomials.
(a) $x^2 + x, x^2 - 1$ **(b)** $x^4 - x, x^2 - 2x + 1, x^4 + 2x^3$

Solution
(a) $x^2 + x = x(x + 1)$
  $x^2 - 1 = (x + 1)(x - 1)$
  LCM $= x(x + 1)(x - 1)$
(b) $x^4 - x = x(x^3 - 1) = x(x - 1)(x^2 + x + 1)$
  $x^2 - 2x + 1 = (x - 1)^2$
  $x^4 + 2x^3 = x^3(x + 2)$
  LCM $= x^3(x - 1)^2(x + 2)(x^2 + x + 1)$

Rules 3 through 8 are stated for two fractions, but they can easily be extended to involve more fractions.

Example 2 Combine and simplify: $\dfrac{3}{2} - \dfrac{4}{3x(x+1)} - \dfrac{x-5}{3x^2}$.

Solution Use the LCM $= 6x^2(x+1)$.

$$\dfrac{3}{2} - \dfrac{4}{3x(x+1)} - \dfrac{x-5}{3x^2} = \dfrac{3 \cdot 3x^2(x+1)}{2 \cdot 3x^2(x+1)} - \dfrac{4 \cdot 2x}{3x(x+1) \cdot 2x} - \dfrac{(x-5) \cdot 2(x+1)}{3x^2 \cdot 2(x+1)}$$

$$= \dfrac{(9x^3 + 9x^2) - 8x - (2x^2 - 8x - 10)}{6x^2(x+1)}$$

$$= \dfrac{9x^3 + 7x^2 + 10}{6x^2(x+1)}$$

Combine and simplify.

1. $\dfrac{x}{5} + \dfrac{2x}{3}$

2. $\dfrac{4x}{3} - \dfrac{x}{2}$

3. $\dfrac{2}{3x^2} - \dfrac{1}{2x}$

4. $\dfrac{3}{2x} + \dfrac{5}{3x} + \dfrac{1}{x}$

5. $\dfrac{7}{x-2} + \dfrac{3}{x+2}$

6. $\dfrac{9}{x-3} + \dfrac{7}{x^2-9}$

7. $\dfrac{5}{(x-1)(x+2)} - \dfrac{8}{x^2-4}$

8. $\dfrac{5x}{(2x+5)^2} + \dfrac{3x-2}{4x^2-25}$

The fundamental properties of fractions can be used to simplify rational expressions whose numerators and denominators may themselves also contain fractions.

Example 3 Simplify: $\dfrac{\dfrac{1}{5+h} - \dfrac{1}{5}}{h}$.

Solution Combine in the numerator and then divide.

$$\dfrac{\dfrac{1}{5+h} - \dfrac{1}{5}}{h} = \dfrac{\dfrac{5-(5+h)}{5(5+h)}}{h} = \dfrac{\dfrac{-h}{5(5+h)}}{h}$$

$$= \dfrac{-h}{5h(5+h)} = -\dfrac{1}{5(5+h)}$$

Example 4 Simplify: $\dfrac{x^{-2} - y^{-2}}{\dfrac{1}{x} - \dfrac{1}{y}}$.

Solution

$$\frac{x^{-2} - y^{-2}}{\dfrac{1}{x} - \dfrac{1}{y}} = \frac{\dfrac{1}{x^2} - \dfrac{1}{y^2}}{\dfrac{1}{x} - \dfrac{1}{y}}$$

$$= \frac{\left(\dfrac{1}{x^2} - \dfrac{1}{y^2}\right)(x^2 y^2)}{\left(\dfrac{1}{x} - \dfrac{1}{y}\right)(x^2 y^2)} \qquad \text{(Rule 2)}$$

$$= \frac{y^2 - x^2}{xy^2 - x^2 y}$$

$$= \frac{(y - x)(y + x)}{xy(y - x)}$$

$$= \frac{y + x}{xy}$$

Working with fractions often creates difficulties for many students. Study the following list; it may help you avoid some common pitfalls.

CAUTION! LEARN TO AVOID MISTAKES LIKE THESE:

WRONG	RIGHT
$\dfrac{2}{3} + \dfrac{x}{5} = \dfrac{2 + x}{3 + 5}$	$\dfrac{2}{3} + \dfrac{x}{5} = \dfrac{2 \cdot 5 + 3 \cdot x}{3 \cdot 5} = \dfrac{10 + 3x}{15}$
$\dfrac{1}{a} + \dfrac{1}{b} = \dfrac{1}{a + b}$	$\dfrac{1}{a} + \dfrac{1}{b} = \dfrac{b + a}{ab}$
$\dfrac{2x + 5}{4} = \dfrac{x + 5}{2}$	$\dfrac{2x + 5}{4} = \dfrac{2x}{4} + \dfrac{5}{4} = \dfrac{x}{2} + \dfrac{5}{4}$
$2 + \dfrac{x}{y} = \dfrac{2 + x}{y}$	$2 + \dfrac{x}{y} = \dfrac{2y + x}{y}$
$3\left(\dfrac{x + 1}{x - 1}\right) = \dfrac{3(x + 1)}{3(x - 1)}$	$3\left(\dfrac{x + 1}{x - 1}\right) = \dfrac{3(x + 1)}{x - 1}$

WRONG	RIGHT
$a \div \dfrac{b}{c} = \dfrac{1}{a} \cdot \dfrac{b}{c}$	$a \div \dfrac{b}{c} = a \cdot \dfrac{c}{b} = \dfrac{ac}{b}$
$\dfrac{1}{a^{-1} + b^{-1}} = a + b$	$\dfrac{1}{a^{-1} + b^{-1}} = \dfrac{1}{\dfrac{1}{a} + \dfrac{1}{b}} = \dfrac{ab}{b + a}$

EXERCISES 2.7

Classify each statement as true or false. If it is false, correct the right side to get a correct equality.

1. $\dfrac{5}{7} - \dfrac{2}{3} = \dfrac{3}{4}$

2. $\dfrac{2x + y}{y - 2x} = -2\left(\dfrac{x + y}{x - y}\right)$

3. $\dfrac{3ax - 5b}{6} = \dfrac{ax - 5b}{2}$

4. $\dfrac{x + x^{-1}}{xy} = \dfrac{x + 1}{x^2 y}$

5. $x^{-1} + y^{-1} = \dfrac{y + x}{xy}$

6. $\dfrac{2}{\frac{3}{4}} = \dfrac{8}{3}$

Simplify, if possible.

7. $\dfrac{8xy}{12yz}$

8. $\dfrac{24abc^2}{36bc^2 d}$

9. $\dfrac{x^2 - 5x}{5 - x}$

10. $\dfrac{n - 1}{n^2 - 1}$

11. $\dfrac{n + 1}{n^2 + 1}$

12. $\dfrac{(x + 1)^2}{x^2 - 1}$

13. $\dfrac{3x^2 + 3x - 6}{2x^2 + 6x + 4}$

14. $\dfrac{x^3 - x}{x^3 - 2x^2 + x}$

15. $\dfrac{4x^2 + 12x + 9}{4x^2 - 9}$

16. $\dfrac{x^2 + 2x + xy + 2y}{x^2 + 4x + 4}$

17. $\dfrac{a^2 - 16b^2}{a^3 + 64b^3}$

18. $\dfrac{a^2 - b^2}{a^2 - 6b - ab + 6a}$

Perform the indicated operations and simplify.

19. $\dfrac{2x^2}{y} \cdot \dfrac{y^2}{x^3}$

20. $\dfrac{3x^2}{2y^2} \div \dfrac{3x^3}{y}$

21. $\dfrac{2a}{3} \cdot \dfrac{3}{a^2} \cdot \dfrac{1}{a}$

22. $\left(\dfrac{a^2}{b^2} \cdot \dfrac{b}{c^2}\right) \div a$

23. $\dfrac{3x}{2y} - \dfrac{x}{2y}$

24. $\dfrac{a + 2b}{a} + \dfrac{3a + b}{a}$

25. $\dfrac{a - 2b}{2} - \dfrac{3a + b}{3}$

26. $\dfrac{7}{5x} - \dfrac{2}{x} + \dfrac{1}{2x}$

27. $\dfrac{x - 1}{3} \cdot \dfrac{x^2 + 1}{x^2 - 1}$

28. $\dfrac{x^2 - x - 6}{x^2 - 3x} \cdot \dfrac{x^3 + x^2}{x + 2}$

29. $\dfrac{x - 1}{x + 2} \div \dfrac{x^2 - x}{x^2 + 2x}$

30. $\dfrac{x^2 + 3x}{x^2 + 4x + 3} \div \dfrac{x^2 - 2x}{x + 1}$

31. $\dfrac{2}{x} - y$

32. $\dfrac{x^2}{x - 1} - \dfrac{1}{x - 1}$

33. $\dfrac{3y}{y + 1} + \dfrac{2y}{y - 1}$

34. $\dfrac{2a}{a^2 - 1} - \dfrac{a}{a + 1}$

35. $\dfrac{2x^2}{x^2 + x} + \dfrac{x}{x + 1}$

36. $\dfrac{3x + 3}{2x^2 - x - 1} + \dfrac{1}{2x + 1}$

37. $\dfrac{5}{x^2 - 4} - \dfrac{3 - x}{4 - x^2}$

38. $\dfrac{1}{a^2 - 4} + \dfrac{3}{a - 2} - \dfrac{2}{a + 2}$

39. $\dfrac{2x}{x^2 - 9} + \dfrac{x}{x^2 + 6x + 9} - \dfrac{3}{x + 3}$

40. $\dfrac{a^2 + 2ab + b^2}{a^2 - b^2} \div \dfrac{a^2 + 3ab + 2b^2}{a^2 - 3ab + 2b^2}$

41. $\dfrac{x^3 + x^2 - 12x}{x^2 - 3x} \cdot \dfrac{3x^2 - 10x + 3}{3x^2 + 11x - 4}$

42. $\dfrac{n^2 + n}{2n^2 + 7n - 4} \cdot \dfrac{4n^2 - 4n + 1}{2n^2 - n - 3} \cdot \dfrac{2n^2 + 5n - 12}{2n^3 - n^2}$

43. $\dfrac{n^3 - 8}{n + 2} \cdot \dfrac{2n^2 + 8}{n^3 - 4n} \cdot \dfrac{n^3 + 2n^2}{n^3 + 2n^2 + 4n}$

44. $\dfrac{a^3 - 27}{a^2 - 9} \div \left(\dfrac{a^2 + 2ab + b^2}{a^3 + b^3} \cdot \dfrac{a^3 - a^2b + ab^2}{a^2 + ab} \right)$

Simplify.

45. $\dfrac{\dfrac{5}{x^2 - 4}}{\dfrac{10}{x - 2}}$

46. $\dfrac{x + \dfrac{3}{2}}{x - \dfrac{1}{2}}$

47. $\dfrac{\dfrac{1}{x} - \dfrac{1}{4}}{x - 4}$

48. $\dfrac{\dfrac{1}{x} + \dfrac{1}{5}}{x + 5}$

49. $\dfrac{\dfrac{1}{4 + h} - \dfrac{1}{4}}{h}$

50. $\dfrac{\dfrac{1}{x^2} - \dfrac{1}{9}}{x - 3}$

Simplify, and express as a single fraction without negative exponents.

*51. $\dfrac{a^{-1} - b^{-1}}{a - b}$

*52. $\dfrac{(a + b)^{-1}}{a^{-1} + b^{-1}}$

*53. $\dfrac{x^{-2} - y^{-2}}{xy}$

*54. There are three tests and a final examination given in a mathematics course. Let a, b, and c be the numerical grades of the tests and let d represent the final examination grade.

   (a) If the final grade is computed by allowing the average of the three tests and the exam to count the same, then show that the final average is given by the expression $\dfrac{a + b + c + 3d}{6}$.

   (b) Assume that the average of the three tests accounts for 60% of the final grade, and that the examination accounts for 40%. Show that the final average is given by the expression $\dfrac{a + b + c + 2d}{5}$.

*55. The new pocket electronic calculators frequently require that certain calculations be performed in a different manner to accommodate the machine. Show that in each case the expression on the left can be computed by using the equivalent expression on the right.

   (a) $\dfrac{A}{B} + \dfrac{C}{D} = \dfrac{\dfrac{A \cdot D}{B} + C}{D}$

   (b) $A \cdot B + C \cdot D + E \cdot F = \left[\dfrac{\left(\dfrac{A \cdot B}{D} + C \right) \cdot D}{F} + E \right] \cdot F$

The factored form of the trinomial square $a^2 + 2ab + b^2$ is $(a + b)^2$. Turning this around, we say that the *expanded form* of $(a + b)^2$ is $a^2 + 2ab + b^2$. And if $(a + b)^2$ is multiplied by $a + b$, we get the expansion of $(a + b)^3$. Here is a list of the expansions of the first five powers of the binomial $a + b$. You can easily verify these results by multiplying the expansion in each row by $a + b$ to get the expansion in the next row.

can use Pascal's △

$$(a + b)^1 = a + b$$
$$(a + b)^2 = a^2 + 2ab + b^2$$
$$(a + b)^3 = a^3 + 3a^2b + 3ab^2 + b^3$$
$$(a + b)^4 = a^4 + 4a^3b + 6a^2b^2 + 4ab^3 + b^4$$
$$(a + b)^5 = a^5 + 5a^4b + 10a^3b^2 + 10a^2b^3 + 5ab^4 + b^5$$

Our objective here is to learn how to find such expansions directly without having to multiply. That is, we want to be able to expand $(a + b)^n$, especially for larger values of n, without having to multiply $a + b$ by itself n times.

Let n represent a positive integer. As seen in the table, each expansion begins with a^n and ends with b^n. Moreover, each expansion has $n + 1$ terms that are all preceded by plus signs. Now look at the case for $n = 5$. Replace the first term a^5 by a^5b^0 and use a^0b^5 in place of b^5. Then

$$(a + b)^5 = a^5b^0 + 5a^4b + 10a^3b^2 + 10a^2b^3 + 5ab^4 + a^0b^5$$

In this form it becomes clear that (from left to right) the exponents for a successively decrease by 1, beginning with 5 and ending with zero. At the same time, the exponents for b increase from zero to 5. Also note that the sum of the exponents for each term is 5. You can easily verify that similar patterns also hold for the other cases shown.

Using the preceding observations we would *expect* the expansion of $(a + b)^6$ to have seven terms that, except for the unknown coefficients, look like this:

$$a^6 + \underline{\quad}a^5b + \underline{\quad}a^4b^2 + \underline{\quad}a^3b^3 + \underline{\quad}a^2b^4 + \underline{\quad}ab^5 + b^6$$

Our list of expansions reveals that the second coefficient, as well as the coefficient of the next to the last term, is the number n. Filling in these coefficients for the case $n = 6$ gives

$$a^6 + 6a^5b + \underline{\quad}a^4b^2 + \underline{\quad}a^3b^3 + \underline{\quad}a^2b^4 + 6ab^5 + b^6$$

To get the remaining coefficients we return to the case $n = 5$ and learn how such coefficients can be generated. Look at the second and third terms.

$$\underbrace{\text{⑤}a^{④}b}_{\text{2nd term}} \qquad \underbrace{10a^3b^{②}}_{\text{3rd term}}$$

If the exponent 4 of *a* in the *second* term is multiplied by the coefficient 5 of the *second* term and then divided by the exponent 2 of *b* in the *third* term, the result is 10, the coefficient of the third term.

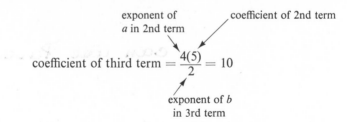

Verify that this procedure works for the next coefficient.

On the basis of the evidence we expect the missing coefficients for the case $n = 6$ to be obtainable in the same way. Here are the computations:

$$\text{Use } \textcircled{6}a^{\textcircled{5}}b + \underline{\quad}a^4b^{\textcircled{2}}: \quad \text{3rd coefficient} = \frac{5(6)}{2} = 15$$

$$\text{Use } \textcircled{15}a^{\textcircled{4}}b^2 + \underline{\quad}a^3b^{\textcircled{3}}: \quad \text{4th coefficient} = \frac{4(15)}{3} = 20$$

$$\text{Use } \textcircled{20}a^{\textcircled{3}}b^3 + \underline{\quad}a^2b^{\textcircled{4}}: \quad \text{5th coefficient} = \frac{3(20)}{4} = 15$$

Finally, we may write the following expansion.

$$(a + b)^6 = a^6 + 6a^5b + 15a^4b^2 + 20a^3b^3 + 15a^2b^4 + 6ab^5 + b^6$$

You can verify that this equality is correct by multiplying the expansion for $(a + b)^5$ by $a + b$.

More labor can be saved by observing the symmetry in the expansions of $(a + b)^n$. For instance, when $n = 6$ the coefficients around the middle term are symmetric. Similarly, when $n = 5$ the coefficients around the two middle terms are symmetric.

Example 1 Write the expansion of $(x + 2)^7$.

Solution Let x and 2 play the role of a and b in $(a + b)^7$ respectively.

$$(x + 2)^7 = x^7 + 7x^62 + \underline{\quad}x^52^2 + \underline{\quad}x^42^3 + \underline{\quad}x^32^4 + \underline{\quad}x^22^5 + 7x2^6 + 2^7$$

Now find the missing coefficients as follows:

$$\text{3rd coefficient} = \frac{6(7)}{2} = 21 = \text{6th coefficient}$$

$$\text{4th coefficient} = \frac{5(21)}{3} = 35 = \text{5th coefficient}$$

The completed expansion may now be given as follows:

$$(x + 2)^7 = x^7 + 7x^62 + 21x^52^2 + 35x^42^3 + 35x^32^4 + 21x^22^5 + 7x2^6 + 2^7$$
$$= x^7 + 14x^6 + 84x^5 + 280x^4 + 560x^3 + 672x^2 + 448x + 128$$

After some experience with these computations you should be able to write the expansion for the general case $(a + b)^n$, the **binomial theorem**:

$$(a + b)^n = a^n + \frac{n}{1}a^{n-1}b + \frac{n(n-1)}{1 \cdot 2}a^{n-2}b^2 + \frac{n(n-1)(n-2)}{1 \cdot 2 \cdot 3}a^{n-3}b^3 + \cdots + \frac{n}{1}ab^{n-1} + b^n$$

To get the expansion of the binomial $a - b$, white $a - b = a + (-b)$ and substitute into the previous form. For example, with $n = 6$,

$$(a - b)^6 = [a + (-b)]^6 = a^6 + 6a^5(-b) + 15a^4(-b)^2 + 20a^3(-b)^3$$
$$+ 15a^2(-b)^4 + 6a(-b)^5 + (-b)^6$$
$$= a^6 - 6a^5b + 15a^4b^2 - 20a^3b^3 + 15a^2b^4 - 6ab^5 + b^6$$

This result indicates that the expansion of $(a - b)^n$ is the same as the expansion of $(a + b)^n$ except that the signs alternate, beginning with plus.

Example 2   Expand: $(x - 1)^7$.

Solution   Use the coefficients found in Example 1, with alternating signs.

$$(x - 1)^7 = x^7 - 7x^61^1 + 21x^51^2 - 35x^41^3 + 35x^31^4 - 21x^21^5 + 7x^11^6 - 1^7$$
$$= x^7 - 7x^6 + 21x^5 - 35x^4 + 35x^3 - 21x^2 + 7x - 1$$

EXERCISES 2.8

Expand and simplify.

1. $(x + 1)^5$ **2.** $(x - 1)^6$ **3.** $(x + 1)^7$ **4.** $(x - 1)^8$

5. $(a - b)^4$ **6.** $(3x - 2)^4$ **7.** $(3x - y)^5$ **8.** $(x + y)^8$

9. $(a^2 + 1)^5$ **10.** $(2 + h)^9$ **11.** $(1 - h)^{10}$ **12.** $(x - 2)^7$

Simplify.

13. $\dfrac{(1 + h)^3 - 1}{h}$ **14.** $\dfrac{(3 + h)^4 - 81}{h}$ **15.** $\dfrac{(c + h)^3 - c^3}{h}$

16. $\dfrac{(x + h)^6 - x^6}{h}$ **17.** $\dfrac{2(x + h)^5 - 2x^5}{h}$ **18.** $\dfrac{\dfrac{1}{(2 + h)^2} - \dfrac{1}{4}}{h}$

19. Evaluate 2^{10} by expanding $(1 + 1)^{10}$.

20. Write the first five terms in the expansion of $(x + 1)^{15}$. What are the last five terms?

21. Write the first five terms and the last five terms in the expansion of $(c + h)^{20}$.

22. Write the first four terms and the last four terms in the expansion of $(a - 1)^{30}$.

23. Study this triangular array of numbers and discover the connection with the expansions of $(a + b)^n$, where $n = 1, 2, 3, 4, 5, 6$.

$$
\begin{array}{ccccccccccc}
 & & & & & 1 & & 1 & & & \\
 & & & & 1 & & 2 & & 1 & & \\
 & & & 1 & & 3 & & 3 & & 1 & \\
 & & 1 & & 4 & & 6 & & 4 & & 1 \\
 & 1 & & 5 & & 10 & & 10 & & 5 & & 1 \\
1 & & 6 & & 15 & & 20 & & 15 & & 6 & & 1
\end{array}
$$

24. Discover how the 6th row of the triangle in Exercise 23 can be obtained from the 5th row by studying the connection between the 4th and 5th rows indicated by this scheme.

$$
\begin{array}{ccccccccc}
1 & & 4 & & 6 & & 4 & & 1 \\
 & + & & + & & + & & + & \\
1 & & 5 & & 10 & & 10 & & 5 & & 1
\end{array}
$$

25. Using the result of Exercise 24, write the 7th, 8th, 9th, and 10th rows of the triangle. (It is called **Pascal's triangle**.)

26. Use the 9th row found in Exercise 25 to expand $(x + h)^9$.

27. Use the 10th row found in Exercise 25 to expand $(x - h)^{10}$.

2.9

Complex Numbers (optional)

There is no real number whose square is -1 because the square of any real number is nonnegative. Another way to state this is to say that $\sqrt{-1}$ is not a real number. (This statement is consistent with Definition 1 of Section 2.2.) We will now make a final extension of our number system in which the square roots of negative real numbers will be defined.

Just as 1 is considered to be the unit for the real numbers, we now define $\sqrt{-1}$ to be the unit for a new set of numbers, the *imaginary numbers*. The symbol i is used to stand for this number and is defined in this way.

$$i = \sqrt{-1} \quad \text{and} \quad i^2 = -1$$

Using i, we can define the square root of a negative number as follows.

$$\sqrt{-n} = \sqrt{-1} \cdot \sqrt{n} = i\sqrt{n} \qquad n > 0$$

Example 1 Simplify: **(a)** $\sqrt{-16} + \sqrt{-25}$ **(b)** $\sqrt{-16} \cdot \sqrt{-25}$

Solution

(a) $\sqrt{-16} = \sqrt{-1} \cdot \sqrt{16} = i \cdot 4 = 4i$

$\sqrt{-25} = \sqrt{-1} \cdot \sqrt{25} = i \cdot 5 = 5i$

Thus

$\sqrt{-16} + \sqrt{-25} = 4i + 5i = 9i$

(b) $\sqrt{-16} \cdot \sqrt{-25} = (4i)(5i) = 20i^2 = 20(-1) = -20$

Note: In the preceding example, $4i$ and $5i$ were combined by using the usual rules of algebra. You will see later that such procedures apply for these new kinds of numbers.

An indicated product of a real number times the imaginary unit i, such as $7i$ or $\sqrt{2}\,i$, is called a **pure imaginary number**. The sum of a real number and a pure imaginary number is called a **complex number**.

A *complex number* has the form $a + bi$, where a and b are real and $i = \sqrt{-1}$.

We say that the real number a is the **real part** of $a + bi$ and the real number b is called the **imaginary part** of $a + bi$. In general, two complex numbers are equal only when both their real parts and their imaginary parts are equal.

The complex numbers $a + bi$ and $c + di$ are equal if and only if $a = c$ and $b = d$.

The collection of complex numbers contains all the real numbers since any real number a can also be written in the form $a = a + 0i$. Similarly, if b is real, $bi = 0 + bi$, so that the complex numbers also contain the pure imaginaries. The following diagram summarizes this point.

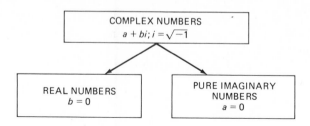

Example 2   Name the sets of numbers to which each of the following belongs:
(a) $\sqrt{\frac{4}{9}}$ (b) $\sqrt{-10}$ (c) $-4 + \sqrt{-12}$

Solution

(a) $\sqrt{\frac{4}{9}} = \frac{2}{3} = \frac{2}{3} + 0i$ is complex, real, and rational.

(b) $\sqrt{-10} = i\sqrt{10}$ is complex and pure imaginary.

(c) $-4 + \sqrt{-12} = -4 + i\sqrt{12} = -4 + 2i\sqrt{3}$ is complex.

How do we add, subtract, multiply, and divide two complex numbers? In answering this question, we must keep in mind that the real numbers are included in the collection of complex numbers and the definitions we construct for the complex numbers must preserve the established operations for the reals. We do not want, for example, to add complex numbers in such a way that would produce inconsistent results for the addition of real numbers.

We add and subtract complex numbers by combining their real and their imaginary parts separately, according to these definitions.

> **Sum and difference of complex numbers**
>
> $$(a + bi) + (c + di) = (a + c) + (b + d)i$$
> $$(a + bi) - (c + di) = (a - c) + (b - d)i$$

Actually, these procedures are quite similar to those used for combining polynomials. For example, compare these two sums.

$$(2 + 3x) + (5 + 7x) = (2 + 5) + (3 + 7)x = 7 + 10x$$
$$(2 + 3i) + (5 + 7i) = (2 + 5) + (3 + 7)i = 7 + 10i$$

Similarly, compare these subtraction problems.

$$(8 + 5x) - (3 + 2x) = (8 - 3) + (5 - 2)x = 5 + 3x$$
$$(8 + 5i) - (3 + 2i) = (8 - 3) + (5 - 2)i = 5 + 3i$$

To learn how to multiply two complex numbers we go back to the set of real numbers and extend the multiplication process. Recall the procedure for multiplying two binomials.

$$(a + b)(c + d) = ac + bc + ad + bd$$

Now compare these two products.

$$(3 + 2x)(5 + 3x) = 15 + 10x + 9x + 6x^2$$
$$(3 + 2i)(5 + 3i) = 15 + 10i + 9i + 6i^2$$

Of course, we can simplify this last expression by noting that $10i + 9i = 19i$ and $6i^2 = -6$. The final result is $9 + 19i$.

In general we can develop a rule for multiplication of two complex numbers by finding the product of $a + bi$ and $c + di$ as follows.

$$(a + bi)(c + di) = ac + adi + bci + bdi^2$$
$$= ac + (ad + bc)i + bd(-1)$$
$$= (ac - bd) + (ad + bc)i$$

Thus we get the following definition for the product of two complex numbers.

Product of complex numbers

$$(a + bi)(c + di) = (ac - bd) + (ad + bc)i$$

Note: It is usually easier to find the product by using the procedure for multiplying binomials rather than by memorizing the formal definition.

Example 3 Multiply: $(5 - 2i)(3 + 4i)$.

Solution

$$(5 - 2i)(3 + 4i) = 15 - 6i + 20i - 8i^2$$
$$= 15 - 6i + 20i + 8$$
$$= 23 + 14i$$

Now consider the quotient of two complex numbers. We wish to express a quotient like the following in the form $a + bi$.

$$\frac{2 + 3i}{3 + i}$$

To do so, we use a method similar to the process of rationalizing a denominator. Note what happens when $3 + i$ is multiplied by its **conjugate**, $3 - i$:

$$(3 + i)(3 - i) = 9 + 3i - 3i - i^2 = 9 - i^2 = 9 + 1 = 10$$

In general, the *conjugate* of $a + bi$ is $a - bi$. Indeed the term *imaginary* is said to have been first applied to these numbers precisely because they seem mysteriously to vanish under certain multiplications.

We are now ready to complete the division problem.

$$\frac{2 + 3i}{3 + i} = \frac{2 + 3i}{3 + i} \cdot \frac{3 - i}{3 - i}$$

$$= \frac{6 + 9i - 2i - 3i^2}{9 - i^2}$$

$$= \frac{6 + 9i - 2i + 3}{9 + 1}$$

$$= \frac{9 + 7i}{10}$$

$$= \frac{9}{10} + \frac{7}{10}i$$

In general, multiplying the numerator and denominator of $\frac{a + bi}{c + di}$ by the conjugate of $c + di$ leads to the following definition for division. (See Exercise 59.)

Quotient of complex numbers

$$\frac{a + bi}{c + di} = \frac{ac + bd}{c^2 + d^2} + \frac{bc - ad}{c^2 + d^2}i \qquad c + di \neq 0$$

Rather than memorize this definition, it is easier to find quotients as was done in the preceding illustration.

Though we will not go into the details here, it can be shown that the basic algebraic rules apply for the complex numbers. For example, the commutative, associative, and distributive laws hold. You will find more on this point in the exercises.

It is also true that the rules for integer exponents apply for complex numbers. For example, $(2 - 3i)^0 = 1$ and $(2 - 3i)^{-1} = \dfrac{1}{2 - 3i}$. In particular, the integral powers of i are easily evaluated. Here are the first four powers of i:

$$i = \sqrt{-1}$$
$$i^2 = -1$$
$$i^3 = i^2 \cdot i = -1 \cdot i = -i$$
$$i^4 = i^2 \cdot i^2 = (-1)(-1) = 1$$

After the first four powers of i a repeating pattern exists, as can be seen from the next four powers of i:

$$i^5 = i^4 \cdot i = 1 \cdot i = i$$
$$i^6 = i^4 \cdot i^2 = 1 \cdot i^2 = -1$$
$$i^7 = i^4 \cdot i^3 = 1 \cdot i^3 = -i$$
$$i^8 = i^4 \cdot i^4 = 1 \cdot i^4 = 1$$

Note the repeating cycle: $i, -1, -i, 1$. For any power of i greater than 4, we merely need to find the preceding power of 4 as in this example.

Example 4   Simplify: **(a)** i^{22} **(b)** i^{39}

Solution
(a) $i^{22} = i^{20} \cdot i^2 = (i^4)^5 \cdot i^2 = 1^5 \cdot i^2 = i^2 = -1$
(b) $i^{39} = i^{36} \cdot i^3 = (i^4)^9 \cdot i^3 = 1^9 \cdot i^3 = i^3 = -i$

The next example illustrates how to operate with a negative integral power of i.

Example 5   Express $2i^{-3}$ as the indicated product of a real number and i.

Solution   First note that $2i^{-3} = \dfrac{2}{i^3}$. Next multiply numerator and denominator by i to obtain a real number in the denominator.

$$2i^{-3} = \frac{2}{i^3} \cdot \frac{i}{i} = \frac{2i}{i^4} = \frac{2i}{1} = 2i$$

Alternate Solution:

$$2i^{-3} = \frac{2}{i^3} = \frac{2}{-i}$$

$$= \frac{2}{-i} \cdot \frac{i}{i} \qquad (i \text{ is the conjugate of } -i)$$

$$= \frac{2i}{-i^2} \qquad (i^2 = -1, \ -i^2 = 1)$$

$$= 2i$$

The general results for rational exponents will not be studied here, although you will see how to find the nth roots of complex numbers in Section 9.7.

An ordered pair of real numbers (x, y) determines the unique complex number $x + yi$ and vice versa. This correspondence between the complex numbers and the ordered pairs of real numbers allows us to give a geometric interpretation of the complex numbers by using the **complex plane**. This plane is a rectangular coordinate system in which the horizontal axis is called the **real axis** and the vertical axis is the **imaginary axis**. The following figure shows how several complex numbers are identified (graphed) in the complex plane.

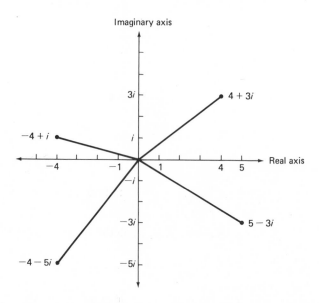

There is an interesting geometric interpretation for the sum of two complex numbers. First locate $a + bi$ and $c + di$. Then complete the parallelogram as shown. The tip of the diagonal of the parallelogram that passes through the origin will represent the sum. (See Exercise 66.) This graphic procedure for finding sums is known as the **parallelogram law.**

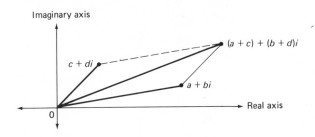

Example 6   Use the parallelogram law to find the sum of $3 + 2i$ and $2 - 4i$.

Solution $(3 + 2i) + (2 - 4i) = 5 - 2i$

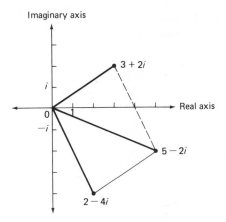

Example 7   Describe the diagonal formed when finding the sum of $3 + 2i$ and $3 - 2i$ by use of the parallelogram law.

Solution   The diagonal is six units in length on the x-axis, from $(0, 0)$ to $(6, 0)$. The sum is $6 + 0i = 6$.

A geometric interpretation of the product of complex numbers will be developed in Section 9.7.

~will not do

EXERCISES 2.9

Classify each statement as true or false.

1. Every real number is a complex number.

2. Every complex number is a real number.

3. Every irrational number is a complex number.

4. Every integer can be written in the form $a + bi$.

5. Every complex number may be expressed as an irrational number.

6. Every negative integer may be written as a pure imaginary number.

Express each of the following numbers in the form $a + bi$.

7. $5 + \sqrt{-4}$ **8.** $7 - \sqrt{-7}$ **9.** -5 **10.** $\sqrt{25}$

Express in the form bi.

11. $3i^3$ **12.** $-5i^5$ **13.** $2i^7$ **14.** $-4i^{19}$

15. $3i^{-3}$ **16.** $\sqrt{-81}$ **17.** $\sqrt{-144}$ **18.** $-\sqrt{-9}$

19. $\sqrt{-\frac{9}{16}}$ **20.** $\dfrac{1}{\sqrt{-25}}$

Simplify.

21. $\sqrt{-9} \cdot \sqrt{-81}$ **22.** $\sqrt{4} \cdot \sqrt{-25}$ **23.** $\sqrt{-3} \cdot \sqrt{-2}$

24. $(2i)(3i)$ **25.** $(-3i^2)(5i)$ **26.** $(i^3)(i^4)$

27. $\sqrt{-9} + \sqrt{-81}$ **28.** $\sqrt{-12} + \sqrt{-75}$ **29.** $\sqrt{-8} + \sqrt{-18}$

30. $2\sqrt{-72} - 3\sqrt{-32}$ **31.** $\sqrt{-9} - \sqrt{-3}$ **32.** $3\sqrt{-80} - 2\sqrt{-20}$

Complete the indicated operation. Express all answers in the form $a + bi$.

33. $(7 + 5i) + (3 + 2i)$ **34.** $(8 + 7i) + (9 - i)$

35. $(8 + 2i) - (3 + 5i)$ **36.** $(7 + 2i) - (4 - 3i)$

37. $(7 + \sqrt{-16}) + (3 - \sqrt{-4})$ **38.** $(8 + \sqrt{-49}) - (2 - \sqrt{-25})$

39. $2i(3 + 5i)$ **40.** $3i(5i - 2)$

41. $(3 + 2i)(2 + 3i)$ **42.** $(\sqrt{5} + 3i)(\sqrt{5} - 3i)$

43. $(5 - 2i)(3 + 4i)$ **44.** $(\sqrt{3} + 2i)^2$

45. $\dfrac{3 + 5i}{i}$ **46.** $\dfrac{5 - i}{i}$

47. $\dfrac{5 + 3i}{2 + i}$ **48.** $\dfrac{7 - 2i}{2 - i}$

49. $\dfrac{3 - i}{3 + i}$ **50.** $\dfrac{8 + 3i}{3 - 2i}$

Simplify and express each answer in the form a + bi.

51. $(3 + 2i)^{-1}$ **52.** $(3 + 2i)^{-2}$

53. Complete the following multiplication table for powers of i.

$\times$	i	i^2	i^3	i^4
i				
i^2				
i^3				
i^4				

54. The set of complex numbers satisfies the associative property for addition. Verify this by completing this problem in two different ways.

$$(3 + 5i) + (2 + 3i) + (7 + 4i)$$

55. Repeat Exercise 54 for multiplication, using this problem.

$$(3 + i)(3 - i)(4 + 3i)$$

56. Find the value of $x^2 + 3x + 5$ when $x = \dfrac{-3 + \sqrt{11}\,i}{2}$.

57. Classify as true or false:
   (a) The sum of a complex number and its conjugate is a real number.
   (b) The product of a complex number and its conjugate is a real number.

58. We say that $0 = 0 + 0i$ is the additive identity for the complex numbers since $0 + z = z$ for any $z = a + bi$. Find the additive inverse (negative) of z.

*59. Write $\dfrac{a + bi}{c + di}$ in the form $x + yi$. (*Hint:* Multiply the numerator and denominator by the conjugate of $c + di$.)

*60. Prove the distributive law by showing that $u(v + w) = uv + uw$, where $u = a + bi$, $v = c + di$, and $w = e + fi$.

For Exercises 61 through 64, find each sum by using the parallelogram law.

61. $(3 + 4i) + (4 + 3i)$ **62.** $(3 - 4i) + (4 - 3i)$

63. $(-3 + 2i) + (-2 - 3i)$ **64.** $(1 + 3i) + (-3 - i)$

*65. Find the sum graphically on a complex plane.

$$[(2 + 5i) + (2 - 2i)] + (3 - 5i)$$

*66. Verify the parallelogram law by showing that $x = a + c$ and $y = b + d$.

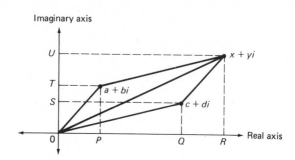

67. Use the parallelogram law to show that the sum of a complex number and its conjugate is a real number.

68. One of the basic rules for operating with radicals is that $\sqrt{ab} = \sqrt{a} \cdot \sqrt{b}$, where a and b are nonnegative real numbers. Prove that this rule does not work when both a and b are negative by showing that $\sqrt{(-4)(-9)} \neq \sqrt{-4} \cdot \sqrt{-9}$.

*69. Use the definition $\sqrt{-x} = i\sqrt{x}$ $(x \geq 0)$ to prove that $\sqrt{ab} = \sqrt{a} \cdot \sqrt{b}$ when $a < 0$ and $b \geq 0$.

REVIEW EXERCISES FOR CHAPTER 2

The solutions to the following exercises can be found within the text of Chapter 2. Try to answer each question without referring to the text.

Section 2.1

1. Simplify: $-(\frac{2}{3})^5(\frac{9}{4})^5$.   **2.** Simplify: $(\frac{1}{2})^3(2)^{-3}$.   **3.** Simplify: $\dfrac{(x^3y)^2y^3}{x^4y^6}$.

4. Simplify and write without negative exponents: $\left(\dfrac{a^{-2}b^3}{a^3b^{-2}}\right)^5$.

5. Prove that $(ab)^m = a^m b^m$, where m is a negative integer. (*Hint:* Let $m = -k$ and apply the rules for positive integral exponents.)

6. What is the motivation behind the definition $b^0 = 1$? Discuss.

Section 2.2

7. What is the definition of $\sqrt[n]{a}$, where $a > 0$ and n is a positive integer ≥ 2?

8. Evaluate: $\sqrt{9 \cdot 25}$.   **9.** Evaluate: $\sqrt[4]{\frac{16}{81}}$.   **10.** Evaluate: $27^{2/3} - 16^{-1/4}$.

Section 2.3

11. Simplify: $\dfrac{6}{\sqrt{3}} + 2\sqrt{75} - \sqrt{3}$.   **12.** Simplify: $2\sqrt{8x^3} + 3x\sqrt{32x} - x\sqrt{18x}$.

Section 2.4

13. Multiply: $(5x - 9)(6x + 2)$.   **14.** Multiply: $(3x^3 - 8x + 4)(2x^2 + 5x - 1)$.

Section 2.5

15. Factor: $24x^9 - 18x^6 - 6x^4 + 90x^2$.

16. Factor: $3x^2 - 75$.

17. Factor: $2x^3 + 128y^3$.

18. Factor: $(x + 1)^2 - 4$.

19. Factor as the difference of squares using irrational numbers: $x^2 - 5$.

20. Factor as the difference of cubes using irrational numbers: $x - 8$.

21. Factor: $3y^5 - 96$.

22. Factor: $ax^2 - 15 - 15ax + 3x$.

Section 2.6

23. Factor: $9x^3 - 42x^2 + 49x$.   **24.** Factor: $24x^2 - 35x + 4$.

25. Factor: $15x^2 + 7x - 8$.   **26.** Factor: $x^4 + x^2 - 12$.

Section 2.7

27. Multiply: $\dfrac{x - 1}{x + 1} \cdot \dfrac{x^2 - x - 2}{5x}$.   **28.** Divide: $\dfrac{(x + 1)^2}{x^2 - 6x + 9} \div \dfrac{3x + 3}{x - 3}$.

29. Add: $\dfrac{3}{x^2 + x} + \dfrac{2}{x^2 - 1}$.   **30.** Combine: $\dfrac{3}{2} - \dfrac{4}{3x(x + 1)} - \dfrac{x - 5}{3x^2}$.

31. Simplify: $\dfrac{\dfrac{1}{5 + h} - \dfrac{1}{5}}{h}$.   **32.** Simplify: $\dfrac{x^{-2} - y^{-2}}{\dfrac{1}{x} - \dfrac{1}{y}}$.

Section 2.8

33. Expand: $(a + b)^6$.   **34.** Expand: $(x - 1)^7$.

Section 2.9

35. Multiply: $(5 - 2i)(3 + 4i)$.

36. Divide: $\dfrac{2 + 3i}{3 + i}$.

37. Simplify: **(a)** i^{22} **(b)** i^{39}

38. Use the parallelogram law to find the sum of $3 + 2i$ and $2 - 4i$.

SAMPLE TEST FOR CHAPTER 2

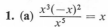

Classify each statement as true or false.

1. (a) $\dfrac{x^3(-x)^2}{x^5} = x$

   (c) $(-27)^{-1/3} = 3$

   (e) $\sqrt{9x^2} = 3|x|$

   (b) $\left(\dfrac{3}{2+a}\right)^{-1} = \dfrac{2}{3} + \dfrac{a}{3}$

   (d) $(x+y)^{3/5} = (\sqrt[3]{x+y})^5$

   (f) $(8+a^3)^{1/3} = 2 + a$

For Exercises 2 through 4 find the correct choice. There is only one correct answer in each case.

2. $\dfrac{4^2(-3)^3 2^0}{6^3(12^2)} =$

   (a) $-\dfrac{1}{216}$ (b) $-\dfrac{1}{72}$ (c) $-\dfrac{1}{36}$ (d) $\left(-\dfrac{1}{6}\right)^5$ (e) $\dfrac{1}{6^3}$

3. $\left(\dfrac{27a^6 b^{-3}}{c^{-2}}\right)^{-2/3} =$

   (a) $\dfrac{b^2}{9a^4 c^{4/3}}$ (b) $\dfrac{b^2}{3a^4 \sqrt[3]{c}}$ (c) $\dfrac{c^2}{18ab}$ (d) $-\dfrac{9c^3 b^2}{a^4}$ (e) $\dfrac{-3b^2}{a^4 c^{4/3}}$

4. $\dfrac{30}{\sqrt{20}} - 2\sqrt{45} =$

   (a) $\dfrac{15 - 3\sqrt{5}}{\sqrt{5}}$ (b) $-3\sqrt{5}$ (c) $-15\sqrt{45}$ (d) $\dfrac{3\sqrt{2}}{2} - 6\sqrt{5}$ (e) $-\dfrac{5}{2}$

5. Simplify.
   (a) $(ab^2 + 5cd) - (3ab^2 - 2cd + a^2) + (2ab^2 - 5a^2)$
   (b) $(5x + 3)(2x - 4) - (3 - x)(3 + x)$

Factor completely.

6. (a) $64 - 27b^3$ (b) $3x^5 - 48x$
7. (a) $x^5 - 32$ (b) $2x^2 - 6xy - 3y^3 + xy^2$
8. Divide and simplify:

$$\dfrac{x^3 + 8}{x^2 - 4x - 12} \div \dfrac{x^3 - 2x^2 + 4x}{x^3 - 6x^2}$$

9. Simplify:

$$\dfrac{\dfrac{1}{x^2} - \dfrac{1}{49}}{x - 7}$$

10. Combine and simplify:

$$\dfrac{1}{x + 3} - \dfrac{2}{x^2 - 9} + \dfrac{x}{2x^2 + x - 15}$$

†11. Write the first four terms in the expansion of $(a - 2)^{20}$.

†12. Express the product $(3 + 7i)(5 - 4i)$ in the form $a + bi$.

†On chapter tests this symbol indicates that the question depends on the optional sections of the chapter.

Page 34

1. 125
2. $-\frac{1}{32}$
3. 0

4. 1,000,000
5. -64
6. 64

7. $\frac{1}{17}$
8. $\frac{1}{23}$
9. 81

10. $a^{11}b^{10}$
11. 64
12. $6x^4$

Page 42

1. 11
2. -4
3. 2

4. 3
5. 35
6. $\frac{3}{5}$

7. 2
8. -70
9. $\frac{4}{3}$

10. 540
11. $-\frac{10}{3}$
12. 4

Page 49

1. $6\sqrt{2}$
2. $6\sqrt{3}$
3. $\sqrt{5}$

4. $6 - 3\sqrt{2}$
5. $5\sqrt[3]{2}$
6. $4\sqrt[3]{2} + 5$

7. $11\sqrt{2}$
8. $13\sqrt{3}$
9. $3\sqrt{5}$

10. $7\sqrt{7}$

Page 54

1. $x^2 + 1$
2. $x^2 + 4x - 13$
3. $2x^3 + 3x^2 + 8x + 1$
4. $-x^3 + 4x - 8$
5. $-3x^4 - 6x^3 + 3x$
6. $-\frac{1}{2}x + 2x^2 + 4x^3$

7. $6x^2 + 13x + 6$
8. $12x^2 + 28x - 5$
9. $16x^2 - 49$
10. $4x^2 - 12x + 9$
11. $x^4 + 9x^3 + 14x^2 - 4x - 8$
12. $4x^6 - 15x^5 + 31x^4 - 22x^3 + 14x^2 - 8x + 5$

Page 58

1. $3(x - 3)$
2. $-5(x - 3)$
3. $5y(x + 5y + 2y^4)$
4. $(x + 6)(x - 6)$

5. $(2x + 7)(2x - 7)$
6. $(a + 2 + 5b)(a + 2 - 5b)$
7. $(x - 3)(x^2 + 3x + 9)$
8. $(x + 3)(x^2 - 3x + 9)$

9. $(2a - 5)(4a^2 + 10a + 25)$
10. $3(x + 4)(x - 4)$
11. $a(x + y)(x^2 - xy + y^2)$
12. $2h(x + 2h)(x - 2h)$

Page 59

1. $(x + \sqrt{2})(x - \sqrt{2})$
2. $(\sqrt{7} + x)(\sqrt{7} - x)$
3. $(\sqrt{x} + 3)(\sqrt{x} - 3)$

4. $(x - \sqrt[3]{4})(x^2 + \sqrt[3]{4}x + (\sqrt[3]{4})^2)$
5. $(1 - \sqrt[3]{h})(1 + \sqrt[3]{h} + (\sqrt[3]{h})^2)$
6. $(\sqrt[3]{x} + 3)[(\sqrt[3]{x})^2 - 3\sqrt[3]{x} + 9]$

Page 66

1. $(a + 3)^2$
2. $(x - 5)^2$
3. $(2x + 3y)^2$

4. $(x + 3)(x + 5)$
5. $(a - 10)(a - 2)$
6. Not factorable.

7. $(5x - 2)(2x - 7)$
8. $(3x + 2)(2x - 5)$
9. Not factorable.

Page 71

1. $x + 2$
2. $\dfrac{x}{x + 2}$
3. $\dfrac{x + 3}{x - 2}$

4. $\dfrac{x + 5}{x - 2}$
5. $\dfrac{1}{x^2 + 4}$
6. $-\dfrac{x + 2}{x}$

7. 1
8. $\dfrac{x - 2}{3x(x + 1)}$
9. $\dfrac{a + b}{a - b}$

10. 1

Page 73

1. $\dfrac{13x}{15}$

2. $\dfrac{5x}{6}$

3. $\dfrac{4 - 3x}{6x^2}$

4. $\dfrac{25}{6x}$

5. $\dfrac{2(5x + 4)}{(x - 2)(x + 2)}$

6. $\dfrac{9x + 34}{x^2 - 9}$

7. $-\dfrac{3x + 2}{(x - 1)(x^2 - 4)}$

8. $\dfrac{16x^2 - 14x - 10}{(2x + 5)^2(2x - 5)}$

Three

Linear
Functions

Suppose that you are riding in a car that is averaging 40 miles per hour. Then the distance traveled is determined by the time traveled; distance = rate × time. Symbolically this relationship can be expressed by the equation $s = 40t$, where t is the time measured in hours. In $t = 2$ hours the distance traveled is $s = 40(2)$ = 80 miles. Likewise, for each specific value of $t \geq 0$ the equation produces *exactly one* value for s. This correspondence between the distance s and the time t is an example of a *functional relationship*. More specifically, we say that the equation $s = 40t$ defines s as a *function* of t.

We say that $s = 40t$ defines s as a function of t because for *each* choice of t there corresponds *exactly one* value for s. We first choose a value of t. Then there is a corresponding value of s that depends on t (s depends on t); s is the *dependent variable* and t is the *independent variable* of the function defined by $s = 40t$.

In contrast, the equation $y^2 = 12x$ does *not* define y as a function of x; for a given value of x there is *more than one* corresponding value for y. If $x = 3$, for example, then $y^2 = 36$ and $y = 6$ or $y = -6$.

Because the variable t represents time in the equation $s = 40t$, it is reasonable to say that $t \geq 0$. This set of allowable values for the independent variable is called the **domain** of the function. The set of corresponding values for the dependent variable is called the **range** of the function.

A **function** is a correspondence between two sets, the domain and the range, such that for each value in the domain there corresponds exactly one value in the range.

The specific letters used for the independent and dependent variables are of no consequence. Usually we will use x for the independent variable and y for the dependent variable. Thus the equation $y = 40x$ can be used to define the same function as $s = 40t$. However, letters that are suggestive, such as t for time, can prove to be helpful.

Most of the expressions we encountered earlier can be used to define functions. Here are some illustrations. Note that in each case *the domain of the function is taken as the largest set of real numbers for which the defining expression in x makes sense.* Unless otherwise indicated, this will be our policy throughout this text.

FUNCTION GIVEN BY	DOMAIN
$y = 6x^4 - 3x^2 + 7x + 1$	All real numbers
$y = \dfrac{2x}{x^2 - 4}$	All reals except 2 and −2
$y = \lvert x \rvert$	All real numbers
$y = \sqrt{x}$	All real $x \geq 0$

Example 1   Explain why the following equation defines y as a function of x and find the domain: $y = \dfrac{1}{\sqrt{x-1}}$.

Solution   For each allowable x the expression $\dfrac{1}{\sqrt{x-1}}$ produces just one y-value.

Therefore the given equation defines a function. To find the domain observe that $x - 1$ cannot be negative because the even root of a negative value is not a real number. Thus $x - 1 \geq 0$, or $x \geq 1$. But $x = 1$ produces division by zero, so we exclude it. Hence the domain consists of all $x > 1$.

$1 - x^2$

Example 2   Find the domain of the function given by $y = \sqrt{1 - x^2}$.

Solution   To avoid square roots of negative nambers we must have $1 - x^2 \geq 0$, or $x^2 - 1 \leq 0$. Now factor $x^2 - 1$ and solve the inequality $(x + 1)(x - 1) \leq 0$. The domain consists of all values of x from -1 through 1; that is, $-1 \leq x \leq 1$.

TEST YOUR UNDERSTANDING

Decide whether the given equation defines y to be a function of x. For each function, find the domain.

1. $y = (x + 2)^2$

2. $y = \dfrac{1}{(x + 2)^2}$

3. $y = \dfrac{1}{x^2 + 2}$

4. $y = \pm 3x$

5. $y = \sqrt{x^2 + x - 6}$

6. $y^2 = x^2$

7. $y = \dfrac{x}{|x|}$

8. $y = \dfrac{1}{\sqrt{x^2 + 2x + 1}}$

Thus far only equations have been used to define functions. One could gain the impression that an equation is a function and that equations are the only way to state functions. This impression is not quite correct. The function is the *correspondence* between the variables, and such correspondences can be stated in many ways. Here, for example, is a *table of values*.

x	1	2	5	-7	23	$\sqrt{2}$
y	6	-6	6	-4	0	6

This table defines y to be a function of x because for each domain value (1, 2, 5, -7, 23, $\sqrt{2}$) there corresponds *exactly one* value for y. The fact that some range values, like 6, are used more than once does not matter. Keep in mind that it is the domain value that can only be used once. A function can also be defined by a graph, as we will see later.

We say that an equation, like $y = 40x$, is a function. Such informal language is commonly used and should not cause any difficulties.

Instead of using a single equation to define a function, there will be times when a function is defined in terms of more than one equation. For instance, the following three equations define a function whose domain is the set of all real numbers.

$$y = \begin{cases} 1 & \text{for } x \leq -6 \\ x^2 & \text{for } -6 < x < 0 \\ 2x + 1 & \text{for } x \geq 0 \end{cases}$$

Note that if $x = 5$ the corresponding y-value comes from $y = 2x + 1$, namely 11; if $x = -5$, $y = (-5)^2 = 25$; and if $x = -7$, $y = 1$.

Example 3 Decide whether these two equations define a function.

this is one equation.
y has 2 diff. values
for same value of x :. not a function

$$y = \begin{cases} 3x - 1 & \text{if } x \leq 1 \\ 2x + 1 & \text{if } x \geq 1 \end{cases}$$

Solution If $x = 1$, the first equation gives $y = 3(1) - 1 = 2$. The second equation gives $y = 2(1) + 1 = 3$. Since we have two different y-values for the same x-value, the equations do *not* define a function.

Example 4 Use the definition of absolute value (Section 1.6) to explain why the equation $y = |x|$ defines a function whose domain consists of all real numbers.

Solution The definition of absolute value says that $|x| = x$ for $x \geq 0$ and $|x| = -x$ for $x < 0$. Thus $y = |x|$ defines a function because each real value x produces just one corresponding y-value.

A useful way to refer to a function is to name it by using a specific letter, such as f, g, F, and the like. For example, the function given by $y = \dfrac{1}{x - 3}$ may be referred to as f. The domain of f is the set of all real numbers not equal to 3; that

$f(x)$ $y = \frac{1}{x-3} = f(x) = \frac{1}{x-3}$

y becomes $f(x)$

$f(\underset{x}{\underbrace{\ })}$, whatever inside parentheses is

is, $x \neq 3$. We write $f(x) = \dfrac{1}{x-3}$ to mean that the value of the function f at x is

$\dfrac{1}{x-3}$. Thus $f(4) = 1$ means that when $x = 4, y = 1$.

$f(4) = 1$ is read as "f of 4 is 1" or "f at 4 is 1."

We use $f(x)$ to represent the range value for the specific value of x given in the parentheses. Note that $f(x)$ does *not* mean that we are to multiply f by x; f does not stand for a number. In this illustration f stands for the function that is

given by $y = \dfrac{1}{x-3}.$ $= f(x) = \dfrac{1}{x-3}$

For $x \neq 3$, $f(x) = \dfrac{1}{x-3}.$ Then $f(0) = \dfrac{1}{0-3} = -\dfrac{1}{3}$ and

$$f(9) = \dfrac{1}{9-3} = \dfrac{1}{6}$$

But $f(3)$ is undefined. Can you explain why?   would be % ∅

Let us explore the function concept with another example. Let g be the function defined by $y = g(x) = x^2$. Then we have

$$g(1) = 1^2 = 1$$
$$g(2) = 2^2 = 4$$
$$g(3) = 3^2 = 9$$

Note that $g(1) + g(2) \neq g(3)$. To write $g(1) + g(2) = g(1+2)$ would be to assume, *incorrectly*, that the distributive property holds for the functional notation. This is not true in general, which comes as no great surprise since g is not a number.

Keep in mind that the variable x in $g(x) = x^2$ is only a *placeholder*. Any letter could serve the same purpose. For example, $g(t) = t^2$ and $g(z) = z^2$ both define the same function.

Example 5   For the function g defined by $g(x) = \dfrac{1}{x}$ find:

(a) $3g(x)$ **(b)** $g(3x)$ **(c)** $3 + g(x)$ **(d)** $g(3) + g(x)$ **(e)** $g(3+x)$ **(f)** $g\left(\dfrac{1}{x}\right)$

Solution   **(a)** $3g(x) = 3 \cdot \dfrac{1}{x} = \dfrac{3}{x}$;   **(b)** $g(3x) = \dfrac{1}{3x}$;   **(c)** $3 + g(x) = 3 + \dfrac{1}{x}$;

(d) $g(3) + g(x) = \dfrac{1}{3} + \dfrac{1}{x}$;   **(e)** $g(3+x) = \dfrac{1}{3+x}$;   **(f)** $g\left(\dfrac{1}{x}\right) = \dfrac{1}{\frac{1}{x}} = x$

Example 6 Let $g(x) = x^2$. Evaluate and simplify: $\dfrac{g(x) - g(4)}{x - 4}$.

Solution $g(x) = x^2$ and $g(4) = 16$

$$\frac{g(x) - g(4)}{x - 4} = \frac{x^2 - 16}{x - 4}$$

$$= \frac{(x - 4)(x + 4)}{x - 4}$$

$$= x + 4$$

Example 7 Let $g(x) = \dfrac{1}{x}$. Evaluate and simplify: $\dfrac{g(4 + h) - g(4)}{h}$.

Solution $g(4 + h) = \dfrac{1}{4 + h}$ and $g(4) = \dfrac{1}{4}$

$$\frac{g(4 + h) - g(4)}{h} = \frac{\dfrac{1}{4 + h} - \dfrac{1}{4}}{h}$$

$$= \frac{\dfrac{4 - (4 + h)}{4(4 + h)}}{h}$$

$$= \frac{-h}{4h(4 + h)}$$

$$= -\frac{1}{4(4 + h)}$$

In the preceding examples we evaluated *difference quotients*. Such forms will be of major importance in the study of calculus.

CAUTION! LEARN TO AVOID MISTAKES LIKE THESE:

In each of the following the function f is defined by $f(x) = 3x^2 - 4$.

WRONG	RIGHT
$f(0) = 0$	$f(0) = 3(0)^2 - 4 = -4$
$f(-2) = -f(2)$	$f(-2) = 3(-2)^2 - 4 = 8$ $-f(2) = -[3(2)^2 - 4] = -8$

WRONG	RIGHT
$f\left(\dfrac{1}{2}\right) = \dfrac{1}{f(2)}$	$f\left(\dfrac{1}{2}\right) = 3\left(\dfrac{1}{2}\right)^2 - 4 = -\dfrac{13}{4}$ $\dfrac{1}{f(2)} = \dfrac{1}{8}$
$[f(2)]^2 = f(4)$	$[f(2)]^2 = 8^2 = 64$ $f(4) = 3(4)^2 - 4 = 44$
$2 \cdot f(5) = f(10)$	$2 \cdot f(5) = 2[3(5)^2 - 4] = 142$ $f(10) = 3(10)^2 - 4 = 296$
$f(5) + f(2) = f(7)$	$f(5) + f(2) = 3(5)^2 - 4 + 3(2)^2 - 4$ $\qquad = 79$ $f(7) = 3(7)^2 - 4 = 143$

EXERCISES 3.1

Decide whether the given equation defines y to be a function of x. For each function, find the domain.

1. $y = x^3$

2. $y = \sqrt[3]{x}$

3. $y = \dfrac{1}{\sqrt{x}}$

4. $y = |2x|$

5. $y^2 = 2x$

6. $y = x \pm 3$

7. $y = \dfrac{1}{x+1}$

8. $y = \dfrac{x-2}{x^2+1}$

9. $y = \dfrac{1}{1 \pm x}$

10. $y = \sqrt{x^2 - 4}$

11. $y = \dfrac{1}{\sqrt{x^2 - 4}}$

12. $y = \dfrac{1}{\sqrt[3]{x^2 - 4}}$

Classify each statement as true or false. If it is false, correct the right side to get a correct equation. For each of these statements use $f(x) = -x^2 + 3$.

13. $f(3) = -6$

14. $f(2)f(3) = -33$

15. $3f(2) = -33$

16. $f(3) + f(-2) = 2$

17. $f(3) - f(2) = -5$

18. $f(2) - f(3) = 11$

19. $f(x) - f(4) = -(x-4)^2 + 3$

20. $f(x) - f(4) = x^2 + 19$

21. $f(4 + h) = -h^2 - 8h - 13$

22. $f(4 + h) = -h^2 - 10$

In Exercises 23 through 34, find (a) $f(-1)$; (b) $f(0)$; (c) $f(\tfrac{1}{2})$, if they exist.

23. $f(x) = 2x - 1$

24. $f(x) = -5x + 6$

25. $f(x) = x^2$

26. $f(x) = x^2 - 5x + 6$

27. $f(x) = x^3 - 1$

28. $f(x) = (x - 1)^2$

29. $f(x) = x^4 + x^2$

30. $f(x) = -3x^3 + \frac{1}{2}x^2 - 4x$

31. $f(x) = \dfrac{1}{x - 1}$

32. $f(x) = \sqrt{x}$

33. $f(x) = \dfrac{1}{\sqrt[3]{x}}$

34. $f(x) = \dfrac{1}{3|x|}$

35. For $g(x) = x^2 - 2x + 1$ find:
   (a) $g(10)$ **(b)** $5g(2)$ **(c)** $g(\frac{1}{2}) + g(\frac{1}{3})$ **(d)** $g(\frac{1}{2} + \frac{1}{3})$

36. Let h be given by $h(x) = x^2 + 2x$. Find $h(3)$ and $h(1) + h(2)$ and compare.

37. Let h be given by $h(x) = x^2 + 2x$. Find $3h(2)$ and $h(6)$ and compare.

38. Let h be given by $h(x) = x^2 + 2x$. Find **(a)** $h(2x)$;   **(b)** $h(2 + x)$;   **(c)** $h\left(\dfrac{1}{x}\right)$;
   (d) $h(x^2)$.

In Exercises 39 through 44 find the difference quotient $\dfrac{f(x) - f(3)}{x - 3}$ and simplify for the given function f.

39. $f(x) = x^2$

40. $f(x) = x^2 - 1$

41. $f(x) = \dfrac{1}{x}$

42. $f(x) = \sqrt{x}$

43. $f(x) = 2x + 1$

44. $f(x) = -x^3 + 1$

In Exercises 45 through 50 find the difference quotient $\dfrac{f(2 + h) - f(2)}{h}$ and simplify.

45. $f(x) = x$

46. $f(x) = -x + 3$

47. $f(x) = -x^2$

*48. $f(x) = \sqrt{x + 2}$

*49. $f(x) = \dfrac{1}{x^2}$

*50. $f(x) = \dfrac{1}{x - 1}$

3.2

The Rectangular Coordinate System

A great deal of information can be learned about a functional relationship by studying its graph. A fundamental objective of this course is to acquaint you with the graphs of some important functions, as well as to impart basic graphing procedures. First we need to introduce a **rectangular coordinate system**, as shown in the figures on the next page.

These are the basic features of a rectangular coordinate system:

1. Two perpendicular number lines intersect at a point called the *origin*, which is the point 0 for each line. The horizontal number line is called the *x-axis*; the vertical number line is called the *y-axis*.
2. Unless otherwise specified, the unit length is the same on both axes. The positive and negative directions are shown in the figure on the upper left.
3. The four regions determined by the axes are called *quadrants* and are labeled counterclockwise as in the figure on the upper left.

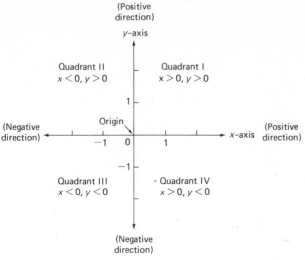

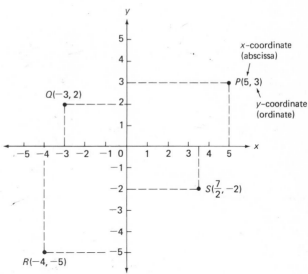

4. The *coordinates* of a point P indicate the direction and distance from the axes to P. The first or x-coordinate (called the *abscissa*) is the distance from the y-axis to P. The second or y-coordinate (called the *ordinate*) is the distance from the x-axis to P. If the first and second coordinates of P are the values x and y respectively, then we write (x, y) or $P(x, y)$ to identify the point. Note the coordinates of the points P, Q, R, and S in the figure on the lower right.

5. The origin has coordinates $(0, 0)$. Other points on the axes have the form $(x, 0)$ for points on the x-axis and $(0, y)$ for points on the y-axis.

If we *begin* with an ordered pair, like (5, 3), the corresponding point may be located by measuring 5 units to the right of the *y*-axis (parallel to the *x*-axis) and then 3 units above the *x*-axis (parallel to the *y*-axis). Do you see that the ordered pair (3, 5) locates a different point? As soon as you realize that (5, 3) and (3, 5) correspond to different points, then it is clear why such a pair is called an *ordered pair of numbers*.

Example 1 Name the ordered pairs indicated by the given letters.

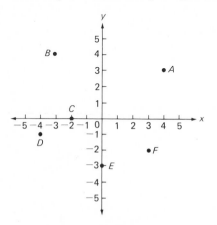

Solution *A*: (4, 3); *B*: (−3, 4); *C*: (−2, 0); *D*: (−4, −1); *E*: (0, −3); *F*: (3, −2)

Example 2 Locate the points corresponding to these ordered pairs: (2, 3); (3, 2); (0, −3); (−3, 0); (−1, 4); (−2, −3).

Solution

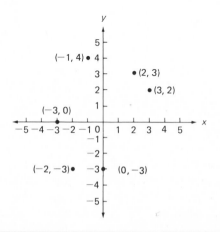

To *graph an equation* stated in the variables x and y means to locate all the points in the rectangular system whose coordinates (x, y) satisfy the given equation. This procedure is shown in Example 2.

Example 3   Graph: $y = 2x - 1$.

Solution   For this example the domain of the function is the set of real numbers. It is helpful to locate a number of specific points first. For example, verify that each of these ordered pairs satisfies the given equation.

$$(-1, -3) \qquad (0, -1) \qquad (\tfrac{1}{2}, 0) \qquad (2, 3)$$

By drawing a straight line through these points we have the graph of $y = 2x - 1$. This is also said to be the graph of the function $f(x) = 2x - 1$.

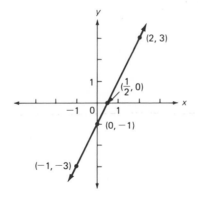

Example 2 illustrates a *linear function*, a function whose graph is a straight line. (Linear functions will be studied in detail in the following section.) A convenient method for graphing such functions is to locate the two points where the graph crosses the axes—the *x-intercept* and the *y-intercept*—and draw a line through these points. In the preceding example, the x-intercept has coordinates $(\tfrac{1}{2}, 0)$ and the y-intercept is at $(0, -1)$.

Note that the graph of $y = 2x - 1$ divides the plane into two *half-planes*. These two half-planes represent the graphs for the two statements of inequality, $y < 2x - 1$ (below the line) and $y > 2x - 1$ (above the line). To show these graphs we would use a dashed line for $y = 2x - 1$ and shade the appropriate half-plane, as in the figures at the top of the next page.

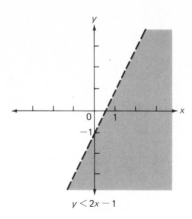

$y < 2x - 1$

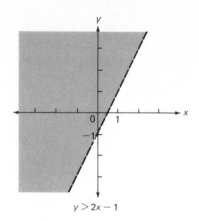

$y > 2x - 1$

To show a graph for an inequality such as $y \leq 2x - 1$ we would draw both the line and the half-plane.

At times we may wish to limit the domain of the variable, as in the next example.

Example 4   Graph: $y = 2x - 1$ for $-2 \leq x \leq 1$.

Solution   The graph is a line segment with endpoints at $(-2, -5)$ and $(1, 1)$.

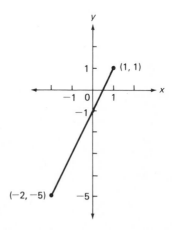

Example 5   Line ℓ is parallel to the x-axis. What can be said about the coordinates of $P(x, y)$ for any P on ℓ?

all y values must be same

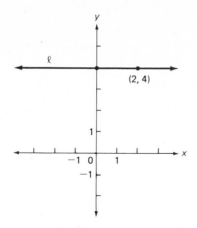

Solution   Consider a specific point on the line, such as $P(2, 4)$. This point is 4 units above the x-axis on a line parallel to the x-axis. It follows that every point (x, y) on the given line must have $y = 4$. That is, for all values x, the point $(x, 4)$ is on ℓ.

EXERCISES 3.2

1. Draw a pair of coordinate axes. Graph the points corresponding to the following ordered pairs of numbers: $(5, 1)$; $(2, \frac{9}{2})$; $(-4, 4)$; $(-4, 0)$; $(-3, -5)$; $(-\frac{1}{2}, -\frac{3}{4})$; $(2, -1)$; $(5, -1)$; $(5, -5)$; $(0, -4)$.

For Exercises 2 through 7 let (x, y) represent any point on the given graph. What conditions on the coordinates must be satisfied?

2.

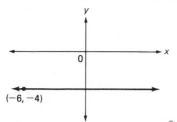

3.

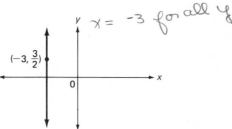

x = -3 for all y

4.

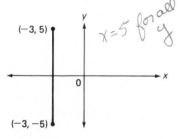

x = 5 for all y

5.

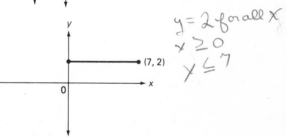

y = 2 for all x
x ≥ 0
x ≤ 7

6.

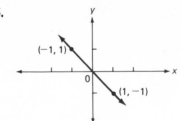

(−1, 1)

(1, −1)

7. $1 \leq y \leq -2$

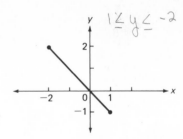

Graph.

8. $y = x + 2$

9. $y = x - 2$

10. $y = -x + 2$

11. $y = 2x + 1$

12. $y = -2x + 3$

13. $y = x$

Graph each line by locating the x- and y-intercepts. Note: *To find the coordinates of the x-intercept, substitute* $y = 0$ *and solve for x; for the y-intercept, let* $x = 0$ *and solve for y.*

14. $y = 3x - 6$

15. $y = -2x + 8$

16. $y = -\frac{1}{2}x + 3$

Graph the points that satisfy each equation for the given values of x.

17. $y = \frac{1}{2}x;\ -6 \leq x \leq 6$

18. $y = -2x + 1;\ -2 \leq x \leq 2$

19. $y = 3x - 5;\ 1 \leq x \leq 4$

20. $y = x^2;\ x = -3, -2, -1, 0, 1, 2, 3$

21. $y = x^2;\ x = -\frac{3}{2}, -1, -\frac{1}{2}, 0, \frac{1}{2}, 1, \frac{3}{2}$

22. $y = x^2;\ -2 \leq x \leq 2$

23. $y = |x|;\ x = -3, -1, 0, 2, 4$

24. $y = |x|;\ -3 \leq x \leq 4$

25. $y = \dfrac{1}{x};\ x = \dfrac{1}{2}, 1, \dfrac{3}{2}, 2, \dfrac{5}{2}, 3$

26. $y = \dfrac{1}{x};\ \dfrac{1}{2} \leq x \leq 3$

27. $y = \sqrt{x}\ ;\ x = 0, \frac{1}{4}, 1, \frac{25}{16}, \frac{9}{4}, 4$

28. $y = \sqrt{x}\ ; 0 \leq x \leq 4$

Graph each inequality.

29. $y > x + 2$

30. $y \leq x + 2$

31. $y \geq x - 1$

32. $y < x - 1$

*33. $2x + y - 4 > 0$

*34. $x - 2y + 2 \leq 0$

3.3

Linear Functions

The graph of every equation of the form $y = mx + b$, where m and b are constant, is a straight line. We say that such an equation defines a *linear function.* (Note that $mx + b$ is a first-degree polynomial in x.) In this section we will learn more about straight lines and their equations.

    The figure at the top of the next page shows two lines ℓ_1 and ℓ_2 intersecting at the point (2, 3). Point (4, 7) is on ℓ_1 and (5, 4) is on ℓ_2.

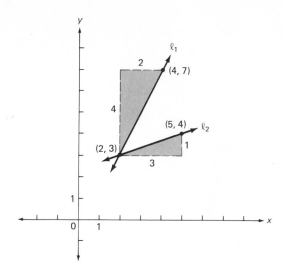

One way these lines differ is in their "steepness"; reading from left to right, ℓ_1 is rising faster than ℓ_2. This notion of steepness can be made more precise by looking at the right triangles determined by the two points indicated on each line.

First look at ℓ_1. Starting at $(2, 3)$ the point $(4, 7)$ can be reached by going 4 units up and 2 units to the right. We say that the change in the y-values (or vertical direction) is 4 units, whereas the change in the x-values (or horizontal direction) is 2 units. We say that the *slope* of ℓ_1 is $\frac{4}{2} = 2$.

For ℓ_2 the change in y is 1 unit up and 3 units right; this is a ratio of 1 to 3, or $\frac{1}{3}$. The *slope* of ℓ_2 is $\frac{1}{3}$.

Do you see that these ratios show numerically that ℓ_1 is steeper than ℓ_2 since $2 > \frac{1}{3}$?

Every nonvertical line has associated with it a unique number called its *slope*, defined as follows.

Definition of slope

If (x_1, y_1) and (x_2, y_2) are any two points on a line ℓ, then the **slope** of ℓ, denoted by m, is defined as

$$m = \frac{y_2 - y_1}{x_2 - x_1}$$

provided $x_1 \neq x_2$.

In the following figure the coordinates of two points A and B have been labeled (x_1, y_1) and (x_2, y_2). The change in the y direction from A to B is given by the difference $y_2 - y_1$; the change in the x direction is $x_2 - x_1$. If a different pair of points is chosen, such as P and Q, then the ratio of these differences is still the same because the resulting triangles (*ABC* and *PQR*) are similar. (Recall that corresponding sides of similar triangles are proportional.)

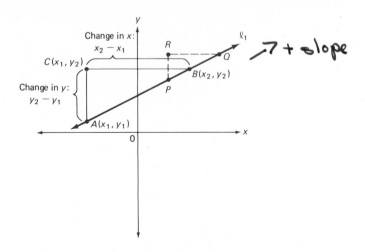

It may be helpful to think of the slope of a line in this way:

$$m = \frac{y_2 - y_1}{x_2 - x_1} = \frac{\text{change in } y}{\text{change in } x} = \frac{\text{vertical change}}{\text{horizontal change}}$$

In the preceding figure both $y_2 - y_1$ and $x_2 - x_1$ are positive. (Why?) Thus the line ℓ_1 has a positive slope. In the following figure $y_2 - y_1$ is negative and $x_2 - x_1$ is positive. Therefore the slope of the line ℓ_2 is negative. <u>Reading from left to right, a rising line has a positive slope and a falling line has a negative slope.</u>

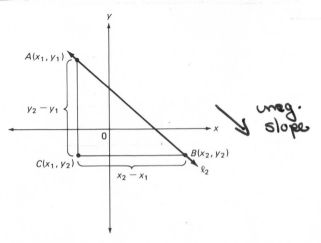

Example 1   Find the slope of line ℓ determined by the points $(-3, 4)$ and $(1, -6)$.

Solution   Use $(x_1, y_1) = (-3, 4)$; $(x_2, y_2) = (1, -6)$.

Then $m = \dfrac{-6 - 4}{1 - (-3)} = -\dfrac{10}{4} = -\dfrac{5}{2}$.

$m = \dfrac{y_2 - y_1}{x_2 - x_1} = \dfrac{-6 - 4}{1 - (-3)} = \dfrac{-10}{4}$

$= \dfrac{-5}{2}$

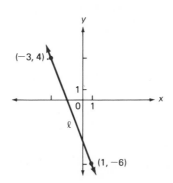

Note: It makes no difference which of the two points is called (x_1, y_1) or (x_2, y_2) since the ratio will still be the same. If $(x_1, y_1) = (1, -6)$ and $(x_2, y_2) = (-3, 4)$, for example, then $\dfrac{y_2 - y_1}{x_2 - y_1} = \dfrac{4 - (-6)}{-3 - 1} = -\dfrac{5}{2} = m$.

Example 2   Graph the line with slope 3 that passes through the point $(-2, -2)$.

Solution   Write $3 = \dfrac{3}{1} = \dfrac{\text{change in } y}{\text{change in } x}$. Now start at $(-2, -2)$ and move 3 units up and 1 unit to the right. This locates the point $(-1, 1)$. Draw the straight line through these two points.

$3 = \dfrac{-2 - y_2}{-2 - x_2} =$

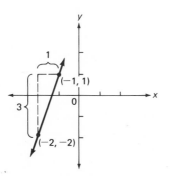

TEST YOUR UNDERSTANDING

Find the slopes of the lines determined by the given pairs of points.

1. $(1, 5); (4, 6)$ **2.** $(3, -5); (-3, 3)$
3. $(-2, -3); (-1, 1)$ **4.** $(-1, 0); (0, 1)$

Draw the line through the given point and with the given slope.

5. $(0, 0); m = 2$ **6.** $(-3, 4); m = -\frac{3}{2}$

Lines with the same slope are parallel. Thus the slope alone does not identify a unique line. As we have seen, however, if just one point on the line is also known, then the line is completely determined.

Let ℓ be a line with slope m that passes through (x_1, y_1). We wish to determine the conditions on the coordinates of any point $P(x, y)$ that is on ℓ.

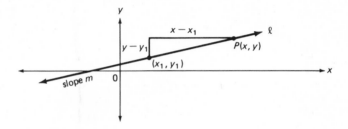

From the figure you can see that $P(x, y)$ will be on ℓ if and only if the ratio $\dfrac{x - y_1}{x - x_1}$ is the same as m. That is, P is on ℓ if and only if

$$m = \frac{y - y_1}{x - x_1}$$

Multiply both sides of this equation by $x - x_1$.

$$m(x - x_1) = y - y_1$$

This step leads to the following form for the equation of a straight line.

Point-slope form of a line

$$y - y_1 = m(x - x_1)$$

where m is the slope and (x_1, y_1) is a point on the line.

Example 3   Find the point-slope form of the line *l* with slope $m = 3$ that passes through the point $(-1, 1)$. Verify that $(-2, -2)$ is on the line.

Solution   Since $m = 3$, any (x, y) on *l* satisfies this equation:

$$y - 1 = 3[x - (-1)] = 3(x + 1)$$

Let $x = -2$:

$$y - 1 = 3(-2 + 1) = -3$$
$$y = -2$$

Thus $(-2, -2)$ is on the line.

TEST YOUR UNDERSTANDING

Write the point-slope form of the line through the given point with slope m.

1. $(2, 6)$; $m = -3$ **2.** $(-1, 4)$; $m = \frac{1}{2}$ **3.** $(5, -\frac{2}{3})$; $m = 1$
4. $(0, 0)$; $m = -\frac{1}{4}$ **5.** $(-3, -5)$; $m = 0$ **6.** $(1, -1)$; $m = -1$

Any horizontal line will have slope zero. This is clear, since for any two points on such a line the distance from the *x*-axis is constant; $y_1 = y_2$.

$$m = \frac{y_2 - y_1}{x_2 - x_1} = \frac{0}{x_2 - x_1} = 0$$

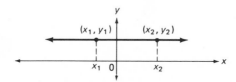

For two points on a vertical line the *x*-values must be the same. (Draw a sketch.) Thus $x_2 - x_1 = 0$ and the definition $m = \dfrac{y_2 - y_1}{x_2 - x_1}$ does not apply. We agree that vertical lines do not have slopes.

Solving $y - y_1 = m(x - x_1)$ for *y*, we have
$$y = mx + (y_1 - mx_1)$$

Since (x_1, y_1) is a fixed point, the quantity $y_1 - mx_1$ is a constant that may be called *b*. Thus we may write $y = mx + b$ as the *slope-intercept form of a line*. In this form the slope is the coefficient of *x*. Since *x* may be any value, we may let

$x = 0$ to find $y = m \cdot 0 + b = b$. Thus $(0, b)$ is on the line. This point is on the y-axis and b is therefore the y-intercept of the line.

Slope-intercept form of a line

$$y = mx + b$$

where m is the slope and b is the y-intercept.

Example 4   Graph the linear function f defined by $y = f(x) = 2x - 1$ by using the slope and y-intercept. Also indicate the domain and range of f, and display $f(2) = 3$ geometrically.

Solution   The y-intercept is -1. Locate $(0, -1)$ and use $m = 2 = \frac{2}{1}$ to reach $(1, 1)$, another point on the line.

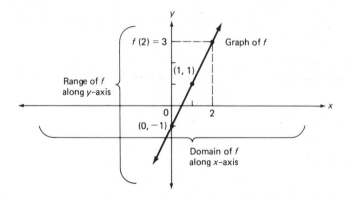

Example 5   Write the slope-intercept form of the line through the two points $(6, -4)$ and $(-3, 8)$.

Solution   First compute the slope.

$$m = \frac{-4 - 8}{6 - (-3)} = \frac{-12}{9} = -\frac{4}{3}$$

Use either point to write the point-slope form, and then convert to the slope-intercept form:

$$y - (-4) = -\tfrac{4}{3}(x - 6)$$
$$y + 4 = -\tfrac{4}{3}x + 8$$
$$y = -\tfrac{4}{3}x + 4$$

Caution:

1. Do not confuse zero slope (for horizontal lines) with no slope (for vertical lines).

2. In computing the slope using the points (x_1, y_1) and (x_2, y_2), observe the same order for the coordinates in the numerator and denominator. Note the following:

$$\frac{y_2 - y_1}{x_2 - x_1} = \frac{y_1 - y_2}{x_1 - x_2} = m$$

But

$$\frac{y_2 - y_1}{x_1 - x_2} = \frac{y_1 - y_2}{x_2 - x_1} = -m$$

EXERCISES 3.3

Compute the slope, if it exists, for the line determined by the given pair of points.

1. $(3, 4); (-5, 2)$
2. $(3, 4); (2, -5)$
3. $(4, 3); (-5, 2)$
4. $(-7, 6); (-7, 106)$
5. $(6, -7); (106, -7)$
6. $(5\sqrt{2}, 3\sqrt{8}); (\sqrt{2}, \sqrt{8})$
7. $(2, -\frac{3}{4}); (-\frac{1}{3}, \frac{2}{3})$
8. $(\frac{1}{2}, 0.1); (-9, 0.1)$
9. $(\sqrt{3}, 8); (-1, 6)$

Draw the line through the indicated point having slope m, as given.

10. $(0, 0); m = 2$
11. $(0, 0); m = -\frac{1}{2}$
12. $(0, 2); m = \frac{3}{4}$
13. $(-\frac{1}{2}, 0); m = -1$
14. $(-2, \frac{3}{2}); m = 0$
15. $(5, -3);$ no slope
16. Draw the two lines through the point $(-3, 4)$ that have slope 4 and $-\frac{1}{4}$.
17. Graph each of the lines with the following slopes through the point $(5, -3)$:

$$m = -2 \qquad m = -1 \qquad m = 0 \qquad m = 1 \qquad m = 2$$

18. In the same coordinate system draw the five lines, each with slope -2, through the points $(-2, 0), (-1, 0), (0, 0), (\frac{1}{2}, 0)$, and $(1, 1)$ respectively.

19. In the same coordinate system draw the lines:
    (a) Through $(1, 0)$ with $m = -1$.
    (b) Through $(0, 1)$ with $m = 1$.
    (c) Through $(-1, 0)$ with $m = -1$.
    (d) Through $(0, -1)$ with $m = 1$.

20. Why is the line determined by the points $(6, -5)$ and $(8, -8)$ parallel to the line through $(-3, 12)$ and $(1, 6)$?

21. Verify that the points $A(1, 2), B(4, -1), C(2, -2)$, and $D(-1, 1)$ are the vertices of a parallelogram. Sketch.

Write the point-slope form of the line through the indicated point with slope m.

22. $(-3, 5)$; $m = -2$ **23.** $(-3, 5)$; $m = 0$

24. $(-3, 5)$; $m = 2$ **25.** $(2, 1)$; $m = \frac{1}{2}$

26. $(8, 0)$; $m = -\frac{2}{3}$ **27.** $(0, 0)$; $m = 5$

28. $(-6, -3)$; $m = \frac{4}{3}$ **29.** $(-6, -3)$; $m = -\frac{3}{4}$

30. $(-\frac{3}{4}, \frac{2}{3})$; $m = 1$ **31.** $(\sqrt{2}, -\sqrt{2})$; $m = 10$

32. (a) Find the slope of line determined by the points $A(-3, 5)$ and $B(1, 7)$, and write its equation in point-slope form, using the coordinates of A.

(b) Do the same as in part (a) using the coordinates of B.

(c) Verify that the equations obtained in parts (a) and (b) give the same slope-intercept form.

Write each equation in slope-intercept form. Identify the slope and y-intercept.

33. $2x + 3y = 6$ **34.** $3x + 2y = 6$

35. $3x - 2y = 6$ **36.** $2x - 3y = 6$

37. $2x + 5y - 10 = 0$ **38.** $5x - 2y + 10 = 0$

39. $4x - 3y - 7 = 0$ **40.** $3x + 4y + 9 = 0$

Use the given pair of points to find m. Then write the equation of the line in slope-intercept form.

41. $(-2, 0)$; $(0, 3)$ **42.** $(\frac{1}{2}, 7)$; $(-4, -\frac{3}{2})$

43. $(-1, -13)$; $(-8, 1)$ **44.** $(-10, -3)$; $(\frac{1}{3}, -3)$

45. $(10, 27)$; $(12, 27)$ **46.** $(\sqrt{2}, 4\sqrt{2})$; $(-3\sqrt{2}, -10\sqrt{2})$

Graph the linear function f by using the slope and y-intercept. Display $y = f(-2)$ on the graph.

47. $y = f(x) = x + 3$ **48.** $y = f(x) = -2x + 1$

49. $y = f(x) = \frac{1}{2}x - 3$ **50.** $y = f(x) = 3x - \frac{1}{2}$

51. (a) Graph the line determined by the points $(5, -2)$ and $(5, 3)$.

(b) For any point (x, y) on this line what must be true about x?

(c) Describe the equation of a vertical line.

52. Graph the lines $y = 5$ and $x = 5$ in the same coordinate system.

53. (a) Graph the two linear functions $y = f(x) = 3x - 2$ and $y = g(x) = -\frac{1}{3}x + 6$ in the same coordinate system.

(b) What is the domain and range for each of these functions?

(c) For how many domain values do these two functions have the same range value?

In Exercises 54 through 57 the domain of the independent variable x is given for the function defined by the equation $y = 3x - 7$. Graph each function.

54. All real numbers.   **55.** All $x \geq 0$.

56. All x where $-1 \leq x \leq 3$.   **57.** $x = -1, 0, 1, 2, 3$

*58. The sides of the parallelogram with vertices $(-1, 1)$, $(0, 3)$, $(2, 1)$, and $(3, 3)$ are the graphs of four different functions. In each case find the equation that defines the function and state the domain.

*59. Any line having a nonzero slope that does not pass through the origin always has both an x- and a y-intercept. Let l be such a line having equation $ax + by = c$.
  (a) Why is $c \neq 0$?
  (b) What are the x- and y-intercepts?
  (c) Derive the equation

$$\frac{x}{q} + \frac{y}{p} = 1$$

  where q and p are the x- and y-intercepts, respectively. This is known as the *intercept form* of a line.
  (d) Use the intercept form to write the equation of the line passing through $(\frac{3}{2}, 0)$ and $(0, -5)$.
  (e) Use the two points in part **(d)** to find the slope, write the slope-intercept form, and compare with the result in part **(d)**.

3.4
Some Special
Functions

Linear functions can be used to define other functions which, in themselves, are not linear but may be described as being "partly" or "piecewise" linear. An important example is the absolute-value function.

$$y = |x| = \begin{cases} x & \text{for } x \geq 0 \\ -x & \text{for } x < 0 \end{cases}$$

To graph this function, first draw the line $y = -x$ and eliminate all those points on it for which x is positive. Then draw the line $y = x$ and eliminate the part for which x is negative. Now join these two parts to get the graph of $y = |x|$.

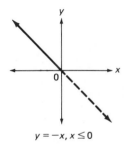
$y = -x, x \leq 0$

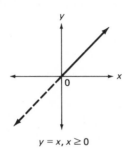
$y = x, x \geq 0$

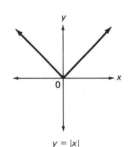
$y = |x|$

The graph of $y = |x|$ consists of two perpendicular half-lines intersecting at the origin. Now $y = |x|$ is not a linear function, but it is linear in parts; the two halves $y = x$, $y = -x$ are linear.

Note that the graph is symmetric about the y-axis. (If the paper were folded along the y-axis, the two parts would coincide.) This symmetry can be observed by noting that the y-values for x and $-x$ are the same. That is,

$$|x| = |-x| \qquad \text{for all } x$$

Example 1   Graph: $y = |x - 2|$.

Solution   For $x \geq 2$, we find $x - 2 \geq 0$, which implies that $y = |x - 2| = x - 2$. This is the half-line through (and to the right of) the point $(2, 0)$, with slope 1. For $x < 2$, we get $x - 2 < 0$, which implies that $y = |x - 2| = -(x - 2) = -x + 2$. This gives the left half of the graph shown.

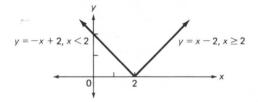

Example 2   Graph the function f defined by the following two equations.

$$y = f(x) = \begin{cases} 2x & \text{if } 0 \leq x \leq 1 \\ -x + 2 & \text{if } 1 < x \end{cases}$$

What are the domain and range of f?

Solution

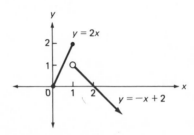

The domain of f is all $x \geq 0$. From the graph we see that the range consists of all $y \leq 2$.

Note: The open dot at (1, 1) means that this point is not part of the graph; the point (1, 2) is on the graph since $f(1) = 2 \cdot 1 = 2$.

Example 3   Graph the function f given by $y = f(x) = \dfrac{|x|}{x}$. What is the domain of f?

Solution   The domain consists of all $x \neq 0$. When $x > 0, |x| = x$ and

$$f(x) = \frac{|x|}{x} = \frac{x}{x} = 1$$

Thus $f(x)$ is the constant 1, for all positive x. Similarly, for $x < 0, |x| = -x$ and

$$f(x) = \frac{|x|}{x} = \frac{-x}{x} = -1$$

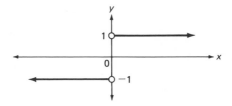

Example 3 is an illustration of <u>a **step function**</u>. Such a function may be described as a function whose graph consists of parts of horizontal lines. Here is the graph of another step function defined for $-2 \leq x < 4$.

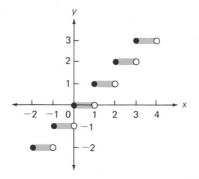

We can find the defining equation for each step. For $-2 \leq x < -1$ the graph shows that the corresponding range value is always -2. That is, $y = -2$ for $-2 \leq x < -1$. In the same way we may write these six defining equations.

$$y = \begin{cases} -2 & \text{for } -2 \le x < -1 \\ -1 & \text{for } -1 \le x < 0 \\ 0 & \text{for } 0 \le x < 1 \\ 1 & \text{for } 1 \le x < 2 \\ 2 & \text{for } 2 \le x < 3 \\ 3 & \text{for } 3 \le x < 4 \end{cases}$$

Observe that in each case, say $2 \le x < 3$, the integer 2 at the left is also the corresponding y-value. Putting it another way, we say that the y-value 2 is the greatest integer less than or equal to x. (It is *the greatest integer not exceeding x itself*.) Using this language, the six equations may be condensed into a single statement.

$$y = \text{greatest integer} \le x \qquad \text{for } -2 \le x < 4$$

For each real number x there must be some integer n (positive, negative, or zero) so that $n \le x < n + 1$. Therefore the greatest integer less than or equal to x equals n. In other words the preceding step function may be extended to a step function with domain of *all* real numbers; its graph would consist of an infinite number of steps.

We use the symbol $[x]$ to mean "greatest integer less than or equal to x." Thus

$$y = [x] = \text{greatest integer} \le x$$

Here are a few illustrations:

   $[2.17] = 2$ because 2 is the greatest integer ≤ 2.17.
   $[0.64] = 0$ because zero is the greatest integer ≤ 0.64.
   $[-2.8] = -3$ because -3 is the greatest integer ≤ -2.8.
   $[-5] = -5$ because -5 is the greatest integer ≤ -5.

Example 4   Graph the step function $y = [2x]$ for the domain $1 \le x < 3$. What is the range of this function?

Solution   In order to evaluate $[2x]$ we need to be able to place $2x$ between two *successive* integers. First note that $1 \le x < 3$ implies that $2 \le 2x < 6$. Therefore $2x$ could be between the integers 2 and 3, 3 and 4, 4 and 5, or 5 and 6. Thus we have the following:

   If $1 \le x < \frac{3}{2}$. then $2 \le 2x < 3$ and $[2x] = 2$.
   If $\frac{3}{2} \le x < 2$, then $3 \le 2x < 4$ and $[2x] = 3$.
   If $2 \le x < \frac{5}{2}$, then $4 \le 2x < 5$ and $[2x] = 4$.
   If $\frac{5}{2} \le x < 3$, then $5 \le 2x < 6$ and $[2x] = 5$.

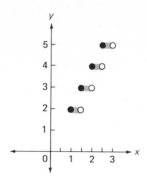

From the graph we see that the range consists of the values 2, 3, 4, and 5.

EXERCISES 3.4

Graph each function and state the domain and range.

1. $y = |x - 1|$ **2.** $y = |x + 3|$ **3.** $y = |2x|$

4. $y = |2x - 1|$ **5.** $y = |3 - 2x|$ **6.** $y = |\frac{1}{2}x + 4|$

7. $y = \begin{cases} 3x & \text{for } -1 \le x \le 1 \\ -x & \text{for } 1 < x \end{cases}$ **8.** $y = \begin{cases} -2x + 3 & \text{for } x < 2 \\ x + 1 & \text{for } x > 2 \end{cases}$

9. $y = \begin{cases} x & \text{for } -2 < x \le 0 \\ 2x & \text{for } 0 < x \le 2 \\ -x + 3 & \text{for } 2 < x \le 3 \end{cases}$

10. $y = \begin{cases} x & \text{if } 0 \le x < 1 \\ x - 1 & \text{if } 1 \le x < 2 \\ x - 2 & \text{if } 2 \le x < 3 \\ x - 3 & \text{if } 3 \le x \le 4 \end{cases}$

Graph each step function for its given domain.

11. $y = \dfrac{x}{|x|}$; all $x \ne 0$ **12.** $y = \dfrac{|x - 2|}{x - 2}$; all $x \ne 2$

13. $y = \dfrac{|x + 3|}{x + 3}$; all $x \ne -3$ **14.** $y = \begin{cases} -1 & \text{for } -1 \le x \le 0 \\ 0 & \text{for } 0 < x \le 1 \\ 1 & \text{for } 1 < x \le 2 \\ 2 & \text{for } 2 < x \le 3 \end{cases}$

15. $y = [x]$; $-3 \le x \le 3$ **16.** $y = [2x]$; $0 \le x \le 2$

*17. $y = 2[x]$; $0 \le x \le 2$ ***18.** $y = [3x]$; $-1 \le x \le 1$

*19. $y = [x - 1]$; $-2 \le x \le 3$

*20. (a) The postage for mailing packages depends on the weight and destination. Let the rates for a certain destination be as follows:

$x =$ WEIGHT (POUNDS)	$y =$ POSTAGE (COST)
under 1	$0.80
1 or more but under 2	$0.90
2 or more but under 3	$1.00
3 or more but under 4	$1.10
4 or more but under 5	$1.20
5 or more but under 6	$1.30
6 or more but under 7	$1.40
7 or more but under 8	$1.50
8 or more but under 9	$1.60
9 or more but under 10	$1.70

This table defines y to be a function of x. If we use P (for postage) we may write $y = P(x)$, and the table gives $P(x) = 1.20$ for $4 \leq x < 5$. Graph this function on its domain $0 < x < 10$. To achieve clarity, you may want to use a larger unit along the vertical axis then on the x-axis.

(b) A formula for $P(x)$ can be given in terms of the greatest integer function. Find such a formula.

3.5

Systems of Linear Equations

The linear equations considered thus far have all been in the slope-intercept form or the point-slope form. There are other ways in which linear equations are often stated. Consider, for example, $y = -\frac{2}{3}x + 4$. Multiply by 3 and isolate the constant term on one side of the equation to obtain $2x + 3y = 12$. A linear equation written in this form is a special case of the **general linear equation** $ax + by = c$, where $a, b,$ and c are constants.

General linear equation

$$ax + by = c$$

In case $a \neq 0$ and $b = 0$, we have $ax + 0y = c$, or

$$x = \frac{c}{a}$$

which is the equation of a vertical line. In case $b \neq 0$, we may solve for y to get

$$y = -\frac{a}{b}x + \frac{c}{b}$$

which is the slope-intercept form; the slope is $-\frac{a}{b}$ and the y-intercept is $\frac{c}{b}$.

It is not worth the effort to memorize these formulas; it is easy enough to take an equation like $2x + 3y = 12$ and solve it for y. Thus $y = -\frac{2}{3}x + 4$, from which we see that the slope is $-\frac{2}{3}$ and the y-intercept is 4.

The equation $y = -\frac{2}{3}x + 4$ shows y to be a function of x *explicitly*; the dependent variable y stands isolated. The form $2x + 3y = 12$ is said to define the same function *implicitly*. In a sense, the equation $2x + 3y = 12$ "contains" a function, but we must solve it for y to see just what its explicit form is.

Let us consider a linear system of two equations as follows:

(1) $2x + 3y = 12$

(2) $3x + 2y = 12$

We may graph this system by graphing each line on the same axes. This may be done by finding the x- and y-intercepts of each line. For instance, letting $x = 0$ in Equation (1) gives the y-intercept $y = 4$. Likewise, letting $y = 0$ produces the x-intercept $x = 6$. The line through $(0, 4)$ and $(6, 0)$ is the graph of Equation (1). Equation (2) may be graphed similarly. Can you find the point of intersection of the two lines?

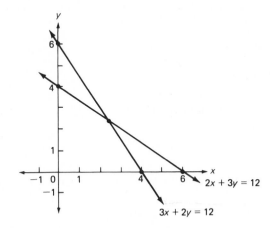

Finding the point of intersection of the two lines is also described as solving the system

$$2x + 3y = 12$$
$$3x + 2y = 12$$

Two procedures will now be described to solve such systems. The underlying idea for each of them is that the coordinates of the point of intersection must satisfy each equation. The **substitution method** will be taken up first.

We begin by letting (x, y) be the coordinates of the point of intersection of the preceding system. Then these x- and y-values fit both equations. Hence either equation may be solved for x or y and then substituted into the other equation. For example, solving the second equation for y produces

$$y = -\tfrac{3}{2}x + 6$$

Substitute this expression into the first and solve for x.

$$2x + 3y = 12$$
$$2x + 3(-\tfrac{3}{2}x + 6) = 12$$
$$2x - \tfrac{9}{2}x + 18 = 12$$
$$-\tfrac{5}{2}x = -6$$
$$x = (-\tfrac{2}{5})(-6) = \tfrac{12}{5}$$

To find the y-value, substitute $x = \tfrac{12}{5}$ into either of the earlier equations. Thus

$$y = -\tfrac{3}{2}(\tfrac{12}{5}) + 6 = \tfrac{12}{5}$$

The given equations may be used for checking. The solution is $(\tfrac{12}{5}, \tfrac{12}{5})$.

TEST YOUR UNDERSTANDING

Use the substitution method to solve each linear system.

1. $y = 3x - 1$
   $y = -5x + 7$

2. $y = 4x + 16$
   $y = -\tfrac{2}{3}x + \tfrac{14}{5}$

3. $4x - 3y = 11$
   $y = 6x - 13$

4. $2x + 2y = \tfrac{4}{5}$
   $-7x + 2y = -1$

5. $x + 7y = 3$
   $5x + 12y = -8$

6. $4x - 2y = 40$
   $-3x + 3y = 45$

Now we will solve the same system as before, this time using the **multiplication-addition** (or **subtraction**) **method.**

$$2x + 3y = 12$$
$$3x + 2y = 12$$

The idea here is to alter the equations so that the coefficients of one of the variables are either negatives of one another or equal to each other. We may, for instance, multiply the first equation by 3 to get $6x + 9y = 36$ and multiply the second by -2 giving $-6x - 4y = -24$. The resulting system looks like this:

$$6x + 9y = 36$$
$$-6x - 4y = -24$$

Keep in mind that we are looking for the pair (x, y) that fits both equations. Thus for these x- and y-values we may add equals to equals to eliminate the variable x and get

$$5y = 12$$
$$y = \tfrac{12}{5}$$

As before, x is now found by substituting $y = \tfrac{12}{5}$ into one of the given equations.

If in the preceding solution the second equation is multiplied by 2 instead of -2, then the system becomes

$$6x + 9y = 36$$
$$6x + 4y = 24$$

Now x can be eliminated by subtracting the equations.

As illustrated in the next example, this method can be condensed into a compact procedure.

Example 1 Solve the system by the multiplication-addition method.

$$\tfrac{1}{3}x - \tfrac{2}{3}y = 4$$
$$7x + 3y = 27$$

Solution

$$15(\tfrac{1}{3}x - \tfrac{2}{3}y = 4) \Longrightarrow \quad 5x - 6y = \quad 60$$
$$2(7x + 3y = 27) \Longrightarrow 14x + 6y = \quad 54$$
$$\text{Add:} \qquad 19x \qquad = 114$$
$$x \qquad = \quad 6$$

Substitute to solve for y.

$$7x + 3y = 27 \Longrightarrow 7(6) + 3y = 27$$
$$3y = -15$$
$$y = -5$$

The solution is $(6, -5)$.

TEST YOUR UNDERSTANDING

Use the multiplication-addition method to solve each linear system.

1. $3x + 4y = 5$
$5x + 6y = 7$

2. $-8x + 5y = -19$
$4x + 2y = -4$

3. $\frac{1}{4}x + \frac{5}{2}y = 2$
$\frac{1}{2}x - 7y = -\frac{17}{4}$

4. $10x + 9y = 0$
$\frac{2}{3}x = -6y$

There is an important feature that all these procedures have in common. They all begin by eliminating one of the two variables. Thus you soon reach one equation in one unknown. This basic strategy of reducing the number of unknowns can be applied to "larger" linear systems (as well as to systems that are not linear). For instance, here is a system of three linear equations in three variables:

$$(1) \qquad 2x - 5y + z = -10$$

$$(2) \qquad x + 2y + 3z = 26$$

$$(3) \quad -3x - 4y + 2z = 5$$

To solve this system we may begin by eliminating the variable x from the first two equations.

$$\left. \begin{array}{c} 2x - 5y + z = -10 \\ -2(x + 2y + 3z = 26) \end{array} \right\} \implies \begin{array}{c} 2x - 5y + z = -10 \\ -2x - 4y - 6z = -52 \\ \hline -9y - 5z = -62 \end{array}$$

Add:

Another equation in y and z can be obtained from Equations (2) and (3) by eliminating x:

$$\left. \begin{array}{c} 3(x + 2y + 3z = 26) \\ -3x - 4y + 2z = 5 \end{array} \right\} \implies \begin{array}{c} 3x + 6y + 9z = 78 \\ -3x - 4y + 2z = 5 \\ \hline 2y + 11z = 83 \end{array}$$

Add:

Now we have this system in two variables:

$$9y + 5z = 62$$

$$2y + 11z = 83$$

You can solve this system as before to find $y = 3$ and $z = 7$. Substituting these values into an earlier equation, say (2), we get

$$x + 2(3) + 3(7) = 26$$

$$x = -1$$

The remaining equations can be used for checking:

$$(1) \qquad 2(-1) - 5(3) + 7 = -10$$
$$(3) \quad -3(-1) - 4(3) + 2(7) = 5$$

The solution is $x = -1$, $y = 3$, $z = 7$.

There are verbal problems that can be solved by using systems of linear equations. In some cases only one equation in one variable can also be used. However, it is worthwhile to learn how to use more than one variable in order to simplify the process of translating from the verbal to the mathematical form.

Example 2 A field goal in basketball is worth 2 points and a foul shot is worth 1 point. In a recent game the school basketball team scored 85 points. If there were twice as many field goals as foul shots, how many of each were there?

Solution Let x be the number of field goals and y the number of foul shots. Then the points due to field goals is $2x$, and y is the number of points due to foul shots. There were twice as many field goals as foul shots, so

$$x = 2y$$

The total number of points in the game was 85; therefore

$$2x + y = 85$$

The answer to the problem is the solution for the system

$$x = 2y$$
$$2x + y = 85$$

which is the ordered pair $(34, 17)$. You can check by returning to the original statement of the problem.

EXERCISES 3.5

Solve each system by the substitution method.

1. $2x + y = -10$
$\quad\;\; 6x - 3y = 6$

2. $-3x + 6y = 0$
$\quad\;\; 4x + y = 9$

3. $v - w = 14$
$\quad\;\; 3v + w = 2$

4. $4x - y = 6$
  $2x + 3y = 10$

5. $x + 5y = -9$
  $4x - 3y = -13$

6. $s + 2t = 5$
  $-3s + 10t = -7$

Solve each system by the multiplication-addition method.

7. $-3x + y = 16$
  $2x - y = 10$

8. $2x + 4y = 24$
  $-3x + 5y = -25$

9. $3y - 9x = 30$
  $8x - 4y = 24$

10. $2u - 6v = -16$
  $5u - 3v = 8$

11. $4x - 5y = 3$
  $16x + 2y = 3$

12. $\frac{1}{3}x + 3y = 6$
  $-x - 8y = 18$

Solve the systems given in Exercises 13 through 28 by using any method.

13. $x - 2y = 3$
  $y - 3x = -14$

14. $2x + y = 6$
  $3x - 4y = 12$

15. $-3x + 8y = 16$
  $16x - 5y = 103$

16. $3x - 8y = -16$
  $7x + 19y = -188$

17. $16x - 5y = 103$
  $7x + 19y = -188$

18. $4x = 7y - 6$
  $9y = -12x + 12$

19. $3s + t - 3 = 0$
  $2s - 3t - 2 = 0$

20. $\frac{1}{2}x - \frac{1}{3}y = 2$
  $\frac{3}{4}x + \frac{2}{5}y = -1$

(*Hint for Exercise 20:*
Multiply the first equation
by 6 and the second by 20.)

21. $\frac{1}{4}x + \frac{1}{3}y = \frac{5}{12}$
  $\frac{1}{2}x + y = 1$

22. $0.1x + 0.2y = 0.7$
  $0.01x - 0.01y = 0.04$

23. $\dfrac{x}{2} + \dfrac{y}{6} = \dfrac{1}{2}$

  $0.2x - 0.3y = 0.2$

24. $2(x + y) = 4 - 3y$

  $\frac{1}{2}x + y = \frac{1}{2}$

25. $2(x - y - 1) = 1 - 2x$

  $6(x - y) = 4 - 3(3y - x)$

26. $\dfrac{x - 2}{5} + \dfrac{y + 1}{10} = 1$

  $\dfrac{x + 2}{3} - \dfrac{y + 3}{2} = 4$

27. $2x = 7$
  $y = 4$

28. $2x - 10c = -y$
  $7x - 2y = 2c$
  (c is constant)

Solve each linear system.

29. $x + y + z = 2$
  $x - y + 3z = 12$
  $2x + 5y + 2z = -2$

30. $x + 2y + 3z = 5$
  $-4x + z = 6$
  $3x - y = -3$

31. $-3x + 3y + z = -10$
  $4x + y + 5z = 2$
  $x - 8y - 2z = 12$

32. $x + 2y + 3z = -4$
  $4x + 5y + 6z = -4$
  $7x - 15y - 9z = 4$

33. Solve the system

$$ax + by = c$$
$$dx + ey = f$$

for x and y, where $a, b, c, d, e,$ and f are constants with $ae - bd \neq 0$.

34. Why do we not allow both *a* and *b* to be zero in the linear equation $ax + by = 0$?

*35. Points $(-8, -16)$, $(0, 10)$, and $(12, 14)$ are three vertices of a parallelogram. Find the coordinates of the fourth vertex if it is located in the third quadrant.

*36. Find the point of intersection for the diagonals of the parallelogram in Exercise 35.

37. The total points that a basketball team scored was 96. If there were $2\frac{1}{2}$ times as many field goals as foul shots, how many of each were there? (Field goals count 2 points; foul shots count 1 point.)

38. During a round of golf a player only scored fours and fives per hole. If he played 18 holes and his total score was 80, how many holes did he play in four strokes and how many in five?

39. The tuition fee at a college plus the room and board comes to $5300 per year. The room and board is $100 less than half the tuition. How much is the tuition, and what does the room and board cost?

40. A college student had a work-study scholarship that paid $2.50 per hour. She also made $1 per hour babysitting. Her income one week was $46 for a total of $23\frac{1}{2}$ hours of employment. How many hours did she spend on each of the two jobs?

*41. An airplane, flying with a tail wind, takes 2 hours and 40 minutes to travel 1120 miles. The plane makes the return trip against the wind in 2 hours and 48 minutes. What is the wind velocity and what is the speed of the plane in still air? (Assume that both velocities are constant; add the velocities for the downwind trip, and subtract them for the return trip.)

42. The perimeter of a rectangle is 72 inches. The length is $3\frac{1}{2}$ times as large as the width. Find the dimensions.

*43. A wholesaler has two grades of oil that ordinarily sell for 64¢ per quart and 44¢ per quart. He wants a blend of the two oils to sell at 55¢ per quart. If he anticipates selling 400 quarts of the new blend, how much of each grade should he use? (One of the equations makes use of the fact that the total income will be $400(0.55) = $220.)

44. A store paid $103.18 for a recent mailing. Some of the letters cost 14¢ postage and the rest needed 10¢ postage. How many letters at each rate were mailed if the total number sent out was 887?

45. The annual return on two investments totals $464. One investment gives 8% interest and the other $7\frac{1}{2}$%. How much money is invested at each rate if the total investment is $6000?

*46. A salesman said that it did not matter to him whether he sold one pair of shoes for $11 or two pairs for $17, because he made the same profit on each sale. How much does one pair of shoes cost the salesman and what is the profit?

*47. To go to work a commuter first averages 36 miles per hour driving his car to the train station and then rides the train, which averages 60 miles per hour. The entire trip takes 1 hour and 22 minutes. It costs the commuter 10¢ per mile to drive the car and 3¢ per mile to ride the train. If the total cost is $2.76, find the distances traveled by car and by train.

3.6

Each of the linear systems considered in the last section had a solution. Such systems are said to be **consistent**. A consistent system of two linear equations in two variables will produce two intersecting lines; their point of intersection is the solution of the system. Here is an illustration of a consistent system:

$$y = 3x - 1$$
$$y = -5x + 7$$

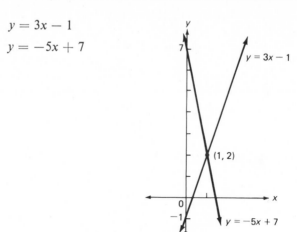

It might happen that a pair of equations really represents the same straight line. The following system gives two different ways of naming the same straight line. You can see this by changing either equation into the form of the other.

$$y = \tfrac{2}{3}x - 5$$
$$2x - 3y = 15$$

Such systems are said to be **dependent**. The graph for a dependent system, therefore, is always a single line.

In a dependent system each pair (x, y) that satisfies one equation must also satisfy the other. Any attempt at "solving" the system will result in some statement that is always true (an *identity*). For example:

$$3(y = \tfrac{2}{3}x - 5) \Longrightarrow 3y = 2x - 15 \Longrightarrow -2x + 3y = -15$$
$$2x - 3y = 15 \Longrightarrow 2x - 3y = 15 \Longrightarrow \underline{2x - 3y = 15}$$
$$\text{Add:} \qquad\qquad 0 = 0$$

In a sense, the given system has an infinite number of solutions.

When a pair of lines is parallel, they cannot have a point of intersection. Such a system has no solution and is called **inconsistent**. Since no (x, y) can fit both equations, any attempt at solving such a system will result in some false statement. For example, try to solve this system:

$$2x - 3y = -3$$
$$3y = 2x - 15$$

The second equation gives $y = \frac{2}{3}x - 5$. Substituting into the first we have

$$2x - 3(\tfrac{2}{3}x - 5) = -3$$
$$2x - 2x + 15 = -3$$
$$15 = -3$$

Note that this "solving" process began by falsely assuming that there was a solution, (x, y). So it should not be surprising that a false result is obtained. In reaching such a false statement we have "solved" the system in the sense that now we know there is no solution. Here is the graph.

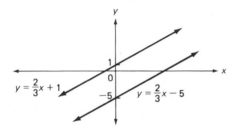

An inconsistent system can also be identified by putting each equation into slope-intercept form and observing that the slopes are equal but the y-intercepts are different.

TEST YOUR UNDERSTANDING

Decide whether the given systems are consistent, inconsistent, or dependent.

1. $x + y = 2$
$x + y = 3$

2. $3x - y = 7$
$-9x + 3y = -21$

3. $2x - 3y = 8$
$-8x + 12y = 33$

4. $\frac{1}{2}x + 5y = -4$
$7x - 3y = 17$

5. $-6x + 3y = 9$
$10x + 5y = -1$

6. $20x + 36y = -27$
$\frac{5}{27}x + \frac{1}{3}y = -\frac{1}{4}$

A special case of a consistent linear system is one for which the two lines are perpendicular. Excluding vertical and horizontal lines, it can be shown that two lines are perpendicular if and only if their slopes are negative reciprocals of one another. (See Exercise 24.)

Example 1   Line ℓ_1 has slope $\frac{5}{2}$ and passes through $P(-2, -6)$. What is the equation of the line perpendicular to ℓ_1 at P? Graph.

Solution   The perpendicular line ℓ_2 has slope $-\dfrac{1}{\frac{5}{2}} = -\dfrac{2}{5}$. Since this line also goes through $P(-2, -6)$, the point-slope form gives

$$y + 6 = -\tfrac{2}{5}(x + 2)$$
$$y = -\tfrac{2}{5}x - \tfrac{34}{5}$$

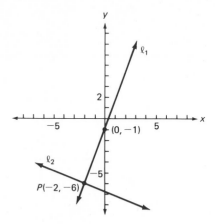

Example 2   Explain why the lines for the given system are perpendicular.

$$\ell_1: \qquad 4x + 3y = 3$$
$$\ell_2: \qquad 3x - 4y = 0$$

Solution   It is convenient to write each equation in slope-intercept form.

$$\ell_1: \qquad y = -\tfrac{4}{3}x + 1$$
$$\ell_2: \qquad y = \tfrac{3}{4}x$$

The slope of ℓ_1 is $-\frac{4}{3}$ and the slope of ℓ_2 is $\frac{3}{4}$. Thus the lines are perpendicular in that $-\dfrac{4}{3} = -\dfrac{1}{\frac{3}{4}}$.

Note that we may also draw the graph for a *system of inequalities.* Consider, for example, this system:

$$2x - y + 4 \leq 0$$
$$x + y - 2 \leq 0$$

It is most convenient to write each in *y-form* first; that is, express y in terms of x. Verify that the following is correct.

$$y \geq 2x + 4$$
$$y \leq -x + 2$$

Now graph each statement of inequality on the same coordinate system. In the graph that follows, the solution for $y \geq 2x + 4$ is shown with horizontal shading; the solution for $y \leq -x + 2$ is shown with vertical shading. The solution for the given system is the region shaded with *both* horizontal and vertical lines, including the boundaries.

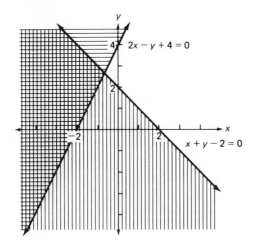

EXERCISES 3.6

In Exercises 1 through 10 decide whether the given systems are consistent, inconsistent, or dependent.

1. $x + 2y = 3$
   $6x + 5y = 4$

2. $x + 2y = 3$
   $10x + 20y = 30$

3. $x + 2y = 3$
   $-x - 2y = 3$

4. $8x + 4y = 12$
   $y = -2x + 3$

5. $4x - 12y = 3$
   $x + \frac{1}{3}y = 3$

6. $-7x + y = 2$
   $28x - 4y = -2$

7. $2x + 5y = -20$
   $x = -\frac{5}{2}y - 10$

8. $x - y = 3$
   $-\frac{1}{3}x + \frac{1}{3}y = 1$

9. $x - 5y = 15$
  $0.01x - 0.05y = 0.5$

10. $4y = 3x + 2$
  $2x = 3y - 3$

In Exercises 11 through 14 write the equation of the line that is perpendicular to the given line and passes through the indicated point.

11. $y = 3x - 1$; $(4, 7)$

12. $y = -10x$; $(0, 0)$

13. $y - 2x = 5$; $(-5, 1)$

14. $3x + 2y = 6$; $(6, 7)$

15. Find the point of intersection of the two lines for Exercise 14 and graph.

16. Line ℓ passes through $(-4, 5)$ and $(8, -2)$.
  (a) Draw the line through $(0, 0)$ perpendicular to ℓ.
  (b) What is the slope of any line perpendicular to ℓ?
  (c) Draw the four lines, each perpendicular to ℓ, through the points $(-4, 5)$, $(4, \frac{1}{3})$, $(0, \frac{8}{3})$, and $(8, -2)$.

Decide whether each of the systems gives parallel or perpendicular lines.

17. $2x - 3y = 4$
  $6y = 4x - 30$

18. $2x - 3y = 4$
  $3x + 2y = 1$

19. (a) Why is the line through $(1, 0)$ and $(0, 1)$ perpendicular to the line through $(0, 0)$ and $(1, 1)$?
  (b) Sketch the lines in part **(a)** and find the coordinates of their point of intersection.

20. Consider the four points $P(5, 11)$, $Q(-7, 16)$, $R(-12, 4)$, and $S(0, -1)$. Show that the four angles of the quadrilateral $PQRS$ are right angles.

21. Find the equations of the diagonals of $PQRS$ (see Exercise 20) and show that the diagonals are perpendicular.

*22. Show that the equation $\frac{1}{2}y = x$ for $x > 2$, and the equation $2x - y = 0$ for x satisfying $3 < x + 1$, define the same function.

*23. Any horizontal line is perpendicular to any vertical line. Why were such lines excluded from the result which states that lines are perpendicular if and only if their slopes are negative reciprocals?

*24. This exercise gives a proof that perpendicular lines have slopes that are negative reciprocals of one another.

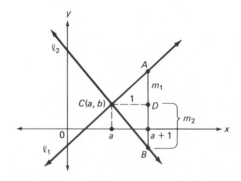

Let line ℓ_1 be perpendicular to line ℓ_2 at the point $C(a, b)$. Use m_1 for the slope of ℓ_1, and m_2 for ℓ_2. We want to show that

$$m_1 m_2 = -1 \quad \text{or} \quad m_1 = -\frac{1}{m_2}$$

Add 1 to the x-coordinate a of point C and draw the vertical line through $a + 1$ on the x-axis. This vertical line will meet ℓ_1 at some point A and ℓ_2 at some point B, forming right triangle ABC with right angle at C. Draw the perpendicular from C to AB meeting AB at D. Then CD has length 1.

(a) Using the right triangle CDA, show that $m_1 = DA$.

(b) Show that $m_2 = DB$. Is m_2 positive or negative?

(c) For right triangle ABC, CD is the mean proportional between segments BD and DA on the hypotenuse. Use this fact to conclude that $\dfrac{m_1}{1} = \dfrac{1}{-m_2}$, or $m_1 m_2 = -1$.

25. (a) Find the coordinate of the midpoint M of each line segment AB.

(i)
```
  -4      0           6
   •——————•———————————•—▸
   A      M           B
```

(ii)
```
  -10             -2   0
   •———————————————•———•—▸
   A               M   B
```

(b) If the coordinates of the endpoints of a line segment are x_1 and x_2, what is the coordinate of its midpoint M?

*26. In the figure PQ is a line segment with midpoint M having coordinates (x', y').

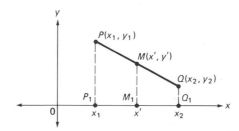

(a) Use the result of Exercise 25 to find x'.

(b) Find y'.

(c) Find the coordinates of the midpoint of the line segment whose endpoints are $(-2, -3)$ and $(6, 2)$.

*27. Find t if the line through $(-1, 1)$ and $(1, \frac{1}{2})$ is perpendicular to the line through $(1, \frac{1}{2})$ and $(7, t)$.

Graph each system of inequalities.

28. $y > 2x + 2$
$\quad y > -x$

29. $y > x + 1$
$\quad y < 1 - x$

*30. $2x - y + 1 \geq 0$
$\quad x - 2y + 2 \leq 0$

*31. $x + 2y - 2 \geq 0$
$\quad x - 2y + 2 \geq 0$

We can solve systems of equations by considering the constants only. Let us do so for this system:

$$x + 4y + 7z = 10$$
$$2x + 5y + 8z = 11$$
$$3x + 6y + 12z = 15$$

First we consider the rectangular array formed by the constants. This array is called a **matrix**; in this case the matrix consists of three rows and four columns.

$$
\begin{array}{c}
& \text{Coefficients} \\
& \text{of } x \quad \text{of } y \quad \text{of } z \\
& \downarrow \quad\quad \downarrow \quad\quad \downarrow \\
\begin{array}{c}
\text{Row 1} \\
\text{Row 2} \\
\text{Row 3}
\end{array}
\left[
\begin{array}{cccc}
1 & 4 & 7 & 10 \\
2 & 5 & 8 & 11 \\
3 & 6 & 12 & 15
\end{array}
\right]
\end{array}
$$

Next we use the same operations we would use with the original set of three equations and attempt to derive a matrix of this form:

$$
\begin{bmatrix}
1 & 0 & 0 & a \\
0 & 1 & 0 & b \\
0 & 0 & 1 & c
\end{bmatrix}
$$

This matrix indicates a solution of $x = a$, $y = b$, and $z = c$. Our goal is to determine a, b, and c for this system. Usually there are a variety of ways to proceed. Here is one sequence of steps that leads to the desired conclusion.

STEP 1. Multiply Row 1 by -2 and add it to Row 2. Then multiply Row 1 by -3 and add it to Row 3.

$$
\begin{bmatrix}
1 & 4 & 7 & 10 \\
0 & -3 & -6 & -9 \\
0 & -6 & -9 & -15
\end{bmatrix}
$$

STEP 2. Look at the second row of this *derived matrix*. If we multiply by $-\frac{1}{3}$ (or divide by -3) we can obtain a 1 in the second position.

$$
\begin{bmatrix}
1 & 4 & 7 & 10 \\
0 & 1 & 2 & 3 \\
0 & -6 & -9 & -15
\end{bmatrix}
$$

STEP 3.   Now multiply the second row by -4 and add this to the first row. Then multiply the second row by 6 and add it to the third row.

$$\begin{bmatrix} 1 & 0 & -1 & -2 \\ 0 & 1 & 2 & 3 \\ 0 & 0 & 3 & 3 \end{bmatrix}$$

STEP 4.   Multiply the third row by $\frac{1}{3}$ (or divide by 3) to obtain 1 in the third position of that row.

$$\begin{bmatrix} 1 & 0 & -1 & -2 \\ 0 & 1 & 2 & 3 \\ 0 & 0 & 1 & 1 \end{bmatrix}$$

STEP 5.   Finally, add the third row to the first row. Then multiply the third row by -2 and add to the second row.

$$\begin{bmatrix} 1 & 0 & 0 & -1 \\ 0 & 1 & 0 & 1 \\ 0 & 0 & 1 & 1 \end{bmatrix}$$

From this last matrix we see that $x = -1$, $y = 1$, and $z = 1$. Check these results by substitution into the original system of equations.

Example 1   Use matrices to solve:

$$2x + 4y = 6$$
$$4x + 5y = 6$$

Solution   Begin with this matrix of constants.

$$\begin{bmatrix} 2 & 4 & 6 \\ 4 & 5 & 6 \end{bmatrix}$$

(a) Multiply Row 1 by $\frac{1}{2}$.

$$\begin{bmatrix} 1 & 2 & 3 \\ 4 & 5 & 6 \end{bmatrix}$$

(b) Multiply the first row by -4 and add to the second row.

$$\begin{bmatrix} 1 & 2 & 3 \\ 0 & -3 & -6 \end{bmatrix}$$

(c) Multiply the second row by $-\frac{1}{3}$.

$$\begin{bmatrix} 1 & 2 & 3 \\ 0 & 1 & 2 \end{bmatrix}$$

(d) Multiply the second row by -2 and add to the first row.

$$\begin{bmatrix} 1 & 0 & -1 \\ 0 & 1 & 2 \end{bmatrix}$$

The last form corresponds to this system of equations.

$$\begin{aligned} 1x + 0y &= -1 \\ 0x + 1y &= 2 \end{aligned}$$

Thus $x = -1$ and $y = 2$. Check this solution.

Frequently shortcuts are possible as you solve a system through the use of matrices. Consider, for example, this system of equations and its corresponding matrix of constants.

$$\begin{aligned} x + y + z &= 1 \\ 2x + y - z &= -3 \\ x - 2y + 2z &= -4 \end{aligned} \qquad \begin{bmatrix} 1 & 1 & 1 & 1 \\ 2 & 1 & -1 & -3 \\ 1 & -2 & 2 & -4 \end{bmatrix}$$

Now multiply the first row by -2 and add it to the third row. Then add the first row to the second row.

$$\begin{bmatrix} 1 & 1 & 1 & 1 \\ 3 & 2 & 0 & -2 \\ -1 & -4 & 0 & -6 \end{bmatrix}$$

Multiply the third row by 3 and add to the second row.

$$\begin{bmatrix} 1 & 1 & 1 & 1 \\ 0 & -10 & 0 & -20 \\ -1 & -4 & 0 & -6 \end{bmatrix}$$

At this point, instead of proceeding with the matrix, note the second row. From this row we deduce that $-10y = -20$, or $y = 2$. Then use this result in the third row:

$$-x - 4y = -6$$
$$-x - 8 = -6$$
$$x = -2$$

Finally, in the first row, we have $x + y + z = 1$. That is, $-2 + 2 + z = 1$ so that $z = 1$. Therefore we have the complete solution: $x = -2$, $y = 2$, and $z = 1$. Thus whenever we get one row of the matrix to contain two zeros in the first three entries, we have found the value of one of the unknowns.

A **square matrix** has the same number of rows as columns, such as this one of two rows and two columns.

$$\begin{bmatrix} a_1 & b_1 \\ a_2 & b_2 \end{bmatrix}$$

Associated with this matrix is a **second-order determinant**, a number written and defined as follows.

$$\begin{vmatrix} a_1 & b_1 \\ a_2 & b_2 \end{vmatrix} = a_1 b_2 - a_2 b_1$$

The arrows below will help you apply this definition. The arrow from top to bottom gives the product $a_1 b_2$ which is the *first* term in $a_1 b_2 - a_2 b_1$. The arrow from bottom to top gives the *second* product $a_2 b_1$.

$$\begin{vmatrix} a_1 & b_1 \\ a_2 & b_2 \end{vmatrix} = a_1 b_2 - a_2 b_1$$

Example 2 Evaluate:

$$\begin{vmatrix} 8 & -20 \\ 4 & 10 \end{vmatrix}$$

Solution

$$\begin{vmatrix} 8 & -20 \\ 4 & 10 \end{vmatrix} = 8(10) - (4)(-20) = 160$$

We can use determinants as a convenient way to represent the solution of a system of equations. To see this, first consider a general system of two equations with two unknowns.

$$a_1 x + b_1 y = c_1$$
$$a_2 x + b_2 y = c_2$$

Assuming that this is a consistent system we can find the unique solution as follows.

Multiply the first equation by b_2 and the second by $-b_1$.

$$a_1 b_2 x + b_1 b_2 y = c_1 b_2$$
$$-a_2 b_1 x - b_1 b_2 y = -c_2 b_1$$

Add to eliminate y.

$$a_1 b_2 x - a_2 b_1 x = c_1 b_2 - c_2 b_1$$

Factor.

$$(a_1 b_2 - a_2 b_1) x = c_1 b_2 - c_2 b_1$$

To solve for x it must be the case that $a_1 b_2 - a_2 b_1 \neq 0$. Then

$$x = \frac{c_1 b_2 - c_2 b_1}{a_1 b_2 - a_2 b_1}$$

Similarly, multiplying the first and second equations by a_2 and $-a_1$, respectively, will produce this solution for y.

$$y = \frac{a_1 c_2 - a_2 c_1}{a_1 b_2 - a_2 b_1}$$

The numerator in the solution for x is $c_1 b_2 - c_2 b_1$. Using our new symbolism, this number is the second-order determinant:

$$\begin{vmatrix} c_1 & b_1 \\ c_2 & b_2 \end{vmatrix} = c_1 b_2 - c_2 b_1$$

Likewise, the numerator for y is

$$\begin{vmatrix} a_1 & c_1 \\ a_2 & c_2 \end{vmatrix} = a_1 c_2 - a_2 c_1$$

Both denominators are

$$\begin{vmatrix} a_1 & b_1 \\ a_2 & b_2 \end{vmatrix} = a_1 b_2 - a_2 b_1$$

In summary, we have the following, known as **Cramer's rule**, for solving a system of linear equations.

Given the consistent system

$$a_1x + b_1y = c_1$$
$$a_2x + b_2y = c_2$$

the solutions for x and y are

$$x = \frac{\begin{vmatrix} c_1 & b_1 \\ c_2 & b_2 \end{vmatrix}}{\begin{vmatrix} a_1 & b_1 \\ a_2 & b_2 \end{vmatrix}} \quad \text{and} \quad y = \frac{\begin{vmatrix} a_1 & c_1 \\ a_2 & c_2 \end{vmatrix}}{\begin{vmatrix} a_1 & b_1 \\ a_2 & b_2 \end{vmatrix}}$$

Example 3   Solve the following system by using determinants.

$$5x - 9y = 7$$
$$-8x + 10y = 2$$

Solution

$$x = \frac{\begin{vmatrix} 7 & -9 \\ 2 & 10 \end{vmatrix}}{\begin{vmatrix} 5 & -9 \\ -8 & 10 \end{vmatrix}} = \frac{70 - (-18)}{50 - 72} = \frac{88}{-22} = -4$$

$$y = \frac{\begin{vmatrix} 5 & 7 \\ -8 & 2 \end{vmatrix}}{\begin{vmatrix} 5 & -9 \\ -8 & 10 \end{vmatrix}} = \frac{10 - (-56)}{-22} = -3$$

The general system

$$a_1x + b_1y = c_1$$
$$a_2x + b_2y = c_2$$

is either dependent or inconsistent when the determinant of the coefficients is zero; that is, when

$$\begin{vmatrix} a_1 & b_1 \\ a_2 & b_2 \end{vmatrix} = 0$$

For example, consider these two systems:

$$(1) \quad \begin{aligned} 2x - 3y &= 5 \\ -10x + 15y &= 8 \end{aligned} \qquad (2) \quad \begin{aligned} 2x - 3y &= 8 \\ -10x + 15y &= -40 \end{aligned}$$

In each case we have the following:

$$\begin{vmatrix} a_1 & b_1 \\ a_2 & b_2 \end{vmatrix} = \begin{vmatrix} 2 & -3 \\ -10 & 15 \end{vmatrix} = 0$$

System (1) is inconsistent and (2) is dependent.

We now define a **third-order determinant** as follows.

$$\begin{vmatrix} a_1 & b_1 & c_1 \\ a_2 & b_2 & c_2 \\ a_3 & b_3 & c_3 \end{vmatrix} = a_1 \begin{vmatrix} b_2 & c_2 \\ b_3 & c_3 \end{vmatrix} - a_2 \begin{vmatrix} b_1 & c_1 \\ b_3 & c_3 \end{vmatrix} + a_3 \begin{vmatrix} b_1 & c_1 \\ b_2 & c_2 \end{vmatrix}$$

$$= a_1(b_2c_3 - b_3c_2) - a_2(b_1c_3 - b_3c_1) + a_3(b_1c_2 - b_2c_1)$$

Gathering terms we have:

$$a_1b_2c_3 + a_2b_3c_1 + a_3b_1c_2 - a_1b_3c_2 - a_2b_1c_3 - a_3b_2c_1$$

Show that the following alternative definition produces the same result.

$$a_1 \begin{vmatrix} b_2 & c_2 \\ b_3 & c_3 \end{vmatrix} - b_1 \begin{vmatrix} a_2 & c_2 \\ a_3 & c_3 \end{vmatrix} + c_1 \begin{vmatrix} a_2 & b_2 \\ a_3 & b_3 \end{vmatrix}$$

This notation can be used to extend Cramer's rule for the solution of three linear equations in three unknowns. Consider this general system.

$$\begin{aligned} a_1x + b_1y + c_1z &= d_1 \\ a_2x + b_2y + c_2z &= d_2 \\ a_3x + b_3y + c_3z &= d_3 \end{aligned}$$

By actually completing the tedious computations involved it can be shown that the solution for this system is the following.

$$x = \frac{\begin{vmatrix} d_1 & b_1 & c_1 \\ d_2 & b_2 & c_2 \\ d_3 & b_3 & c_3 \end{vmatrix}}{D} \qquad y = \frac{\begin{vmatrix} a_1 & d_1 & c_1 \\ a_2 & d_2 & c_2 \\ a_3 & d_3 & c_3 \end{vmatrix}}{D} \qquad z = \frac{\begin{vmatrix} a_1 & b_1 & d_1 \\ a_2 & b_2 & d_2 \\ a_3 & b_3 & d_3 \end{vmatrix}}{D}$$

where $D = \begin{vmatrix} a_1 & b_1 & c_1 \\ a_2 & b_2 & c_2 \\ a_3 & b_3 & c_3 \end{vmatrix}$ and $D \neq 0$.

Note that the determinant D consists of the coefficients of the variables in order. Then the numerators for the solutions for x, y, and z consist of the coefficients, but in each case the constants are used to replace the coefficients of the variable under consideration. Note how this is done in the example that follows.

Example 4   Use Cramer's rule to solve this system.

$$
\begin{aligned}
x + 2y + z &= 3 \\
2x - y - z &= 4 \\
-x - y + 2z &= -5
\end{aligned}
$$

Solution   First we find D to be certain that $D \neq 0$.

$$D = \begin{vmatrix} 1 & 2 & 1 \\ 2 & -1 & -1 \\ -1 & -1 & 2 \end{vmatrix} = 1 \begin{vmatrix} -1 & -1 \\ -1 & 2 \end{vmatrix} - 2 \begin{vmatrix} 2 & 1 \\ -1 & 2 \end{vmatrix} + (-1) \begin{vmatrix} 2 & 1 \\ -1 & -1 \end{vmatrix} = -12$$

Verify each of the following computations.

$$x = \frac{\begin{vmatrix} 3 & 2 & 1 \\ 4 & -1 & -1 \\ -5 & -1 & 2 \end{vmatrix}}{D} = \frac{-24}{-12} = 2 \qquad y = \frac{\begin{vmatrix} 1 & 3 & 1 \\ 2 & 4 & -1 \\ -1 & -5 & 2 \end{vmatrix}}{D} = \frac{-12}{-12} = 1$$

$$z = \frac{\begin{vmatrix} 1 & 2 & 3 \\ 2 & -1 & 4 \\ -1 & -1 & -5 \end{vmatrix}}{D} = \frac{12}{-12} = -1$$

Thus $x = 2$, $y = 1$, and $z = -1$.

At the top of the next page is an alternate procedure for evaluating a third-order determinant. Rewrite the first two columns at the right as shown. Follow the arrows pointing downward to get the three products having a plus sign, and the arrows pointing upward give the three products having a negative sign.

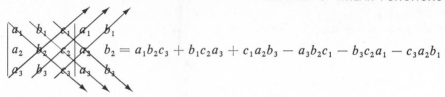

Use this procedure to evaluate the determinants in Example 4.

EXERCISES 3.7

Use matrices to solve each system.

1. $x + 5y = -9$
$\quad 4x - 3y = -13$

2. $4x - y = 6$
$\quad 2x + 3y = 10$

3. $3x + 2y = 18$
$\quad 6x + 5y = 45$

4. $2x + 4y = 24$
$\quad -3x + 5y = -25$

5. $4x - 5y = -2$
$\quad 16x + 2y = 3$

6. $x = y - 7$
$\quad 3y = 2x + 16$

7. $2x = -8y + 2$
$\quad 4y = x - 1$

8. $2x + 5y = 4$
$\quad -10x + 25y = 0$

9. $3x + 4y = 7$
$\quad 4x + 6y = 8$

10. $-3x + 4y = 2$
$\quad 2x - 3y = -3$

11. $3x = 8 - 5y$
$\quad 6x + 9y = 14$

12. $-10x + 5y = 8$
$\quad 15x - 10y = -4$

13. $x - 2y + 3z = -2$
$\quad -4x + 10y + 2z = -2$
$\quad 3x + y + 10z = 7$

14. $2x + 4y + 8z = 14$
$\quad 4x - 2y + 2z = 6$
$\quad -5x + 3y - z = -4$

Evaluate each determinant.

15. $\begin{vmatrix} 5 & -1 \\ -3 & 4 \end{vmatrix}$

16. $\begin{vmatrix} 1 & 2 \\ 3 & 4 \end{vmatrix}$

17. $\begin{vmatrix} 17 & -3 \\ 20 & 2 \end{vmatrix}$

18. $\begin{vmatrix} -7 & 9 \\ -5 & 5 \end{vmatrix}$

19. $\begin{vmatrix} 10 & 5 \\ 6 & -3 \end{vmatrix}$

20. $\begin{vmatrix} 6 & 11 \\ 0 & -9 \end{vmatrix}$

21. $\begin{vmatrix} 16 & 0 \\ -9 & 0 \end{vmatrix}$

22. $\begin{vmatrix} a & b \\ 3a & 3b \end{vmatrix}$

23. $\begin{vmatrix} 2 & 2 & -1 \\ -1 & 3 & -3 \\ 1 & 2 & 3 \end{vmatrix}$

24. $\begin{vmatrix} 2 & 0 & -1 \\ 3 & -2 & 1 \\ -3 & 0 & 4 \end{vmatrix}$

25. $\begin{vmatrix} 1 & -3 & 2 \\ -5 & 2 & 0 \\ 4 & -1 & 3 \end{vmatrix}$

26. $\begin{vmatrix} 1 & 2 & 3 \\ 4 & 5 & 6 \\ 7 & 8 & 9 \end{vmatrix}$

27. $\begin{vmatrix} 1 & 1 & 1 \\ -1 & 1 & 1 \\ -1 & -1 & 1 \end{vmatrix}$

28. $\begin{vmatrix} 1 & 1 & 4 \\ 2 & 2 & -5 \\ 3 & 3 & 6 \end{vmatrix}$

Solve each system by using determinants.

29. $3x + 9y = 15$
    $6x + 12y = 18$

30. $x - y = 7$
    $-2x + 5y = -8$

31. $-4x + 10y = 8$
    $11x - 9y = 15$

32. $7x + 4y = 5$
    $-x + 2y = -2$

33. $5x + 2y = 3$
    $2x + 3y = -1$

34. $3x + 3y = 6$
    $4x - 2y = -1$

35. $x + y + z = 6$
    $2x + y - z = 1$
    $x - 2y + 2z = 3$

36. $2x + y + 2z = 5$
    $x - y - z = -7$
    $3x + y - 5z = -3$

37. $2x + 3y + 2z = 0$
    $x - 2y + z = 7$
    $3x - y + 3z = 11$

38. $3x - y - z = 8$
    $x + 2y + z = 3$
    $2x + 5y + 3z = 7$

39. When variables are used for some of the entries of a determinant, the determinant itself can be used to state equations. Solve for x.

$$\begin{vmatrix} x & 2 \\ 5 & 3 \end{vmatrix} = 8$$

40. Solve the given system.

$$\begin{vmatrix} x & y \\ 2 & 4 \end{vmatrix} = 5$$

$$\begin{vmatrix} 1 & y \\ -1 & x \end{vmatrix} = -\frac{1}{2}$$

*41. (a) Show that if the rows and columns of a second-order determinant are interchanged, the value of the determinant remains the same.
    (b) Do the same for a third-order determinant.

*42. (a) Show that if one of the rows of $\begin{vmatrix} a_1 & b_1 \\ a_2 & b_2 \end{vmatrix}$ is a nonzero multiple of the other, then the determinant is zero.
    (b) Use part (a) and Exercise 41 to demonstrate that the determinant is zero if one column is a nonzero multiple of the other.

*43. Answer parts (a) and (b) of Exercise 42 for a third-order determinant.

*44. (a) Show that if each element of a row (or column) of a second-order determinant is multiplied by the same number k, the value of the determinant is multiplied by k.
    (b) Do the same for a third-order determinant.

*45. Make repeated use of the result in Exercise 44 to show the following.

$$\begin{vmatrix} 27 & 3 \\ 105 & -75 \end{vmatrix} = (45) \begin{vmatrix} 9 & 1 \\ 7 & -5 \end{vmatrix} \quad \text{or} \quad \begin{vmatrix} 27 & 3 \\ 105 & -75 \end{vmatrix} = (45) \begin{vmatrix} 3 & 1 \\ 7 & -15 \end{vmatrix}$$

Then evaluate each side to check.

*46. Prove:
$$\begin{vmatrix} a_1 + t_1 & b_1 \\ a_2 + t_2 & b_2 \end{vmatrix} = \begin{vmatrix} a_1 & b_1 \\ a_2 & b_2 \end{vmatrix} + \begin{vmatrix} t_1 & b_1 \\ t_2 & b_2 \end{vmatrix}$$

*47. Prove that if to each element of a row (or column) of a second-order determinant we add k times the corresponding element of the other row (or column), then the value of the new determinant is the same as that of the original determinant.

REVIEW EXERCISES FOR CHAPTER 3

The solutions to the following exercises can be found within the text of Chapter 3. Try to answer each question without referring to the text.
Section 3.1

1. State the definition of a function.

2. Decide whether the equation $y = \dfrac{1}{\sqrt{x-1}}$ defines y to be a function of x.

3. Decide whether these two equations define y to be a function of x.

$$y = \begin{cases} 3x - 1 & \text{if } x \leq 1 \\ 2x + 1 & \text{if } x \geq 1 \end{cases}$$

4. Find the domain of $y = \dfrac{1}{x^2 + 2}$.

5. Find the domain of $y = \sqrt{x^2 + x - 6}$.

6. For $g(x) = \dfrac{1}{x}$ find $g(3 + x)$.

7. For $g(x) = x^2$ find $\dfrac{g(x) - g(4)}{x - 4}$.

8. For $g(x) = \dfrac{1}{x}$ find $\dfrac{g(4 + h) - g(4)}{h}$.

Section 3.2

9. Graph the function $y = 2x - 1$.

Section 3.3

10. State the definition of the slope of a line.
11. What is the slope of the line determined by the points $(-3, 4)$ and $(1, -6)$?
12. Graph the line through point $(-3, 4)$ with slope $m = -\frac{3}{2}$.
13. Write the point-slope-form equation of a line.
14. Write the point-slope-form equation for the line through $(-1, 4)$ with slope $\frac{1}{2}$.

15. Write the slope-intercept-form equation of a line.

16. Write the slope-intercept form of the line through $(6, -4)$ and $(-3, 8)$.

17. True or false:
   (a) Parallel lines have equal slopes.
   (b) All lines have slope.

Section 3.4

18. Graph $y = |x - 2|$.

19. Graph the step function $y = [2x]$ for the domain $1 \leq x \leq 3$.

20. Graph f, where

$$y = f(x) = \begin{cases} 2x & \text{if } 0 \leq x \leq 1 \\ -x + 2 & \text{if } 1 < x \end{cases}$$

Section 3.5

21. Solve and graph this system:

$$2x + 3y = 12$$
$$3x + 2y = 12$$

22. Solve this system:

$$4x - 2y = 40$$
$$-3x + 3y = 45$$

23. Solve this system:

$$\tfrac{1}{3}x - \tfrac{2}{5}y = 4$$
$$7x + 3y = 27$$

24. Solve this system:

$$2x - 5y + z = -10$$
$$x + 2y + 3z = 26$$
$$-3x - 4y + 2z = 5$$

Section 3.6

25. A line has slope $\tfrac{5}{2}$ and passes through $P(-2, -6)$. Find the equation of the line perpendicular to the given line that also goes through P.

26. True or false:
   (a) Parallel lines represent an inconsistent system.
   (b) Intersecting lines form a dependent system.
   (c) Solving an inconsistent system produces a false conclusion.

27. Classify each of these systems as being consistent, inconsistent, or dependent.
   (a) $3x - y = 7$ (b) $-6x + 3y = 9$ (c) $20x + 36y = -27$
   $-9x + 3y = -21$ $10x + 5y = -1$ $\tfrac{5}{27}x + \tfrac{1}{3}y = -\tfrac{1}{4}$

Section 3.7

28. Use matrices to solve this system:

$$x + 4y + 7z = 10$$
$$2x + 5y + 8z = 11$$
$$3x + 6y + 12z = 15$$

29. Use determinants to solve this system:

$$5x - 9y = 7$$
$$-8x + 10y = 2$$

30. Use Cramer's rule to solve this system:

$$x + 2y + z = 3$$
$$2x - y - z = 4$$
$$-x - y + 2z = -5$$

SAMPLE TEST FOR CHAPTER 3

1. Classify each statement as true or false.
  (a) Horizontal lines do not have slopes.
  (b) A line whose y-intercept is -2 and whose slope is a negative number will intersect the x-axis on the positive side.
  (c) The lines given by $2x + 5y = 7$ and $15 - 10y = 4x$ are parallel.
  (d) A horizontal line cannot be the graph of a function.

2. Graph the line through $(6, -5)$ with slope $-\frac{5}{3}$ and write its point-slope form.

Exercises 3 through 5 refer to the following figure.

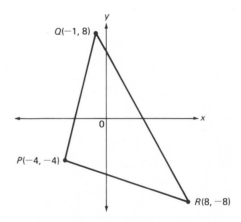

3. Write the equations for:
   (a) The line through P and Q.
   (b) The line through P and R.

4. Why is angle QPR not a right angle?

5. Line segment PQ is the graph of a function f in which x is the independent variable.
   (a) What is the domain of f?
   (b) What is the range of f?
   (c) What is the equation that gives the range value $y = f(x)$ for a domain value x?

6. (a) Write an equation of the line through the point $(2, 8)$ and perpendicular to the line $y = -\frac{2}{3}x + 3$.
   (b) Find the point of intersection of these perpendicular lines.

7. Graph and state the domain and range for $y = |2x + 1|$.

For Exercises 8 through 10 decide whether the system is consistent, inconsistent, or dependent. If it is consistent, find the solution.

8. $4x - 5y = 12$
   $-2x + \frac{5}{2}y = -6$

9. $7x + 4y = 1$
   $-3x + 2y = -19$

10. $-6x + 4y = -24$
    $9x - 6y = 14$

†11. Use matrices to solve this system.

$$x - y + 2z = 5$$
$$2x \qquad - z = -6$$
$$-x - 3y + z = -1$$

Answers To The Test Your Understanding Exercises

Page 97

1. Function; all reals.

2. Function; all real $x \neq -2$.

3. Function; all reals.

4. Not a function.

5. Function; $x \geq 2$ or $x \leq -3$.

6. Not a function.

7. Function; all $x \neq 0$.

8. Function; all $x \neq -1$.

Page 112

1. $\dfrac{6 - 5}{4 - 1} = \dfrac{1}{3}$

2. $\dfrac{3 - (-5)}{-3 - 3} = -\dfrac{8}{6} = -\dfrac{4}{3}$

3. $\dfrac{1 - (-3)}{-1 - (-2)} = \dfrac{4}{1} = 4$

4. $\dfrac{1 - 0}{0 - (-1)} = \dfrac{1}{1} = 1$

5.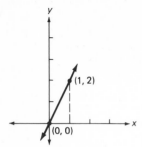

6.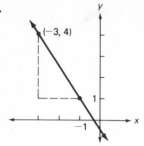

Page 113

1. $y - 6 = -3(x - 2)$

2. $y - 4 = \frac{1}{2}(x + 1)$

.**3.** $y + \frac{2}{3} = 1(x - 5)$

4. $y = -\frac{1}{4}x$

5. $y + 5 = 0(x + 3)$

6. $y + 1 = -(x - 1)$

Page 124

1. $(1, 2)$

2. $(-3, 4)$

3. $(2, -1)$

4. $(\frac{1}{5}, \frac{1}{5})$

5. $(-4, 1)$

6. $(35, 50)$

Page 126

1. $(-1, 2)$

2. $(\frac{1}{2}, -3)$

3. $(\frac{3}{2}, \frac{5}{7})$

4. $(0, 0)$

Page 131

1. Inconsistent.

2. Dependent.

3. Inconsistent.

4. Consistent.

5. Consistent.

6. Dependent.

Four

Quadratic Functions

4.1

Graphing Quadratic Functions

A function defined by a polynomial expression of degree 2 is referred to as a quadratic function in x. Thus the following are all examples of quadratic functions in x:

$$f(x) = -3x^2 + 4x + 1$$
$$g(x) = 7x^2 - 4$$
$$h(x) = x^2$$

The most general form of such a quadratic function is

$$f(x) = ax^2 + bx + c$$

where a, b, and c represent constants, with $a \neq 0$. If $a = 0$, then the resulting polynomial no longer represents a quadratic function; $f(x) = bx + c$ is a linear function.

The simplest quadratic function is given by $f(x) = x^2$. The graph of this quadratic function will serve as the basis for drawing the graph of any quadratic function $f(x) = ax^2 + bx + c$. We can save some labor by noting the *symmetry* that exists. For example, note the following:

$$f(-\tfrac{1}{4}) = f(\tfrac{1}{4}) = \tfrac{1}{16}$$
$$f(-\tfrac{1}{2}) = f(\tfrac{1}{2}) = \tfrac{1}{4}$$
$$f(-\tfrac{3}{4}) = f(\tfrac{3}{4}) = \tfrac{9}{16}$$
$$f(-1) = f(1) = 1$$
$$f(-\tfrac{3}{2}) = f(\tfrac{3}{2}) = \tfrac{9}{4}$$
$$f(-2) = f(2) = 4$$

In general,

$$f(-x) = (-x)^2 = x^2 = f(x)$$

Note: When $f(-x) = f(x)$, the graph is said to be *symmetric* with respect to the y-axis. We can summarize the facts given above in this table of values.

x	0	$\pm\dfrac{1}{4}$	$\pm\dfrac{1}{2}$	$\pm\dfrac{3}{4}$	± 1	$\pm\dfrac{3}{2}$	± 2
$y = f(x)$	0	$\dfrac{1}{16}$	$\dfrac{1}{4}$	$\dfrac{9}{16}$	1	$\dfrac{9}{4}$	4

When these points are located in a rectangular system and connected by a smooth curve, the result looks like the graph at the top of the next page.

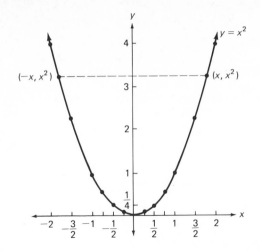

Greater accuracy can be obtained by using more points. But since we can never locate an infinite number of points, we must admit that there is a certain amount of faith involved in connecting the points as we did. The curve is called a **parabola**, and every quadratic function $y = ax^2 + bx + c$ has such a parabola as its graph.

An important feature of a parabola is that it is symmetric about a vertical line called its **axis of symmetry**. The graph of $y = x^2$ is symmetric with respect to the y-axis. This symmetry is due to the fact that $(-x)^2 = x^2$.

The parabola has a *turning point*, called the **vertex**, which is located at the intersection of the parabola with its axis of symmetry. For the preceding graph the coordinates of the vertex are $(0, 0)$.

Next consider $y = 2x^2$. It is clear from this equation that the y-values can be obtained by multiplying x^2 by 2. So we may take the graph of $y = x^2$ and multiply (stretch) its ordinates by 2 to locate the points on the parabola $y = 2x^2$. Similarly, to obtain the graph of $y = \frac{1}{2}x^2$ we divide (shrink) the y-values of $y = x^2$ by 2.

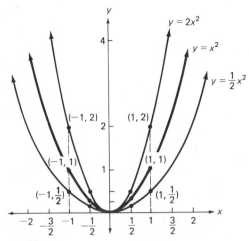

The graph of $y = -x^2$ may be obtained by multiplying each of the ordinates of $y = x^2$ by -1. This step has the effect of "flipping" the parabola $y = x^2$ downward, a *reflection* in the x-axis.

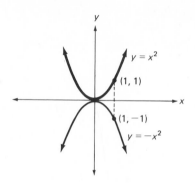

To graph $y = -2x^2$ you may first graph $y = 2x^2$ as before and then draw the reflection in the x-axis, or you may first graph $y = -x^2$ and then multiply by 2.

Let us next consider the quadratic function $y = g(x) = (x - 1)^2$. If we write x^2 as $(x - 0)^2$, we may make a useful comparison between these two functions:

$$y = f(x) = (x - 0)^2$$
$$y = g(x) = (x - 1)^2$$

We know that $x = 0$ is the axis of symmetry for the graph of f. In the same way, $x = 1$ is the axis of symmetry for g. In a similar way, the parabola $y = (x + 1)^2 = [x - (-1)]^2$ has $x = -1$ as an axis of symmetry.

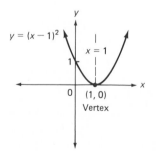

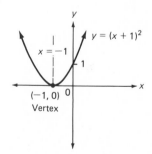

Both these parabolas are congruent to the basic parabola $y = x^2$. Each may be graphed by *translating* (shifting) the parabola $y = x^2$ by 1 unit, to the right for $y = (x - 1)^2$ and to the left for $y = (x + 1)^2$.

The graph of $y = 3(x - 1)^2$ can be obtained from the graph of $y = (x - 1)^2$ by multiplying each ordinate by 3. Also, the graph of $y = 3(x - 1)^2 + 2$ can be drawn by shifting the parabola $y = 3(x - 1)^2$ by 2 units upward. Similarly, shifting $y = 3(x - 1)^2$ by 2 units downward gives the parabola $y = 3(x - 1)^2 - 2$, as in the following figure.

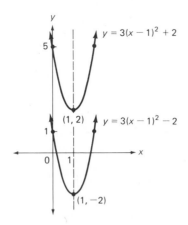

We may now make the following general observation about any parabola, $y = a(x - h)^2 + k$.

1. Vertex: (h, k).
2. Axis of symmetry: $x = h$,
3. Opens up if $a > 0$; down if $a < 0$.
4. To draw the parabola, in case $a > 0$, first sketch a parabola congruent to $y = x^2$ with vertex (h, k) and with $x = h$ as axis of symmetry. Then use the multiplication process for points of this parabola. For $a < 0$, begin with a parabola congruent to $y = -x^2$.

Example   Graph the parabola $y = -2(x - 3)^2 + 4$.

Solution   The graph will be a parabola congruent to $y = -2x^2$, with vertex at $(3, 4)$ and with $x = 3$ as axis of symmetry. A brief table of values, together with the graph, is shown at the top of the next page.

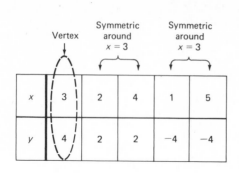

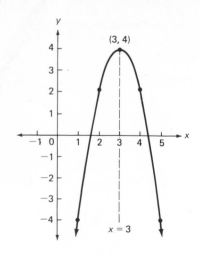

EXERCISES 4.1

Draw each set of graphs in the same coordinate system.

1. (a) $y = x^2$ (b) $y = 2x^2$ (c) $y = 3x^2$
2. (a) $y = x^2$ (b) $y = \frac{1}{4}x^2$ (ç) $y = \frac{1}{4}x^2 + 1$
3. (a) $y = -x^2$ (b) $y = -\frac{1}{2}x^2$ (c) $y = -\frac{1}{2}x^2 + 1$
4. (a) $y = (x - 2)^2$ (b) $y = 2(x - 2)^2$
5. (a) $y = (x + 2)^2$ (b) $y = \frac{1}{3}(x + 2)^2$
6. (a) $y = -(x + \frac{3}{2})^2$ (b) $y = -2(x + \frac{3}{2})^2$

Graph each of the following.

7. $y = (x - 1)^2 + 2$ 8. $y = (x + 1)^2 - 2$
9. $f(x) = -(x + 1)^2 + 2$ 10. $f(x) = -(x + 1)^2 - 2$
11. $y = f(x) = 2(x - 3)^2 - 1$ 12. $y = f(x) = 2(x + \frac{5}{4})^2 + \frac{5}{4}$

13. Graph the function f where $f(x) = \begin{cases} x^2 & \text{if } -2 \le x \le 1 \\ x & \text{if } 1 < x \le 3 \end{cases}$

*14. Graph f where $f(x) = \begin{cases} x^2 - 9 & \text{if } -2 \le x < 4 \\ -3x + 15 & \text{if } 4 \le x < 6 \\ 3 & \text{if } x = 6 \end{cases}$

15. The domain of the functions given in Exercises 1 through 12 consists of all real numbers. Find the range of the functions in the odd-numbered exercises, 1–11.

16. Compare the graphs of $y = x^2$ and $y = |x|$. In what ways are they alike?

17. Graph $x = y^2$. Why is y not a function of x?

18. The graph of $y = ax^2$ passes through the point $(1, -2)$. Find a.

*19. The graph of $y = ax^2 + c$ has its vertex at $(0, 4)$ and passes through the point $(3, -5)$. Find the values for a and c.

*20. Find the value for k so that the graph of $y = (x - 2)^2 + k$ will pass through the point $(5, 12)$.

4.2 Completing the Square

We will now rewrite the general quadratic function $f(x) = ax^2 + bx + c$ into the standard form $f(x) = a(x - h)^2 + k$. In this latter form we can easily graph the function by appropriate translations and reflections of the graph of $y = ax^2$, as in Section 4.1. For example, $y = 3x^2 - 6x + 1$ may be written in standard form as

$$= 3(x-1)^2 + 1$$
$$= 3[(x-1)(x-1)] + 1$$
$$y = 3(x - 1)^2 - 2 \qquad 3[x^2 - 2x + \square] + 1 = 3x^2 - 6x + 1 + 1$$
$$= 3x^2 - 6x + 2 \; ?$$

In this form we recognize a parabola with vertex at $(1, -2)$ and with $x = 1$ as axis of symmetry.

Suppose we wish to write $y = x^2 - 10x$ in standard form. To begin, consider the following identities:

$$(x + h)^2 = x^2 + 2hx + h^2$$
$$(x - h)^2 = x^2 - 2hx + h^2$$

In both these *equations* the third term on the right, h^2, is the square of one-half the coefficient of the x-term (without regard to sign). That is,

$$\text{Third term} = [\tfrac{1}{2}(2h)]^2 = h^2 \qquad \left[\tfrac{1}{2}(10)\right]^2 = 25$$

Using these observations we can convert the expression $x^2 - 10x$ into a trinomial square by adding $[\tfrac{1}{2}(10)]^2 = 25$. Thus

$$x^2 - 10x + 25 = (x - 5)^2$$

But we must retain the original expression by subtracting the same amount, as follows:

$$x^2 - 10x = x^2 - 10x + 25 - 25 = (x - 5)^2 - 25$$

Example 1   Write $x^2 + 3x - 1$ in the form $(x - h)^2 + k$.

Solution   Add and subtract $[\tfrac{1}{2}(3)]^2 = \tfrac{9}{4}$.

$$x^2 + 3x - 1 = x^2 + 3x + \tfrac{9}{4} - \tfrac{9}{4} - 1$$
$$= (x + \tfrac{3}{2})^2 - \tfrac{13}{4}$$

The process used in the preceding example is called **completing the square** or *completing the square in x*. Note that in writing

$$x^2 + 3x - 1 = (x + \tfrac{3}{2})^2 - \tfrac{13}{4}.$$

the right side contains $(x + \tfrac{3}{2})^2$, which is the square of a binomial in x.

TEST YOUR UNDERSTANDING

Convert each quadratic into the standard form $(x - h)^2 + k$.

1. $x^2 + 4x - 3$ 2. $x^2 - 6x + 7$ 3. $x^2 - 2x + 9$
4. $x^2 - 3x + 2$ 5. $x^2 - x + 2$ 6. $x^2 + \tfrac{1}{2}x + 1$

The two general forms $x^2 \pm 2hx + h^2 = (x \pm h)^2$ have 1 as the coefficient of x^2. A modified version of this process allows other coefficients as well. The next two examples illustrate this technique.

Example 2   Write $2x^2 - 12x - 5$ in the form $a(x - h)^2 + k^2$.

Solution   First factor 2 from the first two terms only.

$$2x^2 - 12x - 5 = 2(x^2 - 6x) - 5$$

Complete the square for $x^2 - 6x$ by adding $[\tfrac{1}{2}(6)]^2 = 9$ *inside* the parentheses. Then compensate by subtracting $2(9) = 18$ *outside* the parentheses. Note that the 2 in front of the parentheses must be taken into account. In brief, the work looks like this:

$$2x^2 - 12x + 13 = 2(x^2 - 6x) + 13$$
$$= 2(x^2 - 6x + 9) + 13 - 18$$
$$= 2(x - 3)^2 - 5$$

Example 3   Convert $-\tfrac{1}{3}x^2 - 2x + 1$ into the form $a(x - h)^2 + k$.

Solution   First factor $-\tfrac{1}{3}$ from the first two terms only.

$$-\tfrac{1}{3}x^2 - 2x + 1 = -\tfrac{1}{3}(x^2 + 6x) + 1$$

Next add 9 inside the parentheses to form the perfect square $x^2 + 6x + 9 = (x + 3)^2$. Because of the coefficient in front of the parentheses, however, we will really be adding $-\frac{1}{3}(9) = -3$. Thus 3 must also be added outside the parentheses.

$$-\tfrac{1}{3}x^2 - 2x + 1 = -\tfrac{1}{3}(x^2 + 6x + 9) + 1 + 3$$
$$= -\tfrac{1}{3}(x + 3)^2 + 4$$

To match the general form $a(x - h)^2 + k$, the answer may be written as:

$$-\tfrac{1}{3}[x - (-3)]^2 + 4$$

We have seen that any quadratic expression $ax^2 + bx + c$ may be written in the standard form $a(x - h)^2 + k$. From this form we can identify the vertex, the axis of symmetry, and other information to help us graph the parabola, as illustrated in the next example.

Example 4   Write $y = 3x^2 - 4x - 2$ in the standard form $y = a(x - h)^2 + k$. Find the vertex, axis of symmetry, and graph.

Solution   First complete the square.

$$y = 3x^2 - 4x - 2$$
$$= 3(x^2 - \tfrac{4}{3}x) - 2$$
$$= 3(x^2 - \tfrac{4}{3}x + \tfrac{4}{9}) - 2 - \tfrac{4}{3}$$
$$= 3(x - \tfrac{2}{3})^2 - \tfrac{10}{3}$$

Vertex: $(\tfrac{2}{3}, -\tfrac{10}{3})$
Axis of symmetry: $x = \tfrac{2}{3}$

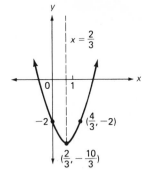

Note: Since the original equation is in the form $y = ax^2 + bx + c$, it is very easy to get the *y*-intercept by letting $x = 0$. This gives the point $(0, -2)$. Since this point is $\frac{2}{3}$ unit to the left of the axis of symmetry $x = \frac{2}{3}$, we quickly find the *symmetric point* $(\frac{4}{3}, -2)$. Can you locate the *x*-intercepts?

Example 5   Graph the function $y = f(x) = |x^2 - 4|$.

Solution   First graph the parabola $y = x^2 - 4$. Then take the part of this curve that is below the x-axis (these are the points for which $x^2 - 4$ is negative) and reflect it through the x-axis.

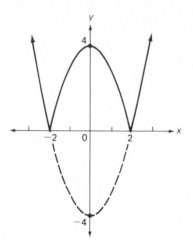

Example 6   State the conditions on the values a and k so that the parabola $y = a(x - h)^2 + k$ opens downward and intersects the x-axis in two points. What are the domain and range of this function?

Solution   In order for the parabola to open downward we must have $a < 0$. If $k > 0$, the parabola will intersect the x-axis in two distinct points. The domain is the set of all real numbers and the range consists of $y \leq k$.

CAUTION! LEARN TO AVOID MISTAKES LIKE THESE:

WRONG	*RIGHT*
$2x^2 - 4x + 3 = 2(x^2 - 2x) + 3$	$2x^2 - 4x + 3 = 2(x^2 - 2x) + 3$
$\quad = 2(x^2 - 2x + 1) + 3 - 1$	$\quad = 2(x^2 - 2x + 1) + 3 - 2$
$\quad = 2(x - 1)^2 + 2$	$\quad = 2(x - 1)^2 + 1$

EXERCISES 4.2

Write in standard form: $a(x - h)^2 + k$.

1. $x^2 + 2x - 5$ **2.** $x^2 - 2x + 5$ **3.** $-x^2 - 6x + 2$

4. $x^2 - 3x + 4$ **5.** $-x^2 + 3x - 4$ **6.** $x^2 - 5x - 2$

7. $x^2 + 5x - 2$ **8.** $2x^2 - 4x + 3$ **9.** $2x^2 + 4x - 3$

10. $5 - 6x + 3x^2$ **11.** $-5 + 6x + 3x^2$ **12.** $-3x^2 - 6x + 5$

13. $x^2 - \frac{1}{2}x + 1$ **14.** $-x^2 - \frac{1}{2}x + 1$ **15.** $\frac{3}{4}x^2 - x - \frac{1}{3}$

16. $-\frac{3}{4}x^2 + x - \frac{1}{3}$ **17.** $-5x^2 - 2x + \frac{4}{5}$ ***18.** $ax^2 + bx + c$

*19. $(x + 1)^2 - 3(x + 1) - \frac{3}{4}$ ***20.** $ax^2 - 2ahx + ah^2 + k$

*21. Compare Exercises 18 and 20 and express h and k in terms of a, b, and c.

Write each of the following in the standard form $y = a(x - h)^2 + k$. *Find the vertex, axis of symmetry, y-intercept, and graph.*

22. $y = 3x^2 - 12x + \frac{29}{2}$ **23.** $y = -\frac{1}{2}x^2 + \frac{5}{2}x - 15$

24. $y = 3x^2 + 3x + \frac{3}{4}$ **25.** $y = -2x^2 - 6x - \frac{9}{2}$

26. Graph the function f where $f(x) = |x^2 - 1|$.

27. Graph the function f where $f(x) = |9 - x^2|$.

28. Graph the function f where $f(x) = |x^2 - x - 6|$.

29. Graph the function f where $f(x) = \begin{cases} 1 & \text{if } -3 \leq x < 0 \\ x^2 - 4x + 1 & \text{if } 0 \leq x < 5 \\ -2x + 16 & \text{if } 5 \leq x < 9 \end{cases}$

30. Graph the function f where $f(x) = \begin{cases} -x^2 + 4 & \text{if } -2 \leq x < 2 \\ x^2 - 10x + 21 & \text{if } 2 \leq x \leq 7 \end{cases}$

For Exercises 31 through 35, state the conditions on the values a and k so that the parabola $y = a(x - h)^2 + k$ *has the following properties.*

31. Opens up and has vertex at $(h, 0)$.

32. Opens down and has range $y \leq 0$.

33. Opens up and has range $y \geq 2$.

34. Opens up and does not intersect the x-axis.

35. Opens down and has two x-intercepts.

*36. This exercise supports (but does not prove) the claim that only three points are needed to determine a parabola. Show that points $(-1, 3)$, $(2, 1)$, and $(4, 8)$

determine a unique parabola with equation $y = ax^2 + bx + c$ by proving that the system

$$a(-1)^2 + b(-1) + c = 3$$
$$a(2)^2 + b(2) + c = 1$$
$$a(4)^2 + b(4) + c = 8$$

produces a unique solution for a, b, and c.

*37. Follow the procedure in Exercise 36 to find the equation of the parabola that is determined by the points $(-2, -7)$, $(1, 8)$, and $(3, -2)$.

4.3

Maximum and Minimum

The graph of the quadratic function defined by $f(x) = ax^2 + bx + c$ is a parabola. There are some special points of parabolas that are useful in solving certain applied problems. In this section the vertex, or turning point, is put to use; the x-intercepts are taken up in the next section.

The parabola with equation

$$y = ax^2 + bx + c = a(x - h)^2 + k$$

opens upward or downward depending on the sign of a. When $a > 0$, the vertex is the lowest point on the parabola; when $a < 0$, it is the highest point.

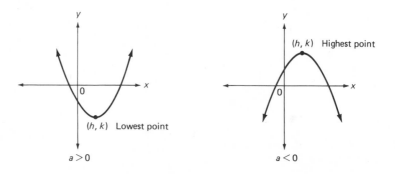

The conversion to the form $y = a(x - h)^2 + k$ instantly identifies (h, k) as this extreme point. We say that the y-value k is the **minimum value** of f when $a > 0$; it is the **maximum value of f when $a < 0$.**

Example 1 Find the maximum value or minimum value of the function $f(x) = 2(x + 3)^2 + 5$.

Solution   Here $a = 2$. Since $a > 0$, $(-3, 5)$ is the lowest point on the parabola. Therefore $f(-3) = 5$ is the minimum value.

Example 2   Find the maximum value of the following quadratic function: $f(x) = -\frac{1}{3}x^2 + x + 2$. At which value x does f achieve this maximum?

Solution   Convert to the form $a(x - h)^2 + k$:

$$
\begin{aligned}
y = f(x) &= -\tfrac{1}{3}(x^2 - 3x) + 2 \\
&= -\tfrac{1}{3}(x^2 - x + \tfrac{9}{4}) + 2 + \tfrac{3}{4} \\
&= -\tfrac{1}{3}(x - \tfrac{3}{2})^2 + \tfrac{11}{4}
\end{aligned}
$$

From this form we have $a = -\frac{1}{3}$. Since $a < 0$, $(\frac{3}{2}, \frac{11}{4})$ is the highest point of the parabola. Thus f has a maximum value of $\frac{11}{4}$ when $x = \frac{3}{2}$; $f(\frac{3}{2}) = \frac{11}{4}$ is the maximum.

Example 3   Find the maximum or minimum value of $f(x) = (x + 2)(x - 6)$ and state the x-value at which this occurs.

Solution

$$
\begin{aligned}
f(x) &= (x + 2)(x - 6) \\
&= x^2 - 4x - 12 \\
&= (x^2 - 4x + 4) - 12 - 4 \\
&= (x - 2)^2 - 16
\end{aligned}
$$

Then -16 is the minimum value, which occurs at $x = 2$.

TEST YOUR UNDERSTANDING

Find the maximum or minimum value of each quadratic function and state the x-value at which this occurs.

1. $f(x) = x^2 - 10x + 21$
2. $f(x) = x^2 + \frac{4}{3}x - \frac{7}{18}$
3. $f(x) = 10x^2 - 20x + \frac{21}{2}$
4. $f(x) = -8x^2 - 64x + 3$
5. $f(x) = -2x^2 - 1$
6. $f(x) = x^2 - 6x + 9$
7. $f(x) = -(2x + 1)(2x - 1)$
8. $f(x) = (x - 3)(x + 4)$
9. $f(x) = 25x^2 + 70x + 49$
10. $f(x) = (5 - 4x)(7 + 4x)$

The following examples demonstrate how the vertex of a parabola can be used in solving applied problems.

Example 4 Suppose that 60 meters of fencing is available to enclose a rectangular garden, one side of which will be against the side of a house. What dimensions of the garden will guarantee a maximum area?

Solution From the sketch you can see that the 60 meters need only be used for three sides, two of which are of the same length x.

The remaining side has length $60 - 2x$, and the area A is given by

$$A(x) = x(60 - 2x)$$
$$= 60x - 2x^2$$

To "maximize" A, convert to the standard form $a(x - h)^2 + k$. Thus

$$A(x) = -2(x^2 - 30x)$$
$$= -2(x^2 - 30x + 225) + 450$$
$$= -2(x - 15)^2 + 450$$

Therefore the maximum area of 450 square meters is obtained when the dimensions are $x = 15$ meters by $60 - 2x = 30$ meters.

Example 5 The sum of two numbers is 24. Find the two numbers if their product is to be a maximum.

Solution Let x represent one of the numbers. Since the sum is 24, the other number is $24 - x$. Now let p represent the product of these numbers.

$$p = x(24 - x)$$
$$= -x^2 + 24x$$
$$= -(x^2 - 24x)$$
$$= -(x^2 - 24x + 144) + 144$$
$$= -(x - 12)^2 + 144$$

Since $a = -1$, the product has a maximum value of 144 when $x = 12$. Hence the numbers are 12 and 12.

EXERCISES 4.3

Find the coordinates of the highest or lowest point of the given quadratic and sketch the graph.

1. $y = x^2 - 4x + 7$ **2.** $y = 1 - 6x - x^2$

3. $y = 1 - 4x + 4x^2$ **4.** $y = -2x^2 + 10x - 5$

Find the maximum or minimum value of the quadratic function and state the x-value at which this occurs.

5. $f(x) = -x^2 + 10x - 18$ **6.** $f(x) = x^2 + 18x + 49$

7. $f(x) = 16x^2 - 64x + 100$ **8.** $f(x) = -\frac{1}{2}x^2 + 3x - 6$

9. $f(x) = 49 - 28x + 4x^2$ **10.** $f(x) = x(x - 10)$

11. $f(x) = -x(\frac{2}{3} + x)$ **12.** $f(x) = (x - 4)(2x - 7)$

13. A manufacturer is in the business of producing small statues called Heros. He finds that the daily cost in dollars, C, of manufacturing n Heros is given by the formula $C = n^2 - 8n + 23$. How many Heros should be produced per day so that the cost will be minimum?

14. The sum of two numbers is 12. Find the two numbers if their product is to be a maximum.

15. The sum of two numbers is n. Find the two numbers such that their product is a maximum.

16. In Exercise 15, are there two numbers that will give a minimum product?

17. A homeowner has 40 feet of wire and wishes to use it to enclose a rectangular garden. What should be the dimensions of the garden so as to enclose the largest possible area?

18. Repeat Exercise 17, but this time assume that the side of the house is to be used as one boundary for the garden. Thus the wire is only needed for the other three sides.

19. A ball is thrown upward with an initial velocity of 16 feet per second. Neglecting air resistance, the formula $h = 32t - 16t^2$ gives its height h after t seconds. What is the maximum height reached by the ball?

*20. Suppose it is known that if 65 apple trees are planted in a certain size orchard, the average yield per tree will be 1500 apples per year. For each additional tree planted in the same orchard, the annual yield per tree drops by 20 apples. How many trees should be planted in order to produce the maximum crop of apples per year? (*Hint:* If n trees are added to the 65 trees, then the yield per tree is $1500 - 20n$.)

*21. The sum of the lengths of the two perpendicular sides of a right triangle is 30 centimeters. What are their lengths if the square of the hypotenuse is a minimum?

*22. Find the maximum or minimum value of $y = ax^2 + bx + c$, $a \neq 0$, and state at which x-value this occurs.

4.4

The Quadratic Formula

The graph of a quadratic function may or may not intersect the *x*-axis. Here are some typical cases:

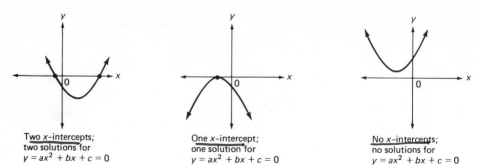

Two *x*-intercepts; two solutions for $y = ax^2 + bx + c = 0$

One *x*-intercept; one solution for $y = ax^2 + bx + c = 0$

No *x*-intercepts; no solutions for $y = ax^2 + bx + c = 0$

It is clear from these figures that if there are *x*-intercepts then these values of *x* are the solutions to the equation $y = ax^2 + bx + c = 0$. If there are no *x*-intercepts then this equation will not have any solutions. In this section we will learn procedures to handle all cases.

As a first example let us find the *x*-intercepts of the parabola $y = f(x) = x^2 - x - 6$. This calls for those values *x* for which $y = 0$. That is, we need to solve the equation $y = f(x) = 0$ for *x*. This can be done by factoring:

$$x^2 - x - 6 = 0$$
$$(x + 2)(x - 3) = 0$$

Since the product of two factors is zero only when one or both of them is zero, it follows that

$$x + 2 = 0 \quad \text{or} \quad x - 3 = 0$$
$$x = -2 \quad \text{or} \quad x = 3$$

The *x*-intercepts are -2 and 3.

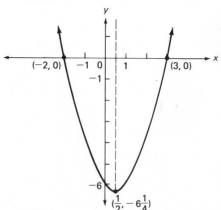

The *x*-intercepts of the parabola are also called the *roots* of the equation $f(x) = 0$.

Example 1   Solve: $25x^2 + 30x + 9 = 0$.

Solution

$$25x^2 + 30x + 9 = 0$$
$$(5x + 3)(5x + 3) = 0$$
$$5x + 3 = 0$$
$$x = -\tfrac{3}{5}$$

Check: $25(\tfrac{9}{25}) + 30(-\tfrac{3}{5}) + 9 = 9 - 18 + 9 = 0$.
Note: Since there is only *one* answer to this quadratic equation, it follows that the parabola $y = 25x^2 + 30x + 9$ has only one *x*-intercept; the parabola is *tangent* to the *x*-axis at its vertex $(-\tfrac{3}{5}, 0)$.

When a quadratic is not factorable, the *x*-intercepts can be found, if there are any, by using an approach that calls on the completing-the-square procedure. The following illustration shows the procedure for solving $x^2 - 2x - 4 = 0$. (Note that $x^2 - 2x - 4$ is not factorable.)

$$x^2 - 2x - 4 = 0$$
$$x^2 - 2x = 4$$
$$x^2 - 2x + 1 = 4 + 1 \qquad \text{(Complete the square by adding 1 to each side.)}$$
$$(x - 1)^2 = 5$$
$$x - 1 = \pm\sqrt{5} \qquad \text{(Take the square root of each side.)}$$
$$x = 1 \pm \sqrt{5}$$

Note that $x = 1 \pm \sqrt{5}$ is used as an abbreviation for $x = 1 + \sqrt{5}$ or $x = 1 - \sqrt{5}$.

Example 2   Find the roots of $2x^2 - 9x - 18 = 0$.

Solution

$$2x^2 - 9x - 18 = 0$$
$$2x^2 - 9x = 18$$
$$x^2 - \frac{9}{2}x = 9$$

$$x^2 - \frac{9}{2}x + \frac{81}{16} = 9 + \frac{81}{16}$$

$$\left(x - \frac{9}{4}\right)^2 = \frac{225}{16}$$

$$x - \frac{9}{4} = \pm\frac{15}{4}$$

$$x = \frac{9}{4} \pm \frac{15}{4}$$

$$x = 6 \text{ or } x = -\frac{3}{2}$$

Example 3 Find the x-intercepts: $y = x^2 + 2x + 5$.

Solution For any x-intercept, $y = 0$.

$$x^2 + 2x + 5 = 0$$
$$x^2 + 2x = -5$$
$$x^2 + 2x + 1 = -4$$
$$(x + 1)^2 = -4$$

Since -4 has no real square roots, there can be no real solution. Therefore the parabola has no x-intercepts.

Note that we could also have obtained this information by placing the equation in standard form.

$$y = f(x) = x^2 + 2x + 5 = (x + 1)^2 + 4$$

We note that $f(-1) = 4$ is the minimum value of f, and since the parabola opens upward, there are no x-intercepts.

TEST YOUR UNDERSTANDING

Find the x-intercepts (if any) of each quadratic function. Use the factoring method when possible; otherwise complete the square.

1. $y = f(x) = 2x^2 + 13x - 24$ **2.** $y = f(x) = 5x^2 - 3x$

3. $y = f(x) = 4x^2 - 1$ **4.** $y = f(x) = -4x^2 + 4x - 1$

5. $y = f(x) = x^2 - 8$ **6.** $f(x) = x^2 + 4$

7. $f(x) = x^2 + 2x - 2$ **8.** $f(x) = 1 + 4x - 3x^2$

9. $f(x) = x^2 + x + 1$ **10.** $f(x) = 9 - 12x + 4x^2$

The factoring method for solving quadratic equations is usually easier than completing the square. But, as we have seen, we cannot always factor. Therefore we will now solve the general quadratic equation $ax^2 + bx + c = 0$ by completing the square. This approach will produce a general solution that can be used for solving *any* quadratic equation.

$$ax^2 + bx + c = 0$$

Add $-c$:

$$ax^2 + bx = -c$$

Divide by a ($a \neq 0$):

$$x^2 + \frac{b}{a}x = -\frac{c}{a}$$

Add $\left[\frac{1}{2}\left(\frac{b}{a}\right)\right]^2 = \frac{b^2}{4a^2}$:

$$x^2 + \frac{b}{a}x + \frac{b^2}{4a^2} = \frac{b^2}{4a^2} - \frac{c}{a}$$

Factor on the left and combine on the right:

$$\left(x + \frac{b}{2a}\right)^2 = \frac{b^2 - 4ac}{4a^2}$$

If $b^2 - 4ac$ is not negative, take square roots and solve for x.

$$x + \frac{b}{2a} = \pm\sqrt{\frac{b^2 - 4ac}{4a^2}}$$

$$x + \frac{b}{2a} = \pm\frac{\sqrt{b^2 - 4ac}}{2a}$$

$$x = -\frac{b}{2a} \pm \frac{\sqrt{b^2 - 4ac}}{2a}$$

Combine terms to obtain the **quadratic formula**.

If $ax^2 + bx + c = 0$ ($a \neq 0$) and $b^2 - 4ac \geq 0$, then

$$x = \frac{-b \pm \sqrt{b^2 - 4ac}}{2a}$$

The values

$$x' = \frac{-b + \sqrt{b^2 - 4ac}}{2a} \quad \text{and} \quad x = \frac{-b - \sqrt{b^2 - 4ac}}{2a}$$

are often called the *roots* of the quadratic equation $ax^2 + bx + c = 0$. They are also the x-intercepts of the parabola $y = ax^2 + bx + c$.

Any quadratic equation may now be solved by identifying the constants a, b, and c and substituting in the quadratic formula. Consider, for example, the equation $x^2 - x - 6 = 0$. Here $a = 1$, $b = -1$, and $c = -6$.

$$x = \frac{-(-1) \pm \sqrt{(-1)^2 - 4(1)(-6)}}{2(1)}$$

$$= \frac{1 \pm \sqrt{25}}{2}$$

$$= \frac{1 \pm 5}{2}$$

Therefore $x = -2$ or $x = 3$.

Example 4   Solve: $2x^2 = -5x + 9$.

Solution   First rewrite the equation in the general form $ax^2 + bx + c = 0$.

$$2x^2 + 5x - 9 = 0$$

Use the quadratic formula with $a = 2$, $b = 5$, and $c = -9$.

$$x = \frac{-5 \pm \sqrt{25 - 4(2)(-9)}}{2(2)}$$

$$= \frac{-5 \pm \sqrt{97}}{4}$$

Thus $x = \dfrac{-5 - \sqrt{97}}{4}$ or $x = \dfrac{-5 + \sqrt{97}}{4}$.

In the derivation of the quadratic formula it was assumed that $b^2 - 4ac$ is not negative. If $b^2 - 4ac < 0$, then no real square roots are possible. Geometrically, this means that the parabola $y = ax^2 + bx + c$ does not meet the x-axis; there are no real solutions for $ax^2 + bx + c = 0$.

When $b^2 - 4ac = 0$, only the solution $x = -\dfrac{b}{2a}$ is possible; the parabola is tangent to the x-axis. Finally, when $b^2 - 4ac > 0$, we get the two solutions as before; these are the x-intercepts of the parabola.

Since $b^2 - 4ac$ tells us how many (if any) solutions $ax^2 + bx + c = 0$ has, it is called the **discriminant**. The use of the discriminant is illustrated in the following table.

QUADRATIC FUNCTION	VALUE OF $b^2 - 4ac$	REAL SOLUTIONS OF $y = 0$	NUMBER OF x-INTERCEPTS
$y = x^2 - 4x + 5$	-4	none	none
$y = x^2 - 4x + 4$	0	one	one
$y = x^2 - x - 6$	25	two	two

Note also that when the discriminant $b^2 - 4ac > 0$, then the two solutions of $ax^2 + bx + c = 0$ will be rational numbers if $b^2 - 4ac$ is a perfect square. In case $b^2 - 4ac$ is not a perfect square, then the roots are irrational as in Example 4.

The x-intercepts of a parabola can also be used to solve a **quadratic inequality**. Consider, for example, the inequality $x^2 - x - 6 < 0$. To solve this inequality first examine the graph of $y = x^2 - x - 6$. This is a parabola with x-intercepts at -2 and 3, as can be seen by writing the equation in factored form as $y = (x + 2)(x - 3)$.

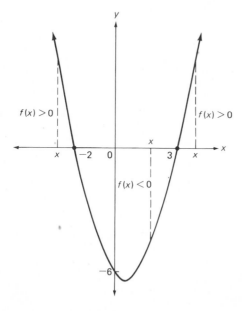

Note that $y = x^2 - x - 6$ is below the x-axis for $-2 < x < 3$. That is, $x^2 - x - 6 < 0$ for $-2 < x < 3$. Also, $x^2 - x - 6 > 0$ for $x < -2$ or $x > 3$.

CAUTION! LEARN TO AVOID MISTAKES LIKE THESE:

When solving a quadratic equation by the factoring method make sure that one side of the equation is 0 before you factor. This will help you to avoid the mistakes shown below when solving an equation such as $x^2 - 6x + 8 = 3$.

WRONG	*RIGHT*
$x^2 - 6x + 8 = 3$	$x^2 - 6x + 8 = 3$
$(x - 4)(x - 2) = 3$	$x^2 - 6x + 5 = 0$ $(x - 5)(x - 1) = 0$
$x - 4 = 3$ or $x - 2 = 1$	$x - 5 = 0$ or $x - 1 = 0$
$x = 7$ or $x = 3$	$x = 5$ or $x = 1$

EXERCISES 4.4

Use the quadratic formula to solve for x.

1. $x^2 - 3x - 10 = 0$ **2.** $2x^2 + 3x - 2 = 0$ **3.** $x^2 - 6x + 9 = 0$

4. $9 - 4x^2 = 0$ **5.** $3x^2 + 7x + 2 = 0$ **6.** $6 - 6x + x^2 = 0$

7. $x^2 = 2x + 4$ **8.** $6x - 14 = x^2$ **9.** $-2x^2 + 3x = -1$

Find the x-intercepts.

10. $y = x^2 - 3x - 4$ **11.** $y = 2x^2 - 5x - 3$ **12.** $y = x^2 - 10x + 25$

13. $y = x^2 - x + 3$ **14.** $y = 9x^2 - 4$ **15.** $y = 2x^2 - 7x + 6$

16. $y = x^2 + 4x + 1$ **17.** $y = -x^2 + 4x - 7$ **18.** $y = 3x^2 + x - 1$

Find the value of $b^2 - 4ac$. Then state if there are (a) no solutions, (b) one solution, (c) two rational solutions, or (d) two irrational solutions.

19. $x^2 - 8x + 16 = 0$ **20.** $x^2 + 3x + 5 = 0$ **21.** $x^2 + 2x - 8 = 0$

22. $4x^2 - 4x + 1 = 0$ **23.** $x^2 + 3x - 1 = 0$ **24.** $2x + 15 = x^2$

25. $2x^2 + x = 5$ **26.** $6x^2 + 7x = 3$ **27.** $1 = x - 2x^2$

Use the discriminant to predict how many times, if any, the parabola will cross the x-axis. Then give the coordinates of (a) the vertex, (b) the y-intercept, and (c) the x-intercepts.

28. $y = x^2 - 6x + 13$ **29.** $y = 9x^2 - 6x + 1$ **30.** $y = -x^2 - 4x + 3$

Solve each quadratic inequality.

31. $x^2 - 2x - 3 < 0$ **32.** $8 + 2x - x^2 < 0$ **33.** $x^2 + 3x - 10 \geq 0$

34. $2x^2 + 3x - 2 \leq 0$ **35.** $x^2 + 6x + 9 < 0$ **36.** $3 - x^2 > 0$

Graph. Show all intercepts.

· **37.** $f(x) = (x - 3)(x + 1)$ **38.** $f(x) = (x - \frac{1}{2})(5 - x)$

· **39.** $f(x) = 4 - 4x - x^2$ **40.** $f(x) = x^2 + 1$

41. $f(x) = 3x^2 - 4x + 1$ **42.** $f(x) = -x^2 + 1$

43. $f(x) = -x^2 + 2x - 4$ **44.** $f(x) = x^2 + 2x$

Find the values of b so that the graph of the parabola will be tangent to the x-axis.

45. $y = x^2 + bx + 9$ **46.** $y = 4x^2 + bx + 25$

Find the values of c so that the equation will have two real roots. (Hint: Let $b^2 - 4ac > 0$.)

47. $-x^2 + 4x + c = 0$ **48.** $2x^2 - 3x + c = 0$

Find the values of a so that the parabola will not intersect the x-axis.

49. $y = ax^2 - x - 1$ **50.** $y = ax^2 + 3x + 7$

*51. Consider the two roots of $ax^2 + bx + c = 0$, where $b^2 - 4ac \geq 0$. Find (a) the sum and (b) the product of these roots.

*52. Solve for x: $x^4 - 5x^2 + 4 = 0$. (*Hint:* Use $u = x^2$ and solve $u^2 - 5u + 4 = 0$.)

*53. Solve for x: $x^3 + 3x^2 - 4x - 12 = 0$ (*Hint:* Factor by grouping.)

*54. Solve for x: $a^3x^2 - 2ax - 1 = 0$ ($a > 0$).

4.5

Circles

Let $A(x_1, y_1)$ and $B(x_2, y_2)$ be two points in a rectangular system. We will use the symbol AB to represent the distance between the points A and B. Then AB is given by the **distance formula**:

$$AB = \sqrt{(x_1 - x_2)^2 + (y_1 - y_2)^2}$$

You can verify this result by studying this figure:

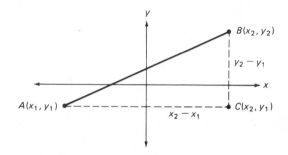

Since AB is the hypotenuse of the right triangle ABC, the Pythagorean theorem gives

$$AB^2 = AC^2 + CB^2$$

But $AC = x_2 - x_1$ and $CB = y_2 - y_1$. Thus

$$AB^2 = (x_2 - x_1)^2 + (y_2 - y_1)^2$$

Taking the positive square root gives the stated result.

Note that the diagram was set up so that $x_2 - x_1 > 0$ and $y_2 - y_1 > 0$. Other situations may have negative values, but it makes no difference because $(x_2 - x_1)^2 = (x_1 - x_2)^2$ and $(y_1 - y_2)^2 = (y_2 - y_1)^2$.

Example 1 Find the length of the line segment determined by points $A(-2, 2)$ and $B(6, -4)$.

Solution

$$
\begin{aligned}
AB &= \sqrt{(-2 - 6)^2 + [2 - (-4)]^2} \\
&= \sqrt{(-8)^2 + 6^2} \\
&= \sqrt{64 + 36} \\
&= \sqrt{100} \\
&= 10
\end{aligned}
$$

We now use the distance formula to obtain the equation of a circle. *A circle consists of all points in the plane, each of which is a fixed distance* r *from a given point called the **center** of the circle;* r *is the **radius** of the circle.*

Here is a circle with center at the origin and radius r.

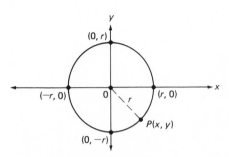

A point will be on this circle if its distance from the origin is precisely equal to r. That is, $P(x, y)$ is on this circle if and only if $OP = r$. Since the origin has coordinates $(0, 0)$, the distance formula gives

$$r = \sqrt{(x - 0)^2 + (y - 0)^2}$$
$$= \sqrt{x^2 + y^2}$$

Squaring produces this result:

$$r^2 = x^2 + y^2$$

We conclude that $P(x, y)$ is on the circle with center O and radius r if and only if the coordinates of P satisfy the preceding equation.

Example 2   What is the center and radius of the circle with equation $x^2 + y^2 = 3$?

Solution   Write $x^2 + y^2 = (\sqrt{3})^2$ and compare to the general result. Then the center is at the origin and the radius $r = \sqrt{3}$.

Now consider any circle of radius r, not necessarily one with the origin as center.

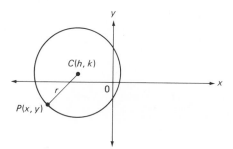

Let the center C have coordinates (h, k). Then, using the distance formula, a point $P(x, y)$ is on this circle if and only if

$$CP = r = \sqrt{(x - h)^2 + (y - k)^2}$$

By squaring we obtain the following:

General form for the equation of a circle

$$(x - h)^2 + (y - k)^2 = r^2$$

with center at (h, k) and radius r.

Example 3   Find the center and radius of the circle with this equation: $(x - 2)^2 + (y + 3)^2 = 4$. Sketch.

Solution   Using $y + 3 = y - (-3)$, rewrite the equation in this form:

$$(x - 2)^2 + [(y - (-3)]^2 = 2^2$$

By comparing to the general form we find the radius $r = 2$ and the center at $(2, -3)$.

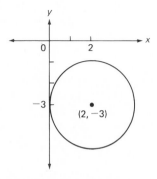

Circles are not graphs of functions. When $x = 2$ is substituted into the equation of the circle in Example 3, we get two y-values, -1 and -5. This is contrary to the definition of a function that calls for just *one* range value y for each domain value x. Even though circles are not functions, we include them in this chapter since they tie in nicely with some of our earlier work and will prove to be useful in later work as well.

Example 4   Write the equation of the circle with center at $(-3, 5)$ and radius 2 in standard form.

Solution   Use $h = -3$, $k = 5$ and $r = 2$ in the general form to obtain

$$[x - (-3)]^2 + (y - 5)^2 = 2^2$$

or

$$(x + 3)^2 + (y - 5)^2 = 4$$

Find the length of the line segment determined by the two points.

1. $(4, 0)$; $(-8, -5)$ **2.** $(9, -1)$; $(2, 3)$ **3.** $(-7, -5)$; $(3, -13)$

Find the center and radius of each circle.

4. $x^2 + y^2 = 100$ **5.** $x^2 + y^2 = 10$

6. $(x - 1)^2 + (y + 1)^2 = 25$ **7.** $(x + \frac{1}{2})^2 + y^2 = 256$

8. $(x + 4)^2 + (y + 4)^2 = 50$

Write the equation of the circle with the given center and radius in standard form.

9. Center at $(0, 4)$; $r = 5$ **10.** Center at $(1, -2)$; $r = \sqrt{3}$

The equation in Example 3 can be written in another form.

$$(x - 2)^2 + (y + 3)^2 = 4$$
$$x^2 - 4x + 4 + y^2 + 6y + 9 = 4$$
$$x^2 - 4x + y^2 + 6y = -9$$

This last equation no longer looks like the equation of a circle. Starting with such an equation we can convert it back into the standard form of a circle by completing the square in both variables, if necessary. For example, let us begin with

$$x^2 - 4x + y^2 + 6y = -9$$

Then complete the squares in x and y:

$$(x^2 - 4x + 4) + (y^2 + 6y + 9) = -9 + 4 + 9$$
$$(x - 2)^2 + (y + 3)^2 = 4$$

Example 5   Find the center and radius of the circle $9x^2 + 12x + 9y^2 = 77$.

Solution   First divide by 9 so that the x^2 and y^2 terms each has a coefficient of 1.

$$x^2 + \tfrac{4}{3}x + y^2 = \tfrac{77}{9}$$

Complete the square in x; add $\frac{4}{9}$ to both sides of the equation.

$$(x^2 + \tfrac{4}{3}x + \tfrac{4}{9}) + y^2 = \tfrac{77}{9} + \tfrac{4}{9}$$
$$(x + \tfrac{2}{3})^2 + y^2 = 9$$

or

$$[x - (-\tfrac{2}{3})]^2 + (y - 0)^2 = 3^2$$

The center is at $(-\tfrac{2}{3}, 0)$ and $r = 3$.

TEST YOUR UNDERSTANDING

Find the center and radius of each circle.

1. $x^2 - 6x + y^2 - 10y = 2$ **2.** $x^2 + y^2 + y = \tfrac{19}{4}$

3. $x^2 - x + y^2 + 2y = \tfrac{23}{4}$ **4.** $16x^2 + 16y^2 - 8x + 32y = 127$

When the equation of a circle is given and the coordinates of a point P on the circle are known, then the equation of the tangent line to the circle at P can be found. For example, the circle $(x + 3)^2 + (y + 1)^2 = 25$ has center $(-3, -1)$ and $r = 5$. The point $P(1, 2)$ is on this circle because its coordinates satisfy the equation of the circle.

$$(1 + 3)^2 + (2 + 1)^2 = 4^2 + 3^2 = 25$$

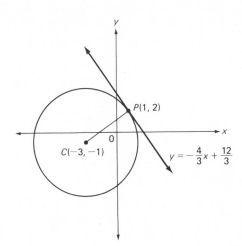

The slope of radius CP is $\dfrac{2 - (-1)}{1 - (-3)} = \dfrac{3}{4}$. Then since the tangent at P is perpendicular to the radius CP, its slope is the negative reciprocal of $\tfrac{3}{4}$, namely $-\tfrac{4}{3}$. Now using the point-slope form we get this equation of the tangent at P:

$$y - 2 = -\tfrac{4}{3}(x - 1)$$

In slope-intercept form this becomes

$$y = -\tfrac{4}{3}x + \tfrac{10}{3}$$

CAUTION! LEARN TO AVOID MISTAKES LIKE THESE:

WRONG	RIGHT
The circle $(x + 3)^2 + (x - 2)^2 = 7$ has center $(3, -2)$ and radius 7.	The circle has center $(-3, 2)$ and radius $\sqrt{7}$.
The equation of the circle with center $(-1, 0)$ and radius 5 has equation $x^2 + (y + 1)^2 = 5$.	The circle has equation $(x + 1)^2 + y^2 = 25$.

EXERCISES 4.5

1. Graph these circles in the same coordinate system.
    (a) $x^2 + y^2 = 25$ (b) $x^2 + y^2 = 16$ (c) $x^2 + y^2 = 4$
    (d) $x^2 + y^2 = 2$ (e) $x^2 + y^2 = 1$

2. Graph these circles in the same coordinate system.
    (a) $(x - 3)^2 + (y - 3)^2 = 9$ (b) $(x + 3)^2 + (y - 3)^2 = 9$
    (c) $(x + 3)^2 + (y + 3)^2 = 9$ (d) $(x - 3)^2 + (y + 3)^2 = 9$
    (e) $x^2 + y^2 = 18$

Write the equations of each circle in standard form. Find the center and radius for each.

3. $x^2 - 4x + y^2 - 10y = -28$ **4.** $x^2 - 10x + y^2 - 14y = -25$
5. $x^2 - 8x + y^2 = -14$ **6.** $x^2 + y^2 + 2y = 7$
7. $x^2 - 20x + y^2 + 20y = -100$ **8.** $4x^2 - 4x + 4y^2 = 15$
9. $16x^2 + 24x + 16y^2 - 32y = 119$ **10.** $36x^2 - 48x + 36y^2 + 180y = -160$

Write the equation of each circle in standard form.

11. Center at $(2, 0)$; $r = 2$ **12.** Center at $(\tfrac{1}{2}, 1)$; $r = 10$
13. Center at $(-3, 3)$; $r = \sqrt{7}$ **14.** Center at $(-1, -4)$; $r = 2\sqrt{2}$
15. Draw the circle $x^2 + y^2 = 25$ and the tangent lines at the points $(3, 4)$, $(-3, 4)$, $(3, -4)$, and $(-3, -4)$. Write the equations of these tangent lines.

Draw the given circle and the tangent line at the indicated point for each of the following. Write the equation of the tangent line.

16. $x^2 + y^2 = 80$; $(-8, 4)$ **17.** $x^2 + y^2 = 9$; $(-2, \sqrt{5})$
18. $(x - 4)^2 + (y + 5)^2 = 45$; $(1, 1)$ **19.** $x^2 + 4x + y^2 - 6y = 60$; $(6, 0)$
20. $x^2 + 14x + y^2 + 18y = 39$; $(5, -4)$

*21. Write the equation of the tangent line to the circle $x^2 + y^2 = 80$ at the point in the first quadrant where $x = 4$.

*22. Write the equation of the tangent line to the circle $x^2 + y^2 = 9$ at the point in the third quadrant where $y = -1$.

*23. Write the equation of the tangent line to the circle $x^2 + 14x + y^2 + 18y = 39$ at the point in the second quadrant where $x = -2$.

*24. Let $x^2 + y^2 = 4$ and $y' = -\dfrac{x}{y}$ for $y \neq 0$. Show that when y' is substituted into the expression $-\dfrac{y - xy'}{y^2}$, the result is $-\dfrac{4}{y^3}$.

4.6

Solving
Nonlinear
Systems

A straight line will intersect a parabola or a circle twice, or once, or not at all. Two parabolas of the form $y = ax^2 + bx + c$ can intersect at most two times; likewise for two circles. A circle and a parabola can intersect at most four times. These diagrams illustrate some of these possibilities.

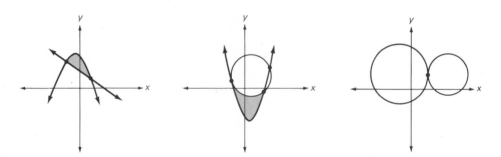

When you study calculus you will learn how to find the areas of the regions between curves. For example, the areas of the shaded regions in the diagrams can be found once the coordinates of the points of intersection are known. Here we will address ourselves only to this part of the problem: finding the points of intersection.

In each of the examples that follow, at least one of the two equations will not be linear. Thus we will be learning how to solve certain types of *nonlinear systems*.

The underlying strategy in solving such systems will be the same as it was for linear systems. We first eliminate one of the two variables to obtain an equation in one unknown. Various possible cases will be illustrated by the examples that follow. All the illustrative examples as well as the exercises have been designed so that the solutions are manageable.

Example 1   (A parabola and a line) Solve the system and graph:

$$y = x^2$$
$$y = -2x + 8$$

Solution   Let (x, y) represent the points of intersection. Since these x- and y-values are the same in both equations, we may set the two values for y equal to each other and solve for x.

$$x^2 = -2x + 8$$
$$x^2 + 2x - 8 = 0$$
$$(x + 4)(x - 2) = 0$$
$$x = -4 \quad \text{or} \quad x = 2$$

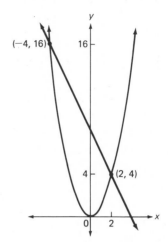

To find the corresponding y-values, either of the original equations may be used. Using $y = -2x + 8$, we have:

For $x = -4$, $y = -2(-4) + 8 = 16$.
For $x = 2$, $y = -2(2) + 8 = 4$.

The solution of the system consists of the two ordered pairs $(-4, 16)$ and $(2, 4)$. The other equation can be used as a check of these results.

Example 2   (A circle and a line) Solve the system and graph:

$$(x + 5)^2 + (y - 2)^2 = 9$$
$$2y - 3x = 4$$

Solution   First solve the linear equation for one of the variables:

$$y = \tfrac{3}{2}x + 2$$

Substitute into the equation for the circle and solve for x.

$$(x + 5)^2 + (\tfrac{3}{2}x + 2 - 2)^2 = 9$$
$$(x^2 + 10x + 25) + \tfrac{9}{4}x^2 = 9$$
$$\tfrac{13}{4}x^2 + 10x + 16 = 0$$
$$13x^2 + 40x + 64 = 0$$

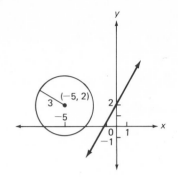

Since $b^2 - 4ac = (40)^2 - 4(13)(64) = -1728$, a negative number, there are no real solutions. Therefore there are no points of intersection; that is, the line and the circle do not meet.

Example 3   (Two parabolas) Solve the system and graph:

$$y = x^2 - 2$$
$$y = -2x^2 + 6x + 7$$

Solution   Set the two values for y equal to each other and solve for x.

$$x^2 - 2 = -2x^2 + 6x + 7$$
$$3x^2 - 6x - 9 = 0$$
$$x^2 - 2x - 3 = 0$$
$$(x + 1)(x - 3) = 0$$
$$x = -1 \text{ or } x = 3$$

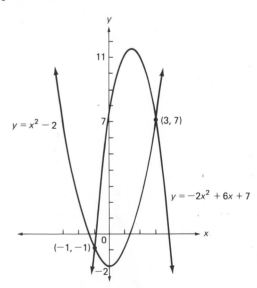

Use $y = x^2 - 2$ to solve for y.

$$y = (-1)^2 - 2 = -1 \qquad y = 3^2 - 2 = 7$$

The points of intersection are $(-1, -1)$ and $(3, 7)$.

Example 4   Solve the system and graph.
$$x^2 + y^2 - 8y = -7$$
$$y - x^2 = 1$$

Solution   Solve the second equation for x^2.

$$x^2 = y - 1$$

Substitute into the first equation and solve for y.

$(y - 1) + y^2 - 8y = -7$

$y^2 - 7y + 6 = 0$

$(y - 1)(y - 6) = 0$

$y = 1$ or $y = 6$

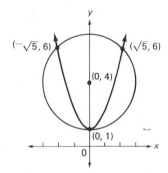

Use $x^2 = y - 1$ to solve for x.

For $y = 1$: $x^2 = 1 - 1 = 0$ $x = 0$

For $y = 6$: $x^2 = 6 - 1 = 5$ $x = \pm\sqrt{5}$

The points of intersection are $(0, 1)$, $(\sqrt{5}, 6)$, and $(-\sqrt{5}, 6)$.

Note: Alternative methods can be used to solve this example (as well as others). Another easy way begins by adding the given equations. Try it. We may also solve $y - x^2 = 1$ for y and substitute into the first equation. You will find that this latter method is more difficult. With practice you will learn how to find the easier methods.

EXERCISES 4.6

Solve each system and graph.

1. $y = -x^2 - 4x + 1$
$y - 2x = 10$

2. $3x - 4y = -5$
$(x + 3)^2 + (y + 1)^2 = 25$

3. $(x + 4)^2 + (y - 1)^2 = 16$
$(x + 4)^2 + (y - 3)^2 = 4$

4. $y = x^2 - 6x + 9$
$(x - 3)^2 + (y - 9)^2 = 9$

5. $y = x^2 + 6x + 6$
$y = -x^2 - 6x + 6$

6. $y = \frac{1}{3}(x - 3)^2 - 3$
$(x - 3)^2 + (y + 2)^2 = 1$

Solve each system.

7. $y = x^2$
$y = -x^2 + 8x - 16$

8. $y = x^2$
$y = x^2 - 8x + 24$

9. $7x + 3y = 42$
$y = -3x^2 - 12x - 15$

10. $y + 2x = 1$
$x^2 + 4x = 6 - 2y$

11. $y - x = 0$
$(x - 2)^2 + (y + 5)^2 = 25$

12. $y - 2x = 0$
$(x - 2)^2 + (y + 5)^2 = 25$

13. $y = -x^2 + 2x$
$x^2 - 2x + y^2 - 2y = 0$
(*Hint:* Complete the square in x in
both equations.)

14. $x^2 + 4x + y^2 - 4y = -4$
$(x - 2)^2 + (y - 2)^2 = 4$

*15. $y = \frac{1}{3}(x - 3)^2 - 3$
$x^2 - 6x + y^2 + 2y = -6$

*16. $y = x^2 - 6x + 9$
$(x - 3)^2 + (y - 2)^2 = 58$

4.7
Difference Quotients and Tangents (optional)

The point $(2, 4)$ is on the parabola $y = f(x) = x^2$. Also, (x, x^2) is on this parabola for any x. If $x \neq 2$ the quotient

$$\frac{f(x) - f(2)}{x - 2} = \frac{x^2 - 4}{x - 2}$$

is called the **difference quotient** of f at 2.

The figure indicates that the difference quotient is the slope of the line through points (x, x^2) and $(2, 4)$. This line may be referred to as a **secant** to the parabola.

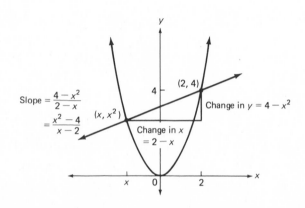

For each $x \neq 2$, there is a secant line having slope $\dfrac{x^2 - 4}{x - 2} = x + 2$. In particular, if $x = -3$, we get $x + 2 = -3 + 2 = -1$ as the slope of the secant.

Example 1

(a) For $g(x) = 9 - x^2$ find the difference quotient $\dfrac{g(x) - g(-2)}{x - (-2)}$ and simplify.

(b) Evaluate the slope of the secant through the points $(-2, g(-2))$ and $(x, g(x))$, where $x = 1$. Draw a figure and label the parts as in the preceding illustration.

Solution

(a) $\dfrac{g(x) - g(-2)}{x - (-2)} = \dfrac{9 - x^2 - [9 - (-2)^2]}{x + 2} = \dfrac{4 - x^2}{x + 2}$

$\qquad = \dfrac{(2 + x)(2 - x)}{x + 2} = 2 - x$

(b) For $x = 1$, $2 - x = 2 - 1 = 1$.

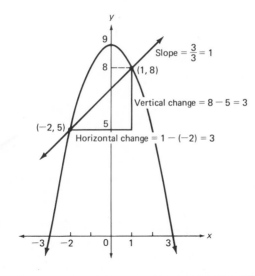

TEST YOUR UNDERSTANDING

For each function f do the following:

(a) Find and simplify the difference quotient $\dfrac{f(x) - f(c)}{x - c}$ for the given value c.

(b) Evaluate the slope of the secant through the point $(c, f(c))$ and $(x, f(x))$ for the given value of x.

1. (a) $f(x) = x^2$; $c = 1$
   (b) $x = -2$

2. (a) $f(x) = -x^2$; $c = 2$
   (b) $x = -1$

The next figure shows the parabola $f(x) = x^2 + 2$ and the three secants through $(1, 3)$ and $(x, f(x))$, where $x = -\frac{1}{2}, 0, \frac{1}{2}$. The darker line is the tangent to the parabola at the point $(1, 3)$.

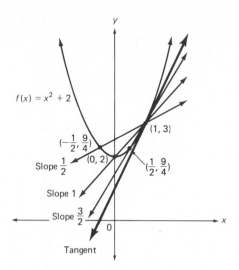

Since $\dfrac{f(x) - f(1)}{x - 1} = x + 1$ is the slope of the secant through $(1, 3)$ and $(x, f(x))$, the three slopes are computed by simply substituting the specific values for x into $x + 1$. Here is a list of slopes of these secants and some others.

	x	$-\frac{1}{2}$	0	$\frac{1}{2}$	$\frac{3}{4}$	$\frac{7}{8}$	$\frac{9}{10}$	$\frac{99}{100}$
Slope of secant $= x + 1$		$\frac{1}{2}$	1	$1\frac{1}{2}$	$1\frac{3}{4}$	$1\frac{7}{8}$	$1\frac{9}{10}$	$1\frac{99}{100}$

What do you *guess* is the slope of the tangent line at $(1, 3)$? The figure and the table *suggest* that the slope of this tangent line is 2. From the table we see that as the values for x are taken closer and closer to 1, the slopes of the corresponding secants get closer and closer to 2. And in the figure we see that as the x-values are taken closer to 1, the secant lines are "turning" counterclockwise and approaching the "limiting" position of the tangent line.

These intuitive observations turn out to be correct. The slope of this tangent line is 2. In the study of calculus such results are obtained by using more precise methods. For our purposes here it will be enough to look at the slopes of the secant, $\dfrac{f(x) - f(1)}{x - 1} = x + 1$, and then observe that since $x + 1$ gets closer to 2 as x is taken closer to 1, the tangent line has slope 2.

In the preceding example, we found that the tangent line through $(1, 3)$ has slope 2. Using the point-slope form for the equation of a line, the equation of this tangent is

$$y - 3 = 2(x - 1)$$

Converting to the slope-intercept form gives $y = 2x + 1$.

Example 2   Evaluate $\dfrac{f(x) - f(5)}{x - 5}$, where $f(x) = -2x^2 + 8x + 3$. Find the slope of the tangent line at $(5, f(5))$ and write its equation in slope-intercept form.

Solution

$$\frac{f(x) - f(5)}{x - 5} = \frac{(-2x^2 + 8x + 3) - (-7)}{x - 5} = \frac{-2x^2 + 8x + 10}{x - 5}$$

$$= \frac{-2(x^2 - 4x - 5)}{x - 5} = \frac{-2(x - 5)(x + 1)}{x - 5}$$

$$= -2(x + 1)$$

$$= -2x - 2$$

As the x-values are taken close to 5, the values $-2x - 2$ get close to -12. The tangent has slope -12. Then its equation is given by

$$y - (-7) = -12(x - 5)$$

or, in slope-intercept form,

$$y = -12x + 53$$

If the quadratic function in Example 2 is converted to the standard form $a(x - h)^2 + k$, we have

$$f(x) = -2(x - 2)^2 + 11$$

Therefore the vertex, or turning point, of the parabola has coordinates $(2, 11)$. At this point the tangent line is horizontal and must have slope zero. The same result is found by using the difference quotient method.

$$\frac{f(x) - f(2)}{x - 2} = \frac{-2x^2 + 8x + 3 - 11}{x - 2}$$

$$= \frac{-2(x^2 - 4x + 4)}{x - 2}$$

$$= \frac{-2(x - 2)^2}{x - 2}$$

$$= -2(x - 2)$$

$$= -2x + 4$$

Then, as the x-values are taken close to 2, $-2x + 4$ is close to zero. The tangent at $(2, 11)$ has slope zero and its equation is $y = 11$.

EXERCISES 4.7

1. Graph $y = f(x) = 4 - x^2$ including a secant line through points $(-1, f(-1))$ and $(x, f(x))$. Use $x \neq -1$ and label the parts of this figure as done on page 184.

2. Follow the instructions of Exercise 1 with $\check{g}(x) = x^2 + 6x + 7$ and the points $(-2, g(-2))$ and $(x, g(x))$, where $x \neq -2$.

3. Follow the instructions of Exercise 1 with $F(x) = \frac{1}{2}x^2 - 4x + 5$ and the points $(2, F(2))$, $(x, F(x))$, where $x \neq 2$.

4. (a) For $f(x) = 4 - x^2$ find $\dfrac{f(x) - f(-1)}{x - (-1)}$ and complete this table of slopes of secants through $(-1, f(-1))$ and $(x, f(x))$.

x	1	$\frac{1}{2}$	0	$-\frac{1}{2}$	$-\frac{3}{4}$	$\frac{9}{10}$	$\frac{99}{100}$
Slope of secant							

   (b) Draw a figure including the tangent at $(-1, f(-1))$ and the two secants through this point and $(x, f(x))$ for $x = \frac{1}{2}, 0$.

   (c) Find the slope of the tangent at $(-1, f(-1))$ and write its equation in slope-intercept form.

5. (a) For $g(x) = x^2 + 6x + 7$ compute the difference quotient $\dfrac{g(x) - g(-2)}{x - (-2)}$ and complete this table of slopes of secants through $(-2, g(-2))$ and $(x, g(x))$.

x	-4	$-3\frac{1}{2}$	-3	$-2\frac{1}{2}$	$-2\frac{1}{10}$	$-2\frac{1}{100}$
Slope of secant						

(b) Draw a figure including the tangent at $(-2, g(-2))$ and the two secants through this point and $(x, g(x))$ for $x = -4, -3\frac{1}{2}$.

(c) Find the slope of the tangent at $(-2, g(-2))$ and write its equation in point-slope form.

6. (a) For $F(x) = \frac{1}{2}x^2 - 4x + 5$ compute $\dfrac{F(x) - F(2)}{x - 2}$ and complete this table of slopes of secants through $(2, F(2))$ and $(x, F(x))$.

x	0	1	$1\frac{1}{2}$	$1\frac{3}{4}$	$1\frac{9}{10}$	$2\frac{1}{10}$	$2\frac{1}{4}$	$2\frac{1}{2}$	3
Slope of secant									

(b) Draw a figure including the tangent through $(2, F(2))$ and the three secants through this point and $(x, F(x))$ for $x = 0, 4,$ and 5.

(c) Find the slope of the tangent through $(2, F(2))$ and write its equation in point-slope form.

For each function f compute the difference quotient $\dfrac{f(x) - f(c)}{x - c}$ *for the indicated value c, and write the equation of the tangent line in slope-intercept form.*

7. $f(x) = x^2 - 4; \; c = 2$ **8.** $f(x) = x^2 - 4; \; c = -2$

9. $f(x) = -x^2; \; c = 3$ **10.** $f(x) = -x^2; \; c = -3$

11. $f(x) = 3x^2 - 5x + 1; \; c = 1$ **12.** $f(x) = 3x^2 - 5x + 1; \; c = -1$

13. $f(x) = -\frac{2}{3}x^2 + x; \; c = 6$ **14.** $f(x) = -\frac{2}{3}x^2 + x; \; c = -6$

For $f(x) = x^2$, $\dfrac{f(x) - f(c)}{x - c} = x + c$. *Then the slope of the tangent through* $(c, f(c))$ *is* $2c$. *In Exercises 15 through 18 find the slope of the tangent to the parabola through point* $(c, f(c))$ *for the given function f.*

15. $f(x) = x^2 - 4$ **16.** $f(x) = -x^2$

17. $f(x) = 3x^2 - 5x + 1$ **18.** $f(x) = -\frac{2}{3}x^2 + x$

*19. Let $h = x - c$ and show that $\dfrac{f(x) - f(c)}{x - c} = \dfrac{f(c + h) - f(c)}{h}$. Give a geometric interpretation by using $f(x) = x^2$.

*20. For $f(x) = x^2$ compute $\dfrac{f(3 + h) - f(3)}{h}$ and use this result to find the slope of the tangent at $x = 3$. (*Hint:* Let the h-values get close to zero.)

*21. Follow the instructions in Exercise 20 but use $f(x) = 4 - x^2$.

*22. Follow the instructions in Exercise 20 but use $f(x) = 3x^2 - 1$.

*23. Follow the instructions in Exercise 20 but use $f(x) = -2x^2 + x + 1$.

*24. Compute $\dfrac{f(x) - f(2)}{x - 2}$ for $f(x) = ax^2 + bx + c$.

*25. Use the result in Exercise 24 to find the slope of the tangent to the parabola $f(x) = ax^2 + bx + c$ at $(2, f(2))$.

4.8

**The Ellipse
and the
Hyperbola
(optional)**

A **conic section** is a curve formed by the intersection of a plane with a double right-circular cone. These curves, also called **conics**, are known as the **circle, ellipse, parabola**, and **hyperbola**. The figure indicates that the inclination of the plane in relation to the axis of the cone determines the nature of the curve. These four curves have played a vital role in mathematics and its applications from the time of the ancient Greeks until the present day.

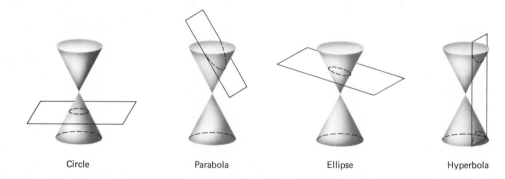

Circle   Parabola   Ellipse   Hyperbola

Since we have already done some work in this chapter with the parabola and the circle, we will now confine our efforts to the study of the ellipse and the hyperbola. Although it is possible to study these curves by using the given geometric interpretations, we will use definitions similar in style to that of a circle as the set of points in a plane equidistant from a fixed point. In the exercises you will also find some work on the parabola along these lines.

By definition, *an ellipse is the set of all points in a plane such that the sum of the distances from two fixed points is a constant.* The two fixed points, F_1 and F_2, are called the **foci** of the ellipse.

We first consider an ellipse whose foci are symmetric about the origin along the x-axis. Thus we let F_1 have coordinates $(-c, 0)$ and F_2 have coordinates $(c, 0)$, where c is some positive number.

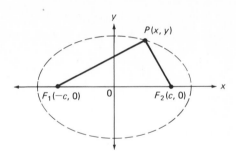

Since the sum of the distances PF_1 and PF_2 must be constant, we choose some positive number a and let this constant equal $2a$. (The form $2a$ will prove to be useful to simplify the algebraic computation.) Thus a point $P(x, y)$ is on the ellipse if and only if $PF_1 + PF_2 = 2a$. (Note $a > c$. Why?)

Using the distance formula gives

$$PF_1 = \sqrt{(x + c)^2 + y^2} \quad \text{and} \quad PF_2 = \sqrt{(x - c)^2 + y^2}$$

Thus

$$PF_1 + PF_2 = \sqrt{(x + c)^2 + y^2} + \sqrt{(x - c)^2 + y^2} = 2a$$

which implies the following:

$$\sqrt{(x + c)^2 + y^2} = 2a - \sqrt{(x - c)^2 + y^2}$$

Squaring both sides and collecting terms we have:

$$a\sqrt{(x - c)^2 + y^2} = a^2 - cx$$

Square both sides again and simplify:

$$(a^2 - c^2)x^2 + a^2y^2 = a^2(a^2 - c^2)$$

Since $a^2 - c^2 > 0$ we may let $b = \sqrt{a^2 - c^2}$ so that $b^2 = a^2 - c^2$. Therefore:

$$b^2x^2 + a^2y^2 = a^2b^2$$

Now divide through by a^2b^2 to obtain the following standard form:

Ellipse with foci at $(-c, 0)$ and $(c, 0)$

$$\frac{x^2}{a^2} + \frac{y^2}{b^2} = 1, \text{ where } b^2 = a^2 - c^2$$

The geometric interpretations of a and b can be found from this equation. Letting $y = 0$ produces the x-intercepts, $x = \pm a$. The points $V_1(-a, 0)$ and $V_2(a, 0)$ are called the **vertices** of the ellipse. The **major axis** of the ellipse is the chord V_1V_2, which has length $2a$. Letting $x = 0$ produces the y-intercepts, $y = \pm b$. The points $(0, -b)$ and $(0, b)$ are the endpoints of the **minor axis**. Note that the minor axis has length $2b$, and $2b < 2a$ since $b = \sqrt{a^2 - c^2} < a$. The intersection of the major and minor axis is the **center** of the ellipse; in this case the center is the origin.

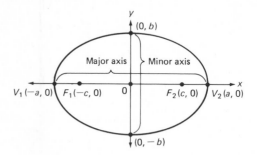

When the foci of the ellipse are on the y-axis, a similar development produces this standard form:

Ellipse with foci at (0, −c) and (0, c)

$$\frac{x^2}{b^2} + \frac{y^2}{a^2} = 1, \text{ where } b^2 = a^2 - c^2$$

The major axis is on the y-axis, and its endpoints are the vertices $(0, \pm a)$. The minor axis is on the x-axis and has endpoints $(\pm b, 0)$. The center is the origin.

Example 1 Find the equation of the ellipse with foci on the x-axis and center $(0, 0)$ if the major axis is 12 and the minor axis is 8. Find the coordinates of the foci.

Solution We are given $2a = 12$ and $2b = 8$. Thus $a = 6$ and $b = 4$ in the standard form.

$$\frac{x^2}{6^2} + \frac{y^2}{4^2} = 1 \qquad \text{or} \qquad \frac{x^2}{36} + \frac{y^2}{16} = 1$$

To locate the foci, recall that $b^2 = a^2 - c^2$ so that $c^2 = a^2 - b^2$.

$$c^2 = 36 - 16 = 20$$
$$c = \pm 2\sqrt{5}$$

The foci are located at the points $(-2\sqrt{5}, 0)$ and $(2\sqrt{5}, 0)$.

Example 2   Change $25x^2 + 16y^2 = 400$ into standard form and graph.

Solution   Divide both sides of $25x^2 + 16y^2 = 400$ by 400 to obtain $\dfrac{x^2}{16} + \dfrac{y^2}{25} = 1$.
Since $a^2 = 25$, $a = 5$ and the major axis is on the y-axis with length $2a = 10$.
Similarly, $b^2 = 16$ gives $b = 4$, and the minor axis has length $2b = 8$. Also,
$c^2 = a^2 - b^2 = 25 - 16 = 9$ so that $c = 3$, which locates the foci at $(0, \pm 3)$.

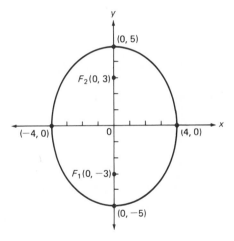

The preceding results can be generalized by allowing the center of the ellipse to be at some point (h, k). If the major axis is horizontal, then the foci have coordinates $(h - c, k)$ and $(h + c, k)$, and it can be shown that the equation has this standard form:

Ellipse with center (h, k)

$$\frac{(x - h)^2}{a^2} + \frac{(y - k)^2}{b^2} = 1, \text{ where } b^2 = a^2 - c^2$$

What is the equation when the major axis is vertical?

Example 3   Write in standard form and graph:

$$4x^2 - 16x - 9y^2 - 18y = 29$$

Solution   We follow a procedure much like that used in Section 4.5 for circles; that is, complete the square in both variables.

$$4x^2 - 16x - 9y^2 - 18y = 29$$
$$4(x^2 - 4x) - 9(y^2 + 2y) = 29$$
$$4(x^2 - 4x + 4) - 9(y^2 + 2y + 1) = 29 + 16 - 9$$
$$4(x - 2)^2 + 9(y + 1)^2 = 36$$

Divide both sides by 36:

$$\frac{(x - 2)^2}{9} + \frac{(y + 1)^2}{4} = 1$$

This is the equation of an ellipse having center at $(2, -1)$, with major axis $2a = 6$ and minor axis $2b = 4$. Since $c^2 = a^2 - b^2 = 5$, $c = \sqrt{5}$, which gives the foci $(2 \pm \sqrt{5}, -1)$.

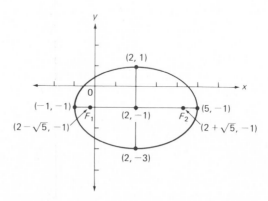

Note: When $a = b$, the standard forms produce $(x - h)^2 + (y - k)^2 = a^2$, which is the equation of a circle with center (h, k) and radius a. Thus a circle may be regarded as a special kind of ellipse, one for which the foci and center coincide.

A hyperbola is defined as the set of all points in a plane such that the difference of the distances from two fixed points is constant. The two fixed points, F_1 and F_2, are called the **foci** of the hyperbola, and its **center** is the midpoint of the **transverse axis** F_1F_2. It turns out that a hyperbola consists of two congruent branches, one of which opens toward the left and the other toward the right.

We begin with a hyperbola with foci on the x-axis at $F_1(-c, 0)$ and $F_2(c, 0)$, where $c > 0$.

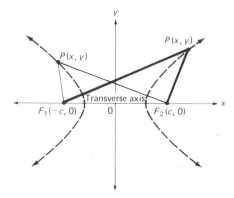

Select a number $a > 0$ so that for any point P on the right branch of the hyperbola we have $PF_1 - PF_2 = 2a$. For any point P on the left branch $PF_2 - PF_1 = 2a$.

Note that $a < c$ since from triangle F_1PF_2 (P on the right branch) we have $PF_1 < PF_2 + F_1F_2$, which gives $PF_1 - PF_2 < F_1F_2$, or $2a < 2c$.

If we now follow the same type of analysis used to derive the equation of an ellipse, it can be shown that the equation of a hyperbola may be written in this standard form:

Hyperbola with foci at (−c, 0) and (c, 0)

$$\frac{x^2}{a^2} - \frac{y^2}{b^2} = 1, \text{ where } b^2 = c^2 - a^2$$

Letting $y = 0$ gives $x = \pm a$; the points $V_1(-a, 0)$ and $V_2(a, 0)$ are the vertices of the hyperbola.

Example 4 Write in standard form and identify the foci and vertices: $16x^2 - 25y^2 = 400$.

Solution Divide through by 400 to place in standard form.

$$\frac{16x^2}{400} - \frac{25y^2}{400} = \frac{400}{400}$$

$$\frac{x^2}{25} - \frac{y^2}{16} = 1$$

Note that $a^2 = 25$ and $b^2 = 16$, so that $a = 5$ and $b = 4$. Then $c^2 = a^2 + b^2 = 25 + 16 = 41$ and $c = \pm\sqrt{41}$. The vertices of the hyperbola are located at $(-5, 0)$ and $(5, 0)$; the foci are at $(-\sqrt{41}, 0)$ and $(\sqrt{41}, 0)$.

Let us return to the standard form for the equation of a hyperbola and solve for y:

$$\frac{x^2}{a^2} - \frac{y^2}{b^2} = 1$$

$$y^2 = \frac{b^2}{a^2}(x^2 - a^2)$$

$$y = \pm\frac{b}{a}\sqrt{x^2 - a^2}$$

Consequently $|x| \geq a$, which means that there are no points of the hyperbola for $-a < x < a$. The geometric interpretation of b is shown in the figure.

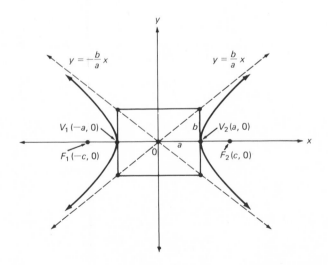

Since $b^2 = c^2 - a^2$, or $a^2 + b^2 = c^2$, it follows that b can be used as a side of a right triangle having hypotenuse c. Thus if we construct a perpendicular at V_2 of length b, the resulting right triangle has hypotenuse c. We also see that this hypotenuse lies on the line $y = \dfrac{b}{a}x$.

For $x \geq a$ we get $\dfrac{b}{a}x > \dfrac{b}{a}\sqrt{x^2 - a^2}$. Therefore in quadrant I the hyperbola is below the line $y = \dfrac{b}{a}x$. Also, since $y = \dfrac{b}{a}\sqrt{x^2 - a^2}$ gets close in value to $y = \dfrac{b}{a}x$ as x gets large, we say that the line $y = \dfrac{b}{a}x$ is an **asymptote** to the hyperbola in quadrant I. Similar observations hold in the remaining quadrants, and we conclude that the lines $y = \pm\dfrac{b}{a}x$ are asymptotes to the hyperbola.

An efficient way to sketch a hyperbola is to use the asymptotes. First draw the rectangle that is $2a$ units wide and $2b$ units high as shown in the preceding figure. Note that the center of the hyperbola is also the center of this rectangle. Draw the diagonals of the rectangle and extend them in both directions; these are the asymptotes. Now sketch the hyperbola by beginning at the vertices $(\pm a, 0)$ so that the lines are asymptotes to the curve and the branches are between the asymptotes.

When the foci of a hyperbola are on the y-axis the equation has this standard form:

Hyperbola with foci at (0, −c) and (0, c)

$$\frac{y^2}{a^2} - \frac{x^2}{b^2} = 1, \text{ where } b^2 = c^2 - a^2$$

The vertices are $(0, \pm a)$, the transverse axis has length $2a$, and the asymptotes are the lines $y = \pm\dfrac{a}{b}x$. The branches of this hyperbola open upward and downward.

When the center of the hyperbola is at some point (h, k) and the transverse axis is horizontal, we have the following standard form:

Hyperbola with center at (h, k)

$$\frac{(x - h)^2}{a^2} - \frac{(y - k)^2}{b^2} = 1, \text{ where } b^2 = c^2 - a^2$$

What is the equation when the transverse axis is vertical?

Example 5   Write in standard form and graph:

$$4x^2 + 16x - 9y^2 + 18y = 29$$

Solution   Complete the square in x and y.

$$4(x^2 + 4x) - 9(y^2 - 2y) = 29$$
$$4(x^2 + 4x + 4) - 9(y^2 - 2y + 1) = 29 + 16 - 9$$
$$4(x + 2)^2 - 9(y - 1)^2 = 36$$

Divide both sides by 36:

$$\frac{(x + 2)^2}{9} - \frac{(y - 1)^2}{4} = 1$$

This is the standard form for a hyperbola with center at $(-2, 1)$. Since $a^2 = 9$, we have $a = 3$ and the vertices are located 3 units from the center at $(-5, 1)$ and $(1, 1)$. Since $c^2 = a^2 + b^2 = 13$, we have the foci located at $\sqrt{13}$ units from the center, namely at $(-2 \pm \sqrt{13}, 0)$.

To sketch the hyperbola first draw the 6 by 4 rectangle with center at $(-2, 1)$ as shown. Draw the asymptotes by extending the diagonals and sketch the branches.

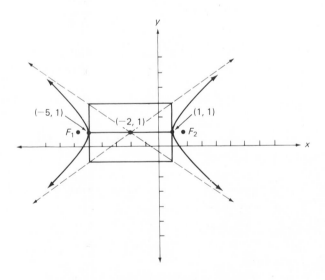

EXERCISES 4.8

Name the conic section and state the coordinates of the center, vertices, and foci. Give the equations of the asymptotes where applicable.

1. $\dfrac{x^2}{25} + \dfrac{y^2}{16} = 1$

2. $\dfrac{x^2}{16} + \dfrac{y^2}{25} = 1$

3. $\dfrac{x^2}{36} - \dfrac{y^2}{25} = 1$

4. $\dfrac{y^2}{36} - \dfrac{x^2}{25} = 1$

5. $25x^2 + 4y^2 = 100$

6. $x^2 - 2y^2 = 6$

7. $x^2 + \dfrac{(y-1)^2}{4} = 1$

8. $\dfrac{(x+2)^2}{16} + \dfrac{(y+5)^2}{25} = 1$

Name the conic and graph.

9. $\dfrac{x^2}{9} + \dfrac{y^2}{16} = 1$

10. $\dfrac{x^2}{16} + \dfrac{y^2}{9} = 1$

11. $16y^2 - 9x^2 = 144$

12. $9x^2 - 16y^2 = 144$

13. $\dfrac{(x-1)^2}{64} + \dfrac{(y-2)^2}{36} = 1$

14. $\dfrac{(y-1)^2}{64} - \dfrac{(x-3)^2}{36} = 1$

15. $16(y-3)^2 - 9(x+2)^2 = -144$

16. $16(y-2)^2 - 9(x+3)^2 = 144$

Write in standard form and identify.

17. $x^2 + y^2 - 2x + 4y + 1 = 0$

18. $x^2 + y^2 + 6x - 4y + 4 = 0$

19. $x^2 + 4y^2 + 2x - 3 = 0$

20. $x^2 - 9y^2 - 2x - 8 = 0$

*21. $9x^2 + 18x - 16y^2 + 96y = 279$

*22. $4x^2 - 16x + y^2 + 8y = -28$

Write the equation of the ellipse in standard form having the given properties.

23. Center $(0, 0)$; horizontal major axis of length 10; minor axis of length 6.

24. Center $(0, 0)$; foci $(\pm 2, 0)$; vertices $(\pm 5, 0)$.

25. Vertices $(0, \pm 5)$; foci $(0, \pm 3)$.

26. Center $(2, 3)$; foci $(-2, 3)$ and $(6, 3)$; minor axis of length 8.

27. Center $(2, -3)$; vertical major axis of length 12; minor axis of length 8.

28. Center $(-5, 0)$; foci $(-5, \pm 2)$; $b = 3$.

Write the equation of the hyperbola in standard form having the given properties.

29. Center $(0, 0)$; foci $(\pm 6, 0)$; vertices $(\pm 4, 0)$.

30. Center $(0, 0)$; foci $(0, \pm 4)$; vertices $(0, \pm 1)$.

31. Center $(-2, 3)$; vertical transverse axis of length 6; $c = 4$.

32. Center $(4, 4)$; vertex $(4, 7)$; $b = 2$.

*33. Center $(0, 0)$; asymptotes $y = \pm \tfrac{1}{2}x$; vertices $(\pm 4, 0)$.

34. The figure describes an instrument that can be used to draw an ellipse. Put thumb-tacks at the points on the paper and place a loop of string around them. Pull the string taut by using a pencil point as indicated. Keeping the string taut, move the pencil around the loop. Why does this motion trace out an ellipse?

35. A parabola can be defined as the set of all points in a plane equidistant from a given fixed line called the **directrix** and a given fixed point called the **focus**. Let focus F have coordinates $(0, p)$, and let the directrix have equation $y = -p$ as indicated.

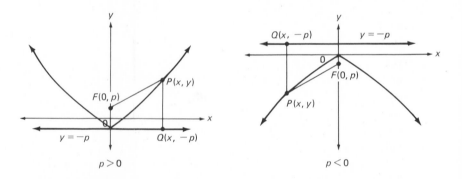

(a) Write the length PF in terms of the coordinates of P and F.

(b) Express the length PQ in terms of y and p.

(c) Equate the results in parts (a) and (b) and derive the form $x^2 = 4py$.

36. Write the equation of the parabola with focus $(0, 3)$ and directrix $y = -3$. Sketch.

37. Write the equation of the parabola with focus $(0, -3)$ and directrix $y = 3$. Sketch.

38. Find the coordinates of the focus and the equation of the directrix for the parabola $y = 2x^2$.

39. Find the coordinates of the focus and the equation of the directrix for the parabola $y = -\frac{1}{4}x^2$.

*40. The origin is the vertex of each parabola having equation $x^2 = 4py$. What is the equation of the parabola with center (h, k) having focus $(h, k + p)$ and directrix $y = k - p$?

*41. A parabola has vertex $(2, -5)$, focus at $(2, -3)$, and directrix $y = -7$. Write its equation in the form of Exercise 40. Then write its equation in the form $y = ax^2 + bx + c$.

*42. Let P be on the right branch of the hyperbola with foci $F_1(-c, 0)$ and $F_2(c, 0)$, and let $PF_1 - PF_2 = 2a$ for $a < c$. Derive the equation $\dfrac{x^2}{a^2} - \dfrac{y^2}{b^2} = 1$, where $b^2 = c^2 - a^2$.

REVIEW EXERCISES FOR CHAPTER 4

The solutions to the following exercises can be found within the text of Chapter 4. Try to answer each question without referring to the text.

Section 4.1

1. Graph $y = (x + 1)^2$.
2. Graph $y = 3(x - 1)^2 + 2$.
3. Graph $y = -2(x - 3)^2 + 4$.

Section 4.2

4. Write in standard form $a(x - h)^2 + k$: $x^2 + 3x - 1$.
5. Write in standard form $a(x - h)^2 + k$: $-\frac{1}{3}x^2 - 2x + 1$.
6. Find the vertex and axis of symmetry of the parabola $y = 3x^2 - 4x - 2$.
7. Graph the function f where $f(x) = |x^2 - 4|$.
8. State the conditions on the values a and k so that the parabola $y = a(x - h)^2 + k$ opens downward and intersects the x-axis in two points.
9. What are the domain and range of the functions in Exercise 8?

Section 4.3

10. What is the highest point for $y = -\frac{1}{3}x^2 + x + 2$?
11. Find the maximum or minimum value of $f(x) = (x + 2)(x - 6)$ and state the x-value at which this occurs.
12. Find the two numbers so that their product is a maximum and their sum is 24.
13. Suppose that 60 meters of fencing is available to enclose a rectangular garden, one side of which will be against the side of a house. What dimensions of the garden will guarantee a maximum area?

Section 4.4

14. Solve for x: $25x^2 + 30x + 9 = 0$.
15. Find the roots: $2x^2 - 9x - 18 = 0$.
16. Find the x-intercepts: $f(x) = x^2 + 2x + 5$.
17. Find the value of the discriminant and decide how many x-intercepts:
    (a) $y = x^2 - 4x + 5$ (b) $y = x^2 - 4x + 4$ (c) $y = x^2 - x - 6$
18. Solve: $x^2 - x - 6 < 0$.

Section 4.5

19. Find the length of the line segment determined by the points $(-2, 2)$ and $(6, -4)$.

20. Find the center and radius of the circle $x^2 + y^2 = 3$.

21. Draw the circle $(x - 2)^2 + (y + 3)^2 = 4$. Where is the center and what is the radius?

22. Find the center and radius of the circle $x^2 - 4x + y^2 + 6y = -9$.

23. Find the center and radius of the circle $9x^2 + 12x + 9y^2 = 77$.

24. Find the equation of the tangent line to the circle $(x + 3)^2 + (y + 1)^2 = 25$ at the point $(1, 2)$. Sketch.

Section 4.6

25. Solve the system and graph:

$$y = x^2$$
$$y = -2x + 8$$

26. Solve the system and graph:

$$y = x^2 - 2$$
$$y = -2x^2 + 6x + 7$$

27. Solve the system and graph:

$$x^2 + y^2 - 8y = -7$$
$$y - x^2 = 1$$

Section 4.7

28. What is the slope of the secant through the points $(-2, g(-2))$ and $(x, g(x))$ on the curve $g(x) = 9 - x^2$?

29. Find the equation of the tangent to the parabola $f(x) = x^2 + 2$ at the point $(1, 3)$ and graph.

30. Find the slope of the tangent line to the parabola $f(x) = -2x^2 + 8x + 3$ at the point $(5, f(5))$ and write its equation in slope-intercept form.

Section 4.8

31. Find the equation of the ellipse with foci on the x-axis and center $(0, 0)$ if the major axis is 12 and the minor axis is 8. Find the coordinates of the foci.

32. Write in standard form and graph:

$$4x^2 - 16x - 9y^2 - 18y = 29$$

33. Write in standard form and identify the foci, vertices, and asymptotes of

$$16x^2 - 25y^2 = 400$$

1. Graph $y = (x - \frac{1}{2})^2 - \frac{9}{4}$.
2. Let $y = -5x^2 + 20x - 1$.
   (a) Write the quadratic in the standard form $y = a(x - h)^2 + k$.
   (b) Give the coordinates of the vertex.
   (c) Write the equation of the axis of symmetry.
   (d) State the domain and range of the quadratic function.
3. (a) Solve for x: $3x^2 - 8x - 3 = 0$.
   (b) Find the x-intercepts of $y = -x^2 + 4x + 7$.

Give the value of the discriminant and use this result to describe the x-intercepts, if any.

4. $y = x^2 + 3x + 1$
5. $y = 6x^2 + 5x - 6$

6. Find the maximum or minimum value of the quadratic function and state the x-value at which this occurs: $f(x) = -\frac{1}{2}x^2 - 6x + 2$.
7. (a) Draw the circle $(x - 3)^2 + (y + 4)^2 = 25$.
   (b) Write the equation of the tangent line to this circle at the point $(6, 0)$.
8. Find the center and radius of the circle $4x^2 + 4x + 4y^2 - 56y = -97$.
9. Solve the system:

$$y = (x - 2)^2 - 2$$
$$(x - 2)^2 + (y - 2)^2 = 4$$

10. In the figure the altitude BC of $\triangle ABC$ is 4 feet. The part of the perimeter $PQRCB$ is to be a total of 28 feet. How long should x be so that the area of rectangle $PCRQ$ is a maximum?

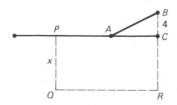

†11. (a) Find and simplify the difference quotient $\dfrac{f(x) - f(5)}{x - 5}$ for $f(x) = x^2 - x$.
   (b) Write the equation of the tangent line to the parabola in (a) at the point $(5, f(5))$.
†12. Write in standard form and identify the curve $4x^2 + 16x - y^2 + 6y = -3$. Find the coordinates of the center, and the vertices.

Answers to the Test Your Understanding Exercises

Page 158

1. $(x + 2)^2 - 7$
2. $(x - 3)^2 - 2$
3. $(x - 1)^2 + 8$
4. $(x - \frac{3}{2})^2 - \frac{1}{4}$
5. $(x - \frac{1}{2})^2 + \frac{7}{4}$
6. $(x + \frac{1}{4})^2 + \frac{15}{16}$

Page 163

1. Minimum value $= -4$ at $x = 5$.
2. Minimum value $= -\frac{5}{6}$ at $x = -\frac{2}{3}$.
3. Minimum value $= \frac{1}{2}$ at $x = 1$.
4. Maximum value $= 131$ at $x = -4$.
5. Maximum value $= -1$ at $x = 0$.
6. Minimum value $= 0$ at $x = 3$.
7. Maximum value $= 1$ at $x = 0$.
8. Minimum value $= -\frac{49}{4}$ at $x = -\frac{1}{2}$.
9. Minimum value $= 0$ at $x = -\frac{7}{5}$.
10. Maximum value $= 36$ at $x = -\frac{1}{4}$.

Page 168

1. $-8, \frac{3}{2}$
2. $0, \frac{3}{5}$
3. $\pm\frac{1}{2}$
4. $\frac{1}{2}$
5. $\pm 2\sqrt{2}$
6. None.
7. $-1 + \sqrt{3}, -1 - \sqrt{3}$
8. $\frac{1}{3}(2 + \sqrt{7}), \frac{1}{3}(2 - \sqrt{7})$
9. None.
10. $\frac{3}{2}$

Page 177

1. 13
2. $\sqrt{65}$
3. $2\sqrt{41}$
4. $(0, 0); 10$
5. $(0, 0); \sqrt{10}$
6. $(1, -1); 5$
7. $(-\frac{1}{2}, 0); 16$
8. $(-4, -4); 5\sqrt{2}$
9. $x^2 + (y - 4)^2 = 25$
10. $(x - 1)^2 + (y + 2)^2 = 3$

Page 178

1. $(3, 5); 6$
2. $(0, -\frac{1}{2}); \sqrt{5}$
3. $(\frac{1}{2}, -1); \sqrt{7}$
4. $(\frac{1}{4}, -1); 3$

Page 185

1. (a) $\dfrac{f(x) - f(1)}{x - 1} = \dfrac{x^2 - 1}{x - 1} = x + 1$
   (b) $-2 + 1 = -1$

2. (a) $\dfrac{f(x) - f(2)}{x - 2} = \dfrac{-x^2 - (-4)}{x - 2} = \dfrac{4 - x^2}{x - 2}$
   $= -(x + 2)$
   (b) $-(-1 + 2) = -1$

Five

Polynomial and Rational Functions

The quadratic function given by $y = x^2$ can also be referred to as the *squaring function*, because for each domain value x the corresponding range value is the square of x. In the same sense, the function f given by $y = f(x) = x^3$ may be called the *cubing function*. A table of values is a helpful aid for drawing the graph of $y = x^3$.

x	-2	-1	$-\dfrac{1}{2}$	0	$\dfrac{1}{2}$	1	2
y	-8	-1	$-\dfrac{1}{8}$	0	$\dfrac{1}{8}$	1	8

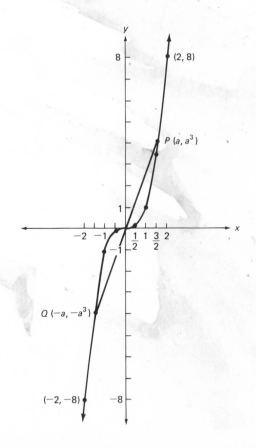

The table and the graph reveal that the curve is symmetric through the origin. Geometrically this means that whenever a line through the origin intersects the curve at a point P, this line will intersect the curve in another point Q (on the opposite side of the origin) so that the lengths of OP and OQ are equal. Algebraically this says that both points (a, a^3) and $(-a, -a^3)$ are on the curve for each value $x = a$. These are said to be **symmetric points**. In particular, since $(2, 8)$ is on the curve, then $(-2, -8)$ is also on the curve.

In general, the graph of a function $y = f(x)$ is said to be *symmetric through the origin* if for all x in the domain of f we have

$$f(-x) = -f(x)$$

Some of our earlier graphing techniques used for parabolas have a natural application to other functions. For example, the graph of $y = 2x^3$ can be obtained from the graph of $y = x^3$ by multiplying each of its ordinates by 2. Here are other illustrations.

Example 1   Graph $y = g(x) = (x - 3)^3$ and $y = h(x) = (x - 3)^3 + 2$.

Solution   The graph of g is obtained by translating $y = x^3$ by 3 units to the right. Adding 2 to each ordinate of $g(x)$ gives the points for the graph of h. That is, shift $y = (x - 3)^3$ by 2 units upward.

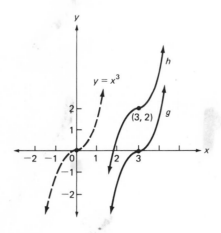

Example 2   Graph $y = f(x) = |(x - 3)^3|$.

Solution   First graph $y = (x - 3)^3$ as in Example 1. Then take the part of this curve that is below the x-axis $[(x - 3)^3 < 0]$ and reflect it through the x-axis. The graph is shown at the top of the next page.

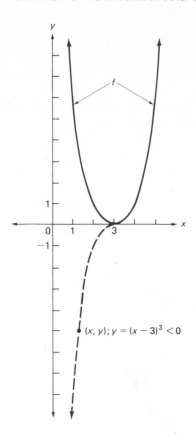

$(x, y); y = (x - 3)^3 < 0$

Before proceeding, it may be helpful to collect some of the general observations that are useful in graphing functions.

Help with graphing

1. If $f(x) = f(-x)$, the curve is symmetric about the y-axis.
2. If $f(-x) = -f(x)$, the curve is symmetric through the origin.
3. The graph of $y = af(x)$ can be obtained by multiplying the ordinates of the curve $y = f(x)$ by the value a. The case $a = -1$ gives $y = -f(x)$, which is the reflection of $y = f(x)$ through the x-axis.
4. The graph of $y = |f(x)|$ can be obtained from the graph of $y = f(x)$ by taking the part of $y = f(x)$ that is below the x-axis and reflecting it through the x-axis.

In each of the following, h is positive.

5. The graph of $y = f(x - h)$ can be obtained by shifting $y = f(x)$ h units to the right.

6. The graph of $y = f(x + h)$ can be obtained by shifting $y = f(x)$ h units to the left.

7. The graph of $y = f(x) + h$ can be obtained by shifting $y = f(x)$ h units upward.

8. The graph of $y = f(x) - h$ can be obtained by shifting $y = f(x)$ h units downward.

Example 3   In the graph the curve C_1 is obtained by shifting $y = x^3$ horizontally and C_2 is obtained by shifting C_1 vertically. What are the equations of C_1 and C_2?

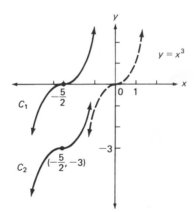

Solution

$$C_1: \qquad y = (x + \tfrac{5}{2})^3$$
$$C_2: \qquad y = (x + \tfrac{5}{2})^3 - 3$$

Example 4   Graph: (a) $y = f(x) = x^4$;   (b) $y = 2x^4$.

Solution

(a) Since $f(-x) = (-x)^4 = x^4 = f(x)$, the graph is symmetric about the y-axis. Use the table of values to locate the right half of the curve; the symmetry gives the rest.

(b) Multiply the ordinates of $y = x^4$ by 2 to get the graph of $y = 2x^4$.

x	0	$\frac{1}{2}$	1	$\frac{3}{2}$	2
y	0	$\frac{1}{16}$	1	$\frac{81}{16}$	16

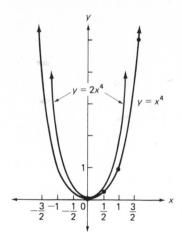

EXERCISES 5.1

In Exercises 1 through 4 graph each set of curves in the same coordinate system. For each exercise use a dotted curve for the first equation and a solid curve for each of the other two.

1. $y = x^2$, $y = (x - 3)^2$, $y = x^2 + 3$

2. $y = x^3$, $y = (x + 2)^3$, $y = x^3 + 4$

3. $y = x^3$, $y = -x^3$, $y = -(x + 2)^3$

4. $y = x^4$, $y = (x - 4)^4$, $y = (x - 4)^4 - 4$

5. Graph $y = |(x + 1)^3|$.

6. Graph $y = 2(x - 1)^3 + 3$.

7. Graph $y = f(x) = |x|$, $y = g(x) = |x - 3|$, and $y = h(x) = |x - 3| + 2$, on the same axes.

Find the equation of the curve C which is obtained from the dotted curve by a horizontal or vertical shift.

8.

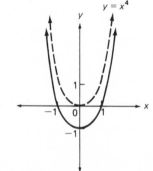

9.

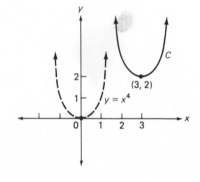

10.

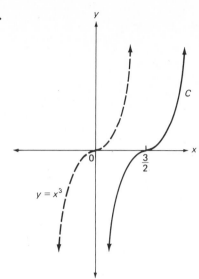

11.

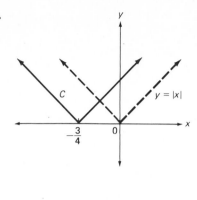

Graph each of the following.

12. $y = |x^4 - 16|$ **13.** $y = |x^3 - 1|$ **14.** $y = |-1 - x^3|$

Graph each of the following. (Hint: Consider the expansion of $(a \pm b)^n$ for appropriate n.)

*15. $y = x^3 + 3x^2 + 3x + 1$ ***16.** $y = x^3 - 6x^2 + 12x - 8$

*17. $y = -x^3 + 3x^2 - 3x + 1$ ***18.** $y = x^4 - 4x^3 + 6x^2 - 4x + 1$

*19. (a) For $y = f(x) = x^3$ evaluate the difference quotient $\dfrac{f(x) - f(2)}{x - 2}$.

   (b) Use the result in part (a) to find the slope of the tangent line to the graph at $(2, f(2))$. (See Section 4.7.)

*20. (a) For $y = f(x) = x^3$ evaluate the difference quotient $\dfrac{f(\frac{1}{2} + h) - f(\frac{1}{2})}{h}$.

   (b) Use the result in part (a) to write the equation of the tangent line to the curve at $(\frac{1}{2}, f(\frac{1}{2}))$. (See Section 4.7.)

5.2

Graphing Some Special Rational Functions

A rational expression is a ratio of polynomials. Such expressions may be used to define functions. For example, $y = f(x) = \dfrac{1}{x}$ gives a rational function whose domain consists of all numbers except zero. The denominator x is a polynomial of degree 1, and the numerator $1 = 1x^0$ is a (constant) polynomial of degree zero.

To draw the graph of this function first observe that it is symmetric through the origin because

$$f(-x) = \frac{1}{-x} = -\frac{1}{x} = -f(x)$$

Moreover, in $y = \dfrac{1}{x}$ both variables must have the same sign since $xy = 1$, a positive number. Hence the graph will only be in quadrants I and III. Now use a table of values to get the curve in the first quadrant, and then use the symmetry to get the remaining half.

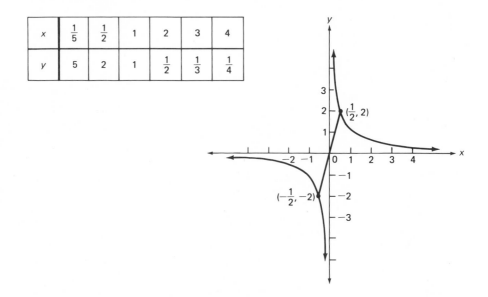

x	$\frac{1}{5}$	$\frac{1}{2}$	1	2	3	4
y	5	2	1	$\frac{1}{2}$	$\frac{1}{3}$	$\frac{1}{4}$

Observe that the curve approaches the x-axis in quadrant I. That is, as the values for x become large, the values for y approach zero. Also, as the values for x approach zero in the first quadrant, the y-values become very large. A similar observation can be made about the curve in the third quadrant. We say that the axes are **asymptotes** for the curve; they are *asymptotic* to the curve.

Example 1   Sketch the graph of $g(x) = \dfrac{1}{x-3}$. Find the asymptotes.

Solution   Using $f(x) = \dfrac{1}{x}$, we have $f(x-3) = \dfrac{1}{x-3} = g(x)$. Therefore the graph of g can be drawn by shifting $f(x) = \dfrac{1}{x}$ by 3 units to the right.

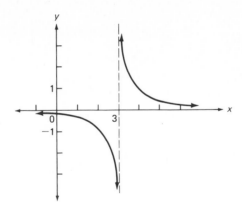

The *x*-axis and the vertical line $x = 3$ are the asymptotes.

Example 2   Graph $y = \dfrac{1}{x^2}$.

Solution   First note that $x \neq 0$. For all other values of x we have $x^2 > 0$ so that the curve will appear in quadrants I and II only. As the values for x become large, the values for $y = \dfrac{1}{x^2}$ become very small. Moreover, as x approaches zero, y becomes very large. Thus the curve is asymptotic to the axes.

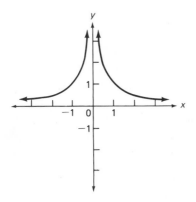

Note that the curve is symmetric about the *y*-axis. That is,

$$f(x) = \frac{1}{x^2} = \frac{1}{(-x)^2} = f(-x)$$

Example 3   Graph $y = \dfrac{1}{(x + 2)^2} - 3$. What are the asymptotes?

Solution   Shift $y = \dfrac{1}{x^2}$ by 2 units left and then shift 3 units down.

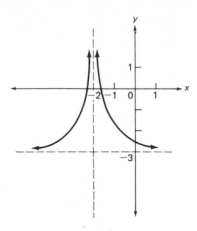

Asymptotic lines: $x = -2$, $y = -3$.

___

The curves studied in this chapter, as well as the parabolas and straight lines in earlier chapters, are very useful in the study of calculus. Having an almost instant recall of the graphs of the following functions will be helpful in future work:

$$y = x \qquad y = |x| \qquad y = x^2 \qquad y = x^3 \qquad y = x^4 \qquad y = \frac{1}{x} \qquad y = \frac{1}{x^2}$$

Not only have you learned what these curves look like, but, just as important, you have also learned how to obtain other curves from them by appropriate translations and reflections.

EXERCISES 5.2

In Exercises 1 through 6 graph each set of curves in the same coordinate system. For each exercise use a dotted curve for the first equation and a solid curve for each of the others.

1. $y = \dfrac{1}{x}$, $y = \dfrac{2}{x}$, $y = \dfrac{1}{2x}$

2. $y = \dfrac{1}{x}$, $y = -\dfrac{1}{x}$

3. $y = \dfrac{1}{x}, y = \dfrac{1}{x+2}$

4. $y = \dfrac{1}{x}, y = \dfrac{1}{x} + 5, y = \dfrac{1}{x} - 5$

5. $y = \dfrac{1}{x^2}, y = \dfrac{1}{(x-3)^2}$

6. $y = \dfrac{1}{x^2}, y = -\dfrac{1}{x^2}$

Graph each of the following. Find all asymptotes, if any.

7. $y = \dfrac{1}{x+4} - 2$

8. $y = \dfrac{1}{(x-2)^2} + 3$

9. $y = \dfrac{1}{|x-2|}$

10. $y = \dfrac{1}{x^3}$

*11. Graph $y = \left| \dfrac{1}{x} - 1 \right|$.

*12. Graph $x = \dfrac{1}{y^2}$. Why is y not a function of x?

5.3
Synthetic Division

In this and in the following two sections we will learn some methods for finding the x-intercepts of polynomial functions and rational functions. First let us briefly review the process used in division of polynomials; it is very much like the division algorithm in arithmetic. Consider the quotient

$$(x^3 - x^2 - 5x + 6) \div (x - 2)$$

STEP 1.   Divide x^3 by x. Place the result in the quotient.

$$
\begin{array}{r}
x^2 \\
x - 2 \overline{\smash{)}\, x^3 - x^2 - 5x + 6}
\end{array}
$$

STEP 2.   Multiply $x - 2$ by x^2 and subtract.

$$
\begin{array}{r}
x^2 \\
x - 2 \overline{\smash{)}\, x^3 - x^2 - 5x + 6} \\
\underline{x^3 - 2x^2 } \\
x^2
\end{array}
$$

STEP 3.   Bring down the next term, $-5x$, and divide again: $x^2 \div x = x$.

$$
\begin{array}{r}
x^2 + x \\
x - 2 \overline{\smash{)}\, x^3 - x^2 - 5x + 6} \\
\underline{x^3 - 2x^2 } \\
x^2 - 5x
\end{array}
$$

STEP 4. Repeat this process until you have used all terms. The completed division follows:

$$\begin{array}{r}
x^2 + x - 3 \\
x - 2\overline{\smash{)}\ x^3 - x^2 - 5x + 6} \\
\underline{x^3 - 2x^2} \\
x^2 - 5x \\
\underline{x^2 - 2x} \\
-3x + 6 \\
-3x + 6
\end{array}$$

There is no remainder. To check, multiply $x^2 + x - 3$ by $x - 2$. The result should be $x^3 - x^2 - 5x + 6$.

When dividing two polynomials, be sure that both the dividend and the divisor are in descending (or ascending) order of powers of the variable. Also, missing powers of the variable in the dividend should be denoted by using a form of zero as in the following example.

Example 1   Divide: $(5x + 3x^3 - 8) \div (x + 3)$.

Solution   First write the dividend in the form $3x^3 + 0x^2 + 5x - 8$; then divide.

$$\begin{array}{r}
3x^2 - 9x + 32 \\
x + 3\overline{\smash{)}\ 3x^3 + 0x^2 + 5x - 8} \\
\underline{3x^3 + 9x^2} \\
-9x^2 + 5x \\
\underline{-9x^2 - 27x} \\
32x - 8 \\
\underline{32x + 96} \\
-104
\end{array}$$

The quotient is $3x^3 - 9x + 32$ and the remainder is -104. To check, recall that the dividend should be equal to the product of the quotient and the divisor, plus the remainder. Verify that the following is correct.

$$(x + 3)(3x^2 - 9x + 32) + (-104) = 3x^3 + 5x - 8$$

We now develop a special procedure for handling long-division problems with polynomials where the divisor is of the form $x - c$. To discover this procedure let us first examine the following long-division problem.

$$\begin{array}{r}
4x^2 + 9x \ - \ \ 3 \\
x - 2\overline{\big)\ 4x^3 + \ \ x^2 - 21x + 13} \\
\underline{4x^3 - 8x^2} \\
9x^2 - 21x \\
\underline{9x^2 - 18x} \\
- \ \ 3x + 13 \\
\underline{- \ \ 3x + \ \ 6} \\
7
\end{array}$$

It should become clear that all the work involves only the coefficients. Thus we could just as easily complete this division by omitting the powers of the variable x^3, x^2, x, as long as we are careful and write the coefficients in the proper places. The division problem would then look like this:

$$\begin{array}{r}
4 + 9 - \ \ 3 \\
1 - 2\overline{\big)\ 4 + 1 - 21 + 13} \\
\underline{④ - 8} \\
9 - 21 \\
\underline{⑨ - 18} \\
- \ \ 3 + 13 \\
\underline{-\ ③ + \ \ 6} \\
7
\end{array}$$

Since the circled numerals are repetitions of those immediately above them, this process can be further shortened by deleting them. Moreover, since these circled numbers are the products of the numbers in the quotient by the 1 in the divisor, we may also eliminate this 1. Then the scheme takes this form:

$$\begin{array}{r}
4 + 9 - \ \ 3 \\
-2\overline{\big)\ 4 + 1 - 21 + 13} \\
\underline{- \ 8} \\
+ 9 - 21 \\
\underline{- 18} \\
- \ \ 3 + 13 \\
\underline{+ \ \ 6} \\
+ \ \ 7
\end{array}$$

Next observe that it is really unnecessary to bring down the -21 and $+13$ because the subtraction can be done just as well without writing these numbers a second time.

$$
\begin{array}{r}
4+9-3 \qquad\qquad \\
-2\,\overline{\big|\,4+1-21+13} \\
-\,8 \\
\hline
+\,9 \\
-\,18 \\
\hline
-\,3 \\
+\,6 \\
\hline
+\,7
\end{array}
$$

A condensed form is obtained by moving the lower numerals upward as follows.

$$
\begin{array}{r}
4+9-3 \qquad\quad \\
-2\,\overline{\big|\,4+1-21+13} \\
-\,8-18+6 \\
\hline
+\,9-3+7
\end{array}
$$

If the lead numeral 4 is brought down from the quotient, then in the last line we have the coefficients of the quotient and the remainder. The top line can therefore be eliminated.

$$
\begin{array}{r}
-2\,\big|\,4+1-21+13 \\
-\,8-18+6 \\
\hline
4+9-3\,\big|+7 \quad \text{remainder} \\
\end{array}
$$

Quotient: $4x^2 + 9x - 3$

To arrive at the final form we change the sign of the divisor, making it $+2$ instead of -2. This change allows us to add rather than subtract throughout, as follows.

$$
\begin{array}{r}
+2\,\big|\,4+1-21+13 \\
+\,8+18-6 \\
\hline
4+9-3\,\big|+7\,=\,\text{remainder} \\
\end{array}
$$

Quotient: $4x^2 + 9x - 3$

As you compare this shortcut to the original long-division process you will see that each essential step there has its counterpart here. That is why the shortcut works. It is called **synthetic division**.

Let us summarize the procedure for synthetic division by using this process with another example.

Example 2   Use synthetic division to find $(2x^3 - 9x^2 + 10x - 7) \div (x - 3)$.

Solution   Write the coefficients of the dividend in descending order. Change the sign of the divisor (change -3 to $+3$).

$$+3\,\underline{|\,2 - 9 + 10 - 7}$$

Now bring down the first term, 2, and multiply by $+3$.

$$
\begin{array}{r}
+3\,\underline{|\,2 - 9 + 10 - 7} \\
\underline{+6 - 9} \\
2
\end{array}
$$

Add -9 and $+6$ to obtain the sum -3. Multiply this sum by $+3$ and repeat the process to the end.

$$
\begin{array}{r}
+3\,\underline{|\,2 - 9 + 10 - 7} \\
\underline{+6 - 9 + 3} \\
2 - 3 + 1\,\underline{|-4}
\end{array}
$$

Since the original dividend is a polynomial of degree 3 and the divisor is a first-degree polynomial, the quotient will have degree 2. Thus we read the last line as implying a quotient of $2x^2 - 3x + 1$ and a remainder of -4. Check this result by using the long-division process.

The synthetic division process has been developed for divisors of the form $x - c$. (Thus, in Example 2, $c = 3$.) A minor adjustment also permits divisors by polynomials of the form $x + c$. For example, a divisor of $x + 2$ may be written as $x - (-2)$; $c = -2$.

Example 3   Use synthetic division to find the quotient and the remainder.

$$(-\tfrac{1}{3}x^4 + \tfrac{1}{6}x^2 - 7x - 4) \div (x + 3)$$

Solution   Write $x + 3$ as $x - (-3)$. Since there is no x^3 term in the dividend, use $0x^3$.

$$
\begin{array}{r}
-3\,\underline{|\,-\tfrac{1}{3} + 0 + \tfrac{1}{6} - 7 - 4} \\
\underline{+1 - 3 + \tfrac{17}{2} - \tfrac{9}{2}} \\
-\tfrac{1}{3} + 1 - \tfrac{17}{6} + \tfrac{3}{2}\,\underline{|-\tfrac{17}{2}} = \text{remainder}
\end{array}
$$

Quotient: $-\tfrac{1}{3}x^3 + x^2 - \tfrac{17}{6}x + \tfrac{3}{2}$ Remainder: $-\tfrac{17}{2}$

Check this result.

Note that the quotient in a synthetic division problem is always a polynomial of degree one less than that of the dividend. This is so because the divisor has degree 1. The bottom line in the synthetic division process, except for the last entry on the right, gives the coefficients of the quotient: a polynomial in standard form.

EXERCISES 5.3

Use synthetic division to find each quotient. For Exercises 1 through 6 check the results by long division.

1. $(x^3 - 2x^2 - 5x + 6) \div (x - 3)$ 2. $(x^3 - x^2 - 5x + 2) \div (x + 2)$
- 3. $(2x^3 + x^2 - 3x + 7) \div (x + 1)$ 4. $(3x^3 - 2x^2 + x - 1) \div (x - 1)$
- 5. $(x^3 + 5x^2 - 7x + 8) \div (x - 2)$ 6. $(x^3 - 3x^2 + x - 5) \div (x + 3)$
- 7. $(x^4 - 3x^3 + 7x^2 - 2x + 1) \div (x + 2)$
8. $(x^4 + x^3 - 2x^2 + 3x - 1) \div (x - 2)$
- 9. $(2x^4 - 3x^2 + 4x - 2) \div (x - 1)$ 10. $(3x^4 + x^3 - 2x + 3) \div (x + 1)$
11. $(x^3 - 27) \div (x - 3)$ 12. $(x^3 - 27) \div (x + 3)$
- 13. $(x^3 + 27) \div (x + 3)$ 14. $(x^3 + 27) \div (x - 3)$
15. $(x^4 - 16) \div (x - 2)$ 16. $(x^4 - 16) \div (x + 2)$
17. $(x^4 + 16) \div (x + 2)$
18. $(4x^5 - x^3 + 5x^2 + \frac{3}{2}x - \frac{1}{2}) \div (x + \frac{1}{2})$
19. $(x^4 - \frac{1}{2}x^3 + \frac{1}{3}x^2 - \frac{1}{4}x + \frac{1}{5}) \div (x - 1)$

Use long division to find each quotient.

*20. $(2x^3 + 9x^2 - 3x - 1) \div (2x - 1)$
*21. $(x^3 - x^2 - x + 10) \div (x^2 - 3x + 5)$
*22. $(3x^5 + 4x^3 - 12x^2) \div (x^2 - x)$

5.4
The Remainder and Factor Theorems

When the polynomial $p(x) = 2x^3 - 9x^2 + 10x - 7$ is divided by $x - 3$, the quotient is the polynomial $q(x) = 2x^2 - 3x + 1$ and the remainder $r = -4$. (See Example 2, page 219.) As a check we see that

$$\underbrace{2x^3 - 9x^2 + 10x - 7}_{p(x)} = \underbrace{(2x^2 - 3x + 1)}_{q(x)} \cdot \underbrace{(x - 3)}_{(x-3)} + \underbrace{(-4)}_{r}$$

In general, whenever a polynomial $p(x)$ is divided by $x - c$ we have

$$p(x) = q(x)(x - c) + r$$

where $q(x)$ is the quotient and r is the (constant) remainder. Since this equation holds for all x, we may let $x = c$ and obtain

$$p(c) = q(c)(c - c) + r$$
$$= q(c) \cdot 0 + r$$
$$= r$$

This result may be summarized as follows:

Remainder theorem

If a polynomial $p(x)$ is divided by $x - c$, the remainder is equal to $p(c)$.

Example 1   Find the remainder when $p(x) = 3x^3 - 5x^2 + 7x + 5$ is divided by $x - 2$.

Solution   By the remainder theorem, the answer is $p(2)$.

$$p(x) = 3x^3 - 5x^2 + 7x + 5$$
$$p(2) = 3(2^3) - 5(2^2) + 7(2) + 5$$
$$= 23$$

Example 2   Let $f(x) = x^3 - 2x^2 + 3x - 1$. Use synthetic division and the remainder theorem to find $f(3)$.

Solution

$$3 \,\underline{\big|\, 1 - 2 + 3 - 1}$$
$$ + 3 + 3 + 18$$
$$\overline{1 + 1 + 6 \big| + 17} = \text{remainder} = f(3)$$

TEST YOUR UNDERSTANDING

Use synthetic division to find the remainder r when $p(x)$ is divided by $x - c$. Verify that $r = p(c)$ by substituting $x = c$ into $p(x)$.

1. $p(x) = x^5 - 7x^4 + 4x^3 + 10x^2 - x - 5$; $x - 1$
2. $p(x) = x^4 + 11x^3 + 11x^2 + 11x + 10$; $x + 10$
3. $p(x) = x^4 + 11x^3 + 11x^2 + 11x + 10$; $x - 10$
4. $p(x) = 6x^3 - 40x^2 + 25$; $x - 6$

Once again, we are going to consider the division of a polynomial $p(x)$ by a divisor of the form $x - c$. First note that

$$p(x) = q(x)(x - c) + r$$

where $q(x)$ is the quotient and r is the (constant) remainder. Now suppose that $r = 0$. Then the remainder theorem gives $p(c) = r = 0$, and the preceding equation becomes

$$p(x) = q(x)(x - c)$$

It follows that $x - c$ is a *factor* of $p(x)$. Conversely, suppose that $x - c$ is a factor of $p(x)$. This means there is another polynomial, say $q(x)$, so that

$$p(x) = q(x)(x - c)$$

or

$$p(x) = q(x)(x - c) + 0$$

which tells us that when $p(x)$ is divided by $x - c$ the remainder is zero. These observations comprise the following result:

Factor theorem

A polynomial $p(x)$ has a factor $x - c$ if and only if $p(c) = 0$.

Example 3   Show that $x - 2$ is a factor of $p(x) = x^3 - 3x^2 + 7x - 10$.

Solution   By the factor theorem we can state that $x - 2$ is a factor of $p(x)$ if $p(2) = 0$.

$$p(2) = 2^3 - 3(2^2) + 7(2) - 10$$
$$= 0$$

Example 4   **(a)** Use the factor theorem to show that $x + 3$ is a factor of $p(x) = x^3 - x^2 - 8x + 12$;   **(b)** Factor $p(x)$ completely.

Solution
(a) First write $x + 3 = x - (-3)$, so that $c = -3$. Then use synthetic division.

$$
\begin{array}{r|rrrr}
-3 & 1 & -1 & -8 & +12 \\
   &   & -3 & +12 & -12 \\
\hline
   & 1 & -4 & +4 & +0 \\
\end{array}
$$

Since $p(-3) = 0$, the factor theorem tells us that $x + 3$ is a factor of $p(x)$.

(b) Synthetic division has produced the quotient $x^2 - 4x + 4$. Therefore, since $x + 3$ is a factor of $p(x)$, we get

$$x^3 - x^2 - 8x + 12 = (x^2 - 4x + 4)(x + 3)$$

To get the complete factored form observe that $x^2 - 4x + 4 = (x - 2)^2$. Thus

$$x^3 - x^2 - 8x + 12 = (x - 2)^2(x + 3)$$

EXERCISES 5.4

For Exercises 1 through 6 use synthetic division and the remainder theorem.

1. $f(x) = x^3 - x^2 + 3x - 2$; find $f(2)$.

2. $f(x) = 2x^3 + 3x^2 - x - 5$; find $f(-1)$.

•**3.** $f(x) = x^4 - 3x^2 + x + 2$; find $f(3)$.

4. $f(x) = x^4 + 2x^3 - 3x - 1$; find $f(-2)$.

5. $f(x) = x^5 - x^3 + 2x^2 + x - 3$; find $f(1)$.

6. $f(x) = 3x^4 + 2x^3 - 3x^2 - x + 7$; find $f(-3)$.

Find the remainder for each division by substitution, using the remainder theorem. That is, in Exercise 7 (for example) let $f(x) = x^3 - 2x^2 + 3x - 5$ and find $f(2) = r$.

7. $(x^3 - 2x^2 + 3x - 5) \div (x - 2)$

8. $(x^3 - 2x^2 + 3x - 5) \div (x + 2)$

•**9.** $(2x^3 + 3x^2 - 5x + 1) \div (x - 3)$

10. $(3x^4 - x^3 + 2x^2 - x + 1) \div (x + 3)$

11. $(4x^5 - x^3 - 3x^2 + 2) \div (x + 1)$

12. $(3x^5 - 2x^4 + x^3 - 7x + 1) \div (x - 1)$

In Exercises 13 through 23 show that the given binomial $x - c$ is a factor of $p(x)$, and then factor $p(x)$ completely.

13. $p(x) = x^3 + 6x^2 + 11x + 6$; $x + 1$

14. $p(x) = x^3 - 6x^2 + 11x - 6$; $x - 1$

•**15.** $p(x) = x^3 + 5x^2 - 2x - 24$; $x - 2$

16. $p(x) = -x^3 + 11x^2 - 23x - 35$; $x - 7$

•**17.** $p(x) = -x^3 + 7x + 6$; $x + 2$

18. $p(x) = x^3 + 2x^2 - 13x + 10$; $x + 5$

19. $p(x) = 6x^3 - 25x^2 - 29x + 20$; $x - 5$

20. $p(x) = 12x^3 - 22x^2 - 100x - 16$; $x + 2$

•**21.** $p(x) = x^4 + 4x^3 + 3x^2 - 4x - 4$; $x + 2$

22. $p(x) = x^4 - 8x^3 + 7x^2 + 72x - 144; x - 4$

23. $p(x) = x^6 + 6x^5 + 8x^4 - 6x^3 - 9x^2; x + 3$

*24. When $x^2 + 5x - 2$ is divided by $x + n$, the remainder is -8. Find all possible values of n and check by division.

• ***25.** Find d so that $x + 6$ is a factor of $x^4 + 4x^3 - 21x^2 + dx + 108$.

*26. Find b so that $x - 2$ is a factor of $x^3 + bx^2 - 13x + 10$.

*27. Find a so that $x - 10$ is a factor of $ax^3 - 25x^2 + 47x + 30$.

5.5

The Rational Root Theorem

Consider the polynomial equation

$$(3x + 2)(5x - 4)(2x - 3) = 0$$

To find the roots, set each factor equal to zero.

$$3x + 2 = 0 \qquad 5x - 4 = 0 \qquad 2x - 3 = 0$$

$$x = -\tfrac{2}{3} \qquad\quad x = \tfrac{4}{5} \qquad\quad x = \tfrac{3}{2}$$

Now multiply the original three factors and keep careful note of the details of this multiplication. Your result should be

$$30x^3 - 49x^2 - 10x + 24 = 0$$

which must have the same three rational roots.

As you analyze this multiplication it becomes clear that the constant 24 is the product of the three constants 2, -4, and -3. Also, the leading coefficient, 30, is the product of the three original coefficients of x, namely 3, 5, and 2. Furthermore, 3, 5, and 2 are the denominators of the roots $-\tfrac{2}{3}$, $\tfrac{4}{5}$, and $\tfrac{3}{2}$. Therefore the denominators of the rational roots are all factors of 30, and their numerators are all factors of 24.

These results are not accidental. It turns out that we have been discussing the following general result:

Rational root theorem

Let $f(x) = a_n x^n + a_{n-1} x^{n-1} + \cdots + a_1 x + a_0$ be an nth-degree polynomial with integer coefficients. If $\dfrac{p}{q}$ is a rational root of $f(x) = 0$, where $\dfrac{p}{q}$ is in lowest terms, then p is a factor of a_0 and q is a factor of a_n.

Let us see how this theorem can be applied to find the rational roots of

$$f(x) = 4x^3 - 16x^2 + 11x + 10 = 0$$

Begin by listing all factors of the constant 10 and of the leading coefficient 4.

Factors of 10: $\pm 1, \pm 2, \pm 5, \pm 10$ (possible numerators)

Factors of 4: $\pm 1, \pm 2, \pm 4$ (possible denominators)

Possible rational roots (take each number in the first row and divide by each number in the second row): $\pm 1, \pm \frac{1}{2}, \pm \frac{1}{4}, \pm 2, \pm 5, \pm \frac{5}{2}, \pm \frac{5}{4}, \pm 10$

To decide which (if any) of these are roots of $f(x) = 0$, we could substitute the values directly into $f(x)$. However, it is easier to use synthetic division because in most cases it leads to easier computations and also makes quotients available. Therefore we proceed by using synthetic division with divisors c, where c is possible rational root. If $f(c) = 0$, then c is a root; if $f(c) \neq 0$, c is not a root.

$$
\begin{array}{r|rrrr}
1 & 4 & -16 & +11 & +10 \\
  &   & +4 & -12 & -1 \\
\hline
  & 4 & -12 & -1 & +9
\end{array}
\qquad
\begin{array}{r|rrrr}
-1 & 4 & -16 & +11 & +10 \\
   &   & -4 & +20 & -31 \\
\hline
   & 4 & -20 & +31 & -21
\end{array}
$$

Since $f(1) = 9 \neq 0$,   Since $f(-1) = -21 \neq 0$,
1 is *not* a root. -1 is *not* a root.

$$
\begin{array}{r|rrrr}
\tfrac{1}{2} & 4 & -16 & +11 & +10 \\
  &   & +2 & -7 & +2 \\
\hline
  & 4 & -14 & +4 & +12
\end{array}
\qquad
\begin{array}{r|rrrr}
-\tfrac{1}{2} & 4 & -16 & +11 & +10 \\
   &   & -2 & +9 & -10 \\
\hline
   & 4 & -18 & +20 & +0
\end{array}
$$

Since $f(\tfrac{1}{2}) = 12 \neq 0$,   Since $f(-\tfrac{1}{2}) = 0$,
$\tfrac{1}{2}$ is *not* a root. $-\tfrac{1}{2}$ *is* a root.

By the factor theorem it follows that $x - (-\tfrac{1}{2}) = x + \tfrac{1}{2}$ is a factor of $f(x)$, and synthetic division gives the other factor, $4x^2 - 18x + 20$.

$$f(x) = (x + \tfrac{1}{2})(4x^2 - 18x + 20)$$

To find other roots of $f(x) = 0$ we could proceed by using the rational root theorem for $4x^2 - 18x + 20 = 0$. But this is unnecessary because the quadratic expression is factorable.

$$
\begin{aligned}
f(x) &= (x + \tfrac{1}{2})(4x^2 - 18x + 20) \\
     &= (x + \tfrac{1}{2})(2)(2x^2 - 9x + 10) \\
     &= 2(x + \tfrac{1}{2})(x - 2)(2x - 5)
\end{aligned}
$$

The solution of $f(x) = 0$ can now be found by setting each factor equal to zero. The solutions are $x = -\tfrac{1}{2}$, $x = 2$, and $x = \tfrac{5}{2}$.

Example 1   Factor $f(x) = x^3 + 4x^2 + x - 6$.

Solution   Since the leading coefficient is 1, whose only factors are ± 1, the possible denominators of a rational root of $f(x) = 0$ can only be ± 1. Hence the possible rational roots must all be factors of -6, namely ± 1, ± 2, ± 3, and ± 6. Use synthetic division to test these cases.

$$\begin{array}{r} -1 \,\underline{\left|\, 1 + 4 + 1 - 6 \right.} \\ \underline{- 1 - 3 + 2} \\ 1 + 3 - 2 \,\underline{\left|- 4\right.} = r \end{array}$$

Since $r = f(-1) \neq 0$, $x - (-1) = x + 1$ is *not* a factor of $f(x)$.

$$\begin{array}{r} 1 \,\underline{\left|\, 1 + 4 + 1 - 6 \right.} \\ \underline{+ 1 + 5 + 6} \\ 1 + 5 + 6 \,\underline{\left|+ 0\right.} = r \end{array}$$

Since $r = f(1) = 0$, $x - 1$ *is* a factor of $f(x)$.

$$x^3 + 4x^2 + x - 6 = (x - 1)(x^2 + 5x + 6)$$

By factoring we have:

$$x^3 + 4x^2 + x - 6 = (x - 1)(x + 2)(x + 3)$$

TEST YOUR UNDERSTANDING

For each $p(x)$ find (a) the possible rational roots of $p(x) = 0$, (b) the factored form of $p(x)$, and (c) the roots of $p(x) = 0$.

1. $p(x) = x^3 - 3x^2 - 10x + 24$ **2.** $p(x) = x^4 + 6x^3 + x^2 - 24x + 16$
3. $p(x) = 4x^3 + 20x^2 - 23x + 6$ **4.** $p(x) = 3x^4 - 13x^3 + 7x^2 - 13x + 4$

Example 2   Solve for x: $p(x) = 2x^5 + 7x^4 - 18x^2 - 8x + 8 = 0$.

Solution   The possible rational roots are ± 1, $\pm \frac{1}{2}$, ± 2, ± 4, and ± 8. Testing these possibilities (left to right), the first root we find is $\frac{1}{2}$.

$$\begin{array}{r} \tfrac{1}{2} \,\underline{\left|\, 2 + 7 + 0 - 10 - 8 + 8 \right.} \\ \underline{\phantom{\tfrac{1}{2}\,|\,2}+ 1 + 4 + 2 - 8 - 8} \\ 2 + 8 + 4 - 16 - 16 \,\underline{\left|+ 0\right.} \end{array}$$

Therefore $x - \frac{1}{2}$ is a factor of $p(x)$.

$$p(x) = (x - \tfrac{1}{2})(2x^4 + 8x^3 + 4x^2 - 16x - 16)$$
$$= 2(x - \tfrac{1}{2})(x^4 + 4x^3 + 2x^2 - 8x - 8)$$

To find other roots of $p(x) = 0$ it now becomes necessary to solve

$$x^4 + 4x^3 + 2x^2 - 8x - 8 = 0$$

The possible rational roots for this equation are ± 1, ± 2, ± 4, and ± 8. However, values like ± 1 that were tried before need not be tried again. Why? We find that $x = -2$ is a root:

$$
\begin{array}{r}
-2\,|\,1 + 4 + 2 - 8 - 8 \\
\underline{-2 - 4 + 4 + 8} \\
1 + 2 - 2 - 4\,|\,{+}\,0
\end{array}
$$

$$
\begin{aligned}
x^4 + 4x^3 + 2x^2 - 8x - 8 &= (x + 2)(x^3 + 2x^2 - 2x - 4) \\
&= (x + 2)[x^2(x + 2) - 2(x + 2)] \\
&= (x + 2)(x^2 - 2)(x + 2) \\
&= (x + 2)^2(x^2 - 2)
\end{aligned}
$$

This gives

$$
\begin{aligned}
p(x) &= 2(x - \tfrac{1}{2})(x^4 + 4x^3 + 2x^2 - 8x - 8) \\
&= 2(x - \tfrac{1}{2})(x + 2)^2(x^2 - 2) \\
&= 2(x - \tfrac{1}{2})(x + 2)^2(x + \sqrt{2})(x - \sqrt{2})
\end{aligned}
$$

Setting each factor equal to zero produces the solutions of $p(x) = 0$:

$$x = \tfrac{1}{2}, \quad x = -2, \quad x = -\sqrt{2}, \quad x = \sqrt{2}$$

The final example of this section shows how the results we have recently learned can be used to solve a nonlinear system.

Example 3 Solve and graph the system:

$$y = x^3$$
$$y = 7x + 6$$

Solution Set the two values for y equal to each other to produce a polynomial equation $p(x) = 0$.

$$x^3 = 7x + 6$$
$$x^3 - 7x - 6 = 0$$

The possible rational roots are ± 1, ± 2, ± 3, and ± 6.

$$-1\,\underline{|1 + 0 - 7 - 6}$$
$$\underline{-1 + 1 + 6}$$
$$1 - 1 - 6\,\underline{|+ 0}$$

Thus

$$x^3 - 7x - 6 = (x + 1)(x^2 - x - 6)$$
$$= (x + 1)(x + 2)(x - 3)$$

Set each factor equal to zero to obtain the values for x: $-2, -1, 3$. Using $y = x^3$ we find the corresponding y-values. (The other equation can be used for checking.)

For $x = -2$, $y = (-2)^3 = -8$.
For $x = -1$, $y = (-1)^3 = -1$.
For $x = 3$, $y = 3^3 = 27$.

The solutions are three ordered pairs of numbers: $(-2, -8)$, $(-1, -1)$, and $(3, 27)$.

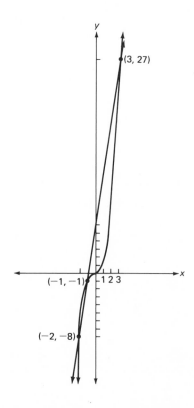

EXERCISES 5.5

Solve for x.

• **1.** $x^3 + 2x^2 - 29x + 42 = 0$

  2. $x^3 + x^2 - 21x - 45 = 0$

• **3.** $2x^3 - 15x^2 + 24x + 16 = 0$

  4. $3x^3 + 2x^2 - 75x - 50 = 0$

• **5.** $x^3 + 3x^2 + 3x + 1 = 0$

  6. $x^4 + 3x^3 + 3x^2 + x = 0$

  7. $x^4 + 6x^3 + 7x^2 - 12x - 18 = 0$

  8. $x^4 + 6x^3 + 2x^2 - 18x - 15 = 0$

• **9.** $x^4 - x^3 - 5x^2 - x - 6 = 0$

  10. $x^4 + 2x^3 - 7x^2 - 18x - 18 = 0$

  11. $x^4 - 5x^3 + 3x^2 + 15x - 18 = 0$

  12. $-x^5 + 5x^4 - 3x^3 - 15x^2 + 18x = 0$

  13. $x^4 + 4x^3 - 7x^2 - 36x - 18 = 0$

  14. $2x^3 - 5x - 3 = 0$

  15. $6x^3 - 25x^2 + 21x + 10 = 0$

  16. $3x^4 - 11x^3 - 3x^2 - 6x + 8 = 0$

Factor.

• **17.** $x^3 - 3x^2 - 10x + 24$

  18. $-x^3 - 3x^2 + 24x + 80$

  19. $x^3 - 28x - 48$

  20. $6x^4 + 9x^3 + 9x - 6$

Solve the system.

• **21.** $y = x^3 - 3x^2 + 3x - 1$
     $y = 7x - 13$

  22. $y = -x^3$
     $y = -3x^2 + 4$

  ***23.** $y = 4x^3 - 7x^2 + 10$
     $y = x^3 + 43x - 5$

  ***24.** $y = x^2 + 4x$
     $(x - 1)^2 + (y - 6)^2 = 37$
     (*Hint:* Substitute $y = x^2 + 4x$ into the second equation.)

• ***25.** Show that $2x^3 - 5x^2 - x + 8 = 0$ has no rational roots.

5.6

Polynomial and Rational Functions

In the last section, as well as in Chapter 2, we learned some procedures for factoring polynomials. We will now use such factored forms to help determine the signs of polynomial and rational functions.

As a first illustration consider the polynomial $f(x) = x^3 + 4x^2 + x - 6$. From Example 1 of Section 5.5 the factored form is

$$f(x) = (x + 3)(x + 2)(x - 1)$$

The individual factors of $f(x)$ are 0 for the values -3, -2, and 1. The x-values around these three numbers can be displayed on a number line:

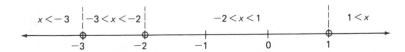

These four cases, left to right, become the four rows, top to bottom, in the table. Specific values for x may be used in each of these four cases to obtain the signs in this table. A detailed analysis follows.

	$x + 3$	$x + 2$	$x - 1$	$f(x) = (x + 3)(x + 2)(x - 1)$
Row 1: $x < -3$	−	−	−	−
Row 2: $-3 < x < -2$	+	−	−	+
Row 3: $-2 < x < 1$	+	+	−	−
Row 4: $1 < x$	+	+	+	+

In Row 1 we have $x < -3$, or $x + 3 < 0$. Since $x + 3$ is negative, there is a minus sign under $x + 3$. Likewise, adding 2 to $x < -3$ gives $x + 2 < -1$. Then $x + 2$ is negative, and there is a minus sign under $x + 2$. Similarly, there is a minus sign under $x - 1$ because adding -1 to $x < -3$ produces the inequality $x - 1 < -4$. Since the product of three negative quantities is negative, there is a minus sign under $f(x)$ in Row 1.

Next consider Row 2 and some value for x such that $-3 < x < -2$. It often helps to think of a specific value, such as $x = -2.5$. For such a value of x, $x + 3$ is positive and both $x + 2$ and $x - 1$ are negative. Since the product of one positive factor and two negative factors is positive, there is a plus sign under $f(x)$ in Row 2. The signs in Rows 3 and 4 are obtained by similar reasoning.

The table shows that $f(x) < 0$ for $x < -3$ (Row 1); also, $f(x) < 0$ for $-2 < x < 1$ (Row 3). We note that $f(x) > 0$ for $-3 < x < -2$ (Row 2) and for $1 < x$ (Row 4). To show an alternate way of stating these results we will now introduce notation for four types of *bounded intervals*.

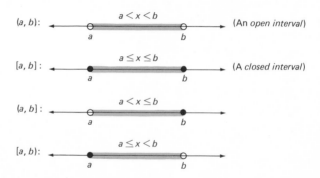

There are also *unbounded intervals*. For example, the set of all $x > 5$ is denoted by $(5, \infty)$. Likewise, $(-\infty, 5]$ represents all $x \le 5$. The symbols ∞ and $-\infty$ are read "plus infinity" and "minus infinity" but do *not* represent numbers. They are symbolic devices used to indicate that *all* x in a given direction, without end, are included, as in the following figure.

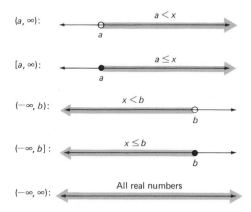

The results of the preceding example can now be stated in this way:

$$f(x) > 0 \text{ on both } (-3, -2) \text{ and } (1, \infty)$$
$$f(x) < 0 \text{ on both } (-\infty, -3) \text{ and } (-2, 1)$$

Example 1 Solve the inequality $p(x) < 0$, where $p(x)$ is defined as: $x^4 + x^3 - 5x^2 + x - 6$.

Solution To factor $p(x)$, begin by using the rational root theorem. The possible rational roots are ± 1, ± 2, ± 3, and ± 6.

$$
\begin{array}{r|rrrrr}
1 & 1 + 1 - 5 + 1 - 6 \\
  & \ \ \ + 1 + 2 - 3 - 2 \\
\hline
  & 1 + 2 - 3 - 2 \ \underline{\underline{|-8}} \ne 0
\end{array}
\qquad
\begin{array}{r|rrrrr}
-1 & 1 + 1 - 5 + 1 - \ \ 6 \\
   & \ \ \ - 1 + 0 + 5 - \ \ 6 \\
\hline
   & 1 + 0 - 5 + 6 \ \underline{\underline{|-12}} \ne 0
\end{array}
$$

$$
\begin{array}{r|rrrrr}
2 & 1 + 1 - 5 + 1 - 6 \\
  & \ \ \ + 2 + 6 + 2 + 6 \\
\hline
  & 1 + 3 + 1 + 3 \ \underline{\underline{|+0}}
\end{array}
\qquad \text{Therefore } x - 2 \text{ is a factor of } p(x).
$$

$$
\begin{aligned}
p(x) &= (x - 2)(x^3 + 3x^2 + x + 3) \\
&= (x - 2)[x^2(x + 3) + (x + 3)] \\
&= (x - 2)(x + 3)(x^2 + 1)
\end{aligned}
$$

Form a table of signs, noting that -3 and 2 are the values for which the factors of $p(x)$ are zero. Note also that $x^2 + 1 > 0$ for all x.

	$x + 3$	$x - 2$	$x^2 + 1$	$p(x)$
$x < -3$	$-$	$-$	$+$	$+$
$-3 < x < 2$	$+$	$-$	$+$	$-$
$2 < x$	$+$	$+$	$+$	$+$

Since $p(x)$ is negative only for $-3 < x < 2$ we find that $p(x) < 0$ on $(-3, 2)$.

TEST YOUR UNDERSTANDING

Use the table method to determine the signs of the polynomial.

1. $p(x) = (x + 6)(x + 5)(x - 6)$
2. $p(x) = 4(x + 2)(x - 1)(x - 2)(x - 3)$
3. $p(x) = -4(x + 2)(x - 1)(x - 2)(x - 3)$
4. $p(x) = (x^2 - 9)(x^2 - 2x + 1)$

Solve the inequality.

5. $x^3 - 3x^2 - 33x + 35 > 0$ **6.** $2x^4 + 5x^3 - 11x^2 - 20x + 12 < 0$

The table method used for polynomials can just as well be used for determining the signs of rational functions. For example, if $f(x) = \dfrac{x^3 - x^2 - 8x + 12}{x - 4}$, then

$$f(x) = \frac{(x + 3)(x - 2)^2}{(x - 4)} \qquad \text{(See Example 4, Section 5.4.)}$$

From this form we are able to solve the inequalities $f(x) > 0$ and $f(x) < 0$.

The signs of $f(x)$ are determined by $x + 3$, $(x - 2)^2$, and $x - 4$. Since -3, 2, and 4 are the values for which these factors are zero, we construct a table as before. Note that the entries under $(x - 2)^2$ are automatically positive since $(x - 2)^2 > 0$ for all $x \neq 2$.

	$x + 3$	$(x - 2)^2$	$x - 4$	$f(x)$
$x < -3$	$-$	$+$	$-$	$+$
$-3 < x < 2$	$+$	$+$	$-$	$-$
$2 < x < 4$	$+$	$+$	$-$	$-$
$4 < x$	$+$	$+$	$+$	$+$

Thus $f(x) > 0$ on both $(-\infty, -3)$ and $(4, \infty)$; and $f(x) < 0$ on both $(-3, 2)$ and $(2, 4)$.

To solve an inequality such as $\dfrac{2x - 5}{x + 1} - \dfrac{3}{x(x + 1)} < 0$, it is helpful to first combine and factor as follows.

$$\frac{2x - 5}{x + 1} - \frac{3}{x(x + 1)} = \frac{2x^2 - 5x - 3}{x(x + 1)}$$
$$= \frac{(2x + 1)(x - 3)}{x(x + 1)}$$

The preceding table method can now be used. For this purpose it is helpful to use $2(x + \frac{1}{2})$ in place of $2x + 1$ since the work is easier when the coefficient for each x is 1. Complete the solution.

We also use the combined form shown to solve an equation such as the following:

$$\frac{(2x + 1)(x - 3)}{x(x + 1)} = 0$$

Recall that a fraction can only be zero when its numerator is equal to zero. Thus we may complete the solution in this way.

$$(2x + 1)(x - 3) = 0$$
$$2x + 1 = 0 \quad \text{or} \quad x - 3 = 0$$
$$x = -\tfrac{1}{2} \quad \text{or} \quad x = 3$$

If we are only interested in solving such an equality, and *not* the inequality, then it is not necessary to combine first, as is shown in Example 2.

Example 2   Solve for x: $\dfrac{2x-5}{x+1} - \dfrac{3}{x^2+x} = 0$.

Solution   Factor the denominator in the second fraction.

$$\frac{2x-5}{x+1} - \frac{3}{x(x+1)} = 0$$

Multiply by the LCD $= x(x+1)$.

$$x(2x-5) - 3 = 0$$
$$2x^2 - 5x - 3 = 0$$
$$(2x+1)(x-3) = 0$$
$$2x + 1 = 0 \quad \text{or} \quad x - 3 = 0$$
$$x = -\tfrac{1}{2} \quad \text{or} \quad x = 3$$

You can check these answers in the original equation.

The final example of this section illustrates how the solution of a rational equation is used in solving a nonlinear system.

Example 3   Solve and graph the system

$$y = \frac{2}{x}$$
$$y = -2x + 5$$

Solution   Set both values for y equal to each other.

$$\frac{2}{x} = -2x + 5$$

Multiply by x.

$$2 = -2x^2 + 5x$$

Solve for x.

$$2x^2 - 5x + 2 = 0$$
$$(2x-1)(x-2) = 0$$
$$x = \tfrac{1}{2} \quad \text{or} \quad x = 2$$

For $x = \frac{1}{2}$, $y = -2(\frac{1}{2}) + 5 = 4$. For $x = 2$, $y = -2(2) + 5 = 1$. The points of intersection are $(\frac{1}{2}, 4)$ and $(2, 1)$.

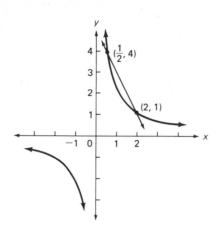

EXERCISES 5.6

Determine the signs of f.

1. $f(x) = (x - 1)(x - 2)(x - 3)$

2. $f(x) = x(x + 2)(x - 2)$

3. $f(x) = \dfrac{(3x - 1)(x + 4)}{x^2(x - 2)}$

4. $f(x) = \dfrac{x(2x + 3)}{(x + 2)(x - 5)}$

Solve each inequality.

5. $(x^2 + 2)(x - 4)(x + 1) > 0$

6. $x(x + 1)(x + 2) < 0$

7. $\dfrac{x - 10}{3(x + 1)(5x - 1)} < 0$

8. $\dfrac{3}{(x + 2)(4x + 3)} > 0$

Determine the signs of f.

9. $f(x) = x^3 + 6x^2 + 8x$

10. $f(x) = x^3 - 8x^2 + x - 8$

11. $f(x) = \dfrac{x}{2x^2 - 5x - 7}$

12. $f(x) = -\dfrac{x^2 - 6x + 9}{x}$

13. $f(x) = x^3 - 6x^2 + 11x - 6$

14. $f(x) = x^3 + 4x^2 - 11x - 30$

15. $f(x) = \dfrac{1}{x + 1} + \dfrac{1}{x - 1}$

16. $f(x) = \dfrac{3}{x + 2} - \dfrac{4}{x - 3}$

17. $f(x) = \dfrac{1}{x^2 + 1} + \dfrac{2}{x - 3}$

18. $f(x) = \dfrac{x - 1}{x^2 + 1} - \dfrac{x - 2}{x^2 + 2}$

Solve for x and check your results.

19. $\dfrac{x}{2} - \dfrac{x}{5} = 6$

20. $\dfrac{2}{x} - \dfrac{5}{x} = 6$

21. $\dfrac{x+3}{5} = 2$

22. $\dfrac{5}{x+3} = 2$

23. $\dfrac{x-1}{2} = \dfrac{x+2}{4}$

24. $\dfrac{2}{x-1} = \dfrac{4}{x+2}$

25. $\dfrac{5}{x} - \dfrac{3}{4x} = 1$

26. $\dfrac{5x}{2} - \dfrac{3x}{4} = 2$

27. $\dfrac{2x+1}{5} - \dfrac{x-2}{3} = 1$

28. $\dfrac{3x-2}{4} - \dfrac{x-1}{3} = \dfrac{1}{2}$

29. $\dfrac{x+1}{x+10} = \dfrac{1}{2x}$

30. $\dfrac{3}{x+1} = \dfrac{9}{x^2 - 3x - 4}$

31. $\dfrac{x(2x+3)(x-5)}{x-4} = 0$

32. $\dfrac{1}{2x-4} + \dfrac{1}{2x+4} = 0$

33. $\dfrac{2}{x^2-4} + \dfrac{1}{x-2} - \dfrac{1}{x+2} = 0$

34. $\dfrac{1}{x^2+4} + \dfrac{1}{x^2-4} = \dfrac{18}{x^4-16}$

*35. Solve $f(x) > 0$, where $f(x) = x^3 - 3x^2 + 3x - 1$, and graph $y = f(x)$.

*36. Solve $f(x) < 0$, where $f(x) = \dfrac{1}{x^3 + 6x^2 + 12x + 8}$, and graph $y = f(x)$.

Solve each system and graph.

*37. $2x + 3y = 7$

$y = \dfrac{1}{x}$

*38. $y = 2x - 5$

$y = -\dfrac{1}{x^2 - 2x + 1}$

5.7

Decomposing Rational Functions (optional)

In Chapter 2 we learned how to combine rational expressions. For example, combining the fractions in

(1) $\quad \dfrac{6}{x-4} + \dfrac{3}{x-2} \quad$ produces $\quad$ (2) $\quad \dfrac{9x-24}{(x-4)(x-2)}$

It is now our goal to start with a rational expression such as (2) and *decompose* it into the form (1). When this is accomplished we say that $\dfrac{9x-24}{(x-4)(x-2)}$ has been decomposed into (simpler) **partial fractions**. We will only consider examples that involve linear factors. Look at this example again.

$$\frac{9x-24}{(x-4)(x-2)} = \frac{6}{x-4} + \frac{3}{x-2}$$

First observe that each factor in the denominator on the left serves as a denominator of a partial fraction on the right. Let us assume, for the moment, that the numerators 6 and 3 are not known. Then it is reasonable to begin by writing

$$\frac{9x - 24}{(x - 4)(x - 2)} = \frac{A}{x - 4} + \frac{B}{x - 2}$$

where A and B are the constants to be found. To find these values we first clear fractions by multiplying both sides by $(x - 4)(x - 2)$.

$$(x - 4)(x - 2) \cdot \frac{9x - 24}{(x - 4)(x - 2)} = (x - 4)(x - 2)\left[\frac{A}{x - 4} + \frac{B}{x - 2}\right]$$

$$9x - 24 = A(x - 2) + B(x - 4)$$

Since we want this equation to hold for all values of x, we may select specific values for x that will produce the constants A and B. Observe that when $x = 4$ the term $B(x - 4)$ will become zero.

$$9(4) - 24 = A(4 - 2) + B(4 - 4)$$

$$12 = 2A + 0$$

$$6 = A$$

Similarly, B can be found by letting $x = 2$.

$$9(2) - 24 = A(2 - 2) + B(2 - 4)$$

$$-6 = 0 - 2B$$

$$3 = B$$

Example 1 Decompose $\dfrac{6x^2 + x - 37}{(x - 3)(x + 2)(x - 1)}$ into partial fractions.

Solution Since there are three linear factors in the denominator we begin with the form:

$$\frac{6x^2 + x - 37}{(x - 3)(x + 2)(x - 1)} = \frac{A}{x - 3} + \frac{B}{x + 2} + \frac{C}{x - 1}$$

Multiply by $(x - 3)(x + 2)(x - 1)$ to clear fractions.

$$6x^2 + x - 37 = A(x + 2)(x - 1) + B(x - 3)(x - 1) + C(x - 3)(x + 2)$$

Since the second and third terms on the right have the factor $x - 3$, the value $x = 3$ will make these two terms zero.

$$6(3)^2 + 3 - 37 = A(3 + 2)(3 - 1) + B(0)(0 - 1) + C(0)(0 - 12)$$
$$54 + 3 - 37 = A(5)(2) + 0 + 0$$
$$20 = 10A$$
$$2 = A$$

To find B use $x = -2$.

$$6(-2)^2 + (-2) - 37 = A(-2 + 2)(-2 - 1) + B(-2 - 3)(-2 - 1) + C(-2 - 3)(-2 + 2)$$
$$24 - 2 - 37 = 0 + B(-5)(-3) + 0$$
$$-15 = 15B$$
$$-1 = B$$

To find C let $x = 1$.

$$6(1)^2 + 1 - 37 = 0 + 0 + C(1 - 3)(1 + 2)$$
$$-30 = -6C$$
$$5 = C$$

Substituting the values for A, B, and C into the original form produces the desired decomposition.

$$\frac{6x^2 + x - 37}{(x - 3)(x + 2)(x - 1)} = \frac{2}{x - 3} - \frac{1}{x + 2} + \frac{5}{x - 1}$$

You can check this result by combining the right side.

TEST YOUR UNDERSTANDING

Decompose into partial fractions.

1. $\dfrac{8x - 19}{(x - 2)(x - 3)}$ **2.** $\dfrac{1}{(x + 2)(x - 4)}$ **3.** $\dfrac{6x^2 - 22x + 18}{(x - 1)(x - 2)(x - 3)}$

Factor the denominator and decompose into partial fractions.

4. $\dfrac{4x + 6}{x^2 + 5x + 6}$ **5.** $\dfrac{23x - 1}{6x^2 + x - 1}$ **6.** $\dfrac{5x^2 - 24x - 173}{x^3 + 4x^2 - 31x - 70}$

Let us look at a somewhat different situation.

$$\frac{7}{x + 3} - \frac{4}{(x + 3)^2} = \frac{7x + 17}{(x + 3)^2}$$

Starting with the right side, we note that the denominator is the square of the linear factor $x + 3$. On the left side both $x + 3$ and $(x + 3)^2$ are denominators of partial fractions. Assuming that the specific numerators are not known, we begin the decomposition process in this way:

$$\frac{7x + 17}{(x + 3)^2} = \frac{A}{x + 3} + \frac{B}{(x + 3)^2}$$

Clear fractions:

(1) $7x + 17 = A(x + 3) + B$

To find B let $x = -3$.

$$7(-3) + 17 = A(0) + B$$
$$-4 = B$$

Substitute this value for B into Equation (1).

(2) $7x + 17 = A(x + 3) - 4$

Now find A by substituting some easy value for x, say $x = 1$, into (2).

$$7(1) + 17 = A(1 + 3) - 4$$
$$24 = 4A - 4$$
$$7 = A$$

Substituting these values for A and B into our original form produces the decomposition.

$$\frac{7x + 17}{(x + 3)^2} = \frac{7}{x + 3} + \frac{-4}{(x + 3)^2} = \frac{7}{x + 3} - \frac{4}{(x + 3)^2}$$

Note: If the original denominator had been $(x + 3)^3$, then we would have used the *additional* fraction $\dfrac{C}{(x + 3)^3}$ to start with.

Example 2   Decompose $\dfrac{6 + 26x - x^2}{(2x - 1)(x + 2)^2}$ into partial fractions.

Solution   Begin with this form:

$$\frac{6 + 26x - x^2}{(2x - 1)(x + 2)^2} = \frac{A}{2x - 1} + \frac{B}{x + 2} + \frac{C}{(x + 2)^2}$$

Clear fractions.

$$(1) \quad 6 + 26x - x^2 = A(x + 2)^2 + B(2x - 1)(x + 2) + C(2x - 1)$$

Find A by substituting $x = \frac{1}{2}$.

$$6 + 13 - \tfrac{1}{4} = A(\tfrac{5}{2})^2 + 0 + 0$$
$$\tfrac{75}{4} = \tfrac{25}{4}A$$
$$3 = A$$

Find C by letting $x = -2$.

$$6 - 52 - 4 = 0 + 0 + C(-5)$$
$$-50 = -5C$$
$$10 = C$$

Substitute $A = 3$ and $C = 10$ into Equation (1).

$$(2) \quad 6 + 26x - x^2 = 3(x + 2)^2 + B(2x - 1)(x + 2) + 10(2x - 1)$$

To find B use a simple value like $x = 1$ in (2).

$$6 + 26 - 1 = 3(9) + B(1)(3) + 10(1)$$
$$-6 = 3B$$
$$-2 = B$$

Then the decomposition is

$$\frac{6 + 26x - x^2}{(2x - 1)(x + 2)^2} = \frac{3}{2x - 1} - \frac{2}{x + 2} + \frac{10}{(x + 2)^2}$$

You can check this by combining the right side.

An alternate method for finding the unknown constants in a decomposition problem makes use of the techniques of solving linear systems studied in Section 3.5. This method begins the same as before. In Example 2, for instance, we again reach this equation after clearing fractions:

$$6 + 26x - x^2 = A(x + 2)^2 + B(2x - 1)(x + 2) + C(2x - 1)$$

Convert each side to a polynomial in standard form (using decreasing exponents). Thus

$$-x^2 + 26x + 6 = Ax^2 + 4Ax + 4A + 2Bx^2 + 3Bx - 2B + 2Cx - C$$
$$= (A + 2B)x^2 + (4A + 3B + 2C)x + (4A - 2B - C)$$

Now call on the fact that *when two polynomials written in standard form are equal, their coefficients are the same.* According to this criterion we may write:

Coefficients of x^2: $-1 = A + 2B$

Coefficients of x: $26 = 4A + 3B + 2C$

Constants: $6 = 4A - 2B - C$

We now have a linear system of three equations in A, B, and C. The solution to this system is $A = 3$, $B = -2$, and $C = 10$.

Which procedure should be used? This method of equating coefficients? Or the earlier procedure in which substitutions were used? Experience is the best teacher here, but as a general working rule always try to find as many of the unknown constants by the substitution method before you equate coefficients.

In every decomposition problem done so far, you will notice that the degree of the polynomial in the numerator is always *less* than the degree in the denominator. Here is an example where this is not the case.

$$\frac{2x^3 + 3x^2 - x + 16}{x^2 + 2x - 3}$$

In such cases the *first* step is to divide.

$$
\begin{array}{r}
2x - 1 \\
x^2 + 2x - 3 \overline{\smash{\big)}\, 2x^3 + 3x^2 - x + 16} \\
\underline{2x^3 + 4x^2 - 6x } \\
- x^2 + 5x \\
\underline{- x^2 - 2x + 3} \\
7x + 13
\end{array}
$$

(Remainder has degree *less* than degree of divisor.)

Now write

$$\frac{2x^3 + 3x^2 - x + 16}{x^2 + 2x - 3} = \text{quotient} + \frac{\text{remainder}}{\text{divisor}}$$

$$= 2x - 1 + \frac{7x + 13}{x^2 + 2x - 3}$$

The problem will be completed by decomposing $\dfrac{7x + 13}{x^2 + 2x - 3}$. For illustrative purposes we will use the alternate procedure described.

$$\frac{7x + 13}{x^2 + 2x - 3} = \frac{7x + 13}{(x - 1)(x + 3)} = \frac{A}{x - 1} + \frac{B}{x + 3}$$

Clear fractions.

$$7x + 13 = A(x + 3) + B(x - 1)$$

$$= (A + B)x + (3A - B)$$

Equate coefficients and solve for A and B.

$$\begin{array}{rcl} A + B &=& 7 \\ 3A - B &=& 13 \qquad \text{(add)} \\ \hline 4A &=& 20 \\ A &=& 5 \end{array}$$

Substitute into $A + B = 7$ to get $B = 7 - 5 = 2$. Therefore the final decomposition is

$$\frac{2x^3 + 3x^2 - x + 16}{x^2 + 2x - 3} = 2x - 1 + \frac{5}{x - 1} + \frac{2}{x + 3}$$

Caution: When the degree of the numerator is *not* less than the degree in the denominator, you *must* divide first. If this step is ignored, the resulting decomposition will be wrong. For example, suppose you started incorrectly in this way:

$$\frac{2x^3 + 3x^2 - x + 16}{(x - 1)(x + 2)} = \frac{A}{x - 1} + \frac{B}{x + 3}$$

This approach will produce the following *incorrect* answer:

$$\frac{2x^3 + 3x^2 - x + 16}{(x - 1)(x + 3)} = \frac{5}{x - 1} + \frac{2}{x + 3}$$

EXERCISES 5.7

Decompose into partial fractions.

1. $\dfrac{2x}{(x + 1)(x - 1)}$ 2. $\dfrac{x}{x^2 - 4}$ 3. $\dfrac{x + 7}{x^2 - x - 6}$

4. $\dfrac{4x^2 + 16x + 4}{(x + 3)(x^2 - 1)}$ 5. $\dfrac{5x^2 + 9x - 56}{(x - 4)(x - 2)(x + 1)}$ 6. $\dfrac{x}{(x - 3)^2}$

7. $\dfrac{3x - 3}{(x - 2)^2}$ 8. $\dfrac{2 - 3x}{x^2 + x}$ 9. $\dfrac{3x - 30}{15x^2 - 14x - 8}$

10. $\dfrac{2x + 1}{(2x + 3)^2}$ 11. $\dfrac{x^2 - x - 4}{(x - 2)^3}$ 12. $\dfrac{x^2 + 5x + 8}{(x - 3)(x + 1)^2}$

In Exercises 13 through 15 first divide and then complete the decomposition into partial fractions.

13. $\dfrac{x^3 - x + 2}{x^2 - 1}$ 14. $\dfrac{4x^2 - 14x + 2}{4x^2 - 1}$

15. $\dfrac{12x^4 - 12x^3 + 7x^2 - 2x - 3}{4x^2 - 4x + 1}$

Decompose into partial fractions.

16. $\dfrac{10x^2 - 16}{x^4 - 5x^2 + 4}$

17. $\dfrac{10x^3 - 15x^2 - 35x}{x^2 - x - 6}$

*18. $\dfrac{25x^3 + 10x^2 + 31x + 5}{25x^2 + 10x + 1}$

*19. $\dfrac{x^5 - 3x^4 + 2x^3 + x^2 + x + 4}{x^3 - 3x^2 + 3x - 1}$

REVIEW EXERCISES FOR CHAPTER 5

The solutions to the following exercises can be found within the text of Chapter 5. Try to answer each question without referring to the text.

Section 5.1

1. Graph $y = (x - 3)^3$.

2. The curve in the diagram can be obtained by shifting the graph of $y = x^3$ both horizontally and vertically. What is its equation?

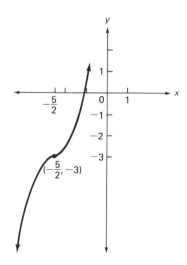

3. Graph $y = 2x^4$.

Section 5.2

4. Graph $y = \dfrac{1}{x^2}$.

5. Graph $y = \dfrac{1}{(x + 2)^2} - 3$.

Section 5.3

6. Use synthetic division to find the quotient and remainder for the division problem $(2x^3 - 9x^2 + 10x - 7) \div (x - 3)$.

7. Use synthetic division to divide $-\frac{1}{3}x^4 + \frac{1}{6}x^2 - 7x - 4$ by $x + 3$.

Section 5.4

8. State the remainder theorem.

9. Use synthetic division to find $p(2)$, where $p(x) = 3x^3 - 5x^2 + 7x + 5$.

10. State the factor theorem.

11. Use the factor theorem to explain why $x + 3$ is a factor of the function $p(x) = x^3 - x^2 - 8x + 12$.

Section 5.5

12. What are the possible rational roots of $p(x) = 4x^3 - 16x^2 + 11x + 10 = 0$?

13. Factor $f(x) = x^3 + 4x^2 + x - 6$.

14. Factor $p(x) = 2x^5 + 7x^4 - 18x^2 - 8x + 8$ and find the roots of $p(x) = 0$.

15. Solve and graph the system

$$y = x^3$$
$$y = 7x + 6$$

Section 5.6

16. Determine the signs of $f(x) = (x + 3)(x + 2)(x - 1)$.

17. Solve the inequality $p(x) < 0$, where $p(x) = x^4 + x^3 - 5x^2 + x - 6$.

18. Determine the signs of $f(x) = \dfrac{x^3 - x^2 - 8x + 12}{x - 4}$.

19. Solve for x: $\dfrac{2x - 5}{x + 1} - \dfrac{3}{x^2 + x} = 0$.

Section 5.7

20. Decompose $\dfrac{6x^2 + x - 37}{(x - 3)(x + 2)(x - 1)}$ into partial fractions.

21. Decompose $\dfrac{6 + 26x - x^2}{(2x - 1)(x + 2)^2}$ into partial fractions.

22. Decompose $\dfrac{2x^3 + 3x^2 - x + 16}{x^2 + 2x - 3}$ into partial fractions.

SAMPLE TEST FOR CHAPTER 5

Graph each function and write the equation of the asymptotes if there are any.

1. $f(x) = (x + 2)^3 - \frac{3}{2}$ 　　　　　　**2.** $g(x) = -\dfrac{1}{x^2} + 2$

3. Use synthetic division to divide $2x^5 + 5x^4 - x^2 - 21x + 7$ by $x + 3$.

4. (a) Let $p(x) = 27x^4 - 36x^3 + 18x^2 - 4x + 1$. Use the remainder theorem to evaluate $p(\frac{1}{3})$.

　　　(b) Use the result of part (a) and the factor theorem to determine whether or not $x - \frac{1}{3}$ is a factor of $p(x)$.

5. Show that $x - 2$ is a factor of $p(x) = x^4 - 4x^3 + 7x^2 - 12x + 12$, and factor $p(x)$ completely.

6. Make use of the rational root theorem to factor $f(x) = x^4 + 5x^3 + 4x^2 - 3x + 9$.

7. Solve: $2x^4 + x^3 - 17x^2 - 4x + 6 = 0$.

8. Determine the signs of $f(x) = \dfrac{x^2 - 2x}{x + 3}$.

9. Solve for x: $\dfrac{6}{x} = 2 + \dfrac{3}{x + 1}$.

10. Solve the system:

$$y = x^3$$
$$y - 19x = -30$$

†11. Decompose $\dfrac{x - 15}{x^2 - 25}$ into partial fractions.

Answers to the Test Your Understanding Exercises

Page 221

1. $1 \big| 1 - 7 + 4 + 10 - 1 - 5$
   $ \underline{+ 1 - 6 - 2 + 8 + 7}$
   $ 1 - 6 - 2 + 8 + 7 \big| + 2 = r$
   $p(1) = 1 - 7 + 4 + 10 - 1 - 5 = 2$

2. $-10 \big| 1 + 11 + 11 + 11 + 10$
   $ \underline{ - 10 - 10 - 10 - 10}$
   $ 1 + 1 + 1 + 1 \big| + 0 = r$
   $p(-10) = 10,000 - 11,000 + 1100 - 110 + 10 = 0$

3. $10 \big| 1 + 11 + 11 + 11 + 10$
   $ \underline{ + 10 + 210 + 2210 + 22210}$
   $ 1 + 21 + 221 + 2221 \big| + 22220 = r$
   $p(10) = 10,000 + 11,000 + 1100 + 110 + 10$
   $ = 22,220$

4. $6 \big| 6 - 40 + 0 + 25$
   $ \underline{ + 36 - 24 - 144}$
   $ 6 - 4 - 24 \big| - 119 = r$
   $p(6) = 6(6)^3 - 40(6)^2 + 25 = -119$

Page 226

1. (a) $\pm 1, \pm 2, \pm 3, \pm 4, \pm 6, \pm 8, \pm 12, \pm 24$
   (b) $(x + 3)(x - 2)(x - 4)$
   (c) $-3, 2, 4$

2. (a) $\pm 1, \pm 2, \pm 4, \pm 8, \pm 16$
   (b) $(x + 4)^2(x - 1)^2$
   (c) $-4, 1$

3. (a) $\pm 1, \pm \frac{1}{2}, \pm \frac{1}{4}, \pm 2, \pm 3, \pm \frac{3}{2}, \pm \frac{3}{4}, \pm 6$
   (b) $(x + 6)(2x - 1)^2$
   (c) $-6, \frac{1}{2}$

4. (a) $\pm 1, \pm \frac{1}{3}, \pm 2, \pm \frac{2}{3}, \pm 4, \pm \frac{4}{3}$
   (b) $(3x - 1)(x - 4)(x^2 + 1)$
   (c) $\frac{1}{3}, 4$

Page 232

1. $p(x) < 0$ on both $(-\infty, -6)$ and $(-5, 6)$; $p(x) > 0$ on both $(-6, -5)$ and $(6, \infty)$.

2. $p(x) < 0$ on both $(-2, 1)$ and $(2, 3)$; $p(x) > 0$ on each of $(-\infty, -2)$, $(1, 2)$, $(3, \infty)$.

3. $p(x) < 0$ on each of $(-\infty, -2)$, $(1, 2)$, $(3, \infty)$; $p(x) > 0$ on both $(-2, 1)$ and $(2, 3)$.

4. $p(x) < 0$ on both $(-3, 1)$ and $(1, 3)$; $p(x) > 0$ on both $(-\infty, -3)$ and $(3, \infty)$.

5. $-5 < x < 1$ or $x > 7$.

6. $-3 < x < -2$ or $\frac{1}{2} < x < 2$.

Page 238

1. $\dfrac{3}{x-2} + \dfrac{5}{x-3}$

2. $-\dfrac{\frac{1}{6}}{x+2} + \dfrac{\frac{1}{6}}{x-4}$

3. $\dfrac{1}{x-1} + \dfrac{2}{x-2} + \dfrac{3}{x-3}$

4. $\dfrac{6}{x+3} - \dfrac{2}{x+2}$

5. $\dfrac{4}{3x-1} + \dfrac{5}{2x+1}$

6. $\dfrac{4}{x+7} - \dfrac{2}{x-5} + \dfrac{3}{x+2}$

Six

Radical Functions

6.1

Graphing Some Special Radical Functions

A radical expression in x, such as $\sqrt{x}$, may be used to define a function f, where $f(x) = \sqrt{x}$. The domain of f consists of all real numbers $x \geq 0$ since the square root of a negative number is not a real number.

To graph $y = \sqrt{x}$, it is helpful to first square both sides to obtain $y^2 = x$; that is, $x = y^2$. Recall the graph of $y = x^2$ and obtain the graph of $x = y^2$ by reversing the role of the variables.

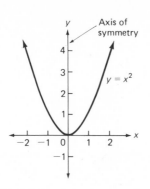

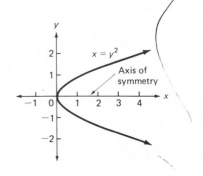

The "sideways" parabola in the diagram is *not* the graph of a function having x as the independent variable because for $x > 0$ there are two corresponding y-values. But if the bottom branch is removed, we have the correct graph of $y = \sqrt{x}$, for $x \geq 0$. (What is the equation for the bottom branch?)

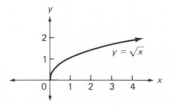

Note that, except for the origin, this graph is in the first quadrant, as could have been predicted by observing that for all $x > 0$, $\sqrt{x} > 0$. You can also verify that the specific points given in the following table are on the graph.

x	0	$\dfrac{1}{4}$	$\dfrac{9}{16}$	1	$\dfrac{9}{4}$	2	3	4
$y = \sqrt{x}$	0	$\dfrac{1}{2}$	$\dfrac{3}{4}$	1	$\dfrac{3}{2}$	$\sqrt{2}$	$\sqrt{3}$	2

Such a table presents us with an alternate method for getting the graph: plot the points in the table and connect with a smooth curve.

Example 1   Find the domain of $y = g(x) = \sqrt{x - 2}$ and graph.

Solution   Since the square root of a negative number is not a real number, the expression $x - 2$ must be nonnegative; therefore the domain of g consists of all $x \geq 2$ ($x - 2 \geq 0$). The graph of g may be found by shifting the graph of $y = \sqrt{x}$ by 2 units to the right.

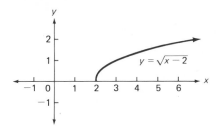

Example 2   Find the domain of $y = f(x) = \sqrt{|x|}$ and graph.

Solution   Since $|x| \geq 0$ for all x, the domain of f consists of all real numbers. To graph f first note that

$$f(-x) = \sqrt{|-x|} = \sqrt{|x|} = f(x)$$

Therefore the graph is symmetric about the y-axis. Thus we first find the graph for $x \geq 0$ and use symmetry to obtain the rest. For $x \geq 0$, we get $|x| = x$ and $f(x) = \sqrt{|x|} = \sqrt{x}$. It is always helpful to locate a few specific points on the curve as an aid to graphing. For example, note that $(1, 1)$, $(4, 2)$, and $(9, 3)$ are all on the curve.

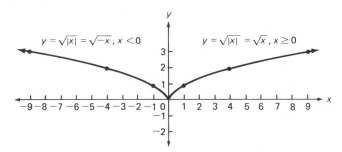

Example 3   Find the domain of $y = h(x) = x^{-1/2}$ and graph by using a table of values.

Solution   Note that $h(x) = x^{-1/2} = \dfrac{1}{x^{1/2}} = \dfrac{1}{\sqrt{x}}$. Thus the domain consists of all $x > 0$. Furthermore, $\dfrac{1}{\sqrt{x}} > 0$ for all x, so we know that the graph must be in the first quadrant only.

Plot the points in the table and connect with a smooth curve.

x	$\frac{1}{9}$	$\frac{1}{4}$	1	4	9
y	3	2	1	$\frac{1}{2}$	$\frac{1}{3}$

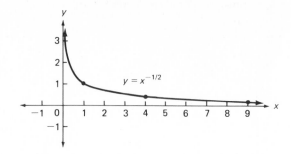

Note that the closer x is to zero, the larger are the corresponding y-values. Moreover, the y-values get closer and closer to zero as x is taken larger and larger. These observations suggest that the coordinate axes are asymptotic to the curve $y = x^{-1/2}$.

TEST YOUR UNDERSTANDING

Graph each function and state the domain.

1. $f(x) = \sqrt{x} - 2$

2. $f(x) = -\sqrt{x - 1}$

3. $f(x) = 2\sqrt{x}$

4. $f(x) = \dfrac{1}{\sqrt{x + 2}}$

The graph of the cube root function $y = \sqrt[3]{x}$ can be found by a process similar to that used for $y = \sqrt{x}$. First take $y = \sqrt[3]{x}$ and cube both sides to get $y^3 = x$. Now recall the graph of $y = x^3$ and obtain the graph of $x = y^3$ by reversing the role of the variables.

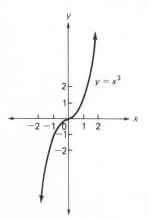

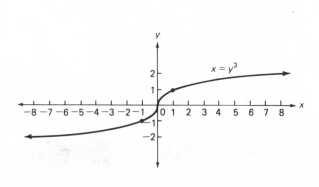

The graph on the right is the desired graph because $x = y^3$ and $y = \sqrt[3]{x}$ are equivalent for all real numbers x.

Example 4 Graph $y = \sqrt[3]{x+1} - 2$ and find the domain.

Solution Consider the graph of $y = \sqrt[3]{x}$ and shift 1 unit to the left and then 2 units down. The domain consists of all real numbers x.

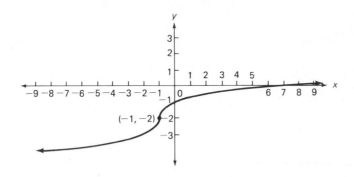

EXERCISES 6.1

For Exercises 1 through 12, find the domain of f, sketch the graph, and give the equations of the asymptotic lines if there are any.

1. $f(x) = \sqrt{x+2}$ 　　　　**2.** $f(x) = x^{1/2} + 2$ 　　　　**3.** $f(x) = \sqrt{x-3} - 1$

4. $f(x) = -\sqrt{x}$ 　　　　**5.** $f(x) = \sqrt{-x}$ 　　　　**6.** $f(x) = \sqrt{(x-2)^2}$

7. $f(x) = 2\sqrt[3]{x}$ 　　　　***8.** $f(x) = |\sqrt[3]{x}|$ 　　　　**9.** $f(x) = -x^{1/3}$

10. $f(x) = \sqrt[3]{-x}$ 　　　　**11.** $f(x) = \dfrac{1}{\sqrt{x}} - 1$ 　　　　**12.** $f(x) = \dfrac{1}{\sqrt{x-2}}$

13. **(a)** Explain why the graph of $f(x) = \dfrac{1}{\sqrt[3]{x}}$ is symmetric through the origin.

　　　(b) What is the domain of f?
　　　(c) Use a table of values to graph f.
　　　(d) What are the equations of the asymptotes?

14. Find the domain of $f(x) = \dfrac{1}{\sqrt[3]{x+1}}$, sketch the graph, and give the equations of the asymptotes.

***15.** Find the graph of the function $y = \sqrt[4]{x}$ by raising both sides of the equation to the fourth power and comparing to the graph of $y = x^4$.

***16.** Reflect the graph of $y = x^2$, for $x \geq 0$, through the line $y = x$. Obtain the equation of this new curve by interchanging variables in $y = x^2$ and solving for y.

*17. Follow the instructions of Exercise 16 with $y = x^3$ for all values x.

*18. For $f(x) = \sqrt{x}$ show that $\dfrac{f(x) - f(4)}{x - 4} = \dfrac{1}{\sqrt{x} + 2}$. (*Hint:* Factor $x - 4$ as the difference of two squares.)

6.2

Radical Equations

Example 4 of Section 6.1 called for the graph of $y = f(x) = \sqrt[3]{x + 1} - 2$. Our sketches of such graphs can be improved by knowing the intercepts. The y-intercept is easy to find: $f(0) = \sqrt[3]{0 + 1} - 2 = 1 - 2 = -1$. Finding the x-intercepts for this graph, as well as for many others, calls for more involved techniques. Since the x-intercepts are points for which $y = 0$, we need to solve this equation:

$$\sqrt[3]{x + 1} - 2 = 0$$

First "isolate" the radical by adding 2 to each side.

$$\sqrt[3]{x + 1} = 2$$

Cube both sides and solve for x.

$$(\sqrt[3]{x + 1})^3 = 2^3$$
$$x + 1 = 8$$
$$x = 7$$

Check this result by substituting into the original equation.

$$\sqrt[3]{7 + 1} - 2 = \sqrt[3]{8} - 2 = 2 - 2 = 0$$

The solution is $x = 7$ and the graph of $y = \sqrt[3]{x + 1} - 2$ crosses the x-axis at (7, 0).

For the remainder of this section we will be concerned with solving radical equations which will enable us to find the x-intercepts of the related graphs.

Example 1   Solve: $\sqrt{x^2 - x - 6} = 0$.

Solution   Square both sides and solve for x.

$$(\sqrt{x^2 - x - 6})^2 = 0^2$$
$$x^2 - x - 6 = 0$$
$$(x + 2)(x - 3) = 0$$
$$x = -2 \quad \text{or} \quad x = 3$$

Check: For $x = -2$: $\sqrt{(-2)^2 - (-2) - 6} = \sqrt{4 + 2 - 6} = \sqrt{0} = 0$
For $x = 3$: $\sqrt{3^2 - 3 - 6} = \sqrt{9 - 9} = \sqrt{0} = 0$

The preceding example, as well as other radical equations, is solved by applying this principle:

If $a = b$, then $a^n = b^n$.

This statement says that every solution of $a = b$ will also be a solution of $a^n = b^n$. Sometimes it is convenient to apply this principle after other changes have been made. For instance, to solve the equation

$$\sqrt{x + 4} + 2 = x$$

first isolate the radical on one side of the equation. Thus

$$\sqrt{x + 4} = x - 2$$

Now square and solve for x.

$$(\sqrt{x + 4})^2 = (x - 2)^2$$
$$x + 4 = x^2 - 4x + 4$$
$$0 = x^2 - 5x$$
$$0 = x(x - 5)$$

Thus we get $x = 0$ or $x = 5$. Check these values in the given equation.

For $x = 0$: $\sqrt{0 + 4} + 2 = \sqrt{4} + 2 = 2 + 2 = 4 \neq 0$; 0 is *not* a solution.

For $x = 5$: $\sqrt{5 + 4} + 2 = \sqrt{9} + 2 = 3 + 2 = 5$; 5 is the *only* solution.

How did the *extraneous* solution zero arise? In going from

$$\sqrt{x + 4} = x - 2 \quad\text{to}\quad (\sqrt{x + 4})^2 = (x - 2)^2$$

we used the basic principle: if $a = b$, then $a^n = b^n$. Therefore every solution of the first equation is also a solution for the second. But this principle is not always reversible. In particular, both 0 and 5 are solutions of the second equation, but only 5 is a solution of the first. In summary, the process of raising both sides of an equation in x to the nth power can produce false solutions. It is therefore vital to check all possible solutions in the original equation.

Note: If the radical in $\sqrt{x + 4} + 2 = x$ is not first isolated, it is still possible to solve the equation, but the work will be more involved. Try it.

TEST YOUR UNDERSTANDING

Solve each equation.

1. $\sqrt{x+1} = 3$ 2. $\sqrt{x} - 2 = 3$ 3. $(x+2)^{1/2} = (2x-5)^{1/2}$

4. $\sqrt{x^2+9} = -5$ 5. $\dfrac{1}{\sqrt{x}} = 3$ 6. $\dfrac{4}{\sqrt{x-1}} = 2$

7. $\sqrt{x^2-5} = 2$ 8. $\sqrt[3]{x+3} - 2 = 0$ 9. $\sqrt{x+16} - x = 4$

10. $2x = 1 + \sqrt{1-2x}$

Radical equations may contain more than one radical. Here are two such equations:

$$(1) \quad x\sqrt{3x} + \sqrt{75x} = 2\sqrt{27x}$$
$$(2) \quad \sqrt{x-7} - \sqrt{x} = 1$$

Equation (1) can be solved by first simplifying each radical and combining. Thus

$$x\sqrt{3x} + \sqrt{75x} = 2\sqrt{27x}$$
$$x\sqrt{3x} + 5\sqrt{3x} = 6\sqrt{3x}$$
$$x\sqrt{3x} + 5\sqrt{3x} - 6\sqrt{3x} = 0$$
$$(x+5-6)\sqrt{3x} = 0$$
$$(x-1)\sqrt{3x} = 0$$

$$x - 1 = 0 \quad \text{or} \quad \sqrt{3x} = 0$$
$$x = 1 \quad \text{or} \quad 3x = 0$$
$$x = 1 \quad \text{or} \quad x = 0$$

Since each of these values checks in the given equation, 0 and 1 are the solutions. For Equation (2), the two radicals cannot be combined. For such cases it is usually best to transform the equation first into one with as few radicals on each side as possible.

$$\sqrt{x-7} - \sqrt{x} = 1$$
$$\sqrt{x-7} = \sqrt{x} + 1$$

Now square:

$$(\sqrt{x-7})^2 = (\sqrt{x} + 1)^2$$
$$x - 7 = x + 2\sqrt{x} + 1$$
$$-8 = 2\sqrt{x}$$
$$-4 = \sqrt{x}$$

Square again:

$$16 = x$$

Check: $\sqrt{16 - 7} - \sqrt{16} = \sqrt{9} - \sqrt{16} = 3 - 4 = -1 \neq 1.$

Therefore this equation has no solution, which could have been observed at an earlier stage as well. For example, $-4 = \sqrt{x}$ has no solution.

The <u>radical signs in the preceding equations may be replaced by appropriate</u> <u>fractional exponents and solved by the same methods.</u> Example 2 shows an equation using fractional exponents.

Example 2   Solve for x:

$$\frac{3(2x - 1)^{1/2}}{x - 3} - \frac{(2x - 1)^{3/2}}{(x - 3)^2} = 0$$

Solution   Multiply through by $(x - 3)^2$ to clear fractions.

$$(x - 3)^2 \left[\frac{3(2x - 1)^{1/2}}{x - 3} \right] - (x - 3)^2 \left[\frac{(2x - 1)^{3/2}}{(x - 3)^2} \right] = (x - 3)^2(0)$$

$$3(x - 3)(2x - 1)^{1/2} - (2x - 1)^{3/2} = 0$$

Note that we may consider $(2x - 1)^{1/2}$ as a common factor.

$$[3(x - 3) - (2x - 1)](2x - 1)^{1/2} = 0$$

$$(x - 8)(2x - 1)^{1/2} = 0$$

Set each factor equal to zero.

$$x - 8 = 0 \quad \text{or} \quad (2x - 1)^{1/2} = 0$$

$$x = 8 \quad \text{or} \quad 2x - 1 = 0$$

$$x = 8 \quad \text{or} \quad x = \tfrac{1}{2}$$

Check in the original equation to show that both $x = \tfrac{1}{2}$ and $x = 8$ are solutions.

As you may have noticed, the algebraic techniques developed earlier involving other kinds of expressions often carry over into this work. Here is an example that uses our knowledge of quadratics in conjunction with radicals.

Example 3   Solve for x: $\sqrt[3]{x^2} + \sqrt[3]{x} - 20 = 0.$

Solution   First write the equation by using rational exponents.

$$x^{2/3} + x^{1/3} - 20 = 0$$

Now think of $x^{2/3}$ as the square of $x^{1/3}$, $x^{2/3} = (x^{1/3})^2$, and use the substitution $u = x^{1/3}$ as follows:

$$0 = x^{2/3} + x^{1/3} - 20$$
$$= (x^{1/3})^2 + x^{1/3} - 20$$
$$= u^2 + u - 20$$

This is a factorable quadratic in u.

$$u^2 + u - 20 = 0$$
$$(u + 5)(u - 4) = 0$$
$$u + 5 = 0 \quad \text{or} \quad u - 4 = 0$$
$$u = -5 \quad \text{or} \quad u = 4$$

Now replace u by $x^{1/3}$.

$$x^{1/3} = -5 \quad \text{or} \quad x^{1/3} = 4$$

Cubing gives

$$x = -125 \quad \text{or} \quad x = 64$$

Check to show that both values are solutions of the given equation.

Example 4 Solve the system and graph:

$$y = \sqrt[3]{x}$$
$$y = \tfrac{1}{4}x$$

Solution For the points of intersection the x- and y-values are the same in both equations. Thus

$$\tfrac{1}{4}x = \sqrt[3]{x}$$

Cube both sides and solve for x.

$$\tfrac{1}{64}x^3 = x$$
$$x^3 = 64x$$
$$x^3 - 64x = 0$$
$$x(x^2 - 64) = 0$$
$$x(x + 8)(x - 8) = 0$$
$$x = 0 \quad \text{or} \quad x = -8 \quad \text{or} \quad x = 8$$

Substitute these values into either of the given equations to obtain the corresponding y-values. The remaining equation can be used for checking. The solutions are $(-8, -2)$, $(0, 0)$, and $(8, 2)$.

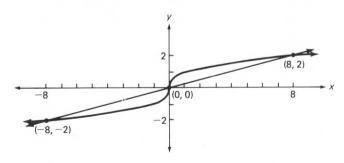

Caution: A common error is to take $x^3 - 64x = 0$ and divide through by x to get $x^2 - 64 = 0$. This step produces the roots ± 8. The root 0 has been lost because we divided by x and 0 is the number for which the factor x in $x(x^2 - 64)$ is zero. You may always divide by a nonzero constant and get an equivalent form of the equation. But when you divide by a variable quantity there is the danger of losing some roots, those for which the divisor is 0.

EXERCISES 6.2

Solve each equation in Exercises 1 through 24.

1. $\sqrt{x - 1} = 4$

2. $\sqrt{3x - 2} - 5 = 0$

3. $\sqrt{4x + 9} - 7 = 0$

4. $x\sqrt{2x} - \sqrt{8x} = -\sqrt{32x}$

5. $\sqrt{20x^3} - \sqrt{45x} = \sqrt{20x}$

6. $\dfrac{12}{\sqrt{x}} = 4$

7. $\dfrac{8}{\sqrt{x + 2}} = 4$

8. $\sqrt{x - 1} = \sqrt{2x - 11}$

9. $\sqrt{3x + 1} = \sqrt{2x + 6}$

10. $\sqrt{x^2 + 2} = 3$

11. $\sqrt{x^2 + 5} + 3 = 0$

12. $\sqrt{x} + \sqrt{x - 7} = 7$

13. $\sqrt{x} + \sqrt{x - 5} = 5$

14. $3\sqrt{x} = 2\sqrt{3}$

15. $\sqrt[3]{2x + 7} = 3$

16. $\left(\dfrac{5x + 4}{2}\right)^{1/3} = 3$

17. $x\sqrt{x - 4} - \sqrt{9x - 36} = 0$

18. $\dfrac{x}{\sqrt{4 - x^2}} - \dfrac{1}{3} = 0$

19. $\dfrac{1}{3(x - 2)^{2/3}} = 0$

20. $x = 8 - 2\sqrt{x}$

21. $3x = \sqrt{3 - 5x - 3x^2}$

22. $\sqrt[3]{x^4} = \sqrt[3]{x}$

23. $4x^{2/3} - 12x^{1/3} + 9 = 0$

24. $2x - 5\sqrt{x} - 3 = 0$

In Exercises 25 through 28, solve the system and graph.

25. $y = \sqrt{x}$

$\quad y = \frac{1}{2}x$

27. $y = \sqrt{-x}$

$\quad y = x^2$

26. $y = \dfrac{2}{\sqrt{x}}$

$\quad x + 3y = 7$

28. $(x - 3)^2 + y^2 = 5$

$\quad y - \sqrt{x} = 0$

*29. The radius of a sphere whose surface area is A is given by the formula $r = \dfrac{1}{2}\sqrt{\dfrac{A}{\pi}}$.
Use this formula to solve for A in terms of π and r. Then find A when $r = 2$ centimeters.

*30. The radius of a right circular cylinder with height h and volume V is given by this
formula: $r = \sqrt{\dfrac{V}{\pi h}}$. Use this formula to solve for V in terms of r and h. Then find V
when $r = 2$ centimeters and $h = 3$ centimeters.

*31. The distance from the point $(3, 0)$ to a point $P(x, y)$ on the curve $y = \sqrt{x}$ is $3\sqrt{5}$.
Find the coordinates of P. (*Hint:* Use the formula for the distance between two
points.)

*32. (a) Let s represent the square of the distance between points $A(6, 0)$ and $P(x, y)$,
where P is on the curve $y = 2\sqrt{x}$. Show that s is a quadratic function of x.
(b) What are the coordinates of P such that s is a minimum value?
(c) Find the slope of the tangent to the curve $y = 2\sqrt{x}$ at the point P.
(d) Prove that the tangent at P and line AP are perpendicular.

6.3
Determining the Signs of Radical Functions

The methods used for determining the signs of rational functions are easily
modified to determine the signs of functions defined in terms of radical expressions. As a first example, consider this function:

$$f(x) = (x - 5)\sqrt{x - 3}$$

Observe that the factor $\sqrt{x - 3}$ indicates $x \geq 3$. (Why?) Therefore the table that
follows does not need to consider any values for $x < 3$.

	$\sqrt{x - 3}$	$x - 5$	$f(x)$
$3 < x < 5$	$+$	$-$	$-$
$5 < x$	$+$	$+$	$+$

Note that the entries in the column headed $\sqrt{x-3}$ are automatically plus because, by definition, the radical sign stands for the positive square root. The table shows that $f(x)$ is negative on the interval $(3, 5)$ and positive on the interval $(5, \infty)$.

Example 1 Find the domain of $f(x) = \dfrac{x}{\sqrt{4-x^2}}$ and determine its signs.

Solution The domain consists of those x for which $4 - x^2 > 0$. To solve this inequality we note that $y = 4 - x^2$ is a parabola that opens downward and crosses the x-axis at ± 2. Hence $4 - x^2 > 0$ on the interval $(-2, 2)$, which is the domain of f. The sign of $f(x)$ depends only on the numerator since the denominator is always positive. Hence $f(x) > 0$ on $(0, 2)$, and $f(x) < 0$ on $(-2, 0)$.

TEST YOUR UNDERSTANDING

Find the domain of each function and determine its signs.

1. $f(x) = \dfrac{x-4}{\sqrt{x}}$ **2.** $f(x) = \dfrac{\sqrt{x+2}}{x}$ **3.** $f(x) = \dfrac{2}{3\sqrt[3]{x}}$

Algebraic procedures developed earlier can be helpful with this type of work. For example, the signs of

$$f(x) = x^{1/2} + \tfrac{1}{2}x^{-1/2}(x-3)$$

can be determined after combining and simplifying as follows:

$$f(x) = x^{1/2} + \tfrac{1}{2}x^{-1/2}(x-3)$$

$$= x^{1/2} + \frac{x-3}{2x^{1/2}}$$

$$= \frac{2x + x - 3}{2x^{1/2}}$$

$$= \frac{3x - 3}{2x^{1/2}}$$

$$= \frac{3(x-1)}{2\sqrt{x}}$$

	$\sqrt{x}$	$x - 1$	$f(x)$
$0 < x < 1$ $1 < x$	$+$ $+$	$-$ $+$	$-$ $+$

Hence $f(x) < 0$ on $(0, 1)$; $f(x) > 0$ on $(1, \infty)$.

Example 2 Find the domain of f, the roots of $f(x) = 0$, and determine the signs of f where

$$f(x) = \frac{x^2 - 1}{2\sqrt{x - 1}} + 2x\sqrt{x - 1}$$

Solution The radical in the denominator calls for $x - 1 > 0$. Therefore the domain consists of all $x > 1$. Now simplify as follows:

$$f(x) = \frac{x^2 - 1}{2\sqrt{x - 1}} + 2x\sqrt{x - 1}$$

$$= \frac{x^2 - 1 + 4x(x - 1)}{2\sqrt{x - 1}}$$

$$= \frac{5x^2 - 4x - 1}{2\sqrt{x - 1}}$$

$$= \frac{(5x + 1)(x - 1)}{2\sqrt{x - 1}}$$

$$= \frac{5(x + \frac{1}{5})(x - 1)}{2\sqrt{x - 1}}$$

The numerator is zero when $x = -\frac{1}{5}$ or $x = 1$. But these values are *not* in the domain of f. Since a fraction can only be zero when the numerator is zero, it follows that $f(x) = 0$ has no solutions. The signs of $f(x)$ are given in this brief table:

	$x + \frac{1}{5}$	$x - 1$	$\sqrt{x - 1}$	$f(x)$
$x > 1$	$+$	$+$	$+$	$+$

Thus $f(x) > 0$ on its domain $(1, \infty)$.

For each exercise (a) find the domain of f, (b) determine the signs of f, and (c) find the roots of $f(x) = 0$.

1. $f(x) = (x + 4)\sqrt{x - 2}$

2. $f(x) = \dfrac{9 + x}{9\sqrt{x}}$

3. $(x) = \sqrt[3]{(x - 2)^2}$

4. $f(x) = \dfrac{1}{\sqrt[3]{x - 2}}$

5. $f(x) = \dfrac{\sqrt{9 - x^2}}{x}$

6. $f(x) = \dfrac{x}{\sqrt{x^2 - 4}}$

7. $f(x) = x^{1/2} - x^{-1/2}$

8. $f(x) = \frac{3}{2}x^{1/2} - \frac{3}{2}x^{-1/2}$

9. $f(x) = (x + 4)\sqrt[3]{x - 2}$

10. $f(x) = \dfrac{(x + 4)\sqrt[3]{x - 2}}{3\sqrt[3]{x}}$

11. $f(x) = x^{1/2} + \frac{1}{2}x^{-1/2}(x - 9)$

12. $f(x) = \dfrac{-5x(x - 4)}{2\sqrt{5 - x}}$

*13. $f(x) = x^{2/3} + \frac{2}{3}x^{-1/3}(x - 10)$

*14. $f(x) = \frac{x^2}{2}(x - 2)^{-1/2} + 2x(x - 2)^{-1/2}$

6.4
Combining
Functions

What does it mean to add two functions f and g? Remember, a function is basically a correspondence; so how do we add correspondences? The answer turns out to be relatively simple.

To illustrate, let us use f and g, where

$$f(x) = \frac{1}{x^3 - 1} \quad \text{and} \quad g(x) = \sqrt{x}$$

The domain of f consists of all $x \neq 1$, and g has domain all $x \geq 0$. The sum of f and g will be symbolized by $f + g$ (it might be helpful if you momentarily used $f \oplus g$ to emphasize that this plus is not our old familiar addition). We will take as the domain of $f + g$ all values x that are in *each* of the domains of f and g simultaneously. Therefore $f + g$ has domain $x \geq 0$ and $x \neq 1$.

Now consider any x in this new domain. What will be the corresponding range value for $f + g$? That is, how should $(f + g)(x)$ be defined? The answer is almost instinctive. For this x, take the two range values $f(x) = \dfrac{1}{x^3 - 1}$ and

$g(x) = \sqrt{x}$, and add them to get $f(x) + g(x)$. Thus

$$(f + g)(x) = f(x) + g(x) = \frac{1}{x^3 - 1} + \sqrt{x}$$

(Note that the plus sign in $f(x) + g(x)$ is addition of real numbers.) Specifically, for $x = 4$ we have

$$(f + g)(4) = f(4) + g(4) = \frac{1}{64 - 1} + \sqrt{4} = \frac{1}{63} + 2 = \frac{127}{63}$$

The difference, product, and quotient are found in a similar manner. Using f and g above, we get

$$(f - g)(x) = f(x) - g(x) = \frac{1}{x^3 - 1} - \sqrt{x}$$

$$(f \cdot g)(x) = f(x)g(x) = \frac{1}{x^3 - 1} \cdot \sqrt{x} = \frac{\sqrt{x}}{x^3 - 1}$$

$$\frac{f}{g}(x) = \frac{f(x)}{g(x)} = \frac{\frac{1}{x^3 - 1}}{\sqrt{x}} = \frac{1}{(x^3 - 1)\sqrt{x}}$$

The domains of $f - g$ and $f \cdot g$ are the same as for $f + g$; namely, $x \geq 0$ and $x \neq 1$. The domain of $\dfrac{f}{g}$ also has those x common to the domains of f and g except those for which $g(x) = 0$; $x > 0$ and $x \neq 1$.

Example 1   Let $f(x) = \dfrac{1}{x^3 - 1}$ and $g(x) = \sqrt{x}$.

(a) Evaluate $f(4)$, and $g(4)$, and compute $f(4) \cdot g(4)$.
(b) Use the expression for $(f \cdot g)(x)$, as in the preceding discussion, to evaluate $(f \cdot g)(4)$.

Solution

(a) $f(4) = \dfrac{1}{4^3 - 1} = \dfrac{1}{63}$, $g(4) = \sqrt{4} = 2$, and $f(4) \cdot g(4) = \dfrac{1}{63} \cdot 2 = \dfrac{2}{63}$

(b) $(f \cdot g)(x) = \dfrac{\sqrt{x}}{x^3 - 1}$

$(f \cdot g)(4) = \dfrac{\sqrt{4}}{4^3 - 1} = \dfrac{2}{63}$

We are ready to state the general definition for forming the sum, difference, product, and quotient of two functions.

For functions f and g the functions $f + g, f - g, f \cdot g$, and $\dfrac{f}{g}$ have range values given by

$$(f + g)(x) = f(x) + g(x)$$
$$(f - g)(x) = f(x) - g(x)$$
$$(f \cdot g)(x) = f(x)g(x)$$
$$\frac{f}{g}(x) = \frac{f(x)}{g(x)}$$

The domains of $f + g, f - g$, and $f \cdot g$, are all the same and consist of all x common to the domains of f and g. The domain of $\dfrac{f}{g}$ has all x common to the domains of f and g except for those x where $g(x) = 0$.

We will study one more way of forming new functions from given functions. It is referred to as the *composition* of functions. Let f and g be given by

$$f(x) = \frac{1}{x - 2} \qquad \text{and} \qquad g(x) = \sqrt{x}$$

Take a specific value in the domain of f, say $x = 6$. Then the corresponding range value is $f(6) = \frac{1}{4}$. Take this range value and use it as a domain value for g to produce $g(\frac{1}{4}) = \sqrt{\frac{1}{4}} = \frac{1}{2}$. This work may be condensed in this way.

$$g(f(6)) = g(\tfrac{1}{4}) = \sqrt{\tfrac{1}{4}} = \tfrac{1}{2}$$

Here are two more illustrations using the condensed notation.

$$g(f(10)) = g\left(\frac{1}{10 - 2}\right) = g\left(\frac{1}{8}\right) = \sqrt{\frac{1}{8}} = \frac{1}{2\sqrt{2}}$$

$$g\left(f\left(\frac{9}{4}\right)\right) = g\left(\frac{1}{\frac{9}{4} - 2}\right) = g\left(\frac{1}{\frac{1}{4}}\right) = g(4) = \sqrt{4} = 2$$

The roles of f and g may be interchanged. Thus

$$f(g(10)) = f(\sqrt{10}) = \frac{1}{\sqrt{10} - 2}$$

and

$$f\left(g\left(\frac{9}{4}\right)\right) = f\left(\sqrt{\frac{9}{4}}\right) = f\left(\frac{3}{2}\right) = \frac{1}{\frac{3}{2} - 2} = -2$$

In some cases this process does not work. For instance, if $x = -3$, then $f(-3) = -\frac{1}{5}$; $g(-\frac{1}{5}) = \sqrt{-\frac{1}{5}}$ is not a real number. We therefore say that $g(f(-3))$ is undefined or that it does not exist.

TEST YOUR UNDERSTANDING

For each pair of functions f and g, evaluate (if possible) each of the following:

   (a) $g(f(1))$ **(b)** $f(g(1))$ **(c)** $f(g(0))$ **(d)** $g(f(-2))$

1. $f(x) = 3x - 1$; $g(x) = x^2 + 4$ **2.** $f(x) = \sqrt{x}$; $g(x) = x^2$

3. $f(x) = \sqrt[3]{3x - 1}$; $g(x) = 5x$ **4.** $f(x) = \dfrac{x + 2}{x - 1}$; $g(x) = x^3$

The preceding computations for specific values of x can be stated in terms of any (allowable) x. For instance, using $f(x) = \dfrac{1}{x - 2}$ and $g(x) = \sqrt{x}$ we have

$$g(f(x)) = g\left(\frac{1}{x - 2}\right) = \sqrt{\frac{1}{x - 2}} = \frac{1}{\sqrt{x - 2}}$$

This new correspondence between a domain value x and the range value $\dfrac{1}{\sqrt{x - 2}}$ is referred to as the **composite function of g by f** (or the *composition of g by f*). This composite function is denoted by $g \circ f$. That is, for the given functions f and g we form the composite of g by f, whose range values $(g \circ f)(x)$, read "g by f of x," are defined by

$$(g \circ f)(x) = g(f(x)) = \frac{1}{\sqrt{x - 2}}$$

The domain of $g \circ f$ will consist of all values x in the domain of f such that $f(x)$ is in the domain of g. Since $f(x) = \dfrac{1}{x - 2}$ has domain all $x \neq 2$ and the domain of $g(x) = \sqrt{x}$ is all $x \geq 0$, the domain of $g \circ f$ is all $x \neq 2$ for which $\dfrac{1}{x - 2}$ is positive; that is, all $x > 2$.

Reversing the role of the two functions gives the composite of f by g, where

$$(f \circ g)(x) = f(g(x)) = f(\sqrt{x}) = \frac{1}{\sqrt{x} - 2}$$

The domain of $f \circ g$ consists of all $x \geq 0$ except $x \neq 4$.

Here is the definition of composite functions.

For functions f and g the composite function g by f, denoted $g \circ f$, has range values defined by

$$(g \circ f)(x) = g(f(x))$$

and domain consisting of all x in the domain of f for which $f(x)$ is in the domain of g.

It may help you to remember the construction of composites by looking at this schematic diagram.

$$x \xrightarrow{\;f\;} f(x) \xrightarrow{\;g\;} g(f(x)) = (g \circ f)(x)$$
$$g \circ f$$

It is also helpful to view the composition $(g \circ f)(x) = g(f(x))$ as consisting of an "inner" function f and an "outer" function g.

Example 2   Form the composite functions $g \circ f$ and $f \circ g$ and give their domains, where $f(x) = \dfrac{1}{x^2 - 1}$ and $g(x) = \sqrt{x}$.

Solution   We find that $f \circ g$ is given by

$$(f \circ g)(x) = f(g(x)) = f(\sqrt{x}) = \frac{1}{(\sqrt{x})^2 - 1} = \frac{1}{x - 1}$$

Domain: $x \geq 0$ and $x \neq 1$
Moreover, $g \circ f$ is given by

$$(g \circ f)(x) = g(f(x)) = g\left(\frac{1}{x^2 - 1}\right) = \sqrt{\frac{1}{x^2 - 1}} = \frac{1}{\sqrt{x^2 - 1}}$$

Domain: $x < -1$ or $x > 1$

From the preceding example it follows that $(f \circ g)(x) \neq (g \circ f)(x)$. We conclude that, in general, the composite of f by g is not equal to the composite of g by f.

The composition of functions may be extended to include more than two functions. For example, if $f(x) = \sqrt{x}$, $g(x) = x^2 + 1$, and $h(x) = \dfrac{1}{x}$, then the composition of f by g by h, denoted $f \circ g \circ h$, is given by

$$(f \circ g \circ h)(x) = f(g(h(x)))$$

$$= f\left(g\left(\frac{1}{x}\right)\right)$$

$$= f\left(\frac{1}{x^2} + 1\right)$$

$$= \sqrt{\frac{1}{x^2} + 1}$$

EXERCISES 6.4

For Exercises 1 through 6, let $f(x) = 2x - 3$ and $g(x) = 3x + 2$.

1. (a) Find $f(1)$, $g(1)$, and $f(1) + g(1)$.
   (b) Find $(f + g)(x)$ and state the domain of $f + g$.
   (c) Use the result in part (b) to evaluate $(f + g)(1)$.

2. (a) Find $g(2)$, $f(2)$, and $g(2) - f(2)$.
   (b) Find $(g - f)(x)$ and state the domain of $g - f$.
   (c) Use the result in part (b) to evaluate $(g - f)(2)$.

3. (a) Find $f(\frac{1}{2})$, $g(\frac{1}{2})$, and $f(\frac{1}{2}) \cdot g(\frac{1}{2})$.
   (b) Find $(f \cdot g)(x)$ and state the domain of $f \cdot g$.
   (c) Use the result in part (b) to evaluate $(f \cdot g)(\frac{1}{2})$.

4. (a) Find $g(-2)$, $f(-2)$, and $\dfrac{g(-2)}{f(-2)}$.

   (b) Find $\dfrac{g}{f}(x)$ and state the domain of $\dfrac{g}{f}$.

   (c) Use the result in part (b) to evaluate $\dfrac{g}{f}(-2)$.

5. (a) Find $g(0)$ and $f(g(0))$.
   (b) Find $(f \circ g)(x)$ and state the domain of $f \circ g$.
   (c) Use the result in part (b) to evaluate $(f \circ g)(0)$.

6. (a) Find $f(0)$ and $g(f(0))$.
   (b) Find $(g \circ f)(x)$ and state the domain of $g \circ f$.
   (c) Use the result in part (b) to evaluate $(g \circ f)(0)$.

For each pair of functions in Exercises 7 through 12, find the following:

   (a) $(f + g)(x)$; domain of $f + g$.

   (b) $\left(\dfrac{f}{g}\right)(x)$; domain $\dfrac{f}{g}$.

   (c) $(f \circ g)(x)$; domain $f \circ g$.

7. $f(x) = x^2; g(x) = \sqrt{x}$ **8.** $f(x) = 5x - 1; g(x) = \dfrac{5}{1 + 3x}$

9. $f(x) = x^3 - 1; g(x) = \dfrac{1}{x}$ **10.** $f(x) = 3x - 1; g(x) = \frac{1}{3}x + \frac{1}{3}$

11. $f(x) = x^2 + 6x + 8; g(x) = \sqrt{x - 2}$

12. $f(x) = \sqrt[3]{x}; g(x) = x^2$

13. Let $f(x) = \dfrac{1}{x}$, $g(x) = 2x - 1$, and $h(x) = x^{1/3}$. Find the following:

    (a) $(f \circ g \circ h)(x)$ **(b)** $(g \circ f \circ h)(x)$ **(c)** $(h \circ f \circ g)(x)$

14. Let $f(x) = x + 2$, $g(x) = \sqrt{x}$, and $h(x) = x^3$. Find the following:

    (a) $(f \circ g \circ h)(x)$ **(b)** $(f \circ h \circ g)(x)$ **(c)** $(g \circ f \circ h)(x)$

    (d) $(g \circ h \circ f)(x)$ **(e)** $(h \circ f \circ g)(x)$ **(f)** $(h \circ g \circ f)(x)$

15. Let $f(x) = \dfrac{1}{x}$. Find $(f \circ f)(x)$.

16. Let $f(x) = x^2$, $g(x) = \dfrac{1}{x - 1}$, and $h(x) = 1 + \dfrac{1}{x}$. Find the following:

    (a) $(f \circ h \circ g)(x)$ **(b)** $(g \circ h \circ f)(x)$ **(c)** $(h \circ g \circ f)(x)$

*17. Let $f(x) = x^3$. Find a function g so that $(f \circ g)(x) = x$ and $(g \circ f)(x) = x$.

*18. Let $f(x) = x$. Find $(f \circ g)(x)$ and $(g \circ f)(x)$ for any function g.

6.5
Decomposition of Composite Functions

One of the most useful skills needed in the study of calculus is the ability to recognize that a given function may be viewed as the composition of two or more functions. For instance, let h be given by

$$h(x) = \sqrt{x^2 + 2x + 2}$$

If we let $f(x) = x^2 + 2x + 2$ and $g(x) = \sqrt{x}$, then the composite g by f is

$$(g \circ f)(x) = g(f(x))$$
$$= g(x^2 + 2x + 2)$$
$$= \sqrt{x^2 + 2x + 2}$$
$$= h(x)$$

Thus the given function h has been *decomposed* into the composition of the two functions f and g.

Such decompositions are not unique. More than one decomposition is possible. For h we may also use

$$t(x) = x^2 + 2x \qquad s(x) = \sqrt{x + 2}$$

Then

$$(s \circ t)(x) = s(t(x))$$
$$= s(x^2 + 2x)$$
$$= \sqrt{x^2 + 2x + 2}$$
$$= h(x)$$

You would most likely agree that the first of these decompositions is more "natural." Just which decomposition one is to choose will, in later work, depend on the situation. For our purposes some additional examples will help to demonstrate the decompositions that are desirable. Other answers for each of these examples are possible.

Example 1 Find f and g such that $h = f \circ g$, where $h(x) = \left(\dfrac{1}{3x - 1}\right)^5$ and the inner function g is rational.

Solution Let $g(x) = \dfrac{1}{3x - 1}$ and $f(x) = x^5$.

$$(f \circ g)(x) = f(g(x))$$
$$= f\left(\frac{1}{3x - 1}\right)$$
$$= \left(\frac{1}{3x - 1}\right)^5$$
$$= h(x)$$

Example 2 Write $h(x) = \dfrac{1}{(3x - 1)^5}$ as the composite of two functions so that the inner function is a binomial.

Solution Let $g(x) = 3x - 1$ and $f(x) = \dfrac{1}{x^5}$.

$$(f \circ g)(x) = f(g(x))$$
$$= f(3x - 1)$$
$$= \frac{1}{(3x - 1)^5}$$
$$= h(x)$$

Example 3   Decompose $h(x) = \sqrt{(x^2 - 3x)^5}$ into two functions so that the inner function is a binomial.

Solution   First write $h(x) = (x^2 - 3x)^{5/2}$. Now let $f(x) = x^2 - 3x$ and let $g(x) = x^{5/2}$.

$$(g \circ f)(x) = g(f(x))$$
$$= g(x^2 - 3x)$$
$$= (x^2 - 3x)^{5/2}$$
$$= h(x)$$

Example 4   Decompose h in Example 3 so that the outer function is a polynomial.

Solution   Write $h(x) = (\sqrt{x^2 - 3x})^5$. Let $f(x) = \sqrt{x^2 - 3x}$ and $g(x) = x^5$.

$$(g \circ f)(x) = g(f(x))$$
$$= g(\sqrt{x^2 - 3x})$$
$$= (\sqrt{x^2 - 3x})^5$$
$$= h(x)$$

TEST YOUR UNDERSTANDING

For each function h find functions f and g so that $h = g \circ f$.

1. $h(x) = (8x - 3)^5$

2. $h(x) = \sqrt[5]{8x - 3}$

3. $h(x) = \sqrt{\dfrac{1}{8x - 3}}$

4. $h(x) = \left(\dfrac{5}{7 + 4x^2}\right)^3$

5. $h(x) = \dfrac{(2x + 1)^4}{(2x - 1)^4}$

6. $h(x) = \sqrt{(x^4 - 2x^2 + 1)^3}$

Examples 3 and 4 showed two different ways of decomposing $h(x) = \sqrt{(x^2 - 3x)^5}$ into the composition of two functions. It is also possible to express h as the composition of three functions. For example, we may use these functions:

$$f(x) = x^2 - 3x \qquad g(x) = x^5 \qquad t(x) = \sqrt{x}$$

Then

$$(t \circ g \circ f)(x) = t(g(f(x)))$$
$$= t(g(x^2 - 3x))$$
$$= t((x^2 - 3x)^5)$$
$$= \sqrt{(x^2 - 3x)^5}$$
$$= h(x)$$

EXERCISES 6.5

For Exercises 1 through 12 find functions f and g so that $h(x) = (f \circ g)(x)$. In each case let the inner function g be a polynomial or a rational function.

1. $h(x) = (3x + 1)^2$

2. $h(x) = (x^2 - 2x)^3$

3. $h(x) = \sqrt{1 - 4x}$

4. $h(x) = \sqrt[3]{x^2 - 1}$

5. $h(x) = \left(\dfrac{x + 1}{x - 1}\right)^2$

6. $h(x) = \left(\dfrac{1 - 2x}{1 + 2x}\right)^3$

7. $h(x) = (3x^2 - 1)^{-3}$

8. $h(x) = \left(1 + \dfrac{1}{x}\right)^{-2}$

9. $h(x) = \sqrt{\dfrac{x}{x - 1}}$

10. $h(x) = \sqrt[3]{\dfrac{x - 1}{x}}$

11. $h(x) = \sqrt{(x^2 - x - 1)^3}$

12. $h(x) = \sqrt[3]{(1 - x^4)^2}$

For Exercises 13 through 18 find three functions f, g, and h such that $k(x) = (h \circ g \circ f)(x)$.

13. $k(x) = (\sqrt{2x + 1})^3$

14. $k(x) = \sqrt[3]{(2x - 1)^2}$

15. $k(x) = \sqrt{\left(\dfrac{x}{x + 1}\right)^5}$

16. $\left(\sqrt[7]{\dfrac{x^2 - 1}{x^2 + 1}}\right)^4$

17. $k(x) = (x^2 - 9)^{2/3}$

18. $k(x) = (5 - 3x)^{5/2}$

6.6

Cross Sections of Solids of Revolution (optional)

When you study calculus you will learn how to find the volume of certain three-dimensional figures. One type is formed by taking a two-dimensional region and revolving it about a line.

In the figure you see the region R whose boundary is the curve $f(x) = \sqrt{x}$, the segment [0, 4] on the x-axis, and a vertical line segment at $x = 4$.

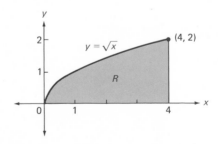

A three-dimensional solid can be generated by letting R revolve 360° around the x-axis. It might help you to visualize this form by thinking of R as a thin, rigid membrane that holds its shape as it spins around the *x*-axis on the interval [0, 4]. In this case *R* "sweeps out" a bowl-shaped three-dimensional figure. (This process can be compared to the rapid movement of a lighted end of a stick at night. Even though the light is always moving, to our eyes it leaves the impression of a complete figure in space.)

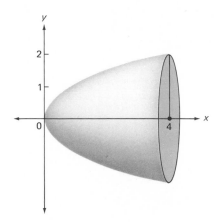

The volume of such a figure can be found (using calculus) by knowing the area of a typical cross section. We will be concerned with just that part of the problem: finding the area of the cross section.

A cross section is formed by cutting the solid with a plane perpendicular to the axis of revolution at any point *P* between 0 and 4.

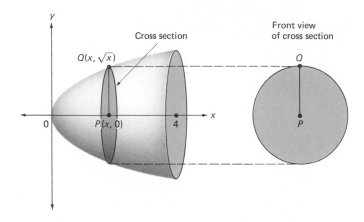

The resulting cross section is also formed by taking the vertical line segment *PQ* as radius and spinning it around the *x*-axis with *P* as center. Therefore the cross section is a circle. If we use *x* as the *x*-coordinate of point *P*, then *Q* has coordinates $(x, \sqrt{x})$ and the radius $PQ = \sqrt{x}$. The area of the cross section is given by

$$A(x) = \pi(PQ)^2 = \pi(\sqrt{x})^2 = \pi x$$

In particular, when $x = 2$, $A(2) = 2\pi$ square units.

A region such as R may be revolved around other lines. For example, we can use the line $y = -1$ as an *axis of revolution*. In such cases it is important to realize that the region R always stays the same distance from the axis. In this case, the inner boundary of R is always 1 unit from the axis $y = -1$.

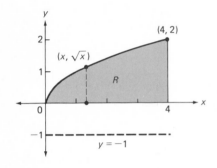

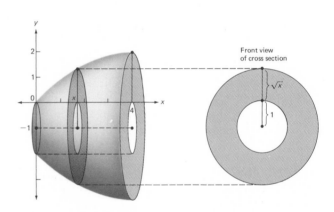

Front view
of cross section

The resulting three-dimensional solid is a bowl-shaped figure with a cylindrical hole through its center. A typical cross section, at x between 0 and 4, is the rim bounded by two concentric circles. The inner circle has radius 1, and the larger circle has radius $1 + \sqrt{x}$. Therefore the area of the cross section is the difference of the areas of these circles:

$$A(x) = \pi(1 + \sqrt{x})^2 - \pi(1)^2$$
$$= \pi[(1 + \sqrt{x})^2 - 1]$$
$$= \pi[1 + 2\sqrt{x} + x - 1]$$
$$= \pi[2\sqrt{x} + x]$$

Specifically, if $x = 1$, we have $A(1) = 3\pi$.

Caution: This is wrong:

$$A(x) = \pi[(1 + \sqrt{x}) - 1]^2$$
$$= \pi(\sqrt{x})^2$$
$$= \pi x$$

The mistake here is that the radii have been subtracted prematurely. The value $\pi(\sqrt{x})^2$ is the area of a circle with radius $\sqrt{x}$.

You can now see that this is not the correct shape of the cross section in question. Since the correct cross section is a rim bounded by two concentric circles, you must subtract *areas* of circles, not their radii.

Example 1   Consider the region R given at the beginning of this section. Find the area of a cross section of the solid generated by revolving the region R around the line $y = 2$.

Solution   The diagram shows the resulting solid with a front view of a cross section at x between 0 and 4.

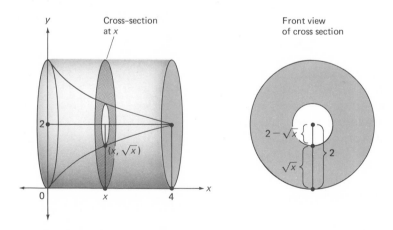

The outer radius of the cross section is 2, and the inner circle has radius $2 - \sqrt{x}$.

Then

$$A(x) = \pi(2)^2 - \pi(2 - \sqrt{x})^2$$
$$= \pi[4 - (2 - \sqrt{x})^2]$$
$$= \pi[4 - (4 - 4\sqrt{x} + x)]$$
$$= \pi(4\sqrt{x} - x)$$

TEST YOUR UNDERSTANDING

Let R be the region whose boundary consists of the curve $y = f(x) = x^2$, the segment [0, 2] on the x-axis, and the vertical line segment at $x = 2$. Find the area of a typical cross section of the solid formed by revolving R about the indicated axes.

1. The *x*-axis.   **2.** The line $y = -1$.   **3.** The line $y = -2$.

4. The line $y = 4$.   **5.** The line $y = 5$.

A two-dimensional region may be completely determined by two intersecting curves. In the next example such a region is used to generate a solid.

Example 2   Let the boundary of region *R* be the curves $y = f(x) = \sqrt{x}$ and $y = g(x) = \frac{1}{2}x$ between their points of intersection. Sketch the region *R* and the resulting solid when *R* is revolved about the *x*-axis. Find the area of a typical cross section.

Solution   At the points where the two curves intersect, the *x*- and *y*-values are the same. Therefore at such points we have

$$f(x) = \sqrt{x} = \tfrac{1}{2}x = g(x)$$

Now solve for *x* by squaring.

$$x = \tfrac{1}{4}x^2$$
$$4x = x^2$$
$$x^2 - 4x = 0$$
$$x(x - 4) = 0$$
$$x = 0 \quad \text{or} \quad x = 4$$

These values produce the points of intersection (0, 0) and (4, 2). The figures show R and the solid of revolution.

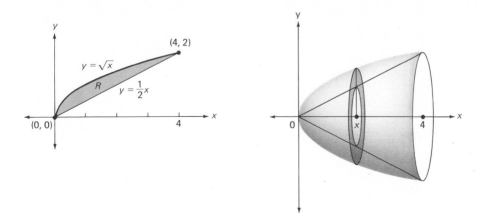

For x between 0 and 4, the cross section is a rim determined by two concentric circles. The inner circle has radius $\frac{1}{2}x$, and the outer circle has radius $\sqrt{x}$. The area of the cross section is

$$A(x) = \pi(\sqrt{x})^2 - \pi(\tfrac{1}{2}x)^2$$
$$= \pi(x - \tfrac{1}{4}x^2)$$

Why is the area of this cross section not equal to $\pi(\sqrt{x} - \frac{1}{2}x)^2$?

EXERCISES 6.6

1. The boundary of the region R is given by the equation $y = \frac{1}{2}x$, $y = 0$, and $x = 4$.
   (a) Draw the region R.
   (b) Draw a figure of the solid obtained by revolving R about the x-axis; include a typical cross section. Draw a separate figure of the front view of the cross section.
   (c) Express the cross-sectional area A as a function of x.

2. Region R of Exercise 1 is revolved about the line $y = 2$. Follow the instructions in parts (b) and (c) of Exercise 1 for this solid.

3. Region R of Exercise 1 is revolved about the line $y = 3$. Follow the instructions in parts (b) and (c) of Exercise 1 for this solid.

4. Region R of Exercise 1 is revolved about the line $y = -2$. Follow the instructions in parts (b) and (c) of Exercise 1 for this solid.

The region R in the figure is bounded by the curve $y = \frac{1}{2}x^2$, the y-axis, and the line $y = 2$.

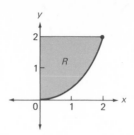

For Exercises 5 through 8, draw a picture of the solid when R is revolved about the indicated axis; include a typical cross section and give its front view. Express the area of the
cross section as a function of x.

5. x-axis   **6.** $y = 2$ **7.** $y = 3$ **8.** $y = -1$

9. The region R is bounded by the curve $y = x^{1/3}$, the x-axis, and the vertical line
$x = 8$. If R is revolved about the x-axis, what is the area of the cross section of the
solid of revolution?

10. Draw a picture of the region bounded by $y = x^2$ and $y = x$.

For Exercises 11 through 14, revolve the region R given in Exercise 10 about the indicated
axis and express the cross-sectional area of the resulting solid as a function of x.

11. x-axis   **12.** $y = 1$ **13.** $y = 2$ **14.** $y = -2$

15. The region R is bounded by $y = \dfrac{1}{x}$ and $y = -x + \dfrac{5}{2}$. A solid is formed by revolving R about the x-axis. What is the area of a typical cross section of this solid?

16. Region R is bounded by the curves $y = \frac{1}{4}x^2$ and $y = 2\sqrt{x}$. If R is revolved about
the x-axis, what is the area of a typical cross section of the resulting solid?

REVIEW EXERCISES FOR CHAPTER 6

*The solutions to the following exercises can be found within the text of Chapter 6. Try to
answer each question without referring to the text.*

Section 6.1

1. Find the domain of $y = \sqrt{x - 2}$ and graph.

2. Find the domain of $y = x^{-1/2}$ and graph.

3. Find the domain of $y = \sqrt[3]{x + 1} - 2$ and graph.

4. Graph $y = |\sqrt[3]{x}|$.

Section 6.2

5. Solve $\sqrt[3]{x+1} - 2 = 0$.

6. Solve $\sqrt{x^2 - x - 6} = 0$.

7. Solve $\sqrt{x+4} + 2 = x$.

8. Solve $x\sqrt{3x} + \sqrt{75x} = 2\sqrt{27x}$.

9. Solve $\sqrt{x-7} - \sqrt{x} = 1$.

10. Solve $\dfrac{3(2x-1)^{1/2}}{x-3} - \dfrac{(2x-1)^{3/2}}{(x-3)^2} = 0$.

11. Solve $\sqrt[3]{x^2} + \sqrt[3]{x} - 20 = 0$.

12. Solve the system and graph:

$$y = \sqrt[3]{x}$$
$$y = \tfrac{1}{4}x$$

Section 6.3

13. Find the domain of $f(x) = (x-5)\sqrt{x-3}$ and determine the signs of f.

14. Find the domain of $f(x) = \dfrac{x}{\sqrt{4-x^2}}$ and determine its signs.

15. Determine the signs of $f(x) = x^{1/2} + \tfrac{1}{2}x^{-1/2}(x-3)$.

16. Find the domain of f, solve $f(x) = 0$, and determine the signs of f, where

$$f(x) = \dfrac{x^2 - 1}{2\sqrt{x-1}} + 2x\sqrt{x-1}$$

Section 6.4

17. Determine $(f+g)(x)$, $(f-g)(x)$, $(f \cdot g)(x)$, and $\dfrac{f}{g}(x)$, where $f(x) = \dfrac{1}{x^3 - 1}$ and $g(x) = \sqrt{x}$.

18. What are the domains of $f+g, f-g$, and $f \cdot g$ given in Exercise 17?

19. What is the domain of $\dfrac{f}{g}$ given in Exercise 17?

20. Evaluate $g\left(f\left(\tfrac{9}{4}\right)\right)$ and $f\left(g\left(\tfrac{9}{4}\right)\right)$, where $f(x) = \dfrac{1}{x-2}$ and $g(x) = \sqrt{x}$.

21. Find $(g \circ f)(x)$ for f and g in Exercise 20 and state the domain.

22. Find $(f \circ g)(x)$ for f and g in Exercise 20 and state the domain.

23. Find $(f \circ g \circ h)(x)$, where $f(x) = \sqrt{x}$, $g(x) = x^2 + 1$, and $h(x) = \dfrac{1}{x}$.

Section 6.5

24. Find functions f and g so that $h(x) = \sqrt{x^2 + 2x + 2} = (g \circ f)(x)$.

25. Find functions f and g so that $h = f \circ g$, where, $h(x) = \left(\dfrac{1}{3x-1}\right)^5$.

26. Decompose $h(x) = \sqrt{(x^2 - 3x)^5}$ into two functions so that the inner function is a binomial.

27. Show that $h(x) = \sqrt{(x^2 - 3x)^5}$ can be written as the composition of three functions.

Section 6.6

For Exercises 28 through 30, let R be the region whose boundary consists of the curve $y = \sqrt{x}$ and the lines $y = 0$ and $x = 4$. Find the area of the cross section of the solid obtained by revolving R about the indicated axes.

28. The x-axis.

29. The line $y = -1$.

30. The line $y = 2$.

31. The region R bounded by the curves $y = \sqrt{x}$ and $y = \frac{1}{2}x$ is revolved about the x-axis to form a solid. What is the cross-sectional area?

SAMPLE TEST FOR CHAPTER 6

Graph each function, state its domain, and give the equations of the asymptotes if there are any.

1. $f(x) = -\sqrt[3]{x - 2}$

2. $g(x) = \dfrac{1}{\sqrt{x}} + 2$

Solve each equation.

3. $\sqrt{18x + 5} - 9x = 1$

4. $6x^{2/3} + 5x^{1/3} - 4 = 0$

For Problems 5 and 6, determine the signs of f.

5. $f(x) = \dfrac{\sqrt[3]{x - 4}}{(x - 2)}$

6. $f(x) = x^{1/2} - \frac{1}{2}x^{-1/2}(x + 4)$

7. Let $f(x) = \dfrac{1}{x^2 - 1}$ and $g(x) = \sqrt{x + 2}$. Find $(f - g)(x)$ and $\dfrac{f}{g}(x)$ and state their domains.

8. For $f(x) = \dfrac{1}{1 - x^2}$ and $g(x) = \sqrt{x}$ find the composites $(f \circ g)(x)$ and $(g \circ f)(x)$ and state their domains.

9. Find functions f and g so that $h(x) = \sqrt[3]{(x - 2)^2} = (f \circ g)(x)$, where g is a binomial.

10. Let $F(x) = \dfrac{1}{(2x - 1)^{3/2}}$. Find functions f, g, and h so that $(f \circ g \circ h)(x) = F(x)$.

†11. Let R be the region whose boundary is formed by $y = \sqrt{x - 2}$, the x-axis, and the vertical line $x = 6$. Let R be revolved about the axis $y = -1$ to form a solid. Express the area of a cross section of this solid as a function of x.

Answers To The Test Your Understanding Exercises

Page 250

1. Domain: $x \geq 0$

3. Domain: $x \geq 0$

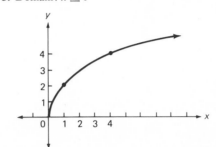

2. Domain: $x \geq 1$

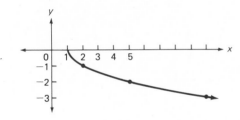

4. Domain: $x > -2$

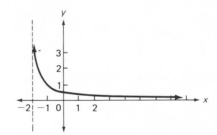

Page 254

1. 8

2. 25

3. 7

4. No solutions.

5. $\frac{1}{9}$

6. 5

7. $-3, 3$

8. 5

9. 0

10. $\frac{1}{2}$

Page 259

1. Domain: $x > 0$
$f(x) > 0$ on $(4, \infty)$
$f(x) < 0$ on $(0, 4)$

2. Domain: $x > -2$ and $x \neq 0$
$f(x) > 0$ on $(0, \infty)$
$f(x) < 0$ on $(-2, 0)$

3. Domain: all $x \neq 0$
$f(x) > 0$ on $(0, \infty)$
$f(x) < 0$ on $(-\infty, 0)$

Page 264

1. (a) $g(f(1)) = g(2) = 8$
(b) $f(g(1)) = f(5) = 14$
(c) $f(g(0)) = f(4) = 11$
(d) $g(f(-2)) = g(-7) = 53$

2. (a) $g(f(1)) = g(1) = 1$
(b) $f(g(1)) = f(1) = 1$
(c) $f(g(0)) = f(0) = 0$
(d) $g(f(-2))$ is undefined.

3. (a) $g(f(1)) = g(\sqrt[3]{2}) = 5\sqrt[3]{2}$
  (b) $f(g(1)) = f(5) = \sqrt[3]{14}$
  (c) $f(g(0)) = f(0) = -1$
  (d) $g(f(-2)) = g(\sqrt[3]{-7}) = -5\sqrt[3]{7}$

4. (a) $g(f(1))$ is undefined.
  (b) $f(g(1))$ is undefined.
  (c) $f(g(0)) = f(0) = -2$
  (d) $g(f(-2)) = g(0) = 0$

Page 269

(Other answers are possible.)
1. $f(x) = 8x - 3 ; g(x) = x^5$
2. $f(x) = 8x - 3 ; g(x) = \sqrt[5]{x}$
3. $f(x) = \dfrac{1}{8x - 3} ; g(x) = \sqrt{x}$

4. $f(x) = \dfrac{5}{7 + 4x^2} ; g(x) = x^3$

5. $f(x) = \dfrac{2x + 1}{2x - 1} ; g(x) = x^4$

6. $f(x) = x^4 - 2x^2 + 1 ; g(x) = \sqrt{x^3}$

Page 274

1. $\pi(x^2)^2 = \pi x^4$
2. $\pi(1 + x^2)^2 - \pi(1)^2 = \pi(2x^2 + x^4)$
3. $\pi(2 + x^2)^2 - \pi(2)^2 = \pi(4x^2 + x^4)$

4. $\pi(4)^2 - \pi(4 - x^2)^2 = \pi(8x^2 - x^4)$
5. $\pi(5)^2 - \pi(5 - x^2)^2 = \pi(10x^2 - x^4)$

Seven

Exponential and Logarithmic Functions

Inverse
Functions
By definition, a function has each domain value x corresponding to exactly one range value y. Some (but not all) functions have the additional property that to every range value y there corresponds exactly one domain value x. Such functions are said to be **one-to-one functions**. To understand this concept, consider the functions $y = x^2$ and $y = x^3$.

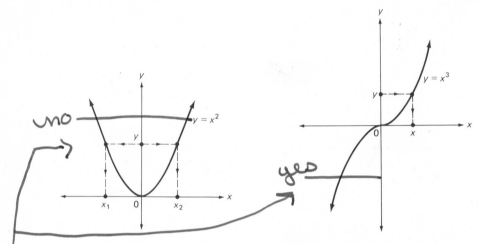

There are two domain values for a range value $y > 0$.	There is exactly one domain value x for each range value y.
$y = x^2$ *is not* a one-to-one function.	$y = x^3$ *is* a one-to-one function.

Once the graph of a function is known there is a simple visual test for determining the one-to-one property. Consider a horizontal line through each range value y, as in the preceding figures. If the line meets the curve exactly once, then we have a one-to-one function; otherwise it is not one-to-one.

TEST YOUR UNDERSTANDING

Use the visual test described above to determine which of the following are one-to-one functions.

1. $y = x^2 - 2x + 1$ **2.** $y = \sqrt{x}$ **3.** $y = \dfrac{1}{x}$

4. $y = |x|$ **5.** $y = 2x + 1$ **6.** $y = \sqrt[3]{x}$

7. $y = -x$ **8.** $y = [x]$

If the variables in $y = x^3$ are interchanged, we obtain $x = y^3$. Here are the graphs for these equations, together with the two graphs on the same axes.

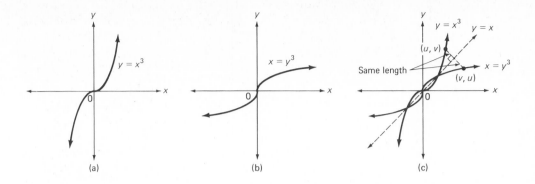

(a) (b) (c)

Due to this interchange of coordinates the two curves are reflections of each other through the line $y = x$, as in part (c) of the figure. Another way to describe this relationship is to say that they are mirror images of one another through the "mirror line" $y = x$; if the paper were folded along $y = x$, the curves would coincide.

The equation $x = y^3$ may be solved for y by taking the cube root of both sides: $y = \sqrt[3]{x}$. The two equations $x = y^3$ and $y = \sqrt[3]{x}$ are equivalent and therefore define the same function. However, since $y = \sqrt[3]{x}$ shows *explicitly* how y depends on x, it is the preferred form.

We began with the one-to-one function $y = f(x) = x^3$ and, by interchanging coordinates, arrived at the new function $y = g(x) = \sqrt[3]{x}$. If the composites $f \circ g$ and $g \circ f$ are formed, something surprising happens.

$$(f \circ g)(x) = f(g(x)) = f(\sqrt[3]{x}) = (\sqrt[3]{x})^3 = x$$

and

$$(g \circ f)(x) = g(f(x)) = g(x^3) = \sqrt[3]{x^3} = x$$

In each case we obtained the same value x that we started with; whatever one of the functions does to a value x, the other function undoes. Whenever two functions act on each other in such a manner, we say they are *inverse functions* or that either function is the inverse of the other.

Definition of inverse functions

Two functions f and g are said to be *inverse functions* if and only if:

1. For each x in the domain of g, $g(x)$ is in the domain of f and

$$(f \circ g)(x) = f(g(x)) = x$$

2. For each x in the domain of f, $f(x)$ is in the domain of g and

$$(g \circ f)(x) = g(f(x)) = x$$

It turns out (as suggested by our work with $y = x^3$) that every one-to-one function f has an inverse g and that their graphs are reflections of each other through the line $y = x$. The technique of interchanging variables, used to obtain $y = \sqrt[3]{x}$ from $y = x^3$, can be applied to many situations, as illustrated in the following examples.

Example 1   Find the inverse g of $y = f(x) = 2x + 3$. Then show that $(f \circ g)(x) = x = (g \circ f)(x)$, and graph both on the same axes.

Solution   Interchange variables in $y = 2x + 3$ and solve for y.

$$x = 2y + 3$$
$$2y = x - 3$$
$$y = \tfrac{1}{2}x - \tfrac{3}{2}$$

Using $y = g(x) = \tfrac{1}{2}x - \tfrac{3}{2}$ we have

$$(f \circ g)(x) = f(g(x)) = f(\tfrac{1}{2}x - \tfrac{3}{2})$$
$$= 2(\tfrac{1}{2}x - \tfrac{3}{2}) + 3$$
$$= x - 3 + 3$$
$$= x$$
$$(g \circ f)(x) = g(f(x)) = g(2x + 3)$$
$$= \tfrac{1}{2}(2x + 3) - \tfrac{3}{2}$$
$$= x + \tfrac{3}{2} - \tfrac{3}{2}$$
$$= x$$

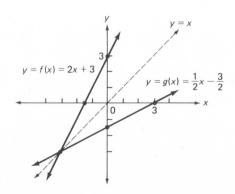

Example 2   Follow the instructions in Example 1 for $y = f(x) = \sqrt{x}$.

Solution   Interchange variables in $y = \sqrt{x}$ to get $x = \sqrt{y}$. At this point we see that x cannot be negative: $x \geq 0$. Solving for y by squaring produces $y = x^2$. Using the letter g for this inverse function, we have $y = g(x) = x^2$ with domain $x \geq 0$.

$$(f \circ g)(x) = f(g(x)) = f(x^2) = \sqrt{x^2} = |x| = x \qquad (\text{since } x \geq 0)$$
$$(g \circ f)(x) = g(f(x)) = g(\sqrt{x}) = (\sqrt{x})^2 = x$$

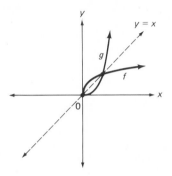

The process of interchanging variables to find the inverse of a function can be explained in terms of the criteria $f(g(x)) = x$. Begin with a function, say $f(x) = 3x - 2$, and let g be its inverse. Thus $f(g(x)) = x$. Substituting $g(x)$ into the equation for f gives

$$x = f(g(x)) = 3g(x) - 2$$

Solving this last equation for $g(x)$ produces

$$g(x) = \tfrac{1}{3}x + \tfrac{2}{3}$$

Now you can see that in taking $y = f(x) = 3x - 2$ and interchanging the variables to get $x = 3y - 2$, we are really letting y stand for the inverse function g. Thus

$$x = 3y - 2 = 3g(x) - 2 = f(g(x))$$

In many cases you will find it more efficient to use the form $x = 3y - 2$ and solve for y. This procedure is notationally simpler than using $x = 3g(x) - 2$ and solving for $g(x)$. But regardless of which notation is used this procedure has its limitations. Thus if the defining expression $y = f(x)$ is complicated it may be algebraically difficult, or even impossible, to do what we did in the special cases above. We will avoid such situations by limiting our work in this section to functions for which the procedure can be applied.

EXERCISES 7.1

In Exercises 1 through 6 show that f and g are inverse functions according to the criteria $(f \circ g)(x) = (g \circ f)(x) = x$. Then graph both functions and the line $y = x$ on the same axes.

1. $f(x) = \frac{1}{3}x - 3$; $g(x) = 3x + 9$

2. $f(x) = 2x - 6$; $g(x) = \frac{1}{2}x + 3$

3. $f(x) = (x + 1)^3$; $g(x) = \sqrt[3]{x} - 1$

4. $f(x) = -(x + 2)^3$; $g(x) = -\sqrt[3]{x} - 2$

5. $f(x) = \dfrac{1}{x - 1}$; $g(x) = \dfrac{1}{x} + 1$

6. $f(x) = x^2 + 2$ for $x \geq 2$; $g(x) = \sqrt{x - 2}$

In Exercises 7 through 18 find the inverse function g of the given function f.

7. $y = f(x) = (x - 5)^3$ **8.** $y = f(x) = x^{1/3} - 3$

9. $y = f(x) = \frac{2}{3}x - 1$ **10.** $y = f(x) = -4x + \frac{2}{5}$

11. $y = f(x) = (x - 1)^5$ **12.** $y = f(x) = -x^5$

13. $y = f(x) = x^{3/5}$ **14.** $y = f(x) = x^{5/3} + 1$

15. $y = f(x) = \dfrac{2}{x - 2}$ **16.** $y = f(x) = -\dfrac{1}{x} - 1$

17. $y = f(x) = x^{-5}$ **18.** $y = f(x) = \dfrac{1}{\sqrt[3]{x - 2}}$

In Exercises 19 through 24 find the inverse g of the given function f, and graph both in the same coordinate system.

19. $y = f(x) = (x + 1)^2; x \geq -1$ **20.** $y = f(x) = x^2 - 4x + 4; x \geq 2$

21. $y = f(x) = \dfrac{1}{\sqrt{x}}$ **22.** $y = f(x) = -\sqrt{x}$

23. $y = f(x) = x^2 - 4; x \geq 0$ ***24.** $y = f(x) = 4 - x^2; 0 \leq x \leq 2$

*25. Aside from the linear function $y = x$, what other linear functions are their own inverses? (*Hint:* Inverse functions are reflections of one another through the line $y = x$.)

*26. In the figure at the top of the next page, curve C_1 is the graph of a one-to-one function $y = f(x)$ and curve C_2 is the graph of $y = g(x)$. The points on C_2 have been obtained by interchanging the first and second coordinates of the points on curve C_1. For a specific value x in the domain of f, the point $P(x, f(x))$ on C_1 produces $Q(f(x), x)$ on C_2. How does this figure demonstrate that $(g \circ f)(x) = x$?

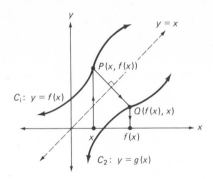

Imagine that a bacterial culture is growing at such a rate that after each hour the number of bacteria has doubled. Thus if there were 10,000 bacteria when the culture started to grow, then after 1 hour the number would have grown to 20,000, after 2 hours there would be 40,000, and so on. Now consider these equalities:

$$20{,}000 = (10{,}000)2^1$$

$$40{,}000 = (10{,}000)2^2$$

$$80{,}000 = (10{,}000)2^3$$

It becomes reasonable to say that

$$y = (10{,}000)2^x$$

gives the number y of bacteria present after x hours. This equation defines an **exponential function** with independent variable x and dependent variable y. We call such a function exponential because its exponent is a variable. In this section we will study exponential functions of the form $y = b^x$ for $b > 0$. The number b is often referred to as the *base* number.

What does the graph of $y = f(x) = 2^x$ look like? We can get a good idea by forming a table of values. Note that in this table we are using domain values x that are integers; also recall that $2^{-x} = \dfrac{1}{2^x}$.

x	-3	-2	-1	0	1	2	3
$y = 2^x$	$\dfrac{1}{8}$	$\dfrac{1}{4}$	$\dfrac{1}{2}$	1	2	4	8

Connecting the points given in the table with a smooth curve produces this graph.

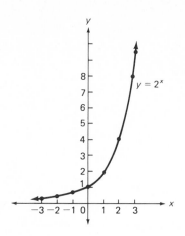

The accuracy of this graph can be improved by using more points. Consider rational values of x, such as $\frac{1}{2}$ or $\frac{3}{2}$:

$$2^{1/2} = \sqrt{2} = 1.4 \qquad \text{(correct to one decimal place from Appendix Table I)}$$

$$2^{3/2} = (\sqrt{2})^3 = (1.4)^3 = 2.7 \qquad \text{(correct to one decimal place)}$$

Using irrational values for x, such as $\sqrt{2}$ or π, is another matter entirely. (Remember that our development of exponents stopped with the rationals.) To give a precise meaning of such numbers is beyond the scope of this course. It does turn out, however, that the indicated shape of the curve $y = 2^x$ is correct and that the formal definitions of values like $2^{\sqrt{2}}$ are made so that they "fit" the curve.

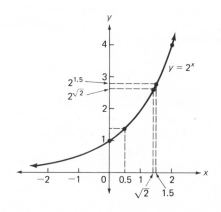

We will go no further into the meaning of numbers like $2^{\sqrt{2}}$ than the preceding geometric description. In advanced work it can be shown that for any positive bases a and b, the following hold for all real numbers r and s.

$$b^r b^s = b^{r+s} \qquad \frac{b^r}{b^s} = b^{r-s} \qquad (b^r)^s = b^{rs}$$

$$a^r b^r = (ab)^r \qquad b^0 = 1 \qquad b^{-r} = \frac{1}{b^r}$$

Our earlier work with the same rules, for rational exponents, can now serve as the basis for accepting these results.

Example 1 Graph the curve $y = 8^x$ on the interval $[-1, 1]$ by using a table of values.

Solution

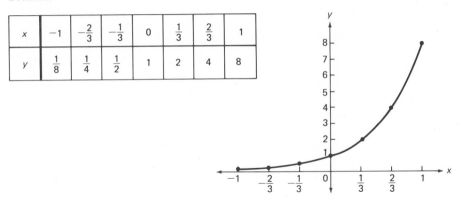

x	-1	$-\frac{2}{3}$	$-\frac{1}{3}$	0	$\frac{1}{3}$	$\frac{2}{3}$	1
y	$\frac{1}{8}$	$\frac{1}{4}$	$\frac{1}{2}$	1	2	4	8

(Note the change in scale on the axes.)

Our attention has been restricted to the exponential functions $y = b^x$, where $b > 1$, all of which have the same shaped graph as $y = 2^x$. For $b = 1$, we get $y = 1^x = 1$ for all x. Since this is a constant function, we do not use base $b = 1$ in the classification of exponential functions. Can you think of reasons why values of $b \leq 0$ are also excluded?

The remaining base values for our purposes are those for which $0 < b < 1$. With $b = \frac{1}{2}$ we get $y = \left(\frac{1}{2}\right)^x = \frac{1}{2^x}$.

x	-2	-1	0	1	2
y	4	2	1	$\frac{1}{2}$	$\frac{1}{4}$

$y \geq b^x$

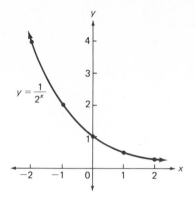

$y = \frac{1}{2^x}$

All the curves $y = b^x$, for $0 < b < 1$, have this same basic shape.

The graph of $y = g(x) = \dfrac{1}{2^x}$ can also be found by comparing it to the graph of $y = f(x) = 2^x$. Since $g(x) = \dfrac{1}{2^x} = 2^{-x} = f(-x)$, the y-values for g are the same as the y-values for f on the opposite side of the y-axis. In other words, the graph of g is the reflection of the graph of f through the y-axis.

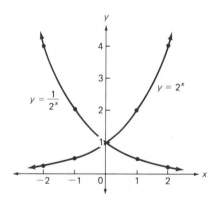

$y = \frac{1}{2^x}$ $y = 2^x$

Example 2 Graph $y = f(x) = 3^x$ by using a table of values. Then sketch $y = \dfrac{1}{3^x}$ by reflecting through the y-axis.

Solution

x	-2	-1	0	1	2
$y = 3^x$	$\frac{1}{9}$	$\frac{1}{3}$	1	3	9

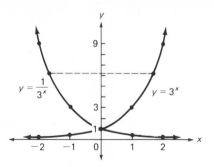
given $y = \frac{1}{b}x$; graph
$y = b^x$ + reflect through
y axis

The graph of 3^x is as follows:

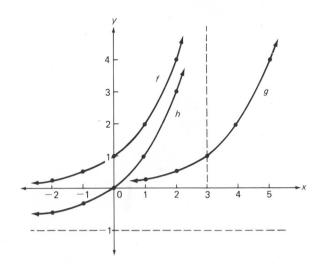

$y = \dfrac{1}{3^x}$ $y = 3^x$

The next example illustrates how to obtain the graphs of exponential functions that are modifications of the form $y = b^x$.

Example 3   Use the graph of $y = f(x) = 2^x$ to sketch the curves $y = g(x) = 2^{x-3}$ and $y = h(x) = 2^x - 1$.

Solution   Since $g(x) = f(x - 3)$ the graph of g can be obtained by shifting $y = 2^x$ by 3 units to the right. Moreover, since $h(x) = f(x) - 1$ the graph of h can be found by shifting $y = 2^x$ down 1 unit.

At the top of the next page are some important properties of the function $y = b^x$, for $b > 0$ and $b \neq 1$.

Properties of $y = f(x) = b^x$

1. The domain consists of all real numbers x.
2. The range consists of all positive numbers y.
3. The function is increasing to the right (the curve is rising) when $b > 1$, and it is decreasing (the curve is falling) when $0 < b < 1$.
4. The curve bends upward for $b > 1$ and downward for $0 < b < 1$.
5. It is a one-to-one function.
6. The point $(0, 1)$ is on the curve. There is no x-intercept.
7. The x-axis is asymptotic to the curve toward the left for $b > 1$ and toward the right for $0 < b < 1$.

For any one-to-one function f we may say that *whenever $f(x_1) = f(x_2)$, then $x_1 = x_2$*. That is, since $f(x_1)$ and $f(x_2)$ represent the same range value there can only be one corresponding domain value; consequently $x_1 = x_2$. Using $f(x) = b^x$, this statement means that if $b^{x_1} = b^{x_2}$, then $x_1 = x_2$.

Example 4 Use the one-to-one property to solve for t in the exponential equation $5^{t^2} = 625$.

Solution Since $625 = 5^4$ we get

$$5^{t^2} = 5^4$$

Now use the one-to-one property of $f(x) = 5^x$ to equate exponents. Thus

$$t^2 = 4$$
$$t = \pm 2$$

Check: $5^{(\pm 2)^2} = 5^4 = 625$.

EXERCISES 7.2

For Exercises 1 through 8 graph the exponential function f by making use of a brief table of values. Then use this curve to sketch the graph of g.

1. $f(x) = 2^x$; $g(x) = 2^{x+3}$
2. $f(x) = 3^x$; $g(x) = 3^x - 2$
3. $f(x) = 4^x$; $g(x) = -(4^x)$
4. $f(x) = 5^x$; $g(x) = (\frac{1}{5})^x$
5. $f(x) = (\frac{3}{2})^x$; $g(x) = (\frac{2}{3})^x$
6. $f(x) = 8^x$; $g(x) = 8^{x-2} + 3$
7. $f(x) = 3^x$; $g(x) = 2(3^x)$
8. $f(x) = 3^x$; $g(x) = \frac{1}{2}(3^x)$

9. Sketch the curves $y = (\frac{3}{2})^x$, $y = 2^x$, and $y = (\frac{5}{2})^x$ on the same axes.

10. Sketch the curves $y = (\frac{1}{4})^x$, $y = (\frac{1}{3})^x$, and $y = (\frac{1}{2})^x$ on the same axes.

In Exercises 11 through 28 use the one-to-one property of an appropriate exponential function to solve the indicated equation.

11. $2^x = 64$ 12. $3^x = 81$ 13. $2^{x^2} = 512$

14. $3^{x-1} = 27$ 15. $5^{2x+1} = 125$ 16. $2^{x^3} = 256$

17. $7^{x^2+x} = 49$ 18. $b^{x^2+x} = 1$ 19. $\dfrac{1}{2^x} = 32$

20. $\dfrac{1}{10^x} = 10,000$ 21. $9^x = 3$ 22. $64^x = 8$

23. $9^x = 27$ 24. $64^x = 16$ 25. $\left(\dfrac{1}{49}\right)^x = 7$

26. $5^x = \dfrac{1}{125}$ 27. $\left(\dfrac{27}{8}\right)^x = \dfrac{9}{4}$ 28. $(0.01)^x = 1000$

*29. Find the equation of the horizontal asymptote to the curve $y = \left(\dfrac{1}{5}\right)^x - 3$.

*30. Find the equation of the horizontal asymptote to the curve $y = 10^x + 2$.

*31. Graph the functions $y = 2^x$ and $y = x^2$ in the same coordinate system for the interval $[0, 5]$. (Use a larger unit on the x-axis than on the y-axis.) What are the points of intersection?

7.3
Logarithmic Functions

It was pointed out in the last section that $y = f(x) = b^x$, for $b > 0$ and $b \neq 1$, is a one-to-one function. Since every one-to-one function has an inverse it follows that f has an inverse. The graph of the inverse function g is the reflection of $y = f(x)$ through the line $y = x$. Here are two typical cases for $b > 1$ and for $0 < b < 1$.

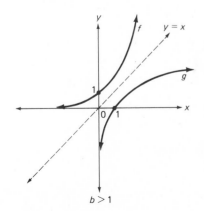

$b > 1$

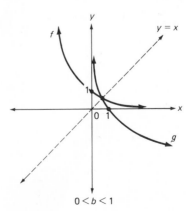

$0 < b < 1$

An equation for the inverse function g can be obtained by interchanging the roles of the variables in $y = b^x$. Thus $x = b^y$ is an equation for g. Unfortunately, we have no way of solving $x = b^y$ to get y explicitly in terms of x. To overcome this difficulty we create some new terminology.

The equation $x = b^y$ tells us that *y is the exponent on b that produces x.* In situations like this the word **logarithm** is used in place of *exponent*. A logarithm, then, is an exponent. Now we may say that *y is the logarithm on b that produces x.* This description can be abbreviated to the form

$$y = \text{logarithm}_b\ x$$

and abbreviating further we reach the final form

$$y = \log_b x$$

which is read "y equals log x to the base b" or simply, "y equals log x base b."

It is important to realize that we are only defining (not proving) the equation $y = \log_b x$ to have the same meaning as $x = b^y$. In other words, the logarithmic form $y = \log_b x$ and the exponential form $x = b^y$ are equivalent. And since they are equivalent they define the same function g: $y = g(x) = \log_b x$.

Now we know that $y = f(x) = b^x$ and $y = g(x) = \log_b x$ are inverse functions. Consequently,

$$f(g(x)) = f(\log_b x) = b^{\log_b x} = x$$

and

$$g(f(x)) = g(b^x) = \log_b(b^x) = x$$

Example 1   Write the equation of the inverse function g of $y = f(x) = 2^x$ and graph both on the same axes.

Solution   The inverse g has equation $y = g(x) = \log_2 x$, and its graph can be obtained by reflecting $y = f(x) = 2^x$ through the line $y = x$.

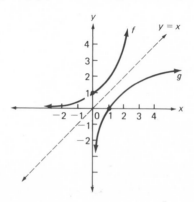

1. Find the equation of the inverse of $y = 3^x$ and graph both on the same axes.
2. Find the equation of the inverse of $y = (\frac{1}{3})^x$ and graph both on the same axes.

Let $y = f(x) = \log_5 x$. Describe how the graph of each of the following can be obtained from the graph of f.

3. $g(x) = \log_5 (x + 2)$ **4.** $g(x) = 2 + \log_5 x$

5. $g(x) = -\log_5 x$ **6.** $g(x) = 2 \log_5 x$

We found $y = \log_b x$ by interchanging the role of the variables in $y = b^x$. As a consequence of this switching, the domains and ranges of the two functions are also interchanged. Thus

Domain of $y = \log_b x$ is the same as the range of $y = b^x$.

Range of $y = \log_b x$ is the same as the domain of $y = b^x$.

These results are incorporated into the following list of important properties of the function $y = \log_b x$, where $b > 0$ and $b \neq 1$.

Properties of y = f(x) = log$_b$ x

1. The domain consists of all positive numbers x.
2. The range consists of all real numbers y.
3. The function increases to the right (the curve is rising) for $b > 1$, and it decreases to the right (the curve is falling) for $0 < b < 1$.
4. The curve bends downward if $b > 1$ and upward if $0 < b < 1$.
5. It is a one-to-one function; if $\log_b (x_1) = \log_b (x_2)$, then $x_1 = x_2$.
6. The point $(1, 0)$ is on the graph. There is no y-intercept.
7. The y-axis is asymptotic to the curve in the downward direction for $b > 1$ and in the upward direction if $0 < b < 1$.

Example 2 Find the domain for $y = \log_2 (x - 3)$.

Solution In $y = \log_2 (x - 3)$ the quantity $x - 3$ plays the role that x does in $\log_2 x$. Thus $x - 3 > 0$ and the domain consists of all $x > 3$.

The equation $y = \log_b x$ is equivalent to $x = b^y$ (not to be confused with the inverse $y = b^x$). Studying the following list of special cases will help you to understand this equivalence.

LOGARITHMIC FORM   EXPONENTIAL FORM

$$\log_5 25 = 2 \qquad\qquad 5^2 = 25$$
$$\log_{27} 9 = \tfrac{2}{3} \qquad\qquad 27^{2/3} = 9$$
$$\log_6 \tfrac{1}{36} = -2 \qquad\qquad 6^{-2} = \tfrac{1}{36}$$
$$\log_b 1 = 0 \qquad\qquad b^0 = 1$$

$3\ \sqrt{27^2} = 3^2$

Of the two forms $y = \log_b x$ and $x = b^y$, the exponential form is usually easier to think with. Consequently when there is a question concerning $y = \log_b x$ it is often useful to convert to the exponential form. For instance, to evaluate $\log_9 27$ we write

$$y = \log_9 27$$

and convert to exponential form. Thus

$$9^y = 27$$

If you happen to recognize that $9^{3/2} = 27$, you will see that $y = \tfrac{3}{2}$. Otherwise, try writing each side of $9^y = 27$ with the same base. Since $27 = 3^3$ and $9^y = (3^2)^y = 3^{2y}$, we have

$$3^{2y} = 3^3$$
$$2y = 3 \qquad (f(t) = 3^t \text{ is one-to-one})$$
$$y = \tfrac{3}{2}$$

Example 3   Solve for b: $\log_b 8 = \tfrac{3}{4}$.

Solution   Convert to exponential form.

$$b^{3/4} = 8$$

Take the $\tfrac{4}{3}$ power of both sides.

$$(b^{3/4})^{4/3} = 8^{4/3}$$
$$b = 16$$

EXERCISES 7.3

Sketch the graph of the function f. Reflect this curve through the line $y = x$ to obtain the graph of the inverse function g, and write the equation for g.

1. $y = f(x) = 4^x$ **2.** $y = f(x) = 5^x$

3. $y = f(x) = (\tfrac{1}{3})^x$ **4.** $y = f(x) = (0.2)^x$

Describe how the graph of h can be obtained from the graph of g. Find the domain of h, and write the equation of the vertical asymptote.

5. $g(x) = \log_3 x$; $h(x) = \log_3 (x + 2)$ **6.** $g(x) = \log_5 x$; $h(x) = \log_5 (x - 1)$

7. $g(x) = \log_8 x$; $h(x) = 2 + \log_8 x$ **8.** $g(x) = \log_{10} x$; $h(x) = 2 \log_{10} x$

Convert from the exponential to the logarithmic form.

9. $2^8 = 256$ **10.** $5^{-3} = \frac{1}{125}$ **11.** $(\frac{1}{3})^{-1} = 3$

12. $81^{3/4} = 27$ **13.** $17^0 = 1$ **14.** $(\frac{1}{49})^{-1/2} = 7$

Convert from the logarithmic form to the exponential form.

15. $\log_{10} 0.0001 = -4$ **16.** $\log_{64} 4 = \frac{1}{3}$ **17.** $\log_{\sqrt{2}} 2 = 2$

18. $\log_{13} 13 = 1$ **19.** $\log_{12} \frac{1}{1728} = -3$ **20.** $\log_{27/8} \frac{9}{4} = \frac{2}{3}$

Solve for the indicated quantity: y, x, or b.

21. $\log_2 16 = y$ **22.** $\log_{1/2} 32 = y$ **23.** $\log_{1/3} 27 = y$

24. $\log_7 x = -2$ **25.** $\log_{1/6} x = 3$ **26.** $\log_8 x = -\frac{2}{3}$

27. $\log_b 125 = 3$ **28.** $\log_b 8 = \frac{3}{2}$ **29.** $\log_b \frac{1}{8} = -\frac{3}{2}$

30. $\log_{100} 10 = y$ **31.** $\log_{27} 3 = y$ **32.** $\log_{1/16} x = \frac{1}{4}$

33. $\log_b \frac{16}{81} = 4$ **34.** $\log_8 x = -3$ **35.** $\log_b \frac{1}{27} = -\frac{3}{2}$

36. $\log_{\sqrt{3}} x = 2$ **37.** $\log_{\sqrt{8}} \frac{1}{8} = y$ **38.** $\log_b \frac{1}{128} = -7$

39. $\log_{0.001} 10 = y$ **40.** $\log_{0.2} 5 = y$ **41.** $\log_9 x = 1$

Evaluate.

*42. $\log_2 (\log_4 256)$ ***43.** $\log_{3/4} (\log_{1/27} \frac{1}{81})$

By interchanging the roles of the variables, find the inverse function g. Show that $(f \circ g)(x) = x$ *and* $(g \circ f)(x) = x$.

*44. $y = f(x) = 2^{x+1}$ ***45.** $y = f(x) = \log_3 (x + 3)$

From the rules of exponents we have

$$2^3 \cdot 2^4 = 2^{3+4} = 2^7$$

Now let us focus on just the exponential part:

$$3 + 4 = 7$$

7.4

The Laws of Logarithms

The three exponents involved here can be expressed as logarithms (exponents are logarithms):

$$3 = \log_2 8 \quad \text{because } 2^3 = 8$$
$$4 = \log_2 16 \quad \text{because } 2^4 = 16$$
$$7 = \log_2 128 \quad \text{because } 2^7 = 128$$

Substituting these expressions into $3 + 4 = 7$ gives

$$\log_2 8 + \log_2 16 = \log_2 128$$

Furthermore, since $128 = 8 \cdot 16$, we have

$$\log_2 8 + \log_2 16 = \log_2 (8 \cdot 16)$$

This is a special case of the first law of logarithms:

Laws of logarithms

If M and N are positive, $b > 0$ and $b \neq 1$, then

LAW 1. $\log_b MN = \log_b M + \log_b N$

LAW 2. $\log_b \dfrac{M}{N} = \log_b M - \log_b N$

LAW 3. $\log_b (N^k) = k \log_b N$

Law (1) says that the log of a product is the sum of the logs of the factors. Can you give similar interpretations for Laws (2) and (3)?

Since logarithms are exponents it is not surprising that these laws can be proved by using the appropriate rules of exponents. Here is a proof of Law (1); the proofs of (2) and (3) are left as exercises:

Let

$$\log_b M = r \quad \text{and} \quad \log_b N = s$$

Convert to exponential form:

$$M = b^r \quad \text{and} \quad N = b^s$$

Multiply the two equations:

$$MN = b^r b^s = b^{r+s}$$

Then convert to logarithmic form:

$$\log_b MN = r + s$$

By substitution, we now have:

$$\log_b MN = \log_b M + \log_b N$$

The laws of logarithms will be useful in the study of calculus, and it will therefore be to our advantage to become familiar with them. The following problems are intended to serve this purpose.

Example 1   For positive numbers A, B, and C show that

$$\log_b \frac{AB^2}{C} = \log_b A + 2 \log_b B - \log_b C$$

Solution

$$\log_b \frac{AB^2}{C} = \log_b (AB^2) - \log_b C \qquad \text{(Law 2)}$$
$$= \log_b A + \log_b B^2 - \log_b C \qquad \text{(Law 1)}$$
$$= \log_b A + 2 \log_b B - \log_b C \qquad \text{(Law 3)}$$

Example 2   Express $\frac{1}{2} \log_b x - 3 \log_b (x - 1)$ as the logarithm of a single expression in x.

Solution

$$\frac{1}{2} \log_b x - 3 \log_b (x - 1) = \log_b x^{1/2} - \log_b (x - 1)^3 \qquad \text{(Law 3)}$$

$$= \log_b \frac{x^{1/2}}{(x - 1)^3} \qquad \text{(Law 2)}$$

$$= \log_b \frac{\sqrt{x}}{(x - 1)^3}$$

TEST YOUR UNDERSTANDING

Convert the given logarithms into expressions involving $\log_b A$, $\log_b B$, *and* $\log_b C$.

1. $\log_b ABC$

2. $\log_b \dfrac{A}{BC}$

3. $\log_b \dfrac{(AB)^2}{C}$

4. $\log_b AB^2C^3$

5. $\log_b \dfrac{A\sqrt{B}}{C}$

6. $\log_b \dfrac{\sqrt[3]{A}}{(BC)^3}$

Change each expression into the logarithm of a single expression in x.

7. $\log_b x + \log_b x + \log_b 3$

8. $2 \log_b (x - 1) + \frac{1}{2} \log_b x$

9. $\log_b (2x - 1) - 3 \log_b (x^2 + 1)$

10. $\log_b x - \log_b (x - 1) - 2 \log_b (x - 2)$

The remaining examples illustrate how the laws of logarithms can be used to solve logarithmic equations.

Example 3   Solve for x: $\log_8 (x - 6) + \log_8 (x + 6) = 2$.

Solution   First note that in $\log_8 (x - 6)$ we must have $x - 6 > 0$, or $x > 6$. Likewise, $\log_8 (x + 6)$ calls for $x > -6$. Therefore the only solutions, if there are any, must satisfy $x > 6$.

$$\log_8 (x - 6) + \log_8 (x + 6) = 2$$
$$\log_8 (x - 6)(x + 6) = 2 \qquad \text{(Law 1)}$$
$$\log_8 (x^2 - 36) = 2$$
$$x^2 - 36 = 8^2 \qquad \text{(converting to exponential form)}$$
$$x^2 - 100 = 0$$
$$(x + 10)(x - 10) = 0$$
$$x = -10 \quad \text{or} \quad x = 10$$

The only possible solutions are -10 and 10. Our initial observation that $x > 6$ automatically eliminates -10. (If that initial observation had not been made, -10 could still have been eliminated by checking in the given equation.) The value $x = 10$ can be checked as follows:

$$\log_8 (10 - 6) + \log_8 (10 + 6) = \log_8 4 + \log_8 16$$
$$= \tfrac{2}{3} + \tfrac{4}{3} = 2$$

Example 4   Solve:

$$\log_{10} (x^3 - 1) - \log_{10} (x^2 + x + 1) = 1.$$

Solution

$$\log_{10} (x^3 - 1) - \log_{10} (x^2 + x + 1) = 1$$

$$\log_{10} \frac{x^3 - 1}{x^2 + x + 1} = 1 \qquad \text{(Law 2)}$$

$$\log_{10} \frac{(x - 1)(x^2 + x + 1)}{x^2 + x + 1} = 1 \qquad \text{(by factoring)}$$

$$\log_{10} (x - 1) = 1$$

$$x - 1 = 10^1 \qquad \text{(Why?)}$$

$$x = 11$$

Check: $\log_{10} (11^3 - 1) - \log_{10} (11^2 + 11 + 1) = \log_{10} 1330 - \log_{10} 133$

$$= \log_{10} \tfrac{1330}{133}$$

$$= \log_{10} 10 = 1$$

Example 5   Solve: $\log_3 2x - \log_3 (x + 5) = 0$.

Solution

$$\log_3 2x - \log_3 (x + 5) = 0$$

$$\log_3 \frac{2x}{x + 5} = 0$$

$$\frac{2x}{x + 5} = 3^0$$

$$\frac{2x}{x + 5} = 1$$

$$2x = x + 5$$

$$x = 5$$

Check: $\log_3 2(5) - \log_3 (5 + 5) = \log_3 10 - \log_3 10 = 0$.

Alternate Solution

$$\log_3 2x - \log_3 (x + 5) = 0$$

$$\log_3 2x = \log_3 (x + 5)$$

$$2x = x + 5 \qquad \text{(Why?)}$$

$$x = 5$$

CAUTION! LEARN TO AVOID MISTAKES LIKE THESE:

WRONG	RIGHT
$\log_b A + \log_b B = \log_b (A + B)$	$\log_b A + \log_b B = \log_b AB$
$\log_b (x^2 - 4) = \log_b x^2 - \log_b 4$	$\log_b (x^2 - 4) = \log_b (x + 2)(x - 2)$ $= \log_b (x + 2) + \log_b (x - 2)$
$(\log_b x)^2 = 2 \log_b x$	$(\log_b x)^2 = (\log_b x)(\log_b x)$
$\log_b A - \log_b B = \dfrac{\log_b A}{\log_b B}$	$\log_b A - \log_b B = \log_b \dfrac{A}{B}$

EXERCISES 7.4

Use the laws of logarithms (as much as possible) to convert the given logarithms into expressions involving sums and differences.

1. $\log_b \dfrac{3x}{x + 1}$ **2.** $\log_b \dfrac{x^2}{x - 1}$ **3.** $\log_b \dfrac{\sqrt{x^2 - 1}}{x}$

4. $\log_b \dfrac{1}{x}$ **5.** $\log_b \dfrac{1}{x^2}$ **6.** $\log_b \sqrt{\dfrac{x + 1}{x - 1}}$

Convert each expression into the logarithm of a single expression in x.

7. $\log_b (x + 1) - \log_b (x + 2)$ **8.** $\log_b x + 2 \log_b (x - 1)$

9. $\frac{1}{2} \log_b (x^2 - 1) - \frac{1}{2} \log_b (x^2 + 1)$ **10.** $\log_b (x + 2) - \log_b (x^2 - 4)$

11. $3 \log_b x - \log_b 2 - \log_b (x + 5)$ **12.** $\frac{1}{3} \log_b (x - 1) + \log_b 3 - \frac{1}{3} \log_b (x + 1)$

Use the appropriate laws of logarithms to explain why each statement is correct.

13. $\log_b 27 + \log_b 3 = \log_b 243 - \log_b 3$

14. $\log_b 16 + \log_b 4 = \log_b 64$

15. $-2 \log_b \frac{4}{9} = \log_b \frac{81}{16}$

16. $\frac{1}{2} \log_b 0.0001 = -\log_b 100$

Solve for x and check.

17. $\log_{10} x + \log_{10} 5 = 2$ **18.** $\log_{10} x + \log_{10} 5 = 1$

19. $\log_{10} 5 - \log_{10} x = 2$ **20.** $\log_{10} (x + 21) + \log_{10} x = 2$

21. $\log_{12} (x - 5) + \log_{12} (x - 5) = 2$ **22.** $\log_3 x + \log_3 (2x + 51) = 4$

23. $\log_{16} x + \log_{16} (x - 4) = \frac{5}{4}$ **24.** $\log_2 (x^2) - \log_2 (x - 2) = 3$

25. $\log_{10} (3 - x) - \log_{10} (12 - x) = -1$

26. $\log_{10}(3x^2 - 5x - 2) - \log_{10}(x - 2) = 1$

27. $\log_{1/7} x + \log_{1/7}(5x - 28) = -2$

28. $\log_{1/3} 12x^2 - \log_{1/3}(20x - 9) = -1$

29. $\log_{10}(x^3 - 1) - \log_{10}(x^2 + x + 1) = -2$

30. $2\log_{10}(x - 2) = 4$

31. $2\log_{25} x - \log_{25}(25 - 4x) = \frac{1}{2}$

32. $\log_3(8x^3 + 1) - \log_3(4x^2 - 2x + 1) = 2$

*33. Prove Law (2). (*Hint:* Follow the proof of Law (1) using $\dfrac{b^r}{b^s} = b^{r-s}$.)

*34. Prove Law (3). (*Hint:* Use $(b^r)^k = b^{rk}$.)

*35. Solve for x: $(x + 2)\log_b b^x = x$.

*36. Solve for x: $\log_{N^2} N = x$.

*37. Solve for x: $\log_x (2x)^{3x} = 4x$.

*38. (a) Explain why $\log_b b = 1$.   **(b)** Show that $(\log_b a)(\log_a b) = 1$. (*Hint:* Use Law (3) and the result $b^{\log_b x} = x$.)

*39. Use $B^{\log_B N} = N$ to derive $\log_B N = \dfrac{\log_b N}{\log_b B}$. (*Hint:* Begin by taking the log base b of both sides.)

7.5

The Base e

The graphs of $y = b^x$ for $b > 1$ all have the same basic shape, as shown in the following figure. Notice that the larger the value of b, the faster the curve rises toward the right and the faster it approaches the x-axis toward the left. You can use your imagination to see that as all possible base values $b > 1$ are considered, the corresponding curves will completely fill in the shaded regions.

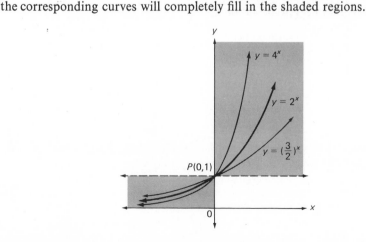

All such curves pass through point $P(0, 1)$. The tangent lines to these curves through P are relatively flat (small positive slope) for values of b close to 1 and very steep for large values of b. The slopes of these tangents consist of all numbers $m > 0$.

Parts (a) and (b) of the next figure show the curves $y = 2^x$ and $y = 3^x$, including the tangent through point $P(0, 1)$.

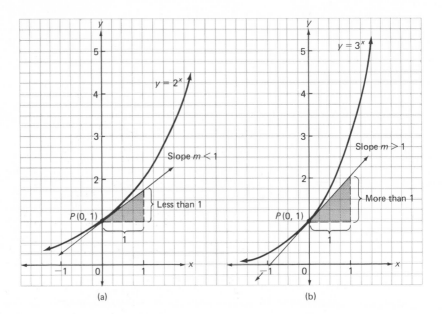

From the grid marks you can observe that the slope of the tangent to $y = 2^x$ is less than 1 because for a horizontal change of 1 unit the vertical change is less than 1 unit. Likewise, you can see that the slope of the tangent to $y = 3^x$ is slightly more than 1. We suspect that there must be a value b so that the slope of the tangent to the corresponding exponential function through point P is exactly equal to 1. In advanced courses it can be shown that such a value does exist. This number plays a very important role in mathematics, and it is designated by the letter e.

Definition of e

The number e is the real number such that the tangent through $P(0, 1)$ to the graph of $y = e^x$ is equal to 1.

Since the curve $y = e^x$ is between $y = 2^x$ and $y = 3^x$, we expect that e satisfies $2 < e < 3$. This is correct; in fact, it turns out that e is an irrational number that is closer to 3 than to 2. To five decimal places, $e = 2.71828$.

For theoretical purposes e is the most important base number for exponential and logarithmic functions. The inverse of $y = e^x$ is given by $y = \log_e x$. In place of $\log_e x$ we will now write $\ln x$, which is called the **natural log of** x. Thus $x = e^y$ and $y = \ln x$ are equivalent.

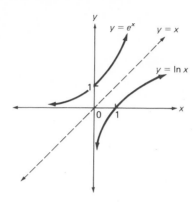

Since $e > 1$, the properties of $y = b^x$ and $y = \log_b x$ $(b > 1)$ carry over to $y = e^x$ and $y = \ln x$. We collect them here for easy reference.

Properties y = e^x	**Properties of y = ln x**
1. Domain: all reals.	1. Domain: all $x > 0$.
2. Range: all $y > 0$.	2. Range: all reals.
3. Increasing to the right.	3. Increasing to the right.
4. Curve bends upward.	4. Curve bends downward.
5. One-to-one function.	5. One-to-one function.
6. $0 < e^x < 1$ for $x < 0$; $\quad e^0 = 1$; $e^x > 1$ for $x > 0$.	6. $\ln x < 0$ for $0 < x < 1$; $\quad \ln 1 = 0$; $\ln x > 0$ for $x > 1$.
7. $e^{x_1}e^{x_2} = e^{x_1+x_2}$ $\quad \dfrac{e^{x_1}}{e^{x_2}} = e^{x_1-x_2}$ $\quad (e^{x_1})^{x_2} = e^{x_1 x_2}$	7. $\ln x_1 x_2 = \ln x_1 + \ln x_2$ $\quad \ln \dfrac{x_1}{x_2} = \ln x_1 - \ln x_2$ $\quad \ln x_1^{x_2} = x_2 \ln x_1$
8. $e^{\ln x} = x$.	8. $\ln e^x = x$.

The examples that follow utilize the base e, and are solved in a manner similar to those done earlier for other bases.

Example 1 Use $y = e^x$ to sketch $y = f(x) = e^{x+2}$ and $y = g(x) = e^x - 2$. What are the equations of the horizontal asymptotes to f and g?

Solution Shift $y = e^x$ left 2 units to graph $y = e^{x+2}$; asymptote: $y = 0$. Shift $y = e^x$ down 2 units to graph $y = e^x - 2$; asymptote: $y = -2$.

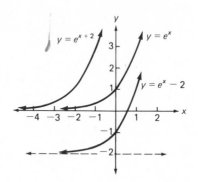

Example 2 (a) Find the domain of $y = \ln(x - 2)$; (b) Sketch $y = \ln x^2$ for $x > 0$.

Solution
(a) Since the domain of $y = \ln x$ is all $x > 0$, the domain of $y = \ln(x - 2)$ has all x for which $x - 2 > 0$; all $x > 2$.
(b) Since $y = \ln x^2 = 2 \ln x$, we obtain the graph by multiplying the ordinates of $y = \ln x$ by 2.

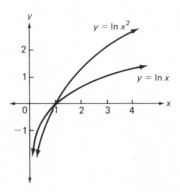

Example 3 Solve for t in $e^{2t-1} = 5$ and express the answer in terms of natural logs.
Solution Either take the natural log of both sides of the given equation or convert it into exponential form.

$$2t - 1 = \ln 5$$
$$2t = 1 + \ln 5$$
$$t = \tfrac{1}{2} + \tfrac{1}{2} \ln 5$$

Check: $e^{2[(1/2)+(1/2) \ln 5]-1} = e^{\ln 5} = 5.$

Example 4   Solve for x: $\ln (x + 1) = 1 + \ln x.$

Solution

$$\ln (x + 1) - \ln x = 1$$
$$\ln \frac{x + 1}{x} = 1$$

Now convert to exponential form:

$$\frac{x + 1}{x} = e$$
$$ex = x + 1$$
$$(e - 1)x = 1$$
$$x = \frac{1}{e - 1}$$

Check this result.

Example 5   **(a)** Show that $h(x) = \ln(x^2 + 5)$ is the composite of two functions.
(b) Show that $F(x) = e^{\sqrt{x^2 - 3x}}$ is the composite of three functions.

Solution
(a) Let $f(x) = \ln x$ and $g(x) = x^2 + 5$. Then

$$(f \circ g)(x) = f(g(x)) = f(x^2 + 5) = \ln(x^2 + 5) = h(x)$$

(b) Let $f(x) = e^x$, $g(x) = \sqrt{x}$, and $h(x) = x^2 - 3x$. Then

$$(f \circ g \circ h)(x) = f(g(h(x)))$$
$$= f(g(x^2 - 3x))$$
$$= f(\sqrt{x^2 - 3x})$$
$$= e^{\sqrt{x^2 - 3x}} = F(x)$$

(Other solutions are possible.)

Example 6   Determine the signs of $f(x) = x^2 e^x + 2xe^x.$

Solution   We find $f(x) = x^2 e^x + 2xe^x = xe^x(x + 2)$. We know that $e^x > 0$ for all x, and the other factors are zero when $x = 0$ or $x = -2$.

	$x + 2$	x	e^x	$f(x)$
$x < -2$	$-$	$-$	$+$	$+$
$-2 < x < 0$	$+$	$-$	$+$	$-$
$0 < x$	$+$	$+$	$+$	$+$

$f(x) > 0$ on the intervals $(-\infty, -2)$ and $(0, \infty)$

$f(x) < 0$ on the interval $(-2, 0)$

EXERCISES 7.5

Sketch each pair of functions on the same axes.

1. $y = e^x$; $y = e^{x-2}$
2. $y = e^x$; $y = 2e^x$
3. $y = \ln x$; $y = \frac{1}{2} \ln x$
4. $y = \ln x$; $y = \ln (x + 2)$
5. $y = \ln x$; $y = \ln (-x)$
6. $y = e^x$; $y = e^{-x}$

Explain how the graph of f can be obtained from the curve $y = \ln x$. (Hint: First apply the appropriate rules of logarithms.)

7. $f(x) = \ln ex$
8. $f(x) = \ln \dfrac{x}{e}$
9. $f(x) = \ln \sqrt{x}$
10. $f(x) = \ln \dfrac{1}{x}$
11. $f(x) = \ln (x^2 - 1) - \ln (x + 1)$
12. $f(x) = \ln x^{-3}$

Find the domain.

13. $f(x) = \ln (x + 2)$
14. $f(x) = \ln |x|$
15. $f(x) = \ln (2x - 1)$

Use the laws of logarithms (as much as possible) to write $\ln f(x)$ as an expression involving sums and differences.

16. $f(x) = x\sqrt{x^2 + 1}$
17. $f(x) = \dfrac{5x}{x^2 - 4}$
18. $f(x) = \sqrt{\dfrac{x + 7}{x - 7}}$
19. $f(x) = \dfrac{(x - 1)(x + 3)^2}{\sqrt{x^2 + 2}}$
20. $f(x) = \dfrac{x}{\sqrt[3]{x^2 - 1}}$
21. $f(x) = \sqrt{x^3(x + 1)}$

Convert each expression into the logarithm of a single expression.

22. $\ln 2 + \ln x - \ln (x - 1)$
23. $\frac{1}{2} \ln x + \ln (x^2 + 5)$
24. $\ln (x^3 - 1) - \ln (x^2 + x + 1)$
25. $3 \ln (x + 1) + 3 \ln (x - 1)$

Solve for x.

26. $e^{3x+5} = 100$ **27.** $e^{-0.01x} = 27$ **28.** $e^{x^2} = e^x e^{3/4}$

29. $e^{\ln(1-x)} = 2x$ **30.** $\ln x + \ln 2 = 1$ **31.** $\ln(x+1) = 0$

32. $\ln(x^2 - 4) - \ln(x + 2) = 0$ **33.** $\ln(x^2 - 4) - \ln(x - 2) = 0$

34. $\ln(x^2 + x - 2) = \ln 4 + \ln(x - 1)$

35. $\ln x = \frac{1}{2}\ln 4 + \frac{2}{3}\ln 8$ **36.** $\ln e^{\sqrt{x+1}} = 3$

37. $\frac{1}{2}\ln(x + 4) = \ln(x - 2)$

Show that each function is the composite of two functions.

38. $h(x) = e^{2x+3}$ **39.** $h(x) = e^{-x^2+x}$

40. $h(x) = \ln(1 - 2x)$ **41.** $h(x) = \ln\dfrac{x}{x+1}$

Show that each function is the composite of three functions.

42. $F(x) = e^{\sqrt{x+1}}$ **43.** $F(x) = e^{(3x-1)^2}$

44. $F(x) = [\ln(x^2 + 1)]^3$ **45.** $F(x) = \ln\sqrt{e^x + 1}$

Determine the signs of each function.

46. $f(x) = xe^x + e^x$ **47.** $f(x) = e^{2x} - 2xe^{2x}$

48. $f(x) = -3x^2 e^{-3x} + 2xe^{-3x}$ **49.** $f(x) = 1 + \ln x$

*50. Solve for x: $\dfrac{e^x + e^{-x}}{2} = 1$.

*51. Show that $(e^x + e^{-x})^2 - (e^x - e^{-x})^2 = 4$.

*52. Show that $\ln\left(\dfrac{x}{4} - \dfrac{\sqrt{x^2 - 4}}{4}\right) = -\ln(x + \sqrt{x^2 - 4})$.

*53. Solve for x in terms of y if $y = \dfrac{e^x}{2} - \dfrac{1}{2e^x}$. (*Hint:* Let $u = e^x$ and solve the resulting quadratic in u.)

7.6
Exponential Growth and Decay

To solve some of the following applied problems it will be helpful to learn how to use natural logarithms to obtain an approximation for x in an equation such as $2^x = 35$. Begin by taking the natural log of both sides. Thus

$$\ln 2^x = \ln 35$$

$$x \ln 2 = \ln 35 \qquad (\text{Why?})$$

$$x = \frac{\ln 35}{\ln 2}$$

An approximation for x can be found by using Appendix Table III, page 470. The entries in this table give values of $\ln x$ to three decimal places. (In most cases $\ln x$ is irrational.) From this table we have $\ln 2 = 0.693$. Even though $\ln 35$ is not given (directly) in the table, it can still be found by applying the second law of logarithms.

$$\ln 35 = \ln (3.5)(10) = \ln 3.5 + \ln 10$$
$$= 1.253 + 2.303 \qquad \text{(Table III)}$$
$$= 3.556$$

Now we have

$$x = \frac{\ln 35}{\ln 2} = \frac{3.556}{0.693} = 5.13$$

As a rough check we see that 5.13 is reasonable since $2^5 = 32$.

Note that the values found in the tables of logarithms are approximations. For the sake of simplicity, however, we will use the equals sign ($=$).

TEST YOUR UNDERSTANDING

Solve each equation for x in terms of natural logarithms. Approximate the answer by using Table III.

1. $4^x = 5$ **2.** $4^{-x} = 5$ **3.** $(\frac{1}{2})^x = 12$

4. $2^{3x} = 10$ **5.** $4^x = 15$ **6.** $67^x = 4$

At the beginning of Section 7.2, we developed the formula $y = (10,000)2^x$, which gives the number of bacteria present after x hours of growth; 10,000 is the initial number of bacteria. How long will it take for this bacterial culture to grow to 100,000? To answer this question we let $y = 100,000$ and solve for x.

$$(10,000)2^x = 100,000$$
$$2^x = 10 \qquad \text{(divide by 10,000)}$$
$$x \ln 2 = \ln 10$$
$$x = \frac{\ln 10}{\ln 2}$$
$$= \frac{2.303}{0.693} = 3.32$$

It will take about 3.3 hours.

In the preceding illustration the exponential and logarithmic functions were used to solve a problem of "exponential growth." Many problems involving exponential growth, or decay, can be solved by using the general formula

$$y = f(x) = Ae^{kx}$$

which shows how the amount of a substance y depends on the time x. Since $f(0) = A$, A represents the initial amount of the substance and k is a constant. In a given situation $k > 0$ signifies that y is growing (increasing) with time. For $k < 0$ the substance is decreasing (compare to the graph of $y = e^x$ and $y = e^{-x}$).

The preceding bacterial problem also fits this general form. This can be seen by substituting $2 = e^{\ln 2}$ into $y = (10{,}000)2^x$:

$$y = (10{,}000)2^x = (10{,}000)(e^{\ln 2})^x = 10{,}000e^{(\ln 2)x}$$

Example 1 A radioactive substance is decreasing (it is changing into another element) according to the formula $y = Ae^{-0.2x}$, where y is the amount of material remaining after x years.

(a) If the initial amount $A = 80$ grams, how much is left after 3 years?

(b) The **half-life** of a radioactive substance is the time it takes for half of it to decompose. Find the half-life of this substance in which $A = 80$ grams.

Solution

(a) Since $A = 80$, $y = 80e^{-0.2x}$. We need to solve for the amount y when $x = 3$.

$$y = 80e^{-0.2x}$$
$$= 80e^{-0.2(3)}$$
$$= 80e^{-0.6}$$
$$= 80(0.549) \qquad \text{(Table II)}$$
$$= 43.920$$

There will be about 43.9 grams after 3 years.

(b) This question calls for the time x at which only half of the initial amount is left. Consequently, the half-life x is the solution to $40 = 80e^{-0.2x}$. Divide each side by 80:

$$\tfrac{1}{2} = e^{-0.2x}$$

Take the natural log of both sides, or change to logarithmic form, to obtain $-0.2x = \ln \tfrac{1}{2}$. Since $\ln \tfrac{1}{2} = \ln 1 - \ln 2 = -\ln 2$, we solve for x as follows.

$$-0.2x = -\ln 2$$
$$x = \frac{\ln 2}{0.2}$$
$$= 3.47$$

The half-life is approximately 3.47 years.

Carbon-14, often written as ^{14}C, is a radioactive isotope of carbon with a half-life of about 5750 years. By finding how much ^{14}C is contained in the remains of a formerly living organism, it becomes possible to determine what percentage this is of the original amount of ^{14}C at the time of death. Once this information is given, the formula $y = Ae^{kx}$ will enable us to date the age of the remains. The dating will be done after we solve for the constant k. Since the amount of ^{14}C after 5750 years will be $\dfrac{A}{2}$, we have:

$$\frac{A}{2} = Ae^{5750k}$$

$$\frac{1}{2} = e^{5750k}$$

$$5750k = \ln \frac{1}{2}$$

$$k = \frac{\ln 0.5}{5750}$$

Substitute this value for k in the general formula:

$$y = Ae^{(\ln 0.5/5750)x}$$

Example 2　An animal skeleton is found to contain one-fourth of its original amount of ^{14}C. How old is the skeleton?

Solution　Let x be the age of the skeleton. Then

$$\frac{1}{4}A = Ae^{(\ln 0.5/5750)x}$$

$$\frac{1}{4} = e^{(\ln 0.5/5750)x}$$

$$\left(\frac{\ln 0.5}{5750}\right)x = \ln \frac{1}{4} = -\ln 4$$

$$x = \frac{(5750)(-\ln 4)}{\ln 0.5}$$

$$= 11,500$$

The skeleton is about 11,500 years old.

EXERCISES 7.6

Estimate the value of y in $y = Ae^{kx}$ for the given values of A, k, and x.

1. $A = 100, k = 0.75, x = 4$　　　　**2.** $A = 25, k = 0.5, x = 10$

3. $A = 1000, k = -1.8, x = 2$　　　　**4.** $A = 12.5, k = -0.04, x = 50$

Solve for k. Leave the answer in terms of natural logarithms.

5. $5000 = 50e^{2k}$
6. $75 = 150e^{10k}$

7. $\dfrac{A}{3} = Ae^{4k}$
8. $\dfrac{A}{2} = Ae^{100k}$

9. A bacterial culture is growing according to the formula $y = 10,000e^{0.6x}$, where x is the time in days. Estimate the number of bacteria after 1 week.

10. Estimate the number of bacteria after the culture in Exercise 9 has grown for 12 hours.

11. How long will it take for the bacterial culture in Exercise 9 to triple in size?

12. How long will it take until the number of bacteria in Exercise 9 reaches 1,000,000?

13. A radioactive substance is decaying according to the formula $y = Ae^{kx}$, where x is the time in years. The initial amount $A = 10$ grams, and 8 grams remains after 5 years.
(a) Find k. Leave the answer in terms of natural logs.
(b) Estimate the amount remaining after 10 years.
(c) Find the half-life.

14. The half-life of radium is approximately 1690 years. A laboratory has 50 milligrams of radium.
(a) Use the half-life to solve for k in $y = Ae^{kx}$. Leave answer in terms of logs.
(b) How long does it take until there are 40 milligrams left?

15. Suppose that 5 grams of a radioactive substance decreases to 4 grams in 30 seconds. What is its half-life?

16. How long does it take for two-thirds of the radioactive material in Exercise 15 to decay?

*17. An Egyptian mummy is found to contain 60% of its ^{14}C. How old is the mummy? (*Hint:* If A is the original amount of ^{14}C, then $\frac{3}{5}A$ is the amount left.)

*18. A skeleton contains one-hundredth of its original amount of ^{14}C. How old is the skeleton?

*19. Answer the question in Exercise 18 if one-millionth of its ^{14}C is left.

**7.7
Common
Logarithms
(optional)**

Logarithms were developed about 350 years ago. Since then they have been widely used to simplify involved numerical computations. Much of this work can now be done more efficiently with the aid of computers and calculators. There is still some use for logarithmic computation, however, especially when an appropriate calculating machine is not available.

For scientific and technical work, numbers are often written in decimal form and we will therefore be using logarithms to the base 10, called **common logarithms**.

      Appendix Table IV, page 471, contains the common logarithms of three-digit numbers from 1.00 to 9.99. To find a logarithm, say $\log_{10} 6.23$, first find the entry 6.2 in the left-hand column under the heading x. Now in the row for 6.2 and in the column headed by the digit 3 you will find the entry .7945. This is the common logarithm of 6.23. We write

$$\log_{10} 6.23 = 0.7945$$

(Note that 0.7945 is really a four-place approximation.) By reversing this process we can begin with $\log_{10} x = 0.7945$ and solve for x.

      Notice that all the common logarithms in Table IV are four-place decimals between 0 and 1. Except for the case $\log_{10} 1 = 0$, they are all approximations. The fact that they are between 0 and 1 will be taken up in the exercises.

Illustrations:

$$\log_{10} 3.07 = 0.4871 \qquad \log_{10} 8.88 = 0.9484$$

      To find $\log_{10} N$, where N may not be between 1 and 10, we first observe that N can be written in this form:

$$N = x10^c$$

where c is an integer and $1 \leq x < 10$. For example:

$$62{,}300 = 6.23(10^4) \qquad 0.00623 = 6.23(10^{-3})$$

This form of N, in conjunction with Table IV, will allow us to find $\log_{10} N$. In general,

$$\begin{aligned}
\log_{10} N &= \log_{10} x10^c \\
&= \log_{10} x + \log_{10} 10^c &\text{(Law 1)} \\
&= \log_{10} x + c &\text{(Why?)}
\end{aligned}$$

The integer c is the **characteristic** of $\log_{10} N$, and the four-place decimal fraction $\log_{10} x$ is its **mantissa**. Using $N = 62{,}300$ we have:

$$\begin{aligned}
\log_{10} 62{,}300 = \log_{10} 6.23(10^4) &= \log_{10} 6.23 + \log_{10} 10^4 \\
&= \log_{10} 6.23 + 4 \\
&= 0.7945 + 4 \qquad \text{(Table IV)} \\
&= 4.7945
\end{aligned}$$

      Since all logarithms considered here are to the base 10 we will simplify the notation and drop the subscript 10 from the logarithmic statements. Thus we write $\log N$ instead of $\log_{10} N$.

Example 1   Find log 0.0419.

Solution

$$\log 0.0419 = \log 4.19(10^{-2})$$
$$= \log 4.19 + \log 10^{-2}$$
$$= 0.6222 + (-2)$$

Suppose, in Example 1, the mantissa 0.6222 and the negative characteristic are combined:

$$0.6222 + (-2) = -1.3778 = -(1 + 0.3778)$$
$$= -1 + (-0.3778)$$

Since Table IV does not have negative mantissas, like -0.3778, we avoid such combining and preserve the form of $\log 0.0419$ so that its mantissa is positive. For computational purposes there are other useful forms of $0.6222 + (-2)$ in which the mantissa 0.6222 is preserved. Note that $-2 = 8 - 10$, $18 - 20$, and so forth. Thus

$$0.6222 + (-2) = 0.6222 + 8 - 10 = 8.6222 - 10 = 18.6222 - 20$$

Likewise,

$$\log 0.00569 = 7.7551 - 10 = 17.7551 - 20$$
$$\log 0.427 = 9.6304 - 10 = 29.6304 - 30$$

An efficient way to find N, if $\log N = 6.1239$, is to find the three-digit number x from Table IV corresponding to the mantissa 0.1239. Then multiply x by 10^6. Thus, since $\log 1.33 = 0.1239$, we have

$$N = 1.33(10^6) = 1{,}330{,}000$$

In the following explanation you can discover why this technique works.

$$\log N = 6.1239$$
$$= 6 + 0.1239$$
$$= 6 + \log 1.33$$
$$= \log 10^6 + \log 1.33$$
$$= \log 10^6 (1.33)$$
$$= \log 1{,}330{,}000$$

Therefore $\log N = \log 1{,}330{,}000$, and we conclude that $N = 1{,}330{,}000$.

TEST YOUR UNDERSTANDING

Find the common logarithm.

1. $\log 267$	**2.** $\log 26.7$	**3.** $\log 2.67$
4. $\log 0.267$	**5.** $\log 0.0267$	**6.** $\log 42{,}000$
7. $\log 0.000813$	**8.** $\log 7990$	**9.** $\log 0.00111$

Find N.

10. $\log N = 2.8248$	**11.** $\log N = 0.8248$
12. $\log N = 9.8248 - 10$	**13.** $\log_{10} N = 0.8248 - 3$
14. $\log N = 7.7126$	**15.** $\log_{10} N = 18.9987 - 20$

For easy reference here are the fundamental laws of logarithms that will be needed in the examples that follow. (Recall $M > 0$ and $N > 0$.)

LAW 1. $\log MN = \log M + \log N$

LAW 2. $\log \dfrac{M}{N} = \log M - \log N$

LAW 3. $\log N^k = k \log N$

Example 2   Estimate $P = (936)(0.00847)$ by using (common) logarithms.

Solution

$$\log P = \log (963)(0.00847)$$
$$= \log 963 + \log 0.00847 \qquad \text{(Law 1)}$$

Now use Table IV.

$$\left. \begin{array}{l} \log 963 = 2.9836 \\ \log 0.00847 = 7.9279 - 10 \end{array} \right\} \text{ (add)}$$
$$\log P = 10.9115 - 10 = 0.9115$$
$$P = 8.16(10^0) = 8.16$$

Note: The mantissa 0.9115 is not in Table IV. In this case we use the closest entry, namely 0.9117, corresponding to $x = 8.16$. Such approximations are good enough for our purposes. (For a more accurate procedure see Exercise 21. Exercise 20 shows how to find $\log x$ when $0 \leq x < 1$ and x has more than three digits.)

Example 3   Use logarithms to estimate $Q = \dfrac{0.00439}{0.705}$.

Solution   We find $\log Q = \log 0.00439 - \log 0.705$ (by Law 2). Now use the table.

$$
\begin{array}{l}
\overbrace{}^{\text{(This form is used to avoid a negative}} \\
\hspace{3.5cm} \text{mantissa when subtracting in the} \\
\hspace{3.5cm} \text{next step.)}
\end{array}
$$

$$
\begin{aligned}
\log 0.00439 &= 7.6425 - 10 = 17.6425 - 20 \\
\log 0.705\ \ &= 9.8482 - 10 =\ \ 9.8482 - 10
\end{aligned}\ \Big\}\ \text{(subtract)}
$$

$$
\begin{aligned}
\log Q &= 7.7943 - 10 \\
Q &= 6.23(10^{-3}) \\
&= 0.00623
\end{aligned}
$$

Example 4   Use logarithms to estimate $R = \sqrt[3]{0.0918}$.

Solution

$$
\begin{aligned}
\log R &= \log (0.0918)^{1/3} \\
&= \tfrac{1}{3} \log 0.0918 \qquad \text{(Law 3)} \\
&= \tfrac{1}{3}(8.9628 - 10) \\
&= \tfrac{1}{3}(28.9628 - 30) \qquad \text{(We avoid the fractional characteristic } -\tfrac{10}{3} \\
&\hspace{3.9cm} \text{by changing to } 28.9628 - 30.) \\
&= 9.6543 - 10 \\
R &= 4.51(10^{-1}) = 0.451
\end{aligned}
$$

Example 5   To determine how much a paint dealer should charge for a gallon of paint, he needs to find out how much the paint cost him per gallon in the first place. The paint is stored in a cylindrical drum $2\tfrac{1}{2}$ feet in diameter and $3\tfrac{3}{4}$ feet high. If he paid \$400 for this quantity of paint, what did it cost him per gallon? (Use 1 cubic foot = 7.48 gallons.)

Solution   The volume of the drum is the area of the base (area of a circle $= \pi r^2$) times the height. Thus there are

$$
\pi(1.25)^2(3.75)
$$

cubic feet of paint in the drum. Then the number of gallons is

$$
\pi(1.25)^2(3.75)(7.48)
$$

Since the total cost was $400, the cost per gallon is given by

$$C = \frac{400}{\pi(1.25)^2(3.75)(7.48)}$$

We use $\pi = 3.14$ to do the computation, using logarithms:

$$\log C = \log 400 - (\log 3.14 + 2 \log 1.25 + \log 3.75 + \log 7.48)$$

$$
\begin{aligned}
&\log 400 = 2.6021 \\
&\log 3.14 = 0.4969 \\
\log 1.25 = 0.0969 \longrightarrow\ &2 \log 1.25 = 0.1938 \\
&\log 3.75 = 0.5740 \\
&\log 7.48 = 0.8739 \\
\hline
&\qquad\qquad\quad 2.1386 \longrightarrow \qquad 2.1386
\end{aligned}
$$

(add)   (subtract)

$$\log C = 0.4635$$
$$C = 2.91 \times 10^0$$
$$= 2.91$$

The paint cost the dealer approximately $2.91 per gallon.

EXERCISES 7.7

In Exercises 1 through 12 estimate the following by using common logarithms.

1. $(512)(84{,}000)$

2. $(906)(2330)(780)$

3. $\dfrac{(927)(818)}{274}$

4. $\dfrac{274}{(927)(818)}$

5. $\dfrac{(0.421)(81.7)}{(368)(750)}$

6. $\dfrac{(579)(28.3)}{\sqrt{621}}$

7. $\dfrac{(28.3)\sqrt{621}}{579}$

8. $\left[\dfrac{28.3}{(579)(621)}\right]^2$

9. $\sqrt{\dfrac{28.3}{(579)(621)}}$

10. $\dfrac{(0.0941)^3(0.83)}{(7.73)^2}$

11. $\dfrac{\sqrt[3]{(186)^2}}{(600)^{1/4}}$

12. $\dfrac{\sqrt[4]{600}}{(186)^{2/3}}$

13. After running out of gasoline, a motorist had her gas tank filled at a cost of $16.93. What was the cost per gallon if the gas tank's capacity is 14 gallons?

14. Suppose that a spaceship takes 3 days, 8 hours, and 20 minutes to travel from the earth to the moon. If the distance traveled was one-quarter of a million miles, what was the average speed of the spaceship in miles per hour?

15. A spaceship, launched from the earth, will travel 432,000,000 miles on its trip to the planet Jupiter. If its average velocity is 21,700 miles per hour, how long will the trip take? Give the answer in years.

16. When P dollars is invested in a bank that pays compound interest at the rate of r percent (expressed as a decimal) per year, the amount A after n years is given by the formula

$$A = P(1 + r)^n$$

(This formula will be derived in Chapter 10.)
 (a) Find A for $P = 2500$, $r = 0.05$ (5%), and $n = 3$.
 (b) An investment of \$3750 earns compound interest at the rate of 6.2% per year. Find the amount A after 5 years.

17. The formula $P = \dfrac{A}{(1 + r)^n}$ gives the initial investment P in terms of the current amount of money A, the annual compound interest rate r, and the number of years n. How much money was invested at 5.8% if after 6 years there is now \$8440 in the bank?

18. If P dollars is invested at an interest rate r and the interest in compounded k times per year, the amount A after n years in given by

$$A = P\left(1 + \frac{r}{k}\right)^{kn}$$

 (a) Use this formula to compute A for $P = \$5000$ and $r = 0.08$ if the interest is compounded semiannually for 3 years.
 (b) Find A in part (a) with interest computed quarterly.
 (c) Find A in part (a) with $k = 8$.

*19. Explain why the mantissas in Table IV are between 0 and 1. (*Hint:* Take $1 \le x < 10$ and now consider the common logarithms of 1, x, and 10.)

*20. It is possible to be more accurate in finding logarithms of four-digit numbers, expressed in thousandths, by *not* first rounding off to hundredths. Here is a computation for finding log 6.477. Study this procedure carefully and then find the logarithms below in the same manner.

$$
\begin{array}{ccc}
N & & \log N \\
\end{array}
$$

$$0.010\left\{0.007\left\{\begin{array}{l}6.470 ------\to 0.8109 \\ 6.477 ------\to\ \ ?\end{array}\right\}d \\ \ \ \ \ \ \ 6.480 -------\to 0.8116\end{array}\right\}0.0007$$

$$\frac{0.007}{0.010} = \frac{d}{0.0007}$$

$$0.7 = \frac{d}{0.0007}$$

$$d = (0.7)(0.0007) = 0.00049$$

$$\log N = 0.8109 + 0.00049$$

$$\quad\quad = 0.8114 \quad\quad \text{(rounded off to four decimal places)}$$

(a) log 3.042   (b) log 7.849   (c) log 1.345   (d) log 5.444
(e) log 6.803   (f) log 2.711   (g) log 4.986   (h) log 9.008
The method used here is called **linear interpolation**. What is the rationale behind this method?

*21. The method in Exercise 20 can be adapted for finding the number when the given logarithm is not an exact table entry. Study the following procedure for finding N in $\log N = 0.734$, and then find the numbers N below in the same manner.

$$
\begin{array}{c}
\log N \qquad\qquad\qquad\qquad N \\
0.0008\left\{0.0006\left\{\begin{array}{l}0.7528 \longrightarrow 5.660 \\ 0.7534 \longrightarrow \quad ? \end{array}\right\}d\right\}0.010 \\
0.7536 \longrightarrow 5.670
\end{array}
$$

$$\frac{0.0006}{0.0008} = \frac{d}{0.01}$$

$$0.75 = \frac{d}{0.01}$$

$$d = (0.01)(0.75) = 0.0075$$

$$N = 5.660 + 0.0075$$

$$= 5.668 \qquad \text{(rounded off to three decimal places)}$$

(a) $\log N = 0.4510$ (b) $\log N = 0.9672$ (c) $\log N = 0.1391$
(d) $\log N = 0.7395$ (e) $\log N = 0.6527$ (f) $\log N = 0.8749$
(g) $\log N = 0.0092$ (h) $\log N = 0.9781$ (i) $\log N = 0.3547$

REVIEW EXERCISES FOR CHAPTER 7

The solutions to the following exercises can be found within the text of Chapter 7. Try to answer each question without referring to the text.

Section 7.1

1. Which of the following are one-to-one functions?

(a) $y = \sqrt{x}$ (b) $y = \dfrac{1}{x}$ (c) $y = |x|$

(d) $y = x^2 - 2x + 1$ (e) $y = \sqrt[3]{x}$

2. Find the inverse of g of $y = f(x) = 2x + 3$ and show that $(f \circ g)(x) = x = (g \circ f)(x)$.

3. Find the inverse g of $f(x) = \sqrt{x}$.

Section 7.2

4. Use a table of values to sketch $y = 8^x$ on the interval $[-1, 1]$.

5. Sketch $y = 3^x$ and $y = (\frac{1}{3})^x$ on the same axes.

6. Explain how to obtain the graphs of $y = 2^{x-3}$ and $y = 2^x - 1$ from $y = 2^x$.

7. Use the one-to-one property of the function $f(x) = 5^x$ to solve the equation $5^{t^2} = 625$ for t.

Section 7.3

8. Which of the following statements are true?
   (a) If $0 < b < 1$, the function $f(x) = b^x$ decreases toward the right.
   (b) The point $(0, 1)$ is on the curve $y = \log_b x$.
   (c) $y = \log_b x$, for $b > 1$, increases toward the right and bends downward.
   (d) The domain of $y = b^x$ is the same as the range of $y = \log_b x$.
   (e) The x-axis is asymptotic to $y = \log_b x$ and the y-axis is asymptotic to $y = b^x$.

9. Write the equation of the inverse of $y = 2^x$ and graph both on the same axes.

10. Explain how the graph of $g(x) = \log_5 (x + 2)$ can be obtained from $f(x) = \log_5 x$.

11. (a) Change to logarithmic form: $27^{2/3} = 9$.
    (b) Change to exponential form: $\log_6 \frac{1}{36} = -2$.

12. Solve for y: $\log_9 27 = y$.

13. Solve for b: $\log_b 8 = \frac{3}{4}$.

Section 7.4

14. Write the three laws of logarithms.

15. Express $\log_b \dfrac{AB^2}{C}$ in terms of $\log_b A$, $\log_b B$, and $\log_b C$.

16. Express $\frac{1}{2} \log_b x - 3 \log_b (x - 1)$ as the logarithm of a single expression in x.

17. Solve for x: $\log_8 (x - 6) + \log_8 (x + 6) = 2$.

18. Solve for x: $\log_{10} (x^3 - 1) - \log_{10} (x^2 + x + 1) = 1$.

Section 7.5

19. Match the columns.

| | SLOPE OF TANGENT TO CURVE |
CURVE	THROUGH $(0, 1)$
(i) $y = 3^x$	(a) Is less than 1.
(ii) $y = 2^x$	(b) Is equal to 1.
(iii) $y = e^x$	(c) Is more than 1.

20. Graph $y = e^x$ and $y = \ln x$ on the same axes.

21. Match the columns:

(i) $\ln x < 0$	(a) $x < 0$
(ii) $\ln x = 0$	(b) $x = 0$
(iii) $\ln x > 0$	(c) $x > 0$
(iv) $0 < e^x < 1$	(d) $0 < x < 1$
(v) $e^x = 1$	(e) $x = 1$
(vi) $e^x > 1$	(f) $x > 1$

22. Explain how the curves $y = e^{x+2}$ and $y = e^x - 2$ can be obtained from $y = e^x$.

23. Explain how the curve $y = \ln x^2$ can be obtained from $y = \ln x$.

24. Solve for t: $e^{2t-1} = 5$.

25. Solve for x: $\ln (x + 1) = 1 + \ln x$.

26. Decompose $k(x) = \ln (x^2 + 5)$ into the composite of two functions.

27. Decompose $F(x) = e^{\sqrt{x^2-3x}}$ into the composite of three functions.

28. Determine the signs of $f(x) = x^2 e^x + 2xe^x$.

Section 7.6

29. Get an approximate solution for x in $2^x = 35$.

30. A radioactive material is decreasing according to the formula $y = Ae^{-0.2x}$, where y is the amount of material remaining after x years. If the initial amount $A = 80$ grams, how much is left after 3 years?

31. Find the half-life of the radioactive substance in Exercise 30.

32. Solve for k in $\dfrac{A}{2} = Ae^{5750k}$. Leave your answer in terms of natural logs.

33. Use the formula $y = Ae^{(\ln 0.5/5750)x}$ to estimate the age of a skeleton that is found to contain one-fourth of its original amount of carbon-14.

Section 7.7

34. Use common logarithms to estimate $P = (963)(0.00847)$.

35. Use common logarithms to estimate $Q = \dfrac{0.00439}{0.750}$.

36. Use common logarithms to estimate $R = \sqrt[3]{0.0918}$.

SAMPLE TEST FOR CHAPTER 7

1. Match each curve with one of the given equations.

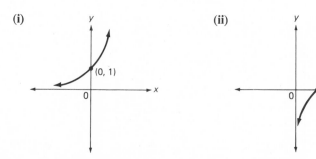

(iii)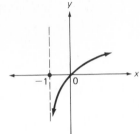

(iv)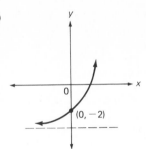

(0, −2)

(v)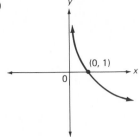

(0, 1)

(a) $y = b^x; \quad b > 1$

(b) $y = b^x; \quad 0 < b < 1$

(c) $y = \log_b x; \quad b > 1$

(d) $y = \log_b x; \quad 0 < b < 1$

(e) $y = \log_b (x + 1); \quad b > 1$

(f) $y = \log_b (x - 1); \quad b > 1$

(g) $y = b^{x+2}; \quad b > 1$

(h) $y = b^x - 2; \quad b > 1$

2. (a) What does it mean to say that a function is one-to-one?

(b) Find the inverse g of $y = f(x) = \sqrt[3]{x} - 1$ and show that $(f \circ g)(x) = x$.

3. (a) Solve for x: (a) $81^x = 9$; (b) $e^{\ln x} = 9$.

4. (a) Solve for b: $\log_b \frac{27}{8} = -3$; (b) evaluate: $\log_{10} 0.01$.

5. Find the domain of each function and give the equation of the vertical or horizontal asymptote.

(a) $y = f(x) = 2^x - 4$

(b) $y = f(x) = \log_3 (x + 4)$

6. Solve for x: $\log_{25} x^2 - \log_{25} (2x - 5) = \frac{1}{2}$.

7. Sketch the graphs of $y = e^{-x}$ and its inverse on the same axes. Write an equation of the inverse in the form $y = g(x)$.

8. Use the laws of logarithms (as much as possible) to write $\ln \dfrac{x^3}{(x + 1)\sqrt{x^2 + 2}}$ as an expression involving sums and differences.

9. Solve for x in $4^{2x} = 5$ and express the answer in terms of natural logs.

10. A radioactive substance decays according to the formula $y = Ae^{-0.04t}$, where t is the time in years. If the initial amount $A = 50$ grams, find the half-life. Leave the answer in terms of natural logs.

†11. Use common logarithms to estimate $N = \dfrac{(2430)^2}{(0.842)\sqrt{27.9}}$.

Answers to the Test Your Understanding Exercises

Page 282

The functions given in 2, 3, 5, 6, and 7 are one-to-one. The others are not.

Page 295

1.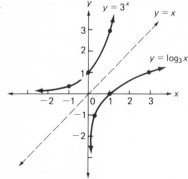

2.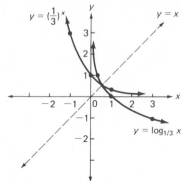

3. Shift 2 units left.

4. Shift 2 units up.

5. Reflect through *x*-axis.

6. Double the size of each ordinate.

Page 300

1. $\log_b A + \log_b B + \log_b C$

2. $\log_b A - \log_b B - \log_b C$

3. $2 \log_b A + 2 \log_b B - \log_b C$

4. $\log_b A + 2 \log_b B + 3 \log_b C$

5. $\log_b A + \frac{1}{2} \log_b B - \log_b C$

6. $\frac{1}{3} \log_b A - 3 \log_b B - 3 \log_b C$

7. $\log_b 3x^2$

8. $\log_b [\sqrt{x}\,(x-1)^2]$

9. $\log_b \dfrac{2x-1}{(x^2+1)^3}$

10. $\log_b \dfrac{x}{(x-1)(x-2)^2}$

Page 310

(Correct to two decimal places.)

1. $\dfrac{\ln 5}{\ln 4} = \dfrac{1.609}{1.386} = 1.16$

2. $-\dfrac{\ln 5}{\ln 4} = -\dfrac{1.609}{1.386} = -1.16$

3. $\dfrac{\ln 12}{\ln 0.5} = \dfrac{2.485}{-0.693} = -3.59$

4. $\dfrac{\ln 10}{3 \ln 2} = \dfrac{2.303}{3(0.693)} = 1.11$

5. $\dfrac{\ln 15}{\ln 4} = \dfrac{2.708}{1.386} = 1.953$

6. $\dfrac{\ln 4}{\ln 67} = \dfrac{1.386}{4.205} = 0.33$

Page 316

1. 2.4265

2. 1.4265

3. 0.4265

4. 9.4265 − 10

5. 8.4265 − 10

6. 4.6232

7. 6.9101 − 10

8. 3.9025

9. 7.0453 − 10

10. 668

11. 6.68

12. 0.668

13. 0.00668

14. 51,600,000

15. 0.0997

Eight

Trigonometry

The word *trigonometry* is based on the Greek words for triangle (*trigōnon*) and measure (*metron*). Thus the study of trigonometry is historically concerned with the measure of triangles, and it is in this manner that we approach the topic here.

When you studied geometry you probably learned to distinguish between a line segment and its measure. Likewise you also learned the difference between an angle and its measure. We will not always emphasize these distinctions. Thus we will often use *AB* to represent the segment with endpoints *A* and *B*, as well as the measure of the segment. Similarly ∠*PQR* will be used to represent the angle with vertex at *Q*, as well as the measure of the angle. Furthermore, ∠*PQR* will be abbreviated to ∠*Q* when it is clear that we are referring to the angle with vertex at *Q*. Thus *Q* can be used to name a vertex, to name an angle with vertex at *Q*, or to denote the measure of the angle.

Consider an acute angle *A*; that is, an angle whose measure in degrees is less than 90. Select a point *C* on one side of angle *A* and construct a perpendicular to form triangle *ABC* (△*ABC*), as in the figure.

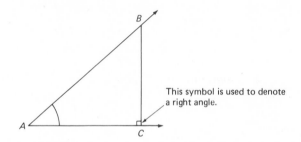

This symbol is used to denote a right angle.

Now consider the ratio of sides *BC* and *AB* of △*ABC*.

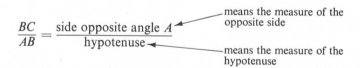

$$\frac{BC}{AB} = \frac{\text{side opposite angle } A}{\text{hypotenuse}}$$

means the measure of the opposite side

means the measure of the hypotenuse

This ratio depends only on the size of angle *A*, not on the location of points *B* and *C*. For example, let us see what happens if a different point *C′* is chosen to determine this ratio. In the next figure, a perpendicular is drawn at point *C′* to form △*AB′C′*.

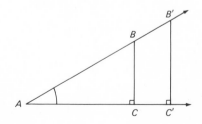

Right triangles ABC and $AB'C'$ are *similar* triangles. They have a common angle A and have equal right angles at C and C'. If two angles of a triangle are equal to two angles of another triangle, then the third angles (B and B') must also be equal. Two triangles are similar if they have their angles respectively equal to one another. Finally, if two triangles are similar, then their corresponding sides are proportional. Thus we may write

$$\frac{BC}{AB} = \frac{B'C}{AB'}$$

Note that C and C' are arbitrary points on the side of angle A. Yet the ratio of the side opposite the angle to the hypotenuse in each right triangle is the same. In every right triangle containing a given acute angle, the ratio of the side opposite that angle to the hypotenuse is *constant*. This ratio is given a a special name—the **sine** of the angle.

sine of $\angle A = \dfrac{\text{side opposite}}{\text{hypotenuse}} = \dfrac{BC}{AB}$

(sine of $\angle A$ is abbreviated as sin A)

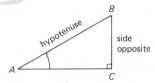

We form another important ratio by considering the side adjacent to angle A and the hypotenuse. We call this ratio the **cosine** of the angle.

cosine of $\angle A = \dfrac{\text{side adjacent}}{\text{hypotenuse}} = \dfrac{AC}{AB}$

(cosine of $\angle A$ is abbreviated as cos A)

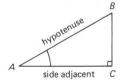

Example 1   Find (a) sin A and (b) cos A.

Solution   Relative to $\angle A$, BC is the side opposite and AC is the side adjacent.

(a) $\sin A = \dfrac{\text{side opposite}}{\text{hypotenuse}} = \dfrac{BC}{AB} = \dfrac{3}{5}$

(b) $\cos A = \dfrac{\text{side adjacent}}{\text{hypotenuse}} = \dfrac{AC}{AB} = \dfrac{4}{5}$

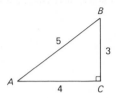

Example 2   For the figure of Example 1 find (a) sin B and (b) cos B.

Solution   Relative to $\angle B$, AC is the side opposite and BC is the side adjacent.

(a) $\sin B = \dfrac{AC}{AB} = \dfrac{4}{5}$ (b) $\cos B = \dfrac{BC}{AB} = \dfrac{3}{5}$

In a right triangle there are six possible ratios that we can form, using one pair of sides at a time. These six ratios, the **trigonometric ratios of an angle**, are of great importance in later work in mathematics. They need to be studied carefully and remembered as definitions. Although they are given in the following table in terms of a specific reference triangle, they should be remembered as ratios of sides that can be applied to any acute angle of a right triangle.

Ratio	Abbreviation	Definition	For $\triangle ABC$	
sine of $\angle A$	sin A	$\dfrac{\text{side opposite}}{\text{hypotenuse}}$	$\dfrac{a}{c}$	
cosine of $\angle A$	cos A	$\dfrac{\text{side adjacent}}{\text{hypotenuse}}$	$\dfrac{b}{c}$	
tangent of $\angle A$	tan A	$\dfrac{\text{side opposite}}{\text{side adjacent}}$	$\dfrac{a}{b}$	
cotangent of $\angle A$	cot A	$\dfrac{\text{side adjacent}}{\text{side opposite}}$	$\dfrac{b}{a}$	
secant of $\angle A$	sec A	$\dfrac{\text{hypotenuse}}{\text{side adjacent}}$	$\dfrac{c}{b}$	
cosecant of $\angle A$	csc A	$\dfrac{\text{hypotenuse}}{\text{side opposite}}$	$\dfrac{c}{a}$	

Example 3 Use the figure and write the six trigonometric ratios for angle B.

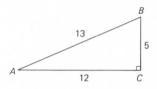

Solution Note that we need to consider sides opposite and adjacent to $\angle B$. The following diagram will help orient you correctly.

$\sin B = \frac{12}{13}$ $\cos B = \frac{5}{13}$ $\tan B = \frac{12}{5}$

$\cot B = \frac{5}{12}$ $\sec B = \frac{13}{5}$ $\csc B = \frac{13}{12}$

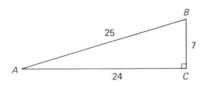

TEST YOUR UNDERSTANDING

Use the accompanying figure and write each of the following trigonometric ratios.

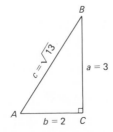

1. $\sin A$ **2.** $\cot B$ **3.** $\sec A$ **4.** $\cos B$

5. Find the value of $(\sin A)^2 + (\cos A)^2$.

6. Find the value of $(\sin A)(\csc A)$.

Note: We usually write an expression such as $(\sin A)^2$ as $\sin^2 A$. Similarly, $\cos^2 A$ will be used to mean $(\cos A)^2$.

It is sometimes useful to name a side of a triangle with a lowercase letter that corresponds to the capital letter used to name the angle opposite that side. For example, the measure of the side opposite angle A may be named a. Moreover, in a right triangle if two sides are given then the third may be found using the Pythagorean theorem, as in the following example.

Example 4   Triangle ABC is a right triangle with right angle C. If $a = 3$ and $b = 2$, find the sine of $\angle B$.

Solution By the Pythagorean theorem $a^2 + b^2 = c^2$. Thus $c^2 = 9 + 4 = 13$ and $c = \sqrt{13}$.

$$\sin B = \frac{AC}{AB} = \frac{2}{\sqrt{13}} \quad \text{or} \quad \frac{2\sqrt{13}}{13}$$

Example 5 Use the Pythagorean theorem to solve for AC in terms of x. Then write each of the six trigonometric ratios of $\angle A$ in terms of x.

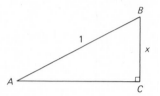

Solution By the Pythagorean theorem $(AC)^2 + x^2 = 1^2 = 1$. Thus $(AC)^2 = 1 - x^2$ and $AC = \sqrt{1 - x^2}$.

$$\sin A = x \qquad\qquad \cot A = \frac{\sqrt{1 - x^2}}{x}$$

$$\cos A = \sqrt{1 - x^2} \qquad \sec A = \frac{1}{\sqrt{1 - x^2}}$$

$$\tan A = \frac{x}{\sqrt{1 - x^2}} \qquad \csc A = \frac{1}{x}$$

Example 6 Use $\triangle ABC$ and show that $(\sin A)(\cos A) = \dfrac{x\sqrt{25 - x^2}}{25}$.

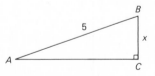

Solution

$$(AC)^2 = 5^2 - x^2 = 25 - x^2$$

$$AC = \sqrt{25 - x^2}$$

$$(\sin A)(\cos A) = \frac{x}{5} \cdot \frac{\sqrt{25 - x^2}}{5} = \frac{x\sqrt{25 - x^2}}{25}$$

EXERCISES 8.1

Use the accompanying figure for Exercises 1 through 12 and give each trigonometric ratio.

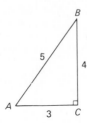

1. cos *A*	**2.** sin *B*	**3.** tan *A*	**4.** sec *B* .
5. csc *A*	**6.** cot *B*	**7.** sin *A*	**8.** cos *B*
9. sec *A*	**10.** tan *B*	**11.** cot *A*	**12.** csc *B*

For each of the following right triangles use the Pythagorean theorem to find the third side. In each case consider a △ABC with right angle C and hypotenuse c. Then name the six trigonometric ratios of (a) angle A and (b) angle B.

13. *a* = 2; *b* = 3	**14.** *a* = 3; *b* = 5	**15.** *a* = 4; *b* = 5
16. *a* = 1; *b* = 1	**17.** *a* = 1; *b* = 2	**18.** *a* = 2; *b* = 5
19. *a* = 3; *c* = 4	**20.** *a* = 2; *c* = 3	**21.** *b* = 2; *c* = 5
22. *b* = 4; *c* = 7		

Use the accompanying figure for △XYZ. Find the value for each of the expressions in Exercises 23 through 30.

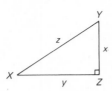

23. (tan *X*)(cot *X*)   **24.** (cos *Y*)(sec *Y*)

25. $(\sin X)\left(\dfrac{1}{\csc X}\right)$ **26.** $(\tan Y)\left(\dfrac{1}{\cot Y}\right)$

27. $\sin^2 X + \cos^2 X$ **28.** $\sin^2 Y + \cos^2 Y$

29. $\sec^2 X - \tan^2 X$ **30.** $\csc^2 Y - \cot^2 Y$

31. Find the six trigonometric ratios of ∠*A* if sin *A* = $\frac{3}{4}$.

32. Find the six trigonometric ratios of ∠*B* if cos *B* = $\frac{2}{5}$.

In Exercises 33 through 36 use the information given for △ABC and verify the indicated equality.

33. $(\sin A)(\cos A) = x\sqrt{1 - x^2}$

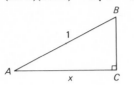

34. $\tan^2 A = \dfrac{x^2}{9 - x^2}$

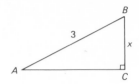

35. $\dfrac{4 \sin^2 A}{\cos A} = \dfrac{x^2}{\sqrt{16 - x^2}}$

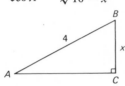

36. $(\sin A)(\cos A) = \dfrac{2x}{4 + x^2}$

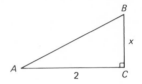

*37. Let x be a positive number. Show that

$$(\sin A)(\tan A) = \dfrac{x^2}{2\sqrt{4 - x^2}}$$

for an appropriate acute angle A.

*38. Let x be a positive number. Show that $\sec^3 A = \dfrac{8}{(4 - x^2)^{3/2}}$ for an appropriate acute angle A.

8.2

Angle Measure

Angles are often measured in degrees, the traditional method inherited from the Babylonians. They used a numeration system based upon groups of 60 and did much to influence the manner in which we measure. We generally consider a complete rotation, a circle, to be divided into 360 parts, each of which is called a *degree*. Each degree is then divided again into 60 parts, each of which is called a *minute*; each minute is divided into 60 parts, each of which is called a *second*.

An angle of 40 degrees, 20 minutes, 45 seconds is abbreviated as 40°20′45″. In much of our work, however, we will consider measures of angles only to the nearest degree or nearest multiple of 10′ (as in Appendix Table V). Addition and subtraction of these angle measures is accomplished by using the conversions of

$1° = 60'$ and $1' = 60''$. For example:

$$40°20'$$
$$+ \underline{35°50'}$$
$$75°70' = 76°10'$$

$$40°20' = 39°80'$$
$$- \underline{35°50'} = \underline{35°50'}$$
$$4°30'$$

Throughout the calculus, as well as in most advanced courses, we generally measure angles in *radians*. Consider an angle whose vertex is at the center of a circle, as in the following figure. The angle θ has a measure of 1 **radian** if it intercepts an arc whose length is equal to the length of the radius of the circle.

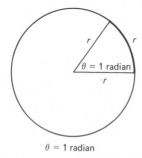

$\theta = 1$ radian

In general, let us consider an angle θ that intercepts an arc of length s in a circle with radius r.

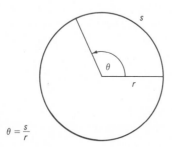

$\theta = \frac{s}{r}$

We then say that the measure of the angle θ is given by the quotient $\frac{s}{r}$; that is,

angle θ has a measure of $\frac{s}{r}$ radians. Note that the size of the circle does not affect the radian measure of the angle; the measure is the ratio of intercepted arc length to the radius and is a real number. Thus for $s = 12$ centimeters and $r = 3$ centimeters, $\theta = \frac{12}{3} = 4$ radians. In particular, when $s = r$ then $\theta = \frac{s}{r} = 1$, and the angle has a measure of 1 radian.

From geometry we know that the circumference of a circle is given by the formula $C = 2\pi r$. Thus $\frac{C}{r} = 2\pi$ and so there are 2π radians in a complete rotation of 360°.

$$2\pi \text{ radians} = 360°$$

If you remember this simple relationship then you can easily convert from radian to degree measure, or from degree to radian measure, by an appropriate division.

$$2\pi \text{ radians} = 360°$$

Divide by 2π: $1 \text{ radian} = \dfrac{360}{2\pi} = \dfrac{180}{\pi} \text{ degrees}$

Divide by 360: $1° = \dfrac{2\pi}{360} = \dfrac{\pi}{180} \text{ radians}$

Example 1 Express **(a)** 30° and **(b)** 120° in radian measure.

Solution

(a) $30° = 30 \times \dfrac{\pi}{180} = \dfrac{\pi}{6} \text{ radians}$

(b) $120° = 120 \times \dfrac{\pi}{180} = \dfrac{2\pi}{3} \text{ radians}$

Example 2 Express **(a)** $\dfrac{\pi}{4}$ radians and **(b)** $\dfrac{5\pi}{6}$ radians in degree measure.

Solution

(a) $\dfrac{\pi}{4} \text{ radians} = \dfrac{\pi}{4} \times \dfrac{180}{\pi} = 45°$

(b) $\dfrac{5\pi}{6} \text{ radians} = \dfrac{5\pi}{6} \times \dfrac{180}{\pi} = 150°$

From now on we will assume an angle to be measured in radians unless degree measure is explicitly stated. For example, an angle of $\dfrac{\pi}{2}$ is interpreted to mean an angle that measures $\dfrac{\pi}{2}$ radians $\left(\dfrac{\pi}{2} \text{ radians} = 90°\right)$.

In general we will leave radian measure in terms of π. However, the following approximations can be used when necessary for computational purposes:

$$1° = \frac{\pi}{180} = 0.0175 \text{ radians}$$

$$1 \text{ radian} = \frac{180}{\pi} = 57.3°$$

We use radian measure to find the area of a sector of a circle. In the figure notice that the area of the shaded sector depends on the central angle θ. That is, $A = k \cdot \theta$, where k is some constant to be determined. To find k, we take the special case where $\theta = 2\pi$. This means that we have the entire circle whose area $A = \pi r^2$.

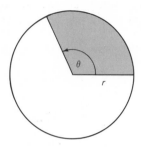

Using this information, we may then proceed as follows:

$$A = k \cdot \theta$$
$$\pi r^2 = k \cdot 2\pi$$
$$k = \tfrac{1}{2} r^2$$

Thus $A = \tfrac{1}{2} r^2 \theta$, where θ is given in radian measure.

Example 3   Find the area of a sector of a circle of radius 6 centimeters if the central angle is 60°.

Solution   First convert 60° to radian measure:

$$60° = 60 \times \frac{\pi}{180} = \frac{\pi}{3}$$

Then $A = \dfrac{1}{2}(6)^2\left(\dfrac{\pi}{3}\right) = 6\pi$ square centimeters.

EXERCISES 8.2

Express each angle in radian measure.

1. 45°	**2.** 60°	**3.** 90°	**4.** 180°	**5.** 270°
6. 360°	**7.** 150°	**8.** 135°	**9.** 225°	**10.** 240°
11. 210°	**12.** 300°	**13.** 330°	**14.** 345°	**15.** 75°

Express each angle in degree measure.

16. $\dfrac{\pi}{2}$	17. π	18. $\dfrac{3\pi}{2}$	19. 2π	20. $\dfrac{\pi}{3}$
21. $\dfrac{5\pi}{9}$	22. $\dfrac{\pi}{6}$	23. $\dfrac{2\pi}{3}$	24. $\dfrac{3\pi}{4}$	25. $\dfrac{5\pi}{4}$
26. $\dfrac{7\pi}{6}$	27. $\dfrac{5\pi}{3}$	28. $\dfrac{\pi}{12}$	29. $\dfrac{5\pi}{18}$	30. $\dfrac{11\pi}{36}$

A circle has a radius of 12 centimeters. Find the area of a sector of this circle for each given central angle.

31. $30°$	32. $90°$	33. $135°$	34. $225°$	35. $315°$

*36. The area of a sector of a circle with radius 6 centimeters is 15 square centimeters. Find the measure of the central angle of the sector in degrees.

*37. Find the area of a sector of a circle whose radius is 2 inches if the length of the intercepted arc is 8 inches.

8.3
Some Special Angles

There are two right triangles that will prove to be of special importance in our work. The first is an isosceles right triangle, that is, one with two equal sides and with acute angles of $45°$. In the following figure $AC = BC = x$. We can then solve for the hypotenuse, c, by using the Pythagorean theorem.

$$x^2 + x^2 = c^2$$
$$2x^2 = c^2$$
$$c^2 = 2x^2$$
$$c = \sqrt{2}\,x \quad \text{or} \quad x\sqrt{2}$$

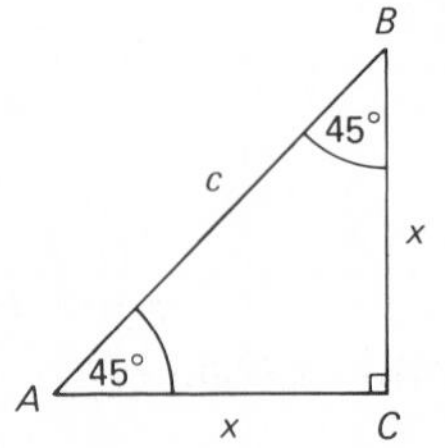

It is easier to remember the relationship of the sides of a $45° - 45° - 90°$ triangle by considering one with the two equal sides of length 1 unit each.

$$c^2 = 1^2 + 1^2 = 2$$
$$c = \sqrt{2}$$

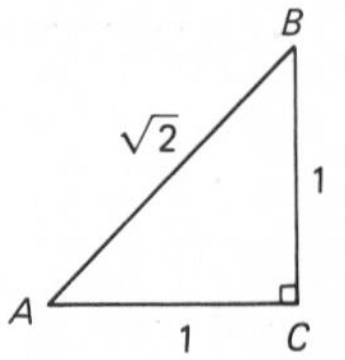

This may be referred to as a $1-1-\sqrt{2}$ triangle. We can now write the six trigonometric ratios for an angle of $45° = \dfrac{\pi}{4}$ radians.

Trigonometric ratios for 45°

$$\sin 45° = \frac{1}{\sqrt{2}} = \frac{\sqrt{2}}{2} \qquad\qquad \cot 45° = \frac{1}{1} = 1$$

$$\cos 45° = \frac{1}{\sqrt{2}} = \frac{\sqrt{2}}{2} \qquad\qquad \sec 45° = \frac{\sqrt{2}}{1} = \sqrt{2}$$

$$\tan 45° = \frac{1}{1} = 1 \qquad\qquad \csc 45° = \frac{\sqrt{2}}{1} = \sqrt{2}$$

Notice that the results are the same if we use the $1-1-\sqrt{2}$ triangle instead of the original one. Consider, for example, the sine ratio from the general triangle. Recall that we found $c = x\sqrt{2}$.

$$\sin 45° = \frac{x}{c}$$

$$= \frac{x}{x\sqrt{2}}$$

$$= \frac{1}{\sqrt{2}} = \frac{\sqrt{2}}{2}$$

A similar observation could have been made for each of the other ratios. Therefore it is advisable to remember the picture of the $1-1-\sqrt{2}$ triangle and use it to determine the trigonometric ratios of $45°$ when needed.

Another very useful triangle to study is the $30°-60°-90°$ triangle. From the equilateral triangle ABD you can see that the perpendicular bisector AC produces a $30° - 60° - 90°$ triangle in which the side opposite the $30°$ angle is one-half the hypotenuse. Thus if we call these lengths x and $2x$, we can determine the third side by the Pythagorean theorem.

$$x^2 + b^2 = (2x)^2$$

$$x^2 + b^2 = 4x^2$$

$$b^2 = 3x^2$$

$$b = x\sqrt{3}$$

As before, it is easiest to remember this triangle if we select the side opposite the 30° angle to be 1 unit in length. We can then use the following 1–$\sqrt{3}$–2 triangle to write the six trigonometric ratios for angles of 30° and 60°.

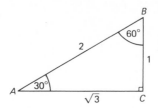

Recall that the trigonometric ratios are constant for a given angle and do not depend upon the size of the triangle. Thus the sine of an angle of 30° is the ratio $\dfrac{BC}{AB} = \dfrac{x}{2x} = \dfrac{1}{2}$. Similar observations can be made for the remaining ratios to show that the preceding triangle can be used to name each ratio. Here are these values for an angle of $30° = \dfrac{\pi}{6}$ radians.

Trigonometric ratios for 30°

$$\sin 30° = \frac{1}{2} \qquad\qquad \cot 30° = \frac{\sqrt{3}}{1} = \sqrt{3}$$

$$\cos 30° = \frac{\sqrt{3}}{2} \qquad\qquad \sec 30° = \frac{2}{\sqrt{3}} = \frac{2\sqrt{3}}{3}$$

$$\tan 30° = \frac{1}{\sqrt{3}} = \frac{\sqrt{3}}{3} \qquad\qquad \csc 30° = \frac{2}{1} = 2$$

TEST YOUR UNDERSTANDING

Find the six trigonometric ratios for an angle of $60° = \dfrac{\pi}{3}$ radians.

At the top of the next page is a table that summarizes the trigonometric ratios for the special angles we have studied thus far.

θ	$\sin \theta$	$\cos \theta$	$\tan \theta$	$\cot \theta$	$\sec \theta$	$\csc \theta$
$\frac{\pi}{6}$ (30°)	$\frac{1}{2}$	$\frac{\sqrt{3}}{2}$	$\frac{\sqrt{3}}{3}$	$\sqrt{3}$	$\frac{2\sqrt{3}}{3}$	2
$\frac{\pi}{4}$ (45°)	$\frac{\sqrt{2}}{2}$	$\frac{\sqrt{2}}{2}$	1	1	$\sqrt{2}$	$\sqrt{2}$
$\frac{\pi}{3}$ (60°)	$\frac{\sqrt{3}}{2}$	$\frac{1}{2}$	$\sqrt{3}$	$\frac{\sqrt{3}}{3}$	2	$\frac{2\sqrt{3}}{3}$

We can use the trigonometric ratios to solve for unknown sides of right triangles. Consider, for example, the following triangle where we wish to determine the length of the side marked x.

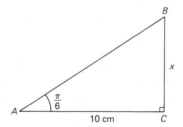

Select an appropriate ratio that includes the side labeled x and the known side AC. The tangent ratio will do the job for us.

$$\tan A = \frac{BC}{AC} = \frac{x}{10}$$

But $\angle A = \frac{\pi}{6}$(or 30°), and we have previously found the value for the tangent of

an angle of $\frac{\pi}{6}$radians. Thus we may complete the solution in this way:

$$\tan \frac{\pi}{6} = \frac{x}{10} = \frac{\sqrt{3}}{3}$$
$$3x = 10\sqrt{3}$$
$$x = \frac{10\sqrt{3}}{3}$$

Using a square root table, or a calculator, we can approximate the answer as 5.8 centimeters. If you use a calculator there may be several ways to compute the answer, depending upon your calculator. In general, it is best to determine the value for $\sqrt{3}$ first. Thus a typical set of steps might be the following:

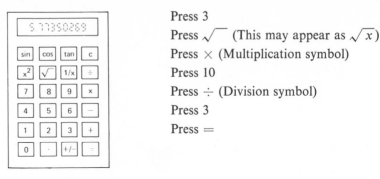

Press 3

Press $\sqrt{}$ (This may appear as $\sqrt{x}$)

Press $\times$ (Multiplication symbol)

Press 10

Press $\div$ (Division symbol)

Press 3

Press $=$

With the use of a table of trigonometric ratios, or an appropriate calculator, such solutions can be completed when the given angle is other than $\frac{\pi}{6}, \frac{\pi}{4}$, or $\frac{\pi}{3}$. We explore such solutions later.

Example Solve the triangle for the side marked x, correct to the nearest tenth of a unit.

Solution

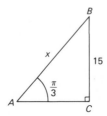

We need a trigonometric ratio that involves sides BC and AB.

$$\sin A = \frac{BC}{AB} = \frac{15}{x}$$

But $\sin \frac{\pi}{3} = \frac{\sqrt{3}}{2}$. Therefore:

$$\sin \frac{\pi}{3} = \frac{15}{x} = \frac{\sqrt{3}}{2}$$

$$x\sqrt{3} = 30$$

$$x = \frac{30}{\sqrt{3}} \qquad \text{(Show that this is equivalent to } 10\sqrt{3}.\text{)}$$

Through use of Appendix Table I or a calculator we find the result to be 17.3, to the nearest tenth of a unit. Show that the same result is obtained if csc A is used instead.

EXERCISES 8.3

Draw a right triangle and find the measure of $\angle A$ in radians if:

1. $\sin A = \dfrac{\sqrt{2}}{2}$ **2.** $\tan A = \sqrt{3}$ **3.** $\csc A = \sqrt{2}$

4. $\cos A = \dfrac{1}{2}$ **5.** $\cot A = \dfrac{\sqrt{3}}{3}$ **6.** $\tan A = 1$

Solve each triangle for the side marked x. Give your answers correct to·the nearest tenth of a unit.

7.

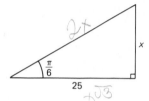

8.

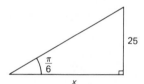

9.

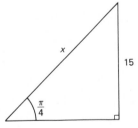

10.

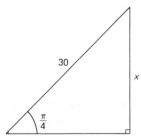

11.

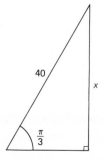

12.

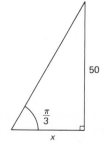

Exercises 13 through 20 refer to △ABC. Find the indicated sides correct to the nearest tenth of a unit. You may use a calculator for your computations.

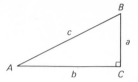

13. $\angle A = \dfrac{\pi}{6}$, $c = 30$; find a.   **14.** $\angle A = \dfrac{\pi}{3}$, $b = 25$; find c.

15. $\angle B = \dfrac{\pi}{4}$, $c = 20$; find b.   **16.** $\angle B = \dfrac{\pi}{6}$, $c = 40$; find b.

17. $\angle A = \dfrac{\pi}{6}$, $b = 15$; find c.   **18.** $\angle A = \dfrac{\pi}{4}$, $a = 35$; find c.

19. $\angle B = \dfrac{\pi}{3}$, $a = 45$; find b.   **20.** $\angle A = \dfrac{\pi}{3}$, $c = 70$; find a.

21. On a sheet of graph paper draw a segment AC that is 20 units in length. At point A use a protractor to construct an angle of 35°. Draw a right angle at C to complete triangle ABC. Measure the length of BC in units, correct to the nearest unit. Use this information to approximate the tangent and cotangent of an angle of 35°.

22. Use the figure constructed for Exercise 21 to approximate the tangent and cotangent of an angle of 55°.

23. Use the figure for Exercise 21 and the Pythagorean theorem to find the length of AB correct to the nearest unit. Use this information to find sin 35° and cos 35°.

8.4

Basic Identities

Before proceeding with further specific angles let us pause to explore some relationships that exist for the trigonometric ratios. Using a right triangle ABC, verify each of the following.

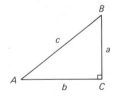

$$\sin A = \cos B = \frac{a}{c} \qquad \cot A = \tan B = \frac{b}{a}$$

$$\cos A = \sin B = \frac{b}{c} \qquad \sec A = \csc B = \frac{c}{b}$$

$$\tan A = \cot B = \frac{a}{b} \qquad \csc A = \sec B = \frac{c}{a}$$

In general, the trigonometric ratio of an acute angle is equal to the *co-named ratio* of its complement. Recall that two angles are said to be complementary if the sum of their angle measures is 90. Using this generalization we may make such statements as the following:

$$\sin 30° = \cos 60° \qquad \tan 40° = \cot 50°$$
$$\sec 70° = \csc 20° \qquad \cos 25° = \sin 65°$$
$$\cot 10° = \tan 80° \qquad \csc 15° = \sec 75°$$

We will now develop a number of basic *identities* through the use of a right triangle.

It is common to use the Greek letter θ (theta) to represent an arbitrary angle. In the triangle shown, let us compare $\sin \theta$ with $\csc \theta$.

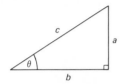

$$\left.\begin{array}{l} \sin \theta = \dfrac{a}{c} \\[2ex] \csc \theta = \dfrac{c}{a} \end{array}\right\} \sin \theta = \dfrac{1}{\csc \theta} = \dfrac{1}{c/a} = 1 \cdot \dfrac{a}{c}$$

$$a/c = a/c$$

We note that the sine and the cosecant of an angle are reciprocals of one another. Using the same figure, we can illustrate two other such pairs of reciprocal ratios.

$$\left.\begin{array}{l} \cos \theta = \dfrac{b}{c} \\[2ex] \sec \theta = \dfrac{c}{b} \end{array}\right\} \cos \theta = \dfrac{1}{\sec \theta} \qquad \left.\begin{array}{l} \tan \theta = \dfrac{a}{b} \\[2ex] \cot \theta = \dfrac{b}{a} \end{array}\right\} \tan \theta = \dfrac{1}{\cot \theta}$$

The preceding results can be summarized by listing these relationships as products. For a given angle θ we have:

$$(\sin \theta) \cdot (\csc \theta) = 1$$
$$(\cos \theta) \cdot (\sec \theta) = 1$$
$$(\tan \theta) \cdot (\cot \theta) = 1$$

Although it is useful to have all six ratios, we can get by with just the sine, cosine, and tangent of an angle. The remaining ratios can then be obtained as reciprocals of these. It is for this reason that major attention will be given to these three basic ratios.

Another very helpful identity relates the sine and cosine functions to the tangent ratio. Again we rely upon our basic right triangle for this development.

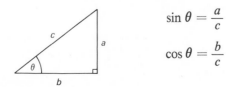

$$\sin \theta = \frac{a}{c}$$

$$\cos \theta = \frac{b}{c}$$

Now consider the quotient of these two ratios.

$$\frac{\sin \theta}{\cos \theta} = \frac{\dfrac{a}{c}}{\dfrac{b}{c}} = \frac{a}{c} \times \frac{c}{b} = \frac{a}{b}$$

But the ratio $\dfrac{a}{b}$ is equal to the tangent of θ. Thus we have the following identity:

$$\tan \theta = \frac{\sin \theta}{\cos \theta}$$

Inasmuch as the tangent and cotangent ratios are reciprocals of one another, we may also write this identity:

$$\cot \theta = \frac{\cos \theta}{\sin \theta}$$

There is another group of identities that will prove helpful in later work. All are derived by considering the Pythagorean relationship:

$$a^2 + b^2 = c^2$$

Now we divide each member of this equation, in turn, by $c^2, b^2,$ and a^2 and express the result as a trigonometric identity.

1. Divide by c^2:

$$\frac{a^2}{c^2} + \frac{b^2}{c^2} = \frac{c^2}{c^2}$$

$$\left(\frac{a}{c}\right)^2 + \left(\frac{b}{c}\right)^2 = \left(\frac{c}{c}\right)^2$$

But $\dfrac{a}{c} = \sin \theta$ and $\dfrac{b}{c} = \cos \theta.$

Thus $(\sin \theta)^2 + (\cos \theta)^2 = 1^2$

$$\sin^2 \theta + \cos^2 \theta = 1$$

2. Divide by b^2:

$$\frac{a^2}{b^2} + \frac{b^2}{b^2} = \frac{c^2}{b^2}$$

$$\left(\frac{a}{b}\right)^2 + \left(\frac{b}{b}\right)^2 = \left(\frac{c}{b}\right)^2$$

But $\dfrac{a}{b} = \tan\theta$ and $\dfrac{c}{b} = \sec\theta$.

Thus
$$(\tan\theta)^2 + 1^2 = (\sec\theta)^2$$
$$\tan^2\theta + 1 = \sec^2\theta$$

3. Divide by a^2:
$$\frac{a^2}{a^2} + \frac{b^2}{a^2} = \frac{c^2}{a^2}$$
$$\left(\frac{a}{a}\right)^2 + \left(\frac{b}{a}\right)^2 = \left(\frac{c}{a}\right)^2$$

But $\dfrac{b}{a} = \cot\theta$ and $\dfrac{c}{a} = \csc\theta$.

Thus
$$1^2 + (\cot\theta)^2 = (\csc\theta)^2$$
$$1 + \cot^2\theta = \csc^2\theta$$

Note that each of these identities may be written in various forms by using the usual laws for working with equations. For example, all of the following are equivalent:
$$\sin^2\theta + \cos^2\theta = 1$$
$$\sin^2\theta = 1 - \cos^2\theta$$
$$\cos^2\theta = 1 - \sin^2\theta$$

Similar observations may be made about the other identities as well.

Here is a summary of the basic identities we have established.

Basic trigonometric identities

$$\sin\theta = \frac{1}{\csc\theta} \qquad \tan\theta = \frac{\sin\theta}{\cos\theta} \qquad \sin^2\theta + \cos^2\theta = 1$$

$$\cos\theta = \frac{1}{\sec\theta} \qquad \cot\theta = \frac{\cos\theta}{\sin\theta} \qquad \tan^2\theta + 1 = \sec^2\theta$$

$$\tan\theta = \frac{1}{\cot\theta} \qquad\qquad\qquad\quad\ \ \cot^2\theta + 1 = \csc^2\theta$$

These basic identities may be used to prove other identities. In general we do so by working on one side of an equation to rewrite it into the form given on the other side. We may also work on both sides to produce equivalent forms. These procedures are shown in the examples that follow.

Example 1   Verify this identity:

$$\frac{\tan^2 \theta + 1}{\tan \theta \csc^2 \theta} = \tan \theta$$

Solution   Supply a reason for each of the steps shown.

$$\frac{\tan^2 \theta + 1}{\tan \theta \csc^2 \theta} = \frac{\sec^2 \theta}{\tan \theta \csc^2 \theta}$$

$$= \frac{\dfrac{1}{\cos^2 \theta}}{\dfrac{\sin \theta}{\cos \theta} \cdot \dfrac{1}{\sin^2 \theta}}$$

$$= \frac{1}{\cos^2 \theta} \cdot \frac{\cos \theta \sin^2 \theta}{\sin \theta}$$

$$= \frac{\sin \theta}{\cos \theta}$$

$$= \tan \theta$$

Example 2   Prove: $(1 - \sin \theta)(1 + \sin \theta) = \dfrac{1}{1 + \tan^2 \theta}.$

Solution   We prove this identity by transforming each side into a third equivalent expression. Supply a reason for each step shown.

$(1 - \sin \theta)(1 + \sin \theta)$	$\dfrac{1}{1 + \tan^2 \theta}$
$1 - \sin^2 \theta$	$\dfrac{1}{\sec^2 \theta}$
$\cos^2 \theta$	$\cos^2 \theta$

When solving identities work on the more complex side and try to reduce it to the form shown on the simpler side. When all else fails, try to write all expressions in terms of the sine and cosine ratios only and then apply appropriate algebraic techniques to simplify. With experience you may find shorter and more creative ways to prove an identity. And occasionally you may find that an "identity" is incorrect; that is, the two sides of the equation are not equivalent. For example, $\sin^2 \theta - \cos^2 \theta = 1$ is *not* an identity. You can verify this by use of a counterexample; find a replacement for θ for which the statement does not hold. One such possibility is $\theta = \dfrac{\pi}{4}$.

$$\sin^2 \frac{\pi}{4} - \cos^2 \frac{\pi}{4} = \left(\frac{1}{\sqrt{2}}\right)^2 - \left(\frac{1}{\sqrt{2}}\right)^2 = 0 \neq 1$$

Until an identity is proved you should not use properties of equality. Therefore you should not perform operations on both sides of an unproved statement of equality at the same time. Squaring both sides of such a statement might produce an incorrect proof that a nonidentity is an identity. For example, $|\sin \theta| = -|\sin \theta|$ is certainly not an identity but squaring both sides gives $\sin^2 \theta = \sin^2 \theta$ which is an identity.

EXERCISES 8.4

Show that each expression is equal to 1.

1. $(\sin \theta)(\cot \theta)(\sec \theta)$ **2.** $(\cos \theta)(\tan \theta)(\csc \theta)$

3. $\cos^2 \theta \, (\tan^2 \theta + 1)$ **4.** $\tan^2 \theta \, (\csc^2 \theta - 1)$

Verify each identity.

5. $\tan \theta + \cot \theta = \dfrac{1}{(\sin \theta)(\cos \theta)}$ **6.** $\tan^2 \theta - \sin^2 \theta = \sin^2 \theta \tan^2 \theta$

7. $(\sec \theta + \tan \theta)(1 - \sin \theta) = \cos \theta$ **8.** $\sec^4 \theta - \tan^4 \theta = 1 + 2 \tan^2 \theta$

9. $\sec \theta - \cos \theta = (\sin \theta)(\tan \theta)$ **10.** $\sin^4 \theta - \cos^4 \theta + \cos^2 \theta = \sin^2 \theta$

11. $\dfrac{\cot \theta - 1}{1 - \tan \theta} = \dfrac{\csc \theta}{\sec \theta}$ **12.** $\dfrac{\tan \theta \sin \theta}{\sec^2 \theta - 1} = \cos \theta$

13. $\dfrac{1 + \tan^2 \theta}{1 + \cot^2 \theta} = \sec^2 \theta - 1$ **14.** $\dfrac{\sin \theta + \tan \theta}{1 + \sec \theta} = \sin \theta$

15. $\dfrac{1 - \cos \theta}{1 + \cos \theta} = (\csc \theta - \cot \theta)^2$ **16.** $\dfrac{1 + \tan \theta}{1 - \tan \theta} = \dfrac{\cot \theta + 1}{\cot \theta - 1}$

17. $(\sin^2 \theta + \cos^2 \theta)^5 = 1$ ***18.** $\dfrac{1 + \sin \theta}{\cos \theta} = \dfrac{\cos \theta}{1 - \sin \theta}$

19. $\dfrac{\tan^2 \theta + 1}{\tan^2 \theta} = \csc^2 \theta$ ***20.** $\dfrac{\cot \theta}{\csc \theta + 1} = \sec \theta - \tan \theta$

*21. $\dfrac{\tan \theta}{\sec \theta - 1} = \dfrac{\sec \theta + 1}{\tan \theta}$ ***22.** $\dfrac{\cos \theta}{\csc \theta - 2 \sin \theta} = \dfrac{\tan \theta}{1 - \tan^2 \theta}$

Show that each of the following is not an identity by finding a specific value of θ for which the statement does not hold.

23. $\dfrac{\cot \theta - 1}{1 - \tan \theta} = \dfrac{\sec \theta}{\csc \theta}$ **24.** $\dfrac{1 + \sin \theta}{\cos \theta} = \dfrac{\cos \theta}{1 + \sin \theta}$

8.5

Right
Triangle
Trigonometry

We are now ready to explore a variety of problems that can be solved through the use of the trigonometric ratios we have studied. Here is a typical problem:

> The angle of elevation of the top of a flagpole is 35° from a point 50 meters from the base of the pole. What is the height of the pole?

First note that a new term is used in this problem: *angle of elevation*. This is the angle between the line of sight to the top of an object and the horizontal. Then we need to solve the following right triangle for the side marked x.

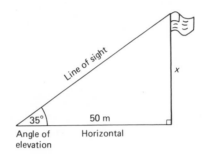

Think of the six trigonometric ratios and consider one that relates the side opposite (x) and the side adjacent (50 meters) to the 35° angle. Clearly, the tangent may be used and we therefore write this ratio:

$$\tan 35° = \frac{x}{50}$$

Solve for x:
$$x = (50)(\tan 35°)$$

We could find x by multiplication if we knew the tangent of 35°. But, to date, we have only found the values of the trigonometric ratios for angles of 30°, 45°, and 60°. So our next task is to determine the trigonometric ratios of other acute angles.

Fortunately there are tables that give the values of trigonometric functions of angles. So let us pause in the solution of this problem and study Appendix Table V, which begins on page 473. A small portion of that table is reproduced at the top of the next page.

First notice that angles are given in both degree and radian measures. The left-hand column of the table gives the measure of angles in units of 10′ from 0° through 45°. The right-hand column, which is read from the bottom of each page to the top, then proceeds from 45° through 90°. Thus to find the ratio of an angle from 0° through 45° you read the angle at the left and go down the appropriate column according to the headings on top. To find the ratio of an angle between

Degrees	Radians	sin	cos	tan	cot	sec	csc		
27° 00′	.4712	.4540	.8910	.5095	1.963	1.122	2.203	1.0996	**63° 00′**
10	741	566	897	132	949	124	190	966	50
20	771	592	884	169	935	126	178	937	40
30	.4800	.4617	.8870	.5206	1.921	1.127	2.166	1.0908	30
40	829	643	857	243	907	129	154	879	20
50	858	669	843	280	894	131	142	850	10
28° 00′	.4887	.4695	.8829	.5317	1.881	1.133	2.130	1.0821	**62° 00′**
10	916	720	816	354	868	134	118	792	50
20	945	746	802	392	855	136	107	763	40
30	.4974	.4772	.8788	.5430	1.842	1.138	2.096	1.0734	30
40	.5003	797	774	467	829	140	085	705	20
50	032	823	760	505	816	142	074	676	10
34° 00′	.5934	.5592	.8290	.6745	1.483	1.206	1.788	.9774	**56° 00′**
10	963	616	274	787	473	209	731	745	50
20	992	640	258	830	464	211	773	716	40
30	.6021	.5664	.8241	.6873	1.455	1.213	1.766	.9687	30
40	050	688	225	916	446	216	758	657	20
50	080	712	208	959	437	218	751	628	10
35° 00′	.6109	.5736	.8192	.7002	1.428	1.221	1.743	.9599	**55° 00′**
10	138	760	175	046	419	223	736	570	50
20	167	783	158	089	411	226	729	541	40
30	.6196	.5807	.8141	.7133	1.402	1.228	1.722	.9512	30
40	225	831	124	177	393	231	715	483	20
50	254	854	107	221	385	233	708	454	10
36° 00′	.6283	.5878	.8090	.7265	1.376	1.236	1.701	.9425	**54° 00′**
		cos	sin	cot	tan	csc	sec	Radians	Degrees

45° and 90°, read the angle in the column on the right and use the headings that appear at the bottom of the table.

In the illustrative portion of the table shown here, we locate 35°00′ (35 degrees and 0 minutes) at the left. Then locate the column headed "tan" at the top. This row and column intersect at the entry that gives the tangent of 35°, namely .7002. Notice that this same entry applies for the cotangent of 55°. Thus locate 55° in the column at the right and the column headed "cot" at the bottom. The intersection here is also .7002, which is what we should expect; that is, tan 35° = cot 55°.

Let us complete our problem now.

$$\tan 35° = \frac{x}{50}$$
$$x = (50)(\tan 35°)$$
$$= (50)(.7002)$$
$$= 35.01$$

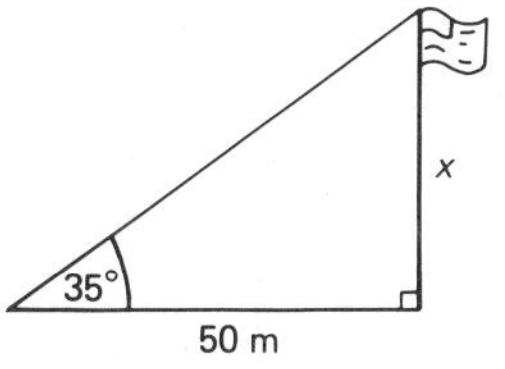

The height of the flagpole is 35 meters, correct to the nearest meter. Note that the entries in Table V are really approximations of the trigonometric ratios to four decimal places. An appropriate calculator will give you the same values to four decimal places.

The preceding problem can be solved with the aid of a calculator, but caution is necessary. You need to determine just what your calculator can do, and in what order. Thus to find tan 35° you must be certain that your calculator can accommodate an angle given in degrees; some instruments have a special setting for degree measure and another for radian measure. Assuming that your calculator will accept degree measure, and has trigonometric functions as well, the steps used might be the following:

$x = (50)(\tan 35°)$

Press 35

Press tan

Press ×

Press 50

Press =

Note the order of pressing keys to find tan 35°. *First* press the degree measure of the given angle; *then* press the desired trigonometric function. Most calculators with trigonometric functions have keys only for the sine, cosine, and tangent function. Can you see how one could find the cotangent of an angle on such a calculator?

Example 1   Use Table V to find (approximate) each trigonometric ratio.
(a) cos 13°;   **(b)** cot 26°;   **(c)** sin 58°;   **(d)** csc 83°

Solution   **(a)** .9744;   **(b)** 2.050;   **(c)** .8480;   **(d)** 1.008

Example 2   Find the measure of angle θ in degrees.
(a) sin θ = .4695;   **(b)** tan θ = 2.145

Solution   It is now necessary to explore the body of the table and locate these values.
(a) We are given the sin θ so that we restrict our attention to the two columns headed "sin" at the top and bottom of the table; these are the first two col-

umns from the left after the measures of the angles are listed. Opposite 28°00′ we find the entry .4695. Thus $\theta = 28°$.

(b) Examine the two columns headed "tan" at the top and bottom of the table. The entry 2.145 is found in a column that has "tan" at the bottom. Therefore we must read the angle measure in the column at the right. You should find $\theta = 65°$; that is, tan 65° = 2.145.

TEST YOUR UNDERSTANDING

Use Table V for these exercises. You may use a calculator to confirm your results. Find each trigonometric ratio.

1. sin 27°   **2.** cos 38°   **3.** tan 72°   **4.** cot 58°

Find each angle θ in both degrees and radians.

5. cos θ = .3907   **6.** tan θ = 1.428   **7.** sin θ = .9744   **8.** sec θ = 1.079

Here are several additional examples that show how the tables may be used to solve problems involving right triangles.

Example 3   From the top of a house the angle of depression of a point on the ground is 25°. The point is 45 meters from the base of the building. How high is the building?

Solution   By the *angle of depression* we mean the angle between the horizontal and the line of sight viewed from the top of an object to a point below. As you will note in the figure, this angle of 25° is not within the triangle. However, the complement of the angle, 65°, is the measure of $\angle B$ inside the triangle. Thus we may write this ratio:

$$\tan 65° = \frac{45}{x}$$

$$x(\tan 65°) = 45$$

$$x = \frac{45}{\tan 65°}$$

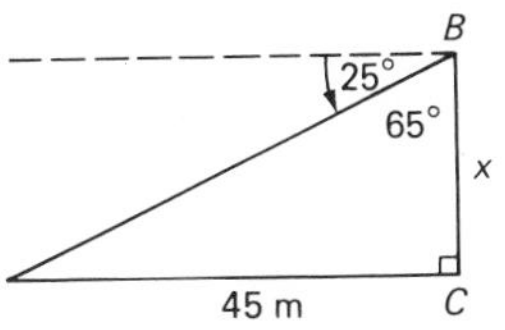

$$= \frac{45}{2.145} \qquad (\tan 65° = 2.145;\ \text{Table V})$$

$$= 21 \text{ meters to the nearest meter}$$

Use the cotangent ratio and solve the problem again with a calculator.

Example 4   A kite is at the end of a 45-meter string that is taut. It is 30 meters above the ground. What is the angle of elevation of the kite?

Solution   We wish to determine $\angle A$ in the figure.

$$\sin A = \frac{30}{45}$$

$$= .6667$$

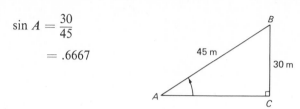

Now look at the body of the table for the *closest entry* in the sin columns to .6667. You should find that the sine of 41°50′ is .6670; therefore to the nearest whole number of degrees and minutes $\angle A = 41°50′$, and to the nearest whole number of degrees $\angle A = 42°$.

Here are some general guidelines for solving trigonometric problems that involve geometric figures:

1. Draw a figure that matches the given information. Try to draw it close to scale.
2. Record any given values directly on the corresponding parts, and label the unknown parts with appropriate letters.
3. Determine the trigonometric ratios and geometric formulas that can be used to solve for an unknown part and solve.

EXERCISES 8.5

1. Find the six trigonometric ratios for $\theta = 35°20′$.
2. Find the six trigonometric ratios for $\theta = 78°50′$.

Find the trigonometric ratios.

3. sin 42°	**4.** cos 48°	**5.** tan 12°	**6.** cot 78°
7. sec 89°	**8.** csc 1°	**9.** sin 2°40′	**10.** tan 45°10′
11. cot 44°50′	**12.** cos 75°30′	**13.** sec 19°20′	**14.** csc 5°50′

Find the measure of θ in degrees and minutes (to the nearest 10′) and also in radians.

15. tan $\theta = 6.314$	**16.** cot $\theta = .4592$	**17.** sin $\theta = .7214$
18. sec $\theta = 14.31$	**19.** cos $\theta = .9940$	**20.** csc $\theta = 2.763$

For Exercises 21 through 27 △ABC is a right triangle with ∠C = 90°. Use an appropriate trigonometric ratio to solve for the indicated part.

21. $\angle A = 70°$, $a = 35$; find b **22.** $\angle A = 70°$, $a = 35$; find c

23. $\angle B = 42°20'$, $a = 20$; find b **24.** $\angle B = 42°20'$, $a = 20$; find c

25. $a = 1$, $b = 3$; find $\angle B$ **26.** $a = 12$, $b = 9.5$; find $\angle A$

27. $b = 9$, $c = 25$; find $\angle B$

28. How high is a building whose horizontal shadow is 50 meters when the angle of elevation is 60°20′ ?

29. At a point 100 feet away from the base of a giant redwood tree a surveyor measures the angle of elevation to the top of the tree to be 70°. How tall is the tree?

30. From the top of a 172-foot-high water tank, the angle of depression to a house is 13°20′. How far away is the house from the water tank?

31. An observation post along a shoreline is 225 feet above sea level. If the angle of depression from this post to a ship at sea is 6°40′, how far is the ship from the shore?

32. A kite is flying at an altitude of 500 feet. The kite string makes an angle of 42°30′ with the ground. Assuming that the string makes a straight line, find its length.

You may use a calculator to solve the following problems.

*33. One side of an inscribed angle of a circle is a diameter of the circle, and the other side is a chord of length 10. If the inscribed angle is 66°, what is the length of the radius?

*34. From the top of a 250-foot cliff, the angle of depression to the far side of a river is 12°, and the angle of depression to a point directly on the opposite side of the river is 62°. How wide is the river between the two points?

*35. A surveyor finds that from point A on the ground the angle of elevation to the top of a mountain is 23°. When he is at a point B that is $\frac{1}{4}$ mile closer to the base of the mountain, the angle of elevation is 43°. How high is the mountain? One mile = 5280 feet. (Assume that the base of the mountain and the two observation points are on the same line.)

8.6

Extending the Definitions

Thus far the six trigonometric ratios have been defined only for acute angles: $0 < \theta < \frac{\pi}{2}$. Now we are going to extend these definitions to include all angles.

To do so, first observe that an acute angle may be placed in a rectangular coordinate system so that its *initial side* lies on the positive part of the x-axis and its *terminal side* is in the first quadrant.

Since the trigonometric ratios do not depend on the size of the sides of the right triangle, we may choose the sides so that the length of the terminal side (the hypotenuse) is 1. Consequently, the intersection of the terminal side of θ with the unit circle centered at the origin is a point $P(x, y)$ whose coordinates satisfy $x^2 + y^2 = 1$.

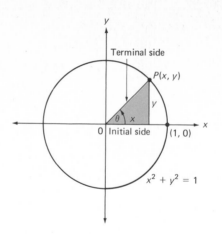

From the right triangle we have $\cos \theta = \dfrac{x}{1} = x$ and $\sin \theta = \dfrac{y}{1} = y$. Therefore we see that the earlier definitions

$$\sin \theta = \frac{\text{opposite}}{\text{hypotenuse}} \quad \text{and} \quad \cos \theta = \frac{\text{adjacent}}{\text{hypotenuse}}$$

are equivalent to

$$\sin \theta = y \quad \text{and} \quad \cos \theta = x$$

where (x, y) are the coordinates of the point of intersection of the terminal side of θ with the unit circle with center at the origin. Note that *sin θ is the second coordinate of point P and cos θ is the first coordinate of P.*

Now look at angles of other sizes. Regardless of the size of θ, its terminal side will always intersect the unit circle at some point $P(x, y)$. For positive angles the terminal side is found by rotating counterclockwise; for negative angles we rotate clockwise. Here are some typical cases.

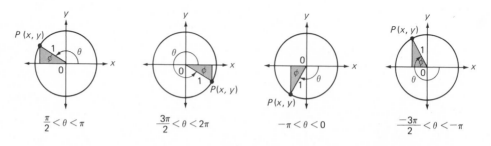

For each case the right triangle formed by drawing the perpendicular from P to the x-axis is called the **reference triangle,** and the acute angle between the terminal side and the x-axis is called the **reference angle.** The reference angle is a positive acute angle: $0 < \phi < \dfrac{\pi}{2}$. For example, the reference angle of $\dfrac{2\pi}{3}$ is $\dfrac{\pi}{3}$, and for $-\dfrac{\pi}{4}$ it is $\dfrac{\pi}{4}$. We use the reference triangle to define the trigonometric ratios of θ as follows:

$$\text{vertical side} = y = \sin \theta$$
$$\text{horizontal side} = x = \cos \theta$$

The remaining trigonometric ratios are formed as before; thus, for example, $\tan \theta = \dfrac{\sin \theta}{\cos \theta}$.

The signs of $x = \cos \theta$ and $y = \sin \theta$ depend upon the quadrant that $P(x, y)$ is in.

Terminal side of θ in	$\sin \theta$	$\cos \theta$
quadrant I	$+$	$+$
quadrant II	$+$	$-$
quadrant III	$-$	$-$
quadrant IV	$-$	$+$

Example 1   Find the six trigonometric ratios for the obtuse angle $\dfrac{5\pi}{6}$ (or 150°).

Solution   Construct the reference triangle and find the values of x and y for the point P.

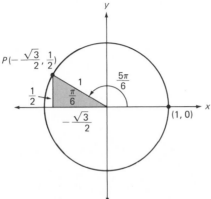

The reference triangle is a 30°–60°–90° triangle in quadrant II. Then $x = -\dfrac{\sqrt{3}}{2}$ and $y = \dfrac{1}{2}$.

$$\sin \frac{5\pi}{6} = y = \frac{1}{2} \qquad\qquad \cot \frac{5\pi}{6} = \frac{x}{y} = -\sqrt{3}$$

$$\cos \frac{5\pi}{6} = x = -\frac{\sqrt{3}}{2} \qquad\qquad \sec \frac{5\pi}{6} = \frac{1}{x} = -\frac{2}{\sqrt{3}}$$

$$\tan \frac{5\pi}{6} = \frac{y}{x} = -\frac{\sqrt{3}}{3} \qquad\qquad \csc \frac{5\pi}{6} = \frac{1}{y} = 2$$

In the preceding example the trigonometric ratios of $\dfrac{5\pi}{6}$ can also be stated in terms of the reference angle $\dfrac{\pi}{6}$, if we take into consideration that in the second quadrant $x < 0$ and $y > 0$. For example:

$$\sin \frac{5\pi}{6} = \sin \frac{\pi}{6} = \frac{1}{2}$$

$$\cos \frac{5\pi}{6} = -\cos \frac{\pi}{6} = -\frac{\sqrt{3}}{2}$$

$$\tan \frac{5\pi}{6} = -\tan \frac{\pi}{6} = -\frac{\sqrt{3}}{3}$$

Example 2   Find $\cos\left(-\dfrac{\pi}{3}\right)$ and $\tan(-225°)$.

Solution   For $-\dfrac{\pi}{3}$ the reference triangle is a $30° - 60° - 90°$ triangle in quadrant IV. Then the coordinates of $P(x, y)$ are $\left(\dfrac{1}{2}, -\dfrac{\sqrt{3}}{2}\right)$, and $\cos\left(-\dfrac{\pi}{3}\right) = \dfrac{1}{2}$.

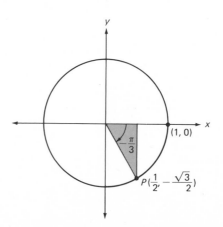

For $-225°$ we have a $45° - 45° - 90°$ reference triangle in quadrant II. Then the coordinates of P are $\left(-\dfrac{\sqrt{2}}{2}, \dfrac{\sqrt{2}}{2}\right)$ and

$$\tan(-225°) = \frac{\dfrac{\sqrt{2}}{2}}{-\dfrac{\sqrt{2}}{2}} = -1$$

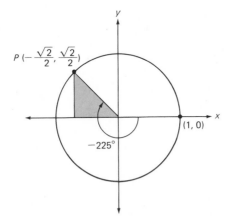

When the reference angle is not one of the special angles Table V can be used, as illustrated next.

Example 3   Find $\cos 200°$.

Solution   The terminal side of $200°$ is in quadrant III. Consequently, $x < 0$. Since the reference angle is $20°$ we have $\cos 200° = -\cos 20° = -.9397$ from Table V.

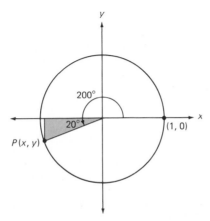

TEST YOUR UNDERSTANDING

Determine each of the following trigonometric ratios. Use Table V whenever the reference angle is not one of the special angles.

1. $\sin 135°$ **2.** $\csc(-135°)$ **3.** $\tan \dfrac{5\pi}{4}$ **4.** $\sec(-330°)$

5. $\cot 147°$ **6.** $\csc 315°$ **7.** $\sin\left(-\dfrac{4\pi}{3}\right)$ **8.** $\tan(-283°)$

If an angle has its terminal side coinciding with one of the coordinate axes, there will not be a reference triangle. Here are two cases. One shows an angle of $\dfrac{\pi}{2}$ and the other an angle of $-\pi$.

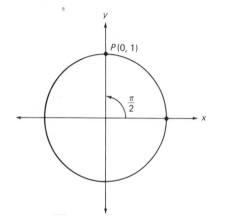
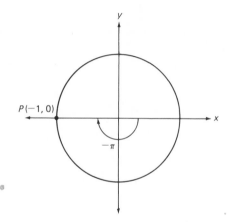

We take the coordinates of P to define the sine and cosine. Thus $\sin \dfrac{\pi}{2} = 1$, $\cos \dfrac{\pi}{2} = 0$, $\sin(-\pi) = 0$, and $\cos(-\pi) = -1$. The remaining trigonometric ratios are defined by forming the appropriate fractions whenever possible. For example, $\tan \dfrac{\pi}{2}$ is undefined since $\dfrac{1}{0}$ is undefined, and $\cot \dfrac{\pi}{2} = \dfrac{x}{y} = \dfrac{0}{1} = 0$.

It is unnecessary to memorize such results. All you need to do is visualize the unit circle with the four points, as shown in the figure at the top of the next page.

The trigonometric ratios, if they exist, are found using the coordinates of the appropriate point. For example, if $\theta = \dfrac{3\pi}{2}$ use the point $(0, -1)$: $\sin \dfrac{3\pi}{2} = -1$, $\cos \dfrac{3\pi}{2} = 0$, $\tan \dfrac{3\pi}{2}$ is undefined, $\cot \dfrac{3\pi}{2} = \dfrac{0}{-1} = 0$, $\sec \dfrac{3\pi}{2}$ is undefined, and

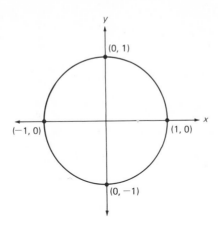

$\csc \dfrac{3\pi}{2} = \dfrac{1}{-1} = -1$. For $\theta = -\dfrac{\pi}{2}$ the trigonometric ratios are exactly the same

as for $\dfrac{3\pi}{2}$ since both angles have the same terminal side.

   To find the trigonometric ratios of angles greater than 2π or less than -2π
we locate the terminal side and then use the coordinates of the point on the unit

circle and the terminal side as before. Consider, for example, an angle of $\dfrac{7\pi}{2}$

radians. Locate the terminal side of this angle by rotating around the unit circle
as shown in this sequence of figures.

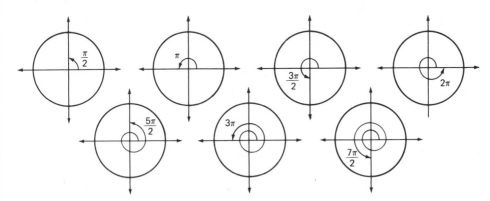

   After 2π the reference angles begin to repeat. Thus the terminal side of $\dfrac{7\pi}{2}$ is

the same as that of $\dfrac{3\pi}{2}$. Therefore the six trigonometric ratios of an angle of $\dfrac{7\pi}{2}$

(630°) are exactly the same as those for an angle of $\dfrac{3\pi}{2}$ (or 270°).

In general, for any angle greater than 2π we *subtract* integral multiples of 2π to locate the terminal side and proceed as before. For angles less than -2π we begin by *adding* integral multiples of 2π.

Example 4   Find cos 932°.

Solution   Since $932° - 2(360°) = 212°$, the terminal side is in quadrant III and the reference angle is 32°. Using Table V, $\cos 932° = -\cos 32° = -.8480$.

Note that the identities established in Section 8.4 still apply, with the added restriction that excludes division by zero. In particular this means we must exclude the situations where a trigonometric ratio fails to exist, such as $\tan \dfrac{\pi}{2}$.

CAUTION! LEARN TO AVOID MISTAKES LIKE THESE:

WRONG	*RIGHT*
$\cos(-50°) = -\cos 50°$	$\cos(-50°) = \cos 50°$
$\cos 125° = \cos 55°$	$\cos 125° = -\cos 55°$
$\sin\left(-\dfrac{\pi}{3}\right) = \dfrac{\sqrt{3}}{2}$	$\sin\left(-\dfrac{\pi}{3}\right) = -\sin \dfrac{\pi}{3} = -\dfrac{\sqrt{3}}{2}$
$\cot \dfrac{\pi}{2} = \dfrac{1}{\tan \dfrac{\pi}{2}}$	$\cot \dfrac{\pi}{2}$ is undefined

EXERCISES 8.6

For Exercises 1 through 6 locate θ in a coordinate system, include the reference triangle, and find the reference angle. Then find the coordinates of $P(x, y)$ on the terminal side and on the unit circle, and write the six trigonometric ratios.

1. $\theta = \dfrac{2\pi}{3}$ **2.** $\theta = \dfrac{29\pi}{6}$ **3.** $\theta = -\dfrac{7\pi}{4}$

4. $\theta = -30°$ **5.** $\theta = 405°$ **6.** $\theta = -930°$

7. What are the coordinates of the point $P(x, y)$ on the unit circle and on the terminal side of $\theta = -\dfrac{\pi}{2}$? Write the trigonometric ratios of $-\dfrac{\pi}{2}$.

8. Follow the instructions in Exercise 7 for $\theta = 3\pi$.

9. Follow the instructions in Exercise 7 for $\theta = \dfrac{-7\pi}{2}$.

10. Complete the table.

θ	$\sin\theta$	$\cos\theta$	$\tan\theta$	$\cot\theta$	$\sec\theta$	$\csc\theta$
0						
$\dfrac{\pi}{2}$						
π						
$\dfrac{3\pi}{2}$						

11. Find θ, where $-2\pi < \theta < 0$, such that θ and $\dfrac{\pi}{2}$ have the same trigonometric ratios.

12. Find θ, where $-2\pi < \theta < 0$, such that θ and $\dfrac{3\pi}{2}$ have the same trigonometric ratios.

Find each trigonometric ratio (if it exists). Use Table V only when necessary.

13. $\tan 220°$

14. $\sec(-72°)$

15. $\sin 261°$

16. $\sec(-\pi)$

17. $\csc(-\pi)$

18. $\tan\dfrac{7\pi}{2}$

19. $\cot\left(\dfrac{7\pi}{2}\right)$

20. $\tan 0$

21. $\cot 0$

22. $\cot\left(-\dfrac{5\pi}{2}\right)$

23. $\tan 8\pi$

24. $\sin\dfrac{9\pi}{2}$

25. $\cos(-275°)$

26. $\csc(-12°)$

27. $\cot 368°$

28. $\sin(242°10')$

29. $\cos(-792°30')$

30. $\tan 120°$

31. $\cot 1200°$

32. $\sec 420°$

33. $\csc(-80°)$

34. $\sin(94°20')$

35. $\cos(-200°50')$

36. $\tan 1°10'$

37. (a) Verify that $P\left(\dfrac{2}{3}, \dfrac{\sqrt{5}}{3}\right)$ is on the unit circle.

   (b) Locate θ, where $0 < \theta < 2\pi$, so that its terminal side intersects the unit circle at P.

   (c) Write the six trigonometric ratios for θ.

   (d) Use Table V to find an approximation for θ.

38. Follow the instructions in Exercise 37 for $P\left(-\dfrac{3}{4}, \dfrac{\sqrt{7}}{4}\right)$.

*39. Find y so that $P\left(\dfrac{\sqrt{3}}{4}, y\right)$ is on the unit circle in the fourth quadrant. Use Table V to find θ, where $-2\pi < \theta < 0$, having OP as terminal side.

*40. The terminal side of θ coincides with the line $y = -x$ and lies in the second quadrant. Find $\cos \theta$.

*41. The terminal side of θ coincides with the line $y = \frac{1}{2}x$ and lies in the third quadrant. Find $\sin \theta$.

42. Explain why $\sin^2 \theta + \cos^2 \theta = 1$, where θ is any angle.

*43. In the figure AB is tangent to the circle at A and meets the terminal side of θ at B. Why is the measure of AB equal to $\tan \theta$? Which segment has measure equal to $\sec \theta$?

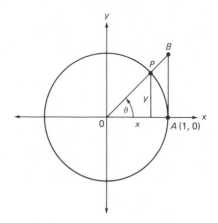

*44. Which segment in the figure has length $\cot \theta$? Which has length $\csc \theta$?

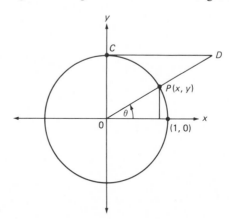

Find all possible values for θ, $0 \leq \theta < 2\pi$. Do not use Table V.

45. $\sin \theta = 1$ 46. $\cos \theta = -1$

47. $\sin \theta = 0$ 48. $\cos \theta = 0$

49. $\sin \theta = \frac{1}{2}$ 50. $\cos \theta = -\frac{1}{2}$

51. $\cos \theta = -\dfrac{\sqrt{2}}{2}$ 52. $\sin \theta = -\dfrac{\sqrt{3}}{2}$

In earlier sections, we learned how to solve for parts of right triangles. That work was based on the trigonometric ratios of acute angles. Now, with the extended definitions of these ratios, we are able to consider other types of triangles. In what follows we will refer to the angles (or vertices) of a triangle by the letters A, B, C and use a, b, c for the opposite sides.

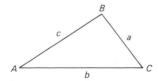

There are some important relationships between the parts of a triangle. We will study two of these, the law of sines and the law of cosines.

For triangle ABC the following equations are referred to as the **law of cosines**.

$$a^2 = b^2 + c^2 - 2bc \cos A$$
$$b^2 = a^2 + c^2 - 2ac \cos B$$
$$c^2 = a^2 + b^2 - 2ab \cos C$$

In verbal form the law of cosines says that *the square of any side of a triangle equals the sum of the squares of the remaining two sides minus twice their product times the cosine of their included angle.*

Note that if $\angle C = 90°$, then $\cos C = 0$ and the third case above reduces to $c^2 = a^2 + b^2$, that is, the Pythagorean theorem is a special case of the law of cosines. To prove this law we place triangle ABC into a rectangular system with A at the origin and AC along the x-axis as shown.

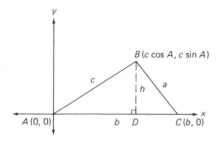

The coordinates of C are $(b, 0)$. To find the coordinates of B, first construct the altitude $BD = h$. From right $\triangle ABD$ we have $\cos A = \dfrac{AD}{c}$, or $AD = c \cos A = x$-coordinate of B. Similarly, $DB = c \sin A = y$-coordinate of B. Now apply the distance formula (page 173) to the points B and C.

$$a^2 = (c \cos A - b)^2 + (c \sin A - 0)^2$$
$$= c^2 \cos^2 A - 2bc \cos A + b^2 + c^2 \sin^2 A$$
$$= b^2 + c^2(\cos^2 A + \sin^2 A) - 2bc \cos A$$

Since $\cos^2 A + \sin^2 A = 1$, we have

$$(1) \quad a^2 = b^2 + c^2 - 2bc \cos A$$

This equation can be solved for $\cos A$:

$$(2) \quad \cos A = \frac{b^2 + c^2 - a^2}{2bc}$$

Equation (1) is used to solve for a side of a triangle when the other two sides and their included angle are given. Equation (2) is used to find the angle once the three sides are known.

Example 1   Solve (approximately) for the remaining parts of triangle ABC if $a = 4$, $b = 7$, and $\angle C = 130°$.

Solution

$$c^2 = 4^2 + 7^2 - 2(4)(7) \cos 130°$$
$$= 65 - 56(-\cos 50°)$$
$$= 65 + 56(.6428)$$
$$= 100.9968$$

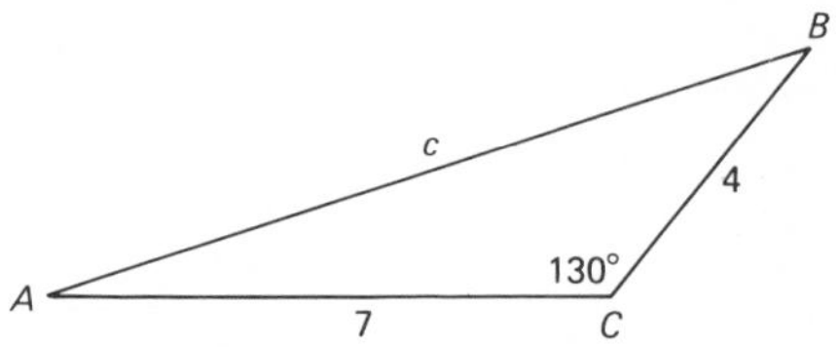

Thus c is approximately 10. To find angle A we may now use

$$\cos A = \frac{b^2 + c^2 - a^2}{2bc}$$
$$= \frac{7^2 + 10^2 - 4^2}{2(7)(10)} = .95$$

We use the nearest entry in Table V to find $A = 18°10'$. The solution is completed by noting that $\angle B = 180° - (\angle A + \angle C) = 31°50'$.

TEST YOUR UNDERSTANDING

In triangle ABC, $b = 9$, $c = 22$, and $\angle A = 42°$.

1. Find a rounded off to one decimal place.

2. Find angle B.

For triangle ABC the following equalities are known as the **law of sines**.

$$\frac{a}{\sin A} = \frac{b}{\sin B} = \frac{c}{\sin C}$$

To prove this, first construct the altitude $CD = h$.

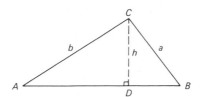

From right triangles ACD and BCD we have $\sin A = \dfrac{h}{b}$ and $\sin B = \dfrac{h}{a}$. Solve each equality for h and set equals to equals.

$$a \sin B = b \sin A$$

Divide by $(\sin A)(\sin B)$: $\dfrac{a}{\sin A} = \dfrac{b}{\sin B}$

Similar reasoning produces: $\dfrac{a}{\sin A} = \dfrac{c}{\sin C}$

The law of sines may be used to solve for the unknown parts of a triangle in the following situations:

1. Given two angles and one side.

2. Given two sides and an opposite angle.

The next example illustrates case 1. Examples 3 and 4 are illustrations of case 2.

Example 2   Let $a = 5$, $\angle B = 75°$, and $\angle C = 41°$. Find $\angle A$, b, and c.

Solution

$$\angle A = 180° - (\angle B + \angle C)$$
$$= 180° - (75° + 41°)$$
$$= 180° - 116°$$
$$= 64°$$

Now use the law of sines to solve for b, and substitute the appropriate values. From $\dfrac{a}{\sin A} = \dfrac{b}{\sin B}$ we obtain $b = \dfrac{a \sin B}{\sin A}$. Thus

$$b = \frac{5 \sin 75°}{\sin 64°} = \frac{5(.9659)}{.8988} = 5.4 \qquad \text{(correct to one decimal place)}$$

We solve for c in a similar manner:

$$c = \frac{a \sin C}{\sin A} = \frac{5 \sin 41°}{\sin 64°} = 3.6 \qquad \text{(correct to one decimal place)}$$

If one is available, use a calculator to confirm these results.

When two sides and an angle opposite one of the sides are given, there are a number of possibilities. The following figures illustrate the four cases in which sides a and b and $\angle A$ are given with $\angle A$ an acute angle.

1. No solution possible since a is too short. This situation will produce the impossible result $\sin B = \dfrac{b \sin A}{a} > 1$.

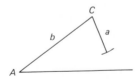

2. Two solutions. This situation occurs when $a < b$ and $\sin B = \dfrac{b \sin A}{a} < 1$.

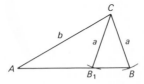

3. One solution. This situation occurs when $a > b$ and $\sin B = \dfrac{b \sin A}{a} < 1$.

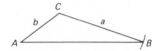

4. One solution in which $\angle B = 90°$. This occurs when $\sin B = \dfrac{b \sin A}{a} = 1$.

See Exercise 25 for an analysis if we are given a, b, and $\angle A$ with $\angle A$ obtuse.

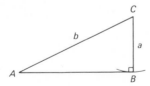

Example 3   Solve triangle ABC if $a = 50$, $b = 65$, and $\angle A = 57°$.

Solution   From $\dfrac{a}{\sin A} = \dfrac{b}{\sin B}$ we obtain $\sin B = \dfrac{b \sin A}{a}$. Thus

$$\sin B = \frac{65(\sin 57°)}{50} = 1.09 > 1$$

Thus there is no solution possible. (Why not?)

Example 4   Approximate the remaining parts of the triangle if $a = 10$, $b = 13$, and $\angle A = 25°$.

Solution   We find that $\sin B = \dfrac{b \sin A}{a} = \dfrac{13(\sin 25°)}{10} = .5494$. Since $\sin B < 1$ and $a < b$, there are two solutions for $\angle B$. Using Table V we find the closest entry to be .5495 and therefore use $\angle B = 33°20'$. The other solution is $180° - \angle B = 146°40'$. If $\angle B = 33°20'$, we have $\angle C = 180° - (\angle A + \angle B) = 121°40'$. Then

$$c = \frac{a \sin C}{\sin A} = \frac{10(\sin 121°40')}{\sin 25°}$$
$$= \frac{10(\sin 58°20')}{\sin 25°} = 20.1$$

Using $\angle B = 146°40'$ we get $\angle C = 8°20'$ and $c = 3.4$. Draw a triangle for each of these solutions.

EXERCISES 8.7

Use the law of cosines to solve for the indicated parts of triangle ABC. Do not use Table V.

1. Solve for a if $b = 4$, $c = 11$, and $\angle A = 60°$.

2. Solve for b if $a = 20$, $c = 8$, and $\angle B = 45°$.

Use the law of sines to solve for the indicated parts of triangle ABC. Do not use Table V.

3. Solve for a and c if $\angle A = 30°$, $\angle B = 120°$, and $b = 54$.

4. Solve for c if $\angle C = 30°$, $\angle A = 135°$, and $a = 100$.

Use the law of cosines to solve for the indicated parts of triangle ABC.

5. Solve for a and $\angle B$ if $\angle A = 20°$, $b = 8$, and $c = 13$.

6. Solve for b and $\angle C$ if $a = 9$, $c = 14$, and $\angle B = 110°$.

7. Solve for $\angle C$ is $a = 12$, $b = 5$, and $c = 13$.

8. Solve for angles A, B, and C if $a = b = c = 10$.

9. Solve for c and $\angle B$ if $a = 18$, $b = 9$, and $\angle C = 30°10'$.

10. Solve for b and $\angle C$ if $a = 15$, $c = 5$, and $\angle B = 157°30'$.

Use the law of sines to solve for the remaining parts of triangle ABC.

11. $\angle A = 25°$, $\angle C = 55°$, $b = 12$ **12.** $\angle C = 110°$, $\angle B = 28°$, $a = 8$

13. $\angle A = 62°20'$, $\angle B = 50°$, $b = 5$ **14.** $\angle A = 155°$, $\angle B = 15°30'$, $c = 20$

Use the law of sines to solve for the indicated parts of triangle ABC whenever possible.

15. Solve for $\angle B$ and c if $\angle A = 53°$, $a = 12$, and $b = 15$.

16. Solve for $\angle C$ and a if $\angle B = 122°$, $b = 20$, and $c = 8$.

17. Solve for $\angle B$ if $b = 25$, $a = 7$, and $\angle A = 75°$.

18. Solve for $\angle B$ if $\angle A = 44°$, $a = 9$, and $b = 12$.

19. Solve for $\angle C$ if $\angle B = 22°40'$, $b = 25$, and $c = 30$.

20. Solve for $\angle C$ if $\angle B = 22°40'$, $b = 8$, and $c = 30$.

21. Two points A and B are on the shoreline of a lake. A surveyor is located at a point C where $AC = 180$ meters and $BC = 120$ meters. He finds that $\angle ACB = 56°20'$. What is the distance between A and B?

22. Two guy wires support a telephone pole. They are attached to the top of the pole and are anchored into the ground on opposite sides of the pole at points A and B. If $AB = 120$ feet and the angles of elevation at A and B are $72°$ and $56°$ respectively, find the length of the wires.

23. An airplane is flying in a straight line toward an airfield at a fixed altitude. At one point the angle of depression to the airfield is $32°$. After flying 2 more miles the angle of depression is $74°$. What is the distance between the airplane and the airfield when the angle of depression is $74°$?

24. A diagonal of a parallelogram has length 80 and makes an angle of 20° with one of the sides. If this side has length 34, find the length of the other side of the parallelogram.

25. Let $\angle A$ and sides a and b be given. If $\angle A$ is obtuse, how many solutions (triangles) are possible when $a \leq b$? How many when $a > b$?

*26. (a) Prove that the area of a triangle equals one-half the product of two of its sides times the sine of the included angle.

　　　(b) Find the area of triangle ABC if $\angle A = 82°$, $b = 14$, and $c = 31$.

REVIEW EXERCISES FOR CHAPTER 8

The solutions to the following exercises can be found within the text of Chapter 8. Try to answer each question without referring to the text.

Section 8.1

1. For a right triangle having sides of length 3, 4, and 5, find the sine and cosine of each of the acute angles.

2. For a right triangle having sides 5, 12, and 13, find the six trigonometric ratios of the angle opposite the side of length 12.

3. Triangle ABC has a right angle at C. If $a = 3$ and $b = 2$, find sin B.

4. Find the six trigonometric ratios for angle A in terms of x.

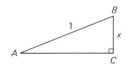

5. Show that $(\sin A)(\cos A) = \dfrac{x\sqrt{25 - x^2}}{25}$ for the given triangle.

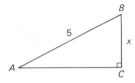

Section 8.2

6. (a) Convert 30° into radian measure.

　　　(b) Convert 120° into radian measure.

7. (a) Convert $\dfrac{\pi}{4}$ radians into degrees.

　　　(b) Convert $\dfrac{5\pi}{6}$ radians into degrees.

8. Find the area of the sector of a circle of radius 6 having a central angle of 60°.

Section 8.3

9. Find the values of the six trigonometric ratios of 45°.

10. Find the values of the six trigonometric ratios of $\dfrac{\pi}{6}$ radians.

11. Solve for x.

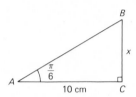

12. Solve for x to the nearest tenth of a unit.

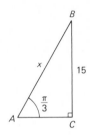

Section 8.4

13. Using the definitions of $\sin \theta$ and $\cos \theta$, derive the identity of $\tan \theta = \dfrac{\sin \theta}{\cos \theta}$.

14. Use the Pythagorean theorem to derive the identity $\sin^2 \theta + \cos^2 \theta = 1$.

15. Verify the identity: $\dfrac{\tan^2 \theta + 1}{\tan \theta \csc^2 \theta} = \tan \theta$.

16. Prove the identity: $(1 - \sin \theta)(1 + \sin \theta) = \dfrac{1}{1 + \tan^2 \theta}$.

Section 8.5

17. Use Table V to find these values:
   (a) $\cos 13°$ (b) $\cot 26°$ (c) $\sin 58°$ (d) $\csc 83°$

18. Use Table V to solve for θ in degrees:
   (a) $\sin \theta = .4695$ (b) $\tan \theta = 2.145$

19. The angle of elevation to the top of a flagpole is 35° from a point 50 meters from the base of the pole. Find the height of the pole.

20. From the top of a house the angle of depression of a point on the ground is 25°. The point is 15 meters from the base of the building. How high is the building?

Section 8.6

21. Find the six trigonometric ratios for $\dfrac{5\pi}{6}$.

22. Find $\cos\left(-\dfrac{\pi}{3}\right)$ and $\tan(-225°)$.

23. Use Table V to find $\cos 200°$.

24. Determine the trigonometric ratios, whenever they exist, for $\dfrac{3\pi}{2}$.

25. Find $\cos 932°$.

Section 8.7

26. Solve for the remaining parts of triangle ABC if $a = 4$, $b = 7$, and $\angle C = 130°$.

27. If $a = 5$, $\angle B = 75°$, and $\angle C = 41°$, find $\angle A$, b, and c.

28. Solve triangle ABC if $a = 50$, $b = 65$, and $\angle A = 57°$.

SAMPLE TEST FOR CHAPTER 8

1. Use the given figure to write the following trigonometric ratios:
  (a) $\tan A$
  (b) $\sin B$
  (c) $\sec A$

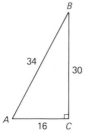

2. If $\sec A = \frac{3}{2}$, find $\sin A$, $\cos A$, and $\tan A$.

3. Use the given triangle to show that

$$3(\sin^2 A)(\sec A) = \frac{x^2}{\sqrt{x^2 + 9}}$$

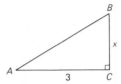

4. Find each ratio (if it exists):

  (a) $\sin \dfrac{\pi}{3}$ (b) $\cot \dfrac{3\pi}{2}$ (c) $\sin(-135°)$

5. Find each ratio (if it exists):

    (a) $\tan 510°$ **(b)** $\cos 3\pi$ **(c)** $\csc\left(-\dfrac{7\pi}{6}\right)$

6. Solve for x in the given triangle.

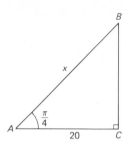

7. A building casts a shadow of 100 feet. The angle of elevation from the tip of the shadow to the top of the building is 75°. Find the height of the building correct to the nearest foot.

8. Prove the identity: $\cot^2\theta - \cos^2\theta = \cos^2\theta \cot^2\theta$.

9. Prove the identity: $\dfrac{1-\sin\theta}{1+\sin\theta} = (\sec\theta - \tan\theta)^2$.

10. **(a)** Solve for a if $\angle A = 60°$, $b = 8$, and $c = 12$. Give $\angle A$ in radians.

    (b) Write an expression for $\sin B$ in terms of other known parts of triangle ABC given in part (a).

Answers To The Test Your Understanding Exercises

Page 329

1. $\frac{7}{25}$ **4.** $\frac{7}{25}$

2. $\frac{7}{24}$ **5.** $(\frac{7}{25})^2 + (\frac{24}{25})^2 = \frac{49}{625} + \frac{576}{625} = \frac{625}{625} = 1$

3. $\frac{25}{24}$ **6.** $(\frac{7}{25})(\frac{25}{7}) = 1$

Page 338

$$\sin 60° = \frac{\sqrt{3}}{2} \qquad \tan 60° = \frac{\sqrt{3}}{1} = \sqrt{3} \qquad \sec 60° = \frac{2}{1} = 2$$

$$\cos 60° = \frac{1}{2} \qquad \cot 60° = \frac{1}{\sqrt{3}} = \frac{\sqrt{3}}{3} \qquad \csc 60° = \frac{2}{\sqrt{3}} = \frac{2\sqrt{3}}{3}$$

Page 351

1. .4540   **3.** 3.078   **5.** $67° = 1.1694$ radians   **7.** $77° = 1.3439$ radians

2. .7880   **4.** .6249   **6.** $55° = .9599$ radians   **8.** $22° = .3840$ radians

Page 358

1. $\dfrac{\sqrt{2}}{2}$ **3.** 1   **5.** -1.540 **7.** $\dfrac{\sqrt{3}}{2}$

2. $-\sqrt{2}$ **4.** $\dfrac{2\sqrt{3}}{3}$ **6.** $-\sqrt{2}$ **8.** 4.331

Page 364

1. 16.5   **2.** $\angle B = 21°30'$

Nine

The Circular Functions

In Section 8.6 we learned that every angle has a sine value. Since angles can be measured in radians and since radians may be regarded as real numbers, for any real number (radian) θ there exists a corresponding value $\sin \theta$. Thus the equation $y = \sin \theta$ defines y to be a *function of* θ whose domain consists of all real numbers. It is called a *circular* or *trigonometric function* due to its connection with the unit circle and trigonometry.

For any angle θ, the range value $y = \sin \theta$ is the second coordinate of the point $P(x, y)$ on the terminal side of θ and on the unit circle. Thus the range consists of all y, where $-1 \leq y \leq 1$.

To graph $y = \sin \theta$ in a rectangular coordinate system, first recall that $\sin \theta$ is positive for θ in quadrants I and II; $0 < \theta < \pi$. Also $\sin \theta$ is negative for $\pi < \theta < 2\pi$. Let us form a table of values by using intervals of $\frac{\pi}{6}$ for $0 \leq \theta \leq 2\pi$.

θ	0	$\frac{\pi}{6}$	$\frac{\pi}{3}$	$\frac{\pi}{2}$	$\frac{2\pi}{3}$	$\frac{5\pi}{6}$	π	$\frac{7\pi}{6}$	$\frac{4\pi}{3}$	$\frac{3\pi}{2}$	$\frac{5\pi}{3}$	$\frac{11\pi}{6}$	2π
$y = \sin \theta$	0	$\frac{1}{2}$	$\frac{\sqrt{3}}{2}$	1	$\frac{\sqrt{3}}{2}$	$\frac{1}{2}$	0	$-\frac{1}{2}$	$-\frac{\sqrt{3}}{2}$	-1	$-\frac{\sqrt{3}}{2}$	$-\frac{1}{2}$	0

We plot these points (using the approximation $\frac{\sqrt{3}}{2} = 0.86$) and connect the points with a smooth curve to find the graph of $y = \sin \theta$ for $0 \leq \theta \leq 2\pi$. (Additional points may be found by using Table V or a calculator.) The segment from zero to 2π on the horizontal axis may be viewed as the circumference of the unit circle after it has been "unrolled."

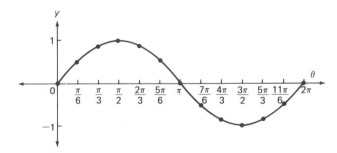

For values of θ, where $2\pi \leq \theta \leq 4\pi$, we know that the terminal sides of θ (in the unit circle) are the same as for those angles between 0 and 2π. Hence everything repeats. Likewise for $-2\pi \leq \theta \leq 0$ and so on. Thus we say that the sine function is *periodic*, with **period** 2π; $\sin(\theta + 2\pi) = \sin \theta$. That is, the sine curve repeats itself to the right and left as shown in the following graph.

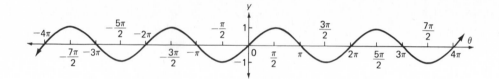

$y = \sin \theta$

Domain: all real numbers

Range: $-1 \leq y \leq 1$

Period: 2π

The graph of $y = \sin \theta$ indicates that the sine function is symmetric through the origin. This symmetry can be verified by returning to the unit circle.

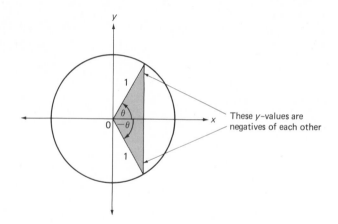

These y-values are negatives of each other

Letting $f(\theta) = \sin \theta$ we have

$$f(-\theta) = \sin (-\theta) = -\sin (\theta) = -f(\theta)$$

or

$$f(-\theta) = -f(\theta)$$

The graph of the cosine function can be obtained from the graph of $y = \sin \theta$ by observing that $\cos \theta = \sin \left(\theta + \dfrac{\pi}{2}\right)$. To see why this is true, consider the two typical situations shown at the top of the next page.

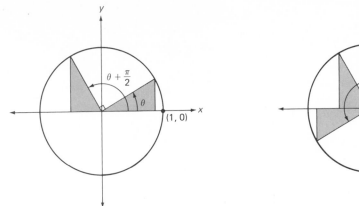

In each case the reference triangles for θ and $\theta + \dfrac{\pi}{2}$ are congruent. Consequently, the side adjacent to the reference angle for θ is the same length as the side opposite to the reference angle for $\theta + \dfrac{\pi}{2}$. It follows that $\cos \theta = \sin \left(\theta + \dfrac{\pi}{2}\right)$.

The graph of $y = \sin \left(\theta + \dfrac{\pi}{2}\right)$ can be obtained by shifting the graph of $y = \sin \theta$ by $\dfrac{\pi}{2}$ units to the left. Thus the cosine curve can be obtained by shifting the sine curve $\dfrac{\pi}{2}$ units to the left.

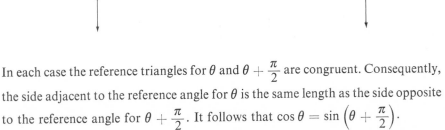

$y = \cos \theta$

Domain: all real numbers

Range: $-1 \le y \le 1$

Period: 2π

In the preceding chapter we used the unit circle to arrive at the definition $\cos \theta = x$. However, to be consistent with the usual labeling of the vertical axis in a rectangular system, we have written the equation in the form $y = \cos \theta$.

Furthermore, from now on we will use x instead of θ for the horizontal axis. We will use the letter θ only when making direct reference to the unit circle.

The cosine curve may be regarded as being $\dfrac{\pi}{2}$ units ahead (or behind) the sine curve and vice versa. Both functions have the same period, 2π. Note that the cosine is symmetric around the y-axis. This symmetry can be verified by studying the following figure.

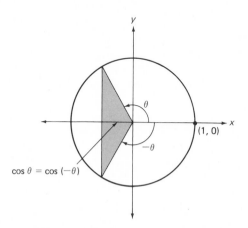

$$\cos \theta = \cos (-\theta)$$

The sine function has a maximum value of 1 and a minimum value of -1. One-half of the maximum value minus the minimum value is the **amplitude**. Thus $\frac{1}{2}[1 - (-1)] = 1$ is the amplitude of $y = \sin x$. Since $-1 \le \sin x \le 1$, multiplying by 2 produces $-2 \le 2 \sin x \le 2$. Thus 2 is the maximum value of $y = 2 \sin x$, -2 is the minimum value, and $\frac{1}{2}[2 - (-2)] = 2$ is the amplitude. Similarly, the amplitude of $y = -2 \sin x$ is $\frac{1}{2}[2 - (-2)] = 2$. In general, the amplitude of $y = a \sin x$ or of $y = a \cos x$ is equal to $|a|$.

In the next figure the idea of amplitude is illustrated for a few cases. Notice that each of these functions has period 2π.

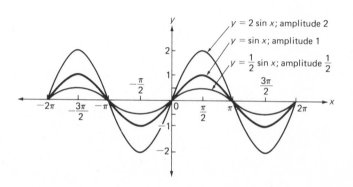

TEST YOUR UNDERSTANDING

Graph the curve for the values $-2\pi \le x \le 2\pi$.

1. $y = -\sin x$

2. $y = 3 \cos x$

Find the amplitude.

3. $y = 10 \sin x$

4. $y = -\frac{2}{3} \cos x$

Explain how the graph of g can be obtained from the graph of f.

5. $f(x) = \sin x$; $g(x) = \sin(x - 2)$

6. $f(x) = \cos x$; $g(x) = 2 + \cos x$

For $y = \sin x$, the coefficient of x is 1. By changing the coefficient we alter the period of the function. Consider, for example, the graph of $y = \sin 2x$. As x assumes values from 0 through π, $2x$ takes on values from 0 through 2π. Thus the graph goes through a complete cycle for $0 \le x \le \pi$ and is said to have a period of π. This information is shown in the following table of values and graph. Note that the graph completes two full cycles in the interval $0 \le x \le 2\pi$.

x	0	$\frac{\pi}{4}$	$\frac{\pi}{2}$	$\frac{3\pi}{4}$	π
$2x$	0	$\frac{\pi}{2}$	π	$\frac{3\pi}{2}$	2π
$y = \sin 2x$	0	1	0	-1	0

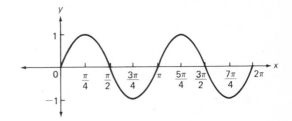

Example 1 Graph $y = \cos x$ and $y = 2 \cos \frac{1}{2}x$ on the same set of axes. Use a dashed curve for $y = \cos x$.

Solution The amplitude of $y = 2 \cos \frac{1}{2}x$ is 2, and the period is 4π since $0 \le \frac{1}{2}x \le 2\pi$ is equivalent to $0 \le x \le 4\pi$.

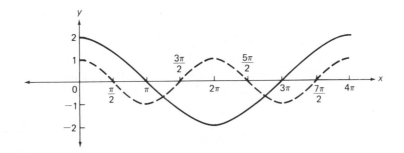

Note that on the interval $[0, 4\pi]$, $y = \cos x$ completes two full cycles whereas $y = 2 \cos \frac{1}{2}x$ completes one cycle.

We can summarize the results about amplitude and length of period by saying that both $y = a \sin bx$ and $y = a \cos bx$ have amplitude $= |a|$ and period $= \dfrac{2\pi}{|b|}$. (See Exercise 28.)

Earlier results involving the composition of functions can be applied to circular functions, as illustrated in the following examples.

Example 2 Let $g(x) = x^2 - 3x$ and $f(x) = \sin x$. Form the composite function $h = f \circ g$.

Solution

$$
\begin{aligned}
h(x) &= (f \circ g)(x) \\
&= f(g(x)) \\
&= f(x^2 - 3x) \\
&= \sin (x^2 - 3x)
\end{aligned}
$$

Example 3 Let $h(x) = \cos \dfrac{x}{x + 1}$. Show that h is the composition of two functions such that the inner function is rational.

Solution Let $g(x) = \dfrac{x}{x + 1}$ and $f(x) = \cos x$. Then

$$
\begin{aligned}
(f \circ g)(x) &= f(g(x)) \\
&= f\left(\frac{x}{x + 1}\right) \\
&= \cos \frac{x}{x + 1} = h(x)
\end{aligned}
$$

EXERCISES 9.1

1. Complete the table and use these points and the symmetry through the origin to graph $y = \sin x$ for $-\pi \le x \le \pi$.

x	$-\pi$	$-\dfrac{5\pi}{6}$	$-\dfrac{2\pi}{3}$	$-\dfrac{\pi}{2}$	$-\dfrac{\pi}{3}$	$-\dfrac{\pi}{6}$	0
$y = \sin x$							

2. Complete the table and use these points and the symmetry through the y-axis to graph $y = \cos x$ for $-\pi \le x \le \pi$.

x	0	$\dfrac{\pi}{4}$	$\dfrac{\pi}{2}$	$\dfrac{3\pi}{4}$	π
$y = \cos x$					

3. Graph $y = 3 \sin x$, $y = \frac{1}{3} \sin x$, and $y = -3 \sin x$ on the same axes for $0 \le x \le 2\pi$. Find the amplitudes.

4. Graph $y = 2 \cos x$, $y = \frac{1}{2} \cos x$, and $y = -2 \cos x$ on the same axes for $-\pi \le x \le \pi$. Find the amplitudes.

Sketch the curve on the interval $0 \le x \le 2\pi$. Find the amplitude and period.

5. $y = \cos 2x$ **6.** $y = -\sin 2x$ **7.** $y = -\frac{3}{2} \sin 4x$

8. $y = \cos 4x$ **9.** $y = -\cos \frac{1}{2}x$ **10.** $y = -2 \sin \frac{1}{2}x$

11. Find the period p of $y = \frac{1}{2} \cos \frac{1}{4}x$. Graph this curve and the curve $y = \cos x$ for $-p \le x \le p$ on the same axes.

12. Find the period p of $y = 3 \sin \frac{1}{3}x$. Graph this curve and the curve $y = \sin x$ for $-p \le x \le p$ on the same axes.

13. Find the period p of $y = 2 \sin \pi x$ and sketch the curve for $0 \le x \le p$.

14. Find the period p of $y = -\dfrac{3}{4} \cos \dfrac{\pi}{2}x$ and sketch the curve for $0 \le x \le p$.

15. Find a and b if the given curve has equation $y = a \sin bx$. (*Hint:* There are three full cycles for $0 \le x \le 2\pi$.)

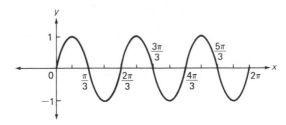

16. Find a and b if the given curve has equation $y = a \cos bx$. (*Hint:* There is one full cycle for $0 \le x \le 3\pi$.)

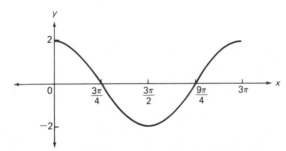

17. Graph $y = 2 + \sin x$ for $0 \le x \le 2\pi$. What is the amplitude?

18. Graph $y = -5 + 2 \cos x$ for $0 \le x \le 2\pi$. What is the amplitude?

19. How can the graph of $y = \sin(2x - 1)$ be obtained from the graph of $y = \sin 2x$? (*Hint:* $2x - 1 = 2(x - \frac{1}{2})$.)

*20. Generalize the observation made in Exercise 19 and describe how the graph of $y = \sin(bx + c)$ can be obtained from $y = \sin bx$ when $c < 0$ and when $c > 0$. (Assume $b > 0$.)

*21. Explain the steps that may be taken to obtain the graph of $y = a \sin(bx + c) + d$ from $y = \sin x$, where a, b, c, and d are positive constants.

*22. Graph $y = \sin x + \cos x$ for $0 \le x \le 2\pi$. (*Hint:* Graph $y = \sin x$ and $y = \cos x$ on the same same axes and then add the ordinates.)

23. Let $f(x) = 2x + 5$ and $g(x) = \cos x$. Form the composites $f \circ g$ and $g \circ f$.

24. Let $f(x) = \sqrt{x^2 + 1}$ and $g(x) = \sin x$. Form the composites $f \circ g$ and $g \circ f$.

25. Let $h(x) = \cos(5x^2)$. Find f and g so that $h = f \circ g$, where the inner function g is quadratic.

26. Let $h(x) = \sin(\ln x)$. Find f and g so that $h = f \circ g$, where the outer function is trigonometric.

27. Let $F(x) = \cos \sqrt[3]{1 - 2x}$. Find f, g, and h so that $F = f \circ g \circ h$, where h is linear.

*28. (a) Let $y = a \sin bx$, $a \ne 0$ and $b \ne 0$. Prove that the amplitude is $|a|$.

(b) For $b > 0$, $y = a \sin bx$ has one complete cycle for $0 \le bx \le 2\pi$. How does this give the period $\dfrac{2\pi}{|b|}$?

(c) For $b < 0$, $y = a \sin bx = -a \sin(-bx)$, where $-b > 0$. How does this give the period $\dfrac{2\pi}{|b|}$?

9.2
Graphing the Other Trigonometric Functions

Since $\tan x = \dfrac{\sin x}{\cos x}$ the properties of the tangent function depend upon the sine and cosine functions. We repeat their graphs for easy reference.

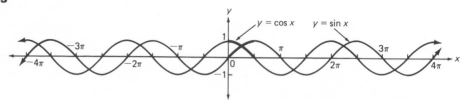

Observe that $\cos x = 0$ for $x = \pm \dfrac{\pi}{2}$, $\pm \dfrac{3\pi}{2}$, $\pm \dfrac{5\pi}{2}$, ... Therefore the domain of $\tan x = \dfrac{\sin x}{\cos x}$ consists of all real numbers x except those of the form $x = \dfrac{\pi}{2} + k\pi$, where k is any integer.

The graph of $y = \tan x$ is symmetric through the origin since for any x in the domain we have

$$\tan(-x) = \frac{\sin(-x)}{\cos(-x)} = \frac{-\sin x}{\cos x} = -\frac{\sin x}{\cos x} = -\tan x$$

The period of $y = \tan x$ is π; that is, $\tan(x + \pi) = \tan x$, as can be observed by returning to the unit circle. Consider, for example, the case where the terminal side of an angle θ is in quadrant I; in particular, we assume that $0 < \theta < \frac{\pi}{2}$. Adding π to θ gives the angle $\theta + \pi$, whose terminal side is in quadrant III. We find that $\tan(\theta + \pi) = \tan\theta$ since the two terminal sides are on the same line, the reference triangles are congruent, and $\frac{\text{opposite}}{\text{adjacent}}$ is positive in each case.

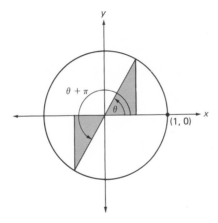

Here is a table of values to help in graphing $y = \tan x$ for $0 \le x < \frac{\pi}{2}$. The symmetry through the origin produces the curve for $-\frac{\pi}{2} < x < 0$.

x	0	$\frac{\pi}{6}$	$\frac{\pi}{4}$	$\frac{\pi}{3}$	1.3	1.4
$y = \tan x$	0	$\frac{\sqrt{3}}{3}$	1	$\sqrt{3}$	3.6	5.8

From Table V

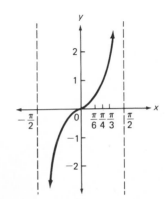

Notice that the vertical lines $x = \pm\frac{\pi}{2}$ are asymptotic to the curve. You can observe this "growth" of $y = \tan x$ as x gets close to $\frac{\pi}{2}$ from Appendix Table V, as well as by studying the sine and cosine curves for $0 \leq \theta < \frac{\pi}{2}$. There you can see that as x gets close to $\frac{\pi}{2}$, the sine gets close to 1 and the cosine gets close to 0. Hence the fraction $\frac{\sin x}{\cos x}$ gets very large.

Since the period of $\tan x$ is π, the preceding figure shows one cycle of the tangent function which repeats to the left and right. The range of $y = \tan x$ consists of all real numbers.

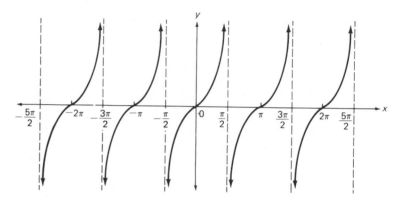

$y = \tan x$

Domain: all $x \neq \frac{\pi}{2} + k\pi$ where k is an integer

Range: all real numbers y

Period: π

A similar analysis leads to the graph of $y = \cot x = \dfrac{\cos x}{\sin x}$.

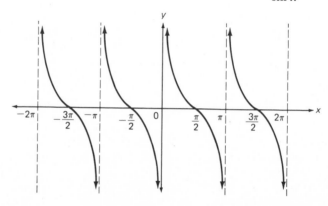

$y = \cot x$

Domain: all $x \neq k\pi$ where k is an integer

Range: all real numbers y

Period: π

Example 1   Graph $y = -\tan 3x$.

Solution   Since $0 \leq 3x < \pi$ is equivalent to $0 \leq x < \dfrac{\pi}{3}$, the period is $\dfrac{\pi}{3}$. To graph $y = -\tan 3x$, first graph $y = \tan 3x$ and then reflect through the x-axis.

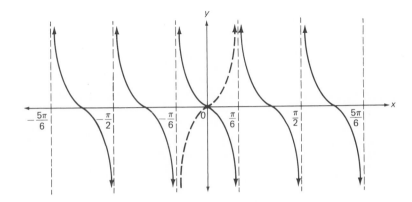

Example 2   Let $F(x) = \tan^2 3x$. Find functions f, g, h so that $F(x) = h(g(f(x)))$.

Solution   Let $f(x) = 3x$, $g(x) = \tan x$, and $h(x) = x^2$. Then

$$h(g(f(x))) = h(g(3x))$$
$$= h(\tan 3x)$$
$$= \tan^2 3x = F(x)$$

The secant function can be graphed by making use of the cosine because $\sec x = \dfrac{1}{\cos x}$ for $\cos x \neq 0$. We need only consider $x \geq 0$ since

$$\sec(-x) = \frac{1}{\cos(-x)} = \frac{1}{\cos x} = \sec x$$

shows that the secant is symmetric with respect to the y-axis. Now consider $0 \leq x < \dfrac{\pi}{2}$. For such x take the reciprocal of $\cos x$ to get $y = \dfrac{1}{\cos x} = \sec x$. On the next page are some specific cases to help you graph the curve.

x	0	$\dfrac{\pi}{6}$	$\dfrac{\pi}{4}$	$\dfrac{\pi}{3}$
$\cos x$	1	$\dfrac{\sqrt{3}}{2}$	$\dfrac{1}{\sqrt{2}}$	$\dfrac{1}{2}$
$\sec x$	1	$\dfrac{2}{\sqrt{3}}$	$\sqrt{2}$	2

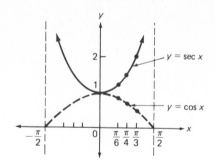

You can see that as x gets close to $\dfrac{\pi}{2}$, $\cos x$ gets close to 0, and therefore the reciprocals get very large. It follows that $x = \dfrac{\pi}{2}$ is a vertical asymptote and, by symmetry, so is $x = -\dfrac{\pi}{2}$. By similar analysis the graph of $y = \sec x$ can be found for $\dfrac{\pi}{2} < x < \dfrac{3\pi}{2}$. Then the periodicity gives the rest. Note that for all x in the domain of the secant we have

$$\sec(x + 2\pi) = \frac{1}{\cos(x + 2\pi)} = \frac{1}{\cos x} = \sec x$$

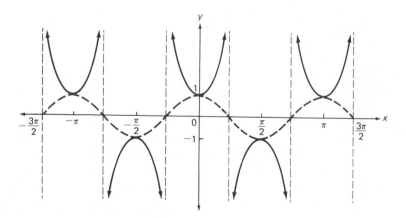

$y = \sec x$

Domain: All $x \neq \pm\dfrac{\pi}{2},\ \pm\dfrac{3\pi}{2},\ \pm\dfrac{5\pi}{2}, \ldots$

Range: All $y \geq 1$ or all $y \leq -1$

Period: 2π

Symmetric about the y-axis

You will be asked to sketch the curve $y = \csc x$ as an exercise.

EXERCISES 9.2

1. Complete the following table (using Appendix Table V only if necessary) and sketch the curve $y = \tan x$ for $-\dfrac{\pi}{2} < x \le 0$. Then use the symmetry through the origin to obtain the graph for $-\dfrac{\pi}{2} < x < \dfrac{\pi}{2}$.

x	-1.4	-1.3	$-\dfrac{\pi}{3}$	$-\dfrac{\pi}{4}$	$-\dfrac{\pi}{6}$	0
$y = \tan x$						

In Exercises 2 through 7 sketch the curve, including at least two full cycles; indicate the period and the vertical asymptotes.

2. $y = \tan 2x$

3. $y = \cot 3x$

4. $y = \dfrac{1}{2}\tan\dfrac{x}{2}$

5. $y = -2\cot\dfrac{x}{2}$

6. $y = -\sec x$

7. $y = \sec 4x$

8. (a) State the domain of $y = \csc x$.
   (b) Prove that the cosecant is symmetric through the origin. Do this by showing $\csc(-x) = -\csc x$.
   (c) Verify that the cosecant has period 2π by showing $\csc(x + 2\pi) = \csc x$.

9. Sketch $y = \sin x$ for $0 \le x \le 2\pi$. Use this curve to obtain the graph of $y = \csc x$ for $0 < x < \pi$ and for $\pi < x < 2\pi$. What are the asymptotes?

10. Find the period and graph $y = \csc 2x$. Include at least two full cycles.

11. Find the period and graph $y = -\csc\frac{1}{3}x$. Include at least two full cycles.

To graph $y = |f(x)|$ we may reflect the negative parts of $y = f(x)$ through the x-axis. Use this procedure to graph the curves in Exercises 12 through 14.

12. $y = |\sin x|$

13. $y = |\tan x|$

14. $y = |\sec x|$

15. Let $f(x) = x^2$ and $g(x) = \tan x$. Form the composites $f \circ g$ and $g \circ f$.

16. Let $f(x) = \dfrac{x}{x+1}$ and $g(x) = \sec x$. Form the composites $f \circ g$ and $g \circ f$.

17. Let $h(x) = \cot^3 x$. Find f and g so that $h = f \circ g$, where the outer function f is a polynomial.

18. Let $F(x) = \tan^2\dfrac{x+1}{x-1}$. Find f, g, and h so that $F = f \circ g \circ h$, where h is rational and f is quadratic.

*19. Refer to the figure and explain why $0 < \dfrac{\sin \theta}{\theta} < 1$ for $0 < \theta < \dfrac{\pi}{2}$. (*Hint:* In a unit circle the central angle in radians is the same as the length of the intercepted arc.)

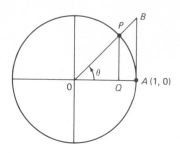

*20. Refer to the figure in Exercise 19 and prove that $\cos \theta < \dfrac{\sin \theta}{\theta} < \dfrac{1}{\cos \theta}$ for $0 < \theta < \dfrac{\pi}{2}$. (*Hint:* Use the areas of the two right triangles and the sector of the circle.)

Stop here

9.3

The Addition Formulas

Is it true that $\cos (30° + 60°) = \cos 30° + \cos 60°$? Some quick calculations show that it is not.

$$\cos (30° + 60°) = \cos 90° = 0 \quad \text{and} \quad \cos 30° + \cos 60° = \frac{\sqrt{3}}{2} + \frac{1}{2}$$

To write $\cos (\theta_1 + \theta_2) = \cos \theta_1 + \cos \theta_2$ would be to assume, incorrectly, that the cosine function obeys the distributive property. We emphasize that cos is the *name* of a function; it is not a number.

The cosine of the sum of two angles can be correctly evaluated by using Formula (3) below; it is one of four very important trigonometric identities known as the *addition formulas*.

Addition formulas

(1) $\qquad \sin (\alpha + \beta) = \sin \alpha \cos \beta + \cos \alpha \sin \beta$
(2) $\qquad \sin (\alpha - \beta) = \sin \alpha \cos \beta - \cos \alpha \sin \beta$
(3) $\qquad \cos (\alpha + \beta) = \cos \alpha \cos \beta - \sin \alpha \sin \beta$
(4) $\qquad \cos (\alpha - \beta) = \cos \alpha \cos \beta + \sin \alpha \sin \beta$

Using Formula (3) we have

$$\cos (30° + 60°) = \cos 30° \cos 60° - \sin 30° \sin 60°$$

$$= \frac{\sqrt{3}}{2} \cdot \frac{1}{2} - \frac{1}{2} \cdot \frac{\sqrt{3}}{2} = 0 = \cos 90°$$

The addition formulas are stated by using the variables α and β, which can represent angles measured in either degrees or radians. We now prove Formula (4) by making use of the unit circle.

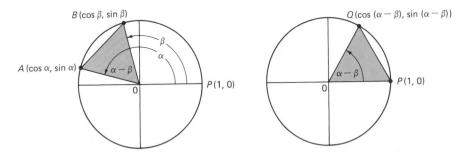

In the unit circle at the left we represent a typical situation in which $\alpha > \beta$. By using the reference triangle for α, we find that the coordinates of A are $(\cos \alpha, \sin \alpha)$. Likewise B has coordinates $(\cos \beta, \sin \beta)$. Then, by the distance formula,

$$(AB)^2 = (\cos \alpha - \cos \beta)^2 + (\sin \alpha - \sin \beta)^2$$

$$= (\cos^2 \alpha - 2 \cos \alpha \cos \beta + \cos^2 \beta) + (\sin^2 \alpha - 2 \sin \alpha \cos \beta + \sin^2 \beta)$$

$$= (\cos^2 \alpha + \sin^2 \alpha) + (\cos^2 \beta + \sin^2 \beta) - 2(\cos \alpha \cos \beta + \sin \alpha \sin \beta)$$

$$= 2 - 2(\cos \alpha \cos \beta + \sin \alpha \sin \beta)$$

In the unit circle at the right the central angle $\alpha - \beta$ has its initial side on the horizontal axis; therefore the coordinates of Q are $(\cos (\alpha - \beta), \sin (\alpha - \beta))$. Using the distance formula, we get

$$(PQ)^2 = [\cos (\alpha - \beta) - 1]^2 + [\sin (\alpha - \beta) - 0]^2$$

$$= \cos^2 (\alpha - \beta) - 2 \cos (\alpha - \beta) + 1 + \sin^2 (\alpha - \beta)$$

$$= 2 - 2 \cos (\alpha - \beta)$$

Triangles AOB and QOP are congruent. (Why?) Then $AB = PQ$ and this implies $(AB)^2 = (PQ)^2$. We may therefore equate the preceding results and then simplify.

$$2 - 2 \cos (\alpha - \beta) = 2 - 2(\cos \alpha \cos \beta + \sin \alpha \sin \beta)$$

$$\cos (\alpha - \beta) = \cos \alpha \cos \beta + \sin \alpha \sin \beta$$

Formula (3) is now easy to prove by recalling that $\cos(-\beta) = \cos\beta$ and $\sin(-\alpha) = -\sin\alpha$. The trick is to write $\alpha + \beta = \alpha - (-\beta)$. Then

$$\cos(\alpha + \beta) = \cos[\alpha - (-\beta)] = \cos\alpha\cos(-\beta) + \sin\alpha\sin(-\beta)$$
$$= \cos\alpha\cos\beta - \sin\alpha\sin\beta$$

The proofs of Formulas (1) and (2) will be called for in the exercises.

Example 1   Evaluate $\sin\frac{7}{12}\pi$ by using $\frac{7}{12} = \frac{1}{4} + \frac{1}{3}$.

Solution   Use Formula (1) with $\alpha = \frac{\pi}{4}$, $\beta = \frac{\pi}{3}$.

$$\sin\frac{7}{12}\pi = \sin\left(\frac{\pi}{4} + \frac{\pi}{3}\right)$$
$$= \sin\frac{\pi}{4}\cos\frac{\pi}{3} + \cos\frac{\pi}{4}\sin\frac{\pi}{3}$$
$$= \frac{\sqrt{2}}{2}\cdot\frac{1}{2} + \frac{\sqrt{2}}{2}\cdot\frac{\sqrt{3}}{2}$$
$$= \frac{1}{4}(\sqrt{2} + \sqrt{6})$$

TEST YOUR UNDERSTANDING

1. Use $15° = 45° - 30°$ and appropriate addition formulas to evaluate $\sin 15°$ and $\cos 15°$.

2. Repeat Exercise 1 with $15° = 60° - 45°$.

3. Use $\frac{11\pi}{12} = \frac{\pi}{6} + \frac{3\pi}{4}$ and appropriate addition formulas to evaluate $\sin\frac{11\pi}{12}$ and $\cos\frac{11\pi}{12}$.

4. Repeat Exercise 3 with $\frac{11\pi}{12} = \frac{7\pi}{6} - \frac{\pi}{4}$.

5. Use an addition formula to evaluate $\sin 195°$.

6. Use an addition formula to evaluate $\cos\frac{5\pi}{12}$.

Formulas (1) and (3) can be used to find an addition formula for the tangent function.

$$\tan(\alpha + \beta) = \frac{\sin(\alpha + \beta)}{\cos(\alpha + \beta)} = \frac{\sin\alpha\cos\beta + \cos\alpha\sin\beta}{\cos\alpha\cos\beta - \sin\alpha\sin\beta}$$

Divide the numerator and denominator by $\cos \alpha \cos \beta$.

$$\tan (\alpha + \beta) = \dfrac{\dfrac{\sin \alpha \cos \beta}{\cos \alpha \cos \beta} + \dfrac{\cos \alpha \sin \beta}{\cos \alpha \cos \beta}}{\dfrac{\cos \alpha \cos \beta}{\cos \alpha \cos \beta} - \dfrac{\sin \alpha \sin \beta}{\cos \alpha \cos \beta}}$$

Thus

(5)
$$\tan (\alpha + \beta) = \frac{\tan \alpha + \tan \beta}{1 - \tan \alpha \tan \beta}$$

By a similar analysis we also get

(6)
$$\tan (\alpha - \beta) = \frac{\tan \alpha - \tan \beta}{1 + \tan \alpha \tan \beta}$$

With the aid of the addition formulas we will be able to derive other useful trigonometric identities. Some of these identities will be considered now and others will be studied in the next section.

The result $\cos \theta = \sin \left(\theta + \dfrac{\pi}{2} \right)$ was derived in Section 9.1 by using the unit circle. This formula can be described as a *reduction formula* in the sense that a trigonometric function of $\theta + \dfrac{\pi}{2}$ is "reduced" to a trigonometric function of just θ. Here is a list of some reduction formulas.

Reduction formulas

$$\sin (-\theta) = -\sin \theta \qquad \cos (-\theta) = \cos \theta$$

$$\sin (\theta - 2\pi) = \sin \theta \qquad \sin (\theta + 2\pi) = \sin \theta$$
$$\cos (\theta - 2\pi) = \cos \theta \qquad \cos (\theta + 2\pi) = \cos \theta$$

$$\sin (\pi - \theta) = \sin \theta \qquad \sin (\pi + \theta) = -\sin \theta$$
$$\cos (\pi - \theta) = -\cos \theta \qquad \cos (\pi + \theta) = -\cos \theta$$
$$\tan (\pi - \theta) = -\tan \theta \qquad \tan (\pi + \theta) = \tan \theta$$

$$\sin \left(\frac{\pi}{2} - \theta \right) = \cos \theta \qquad \sin \left(\frac{\pi}{2} + \theta \right) = \cos \theta$$

$$\cos \left(\frac{\pi}{2} - \theta \right) = \sin \theta \qquad \cos \left(\frac{\pi}{2} + \theta \right) = -\sin \theta$$

$$\tan \left(\frac{\pi}{2} - \theta \right) = \cot \theta \qquad \tan \left(\frac{\pi}{2} + \theta \right) = -\cot \theta$$

Note that each of these results may be restated in degrees; for example, $\cos(180° - \theta) = -\cos\theta$.

Some of these results have been encountered earlier in our work. For instance, we know that the sine of an acute angle equals the cosine of its complement. This fact is included in $\sin\theta = \cos\left(\dfrac{\pi}{2} - \theta\right)$, which holds for all values θ.

Example 2 Explain how the reduction formula $\cot x = \tan\left(\dfrac{\pi}{2} - x\right)$ can be used to graph the cotangent if the graph of the tangent is given.

Solution We find that

$$\cot x = \tan\left(\frac{\pi}{2} - x\right) = \tan\left[-\left(x - \frac{\pi}{2}\right)\right] = -\tan\left(x - \frac{\pi}{2}\right).$$

Thus the graph of the cotangent may be found by shifting the tangent curve $\dfrac{\pi}{2}$ units to the right and then reflecting through the x-axis.

We use addition formulas (3) and (4) to prove two of the preceding reduction formulas. Others will be taken up in the exercises.

$$\cos\left(\frac{\pi}{2} - \theta\right) = \cos\frac{\pi}{2}\cos\theta + \sin\frac{\pi}{2}\sin\theta \qquad \text{(by Formula (4))}$$

$$= 0 \cdot \cos\theta + (1)\sin\theta$$

$$= \sin\theta$$

$$\cos(\pi + \theta) = \cos\pi\cos\theta - \sin\pi\sin\theta \qquad \text{(by Formula (3))}$$

$$= (-1)\cos\theta - 0 \cdot \sin\theta$$

$$= -\cos\theta$$

The reduction formulas involving the tangent can be obtained from the results for the sine and cosine. Thus

$$\tan\left(\frac{\pi}{2} - \theta\right) = \frac{\sin\left(\dfrac{\pi}{2} - \theta\right)}{\cos\left(\dfrac{\pi}{2} - \theta\right)} = \frac{\cos\theta}{\sin\theta} = \cot\theta$$

Why can't the formula for $\tan(\alpha - \beta)$ be used here?

$$EXERCISES\ 9.3$$

In Exercises 1 through 4 use the addition formulas to evaluate $\sin\theta$, $\cos\theta$, *and* $\tan\theta$ *for the specified value of* θ.

1. $\theta = 75°$ **2.** $\theta = 105°$ **3.** $\theta = \dfrac{\pi}{12}$ **4.** $\theta = \dfrac{19\pi}{12}$

5. Use an addition formula to express $\sin\left(\theta + \dfrac{\pi}{3}\right)$ in terms of $\sin\theta$ and $\cos\theta$.

6. Prove: $\cos\left(\theta - \dfrac{\pi}{4}\right) = \dfrac{\sqrt{2}}{2}(\cos\theta + \sin\theta)$.

7. Let $x = \dfrac{\pi}{2} - \theta$ in the equation $\cos\left(\dfrac{\pi}{2} - x\right) = \sin x$. Then prove the following:

$\sin\left(\dfrac{\pi}{2} - \theta\right) = \cos\theta$.

8. Prove addition formula (1). $\left(Hint:\ \text{Begin with } \sin(\alpha + \beta) = \cos\left[\dfrac{\pi}{2} - (\alpha + \beta)\right]\right.$

and note that $\dfrac{\pi}{2} - (\alpha + \beta) = \left(\dfrac{\pi}{2} - \alpha\right) - \beta.\Big)$

9. Use addition formula (1) to prove (2). (*Hint:* Use $\alpha - \beta = \alpha + (-\beta)$.)

10. Use addition formula (5) to prove (6). (*Hint:* Use $\alpha - \beta = \alpha + (-\beta)$.)

For Exercises 11 through 14 prove the reduction formula by using an appropriate addition formula.

11. $\sin(\pi - \theta) = \sin\theta$ **12.** $\cos\left(\dfrac{\pi}{2} + \theta\right) = -\sin\theta$

13. $\tan(\pi - \theta) = -\tan\theta$ **14.** $\cos(\theta - 2\pi) = \cos\theta$

15. Prove: $\sec\left(\dfrac{\pi}{2} - \theta\right) = \csc\theta$.   **16.** Prove: $\csc(\pi - \theta) = \csc\theta$.

17. Derive $\cot(\alpha + \beta) = \dfrac{\cot\alpha\cot\beta - 1}{\cot\alpha + \cot\beta}$ by forming $\dfrac{\cos(\alpha + \beta)}{\sin(\alpha + \beta)}$ and using

Formulas (1) and (3).

18. Derive the formula in Exercise 17 by forming $\dfrac{1}{\tan(\alpha + \beta)}$ and using Formula (5).

19. Explain how the graph of the sine can be obtained from the cosine curve by using the reduction formula $\cos\left(\dfrac{\pi}{2} + \theta\right) = -\sin\theta$.

20. Explain how the graph of the sine can be obtained from the cosine curve by using the reduction formula $\cos\left(\dfrac{\pi}{2} - \theta\right) = \sin\theta$.

*21. Let $f(x) = \sin x$. Prove:

$$\dfrac{f(x + h) - f(x)}{h} = \left(\dfrac{\cos h - 1}{h}\right)\sin x + \left(\dfrac{\sin h}{h}\right)\cos x$$

The formula $\sin(\alpha + \beta) = \sin\alpha\cos\beta + \cos\alpha\sin\beta$ is true for all values of α and β. If, in particular, we use any value α and let $\beta = \alpha$, then

$$\sin 2\alpha = \sin(\alpha + \alpha) = \sin\alpha\cos\alpha + \cos\alpha\sin\alpha$$

Consequently, we have the following *double-angle formula*:

(7) $$\sin 2\alpha = 2\sin\alpha\cos\alpha$$

Next use $\alpha = \beta$ in the formula for $\cos(\alpha + \beta)$:

$$\cos(\alpha + \beta) = \cos\alpha\cos\beta - \sin\alpha\sin\beta$$
$$\cos 2\alpha = \cos(\alpha + \alpha) = \cos\alpha\cos\alpha - \sin\alpha\sin\alpha$$

Thus a double-angle formula for the cosine is

(8) $$\cos 2\alpha = \cos^2\alpha - \sin^2\alpha$$

Since $\cos^2\alpha = 1 - \sin^2\alpha$, Formula (8) may be written as

(9) $$\cos 2\alpha = 1 - 2\sin^2\alpha$$

Similarly, using $\sin^2\alpha = 1 - \cos^2\alpha$, we have

(10) $$\cos 2\alpha = 2\cos^2\alpha - 1$$

Substituting $\alpha = \beta$ in $\tan(\alpha + \beta) = \dfrac{\tan\alpha + \tan\beta}{1 - \tan\alpha\tan\beta}$ gives the following double-angle formula:

(11) $$\tan 2\alpha = \frac{2\tan\alpha}{1 - \tan^2\alpha}$$

Here is a summary of the double-angle formulas in terms of an angle θ.

Double-angle formulas

$$\sin 2\theta = 2\sin\theta\cos\theta$$
$$\cos 2\theta = \cos^2\theta - \sin^2\theta$$
$$\cos 2\theta = 1 - 2\sin^2\theta$$
$$\cos 2\theta = 2\cos^2\theta - 1$$
$$\tan 2\theta = \frac{2\tan\theta}{1 - \tan^2\theta}$$

Example 1 If θ is acute and $\sin \theta = \frac{3}{5}$, find $\sin 2\theta$.

Solution From the right triangle, we find $\cos \theta = \frac{4}{5}$. Then

$$\sin 2\theta = 2 \sin \theta \cos \theta$$
$$= 2(\tfrac{3}{5})(\tfrac{4}{5}) = \tfrac{24}{25}$$

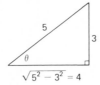

$$\sqrt{5^2 - 3^2} = 4$$

Example 2 Show that $\cos 3\theta = 4 \cos^3 \theta - 3 \cos \theta$.

Solution

$$\cos 3\theta = \cos (2\theta + \theta)$$
$$= \cos 2\theta \cos \theta - \sin 2\theta \sin \theta$$
$$= (2 \cos^2 \theta - 1) \cos \theta - (2 \sin \theta \cos \theta) \sin \theta$$
$$= 2 \cos^3 \theta - \cos \theta - 2 \cos \theta \sin^2 \theta$$
$$= 2 \cos^3 \theta - \cos \theta - 2 \cos \theta (1 - \cos^2 \theta)$$
$$= 4 \cos^3 \theta - 3 \cos \theta$$

Example 3 Let θ be an acute angle of a right triangle where $\sin \theta = \dfrac{x}{2}$. Show that $\sin 2\theta = \dfrac{x\sqrt{4 - x^2}}{2}$.

Solution

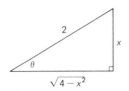

By the Pythagorean theorem, the side adjacent to θ has length $\sqrt{4 - x^2}$. Then, using (7), we have

$$\sin 2\theta = 2 \sin \theta \cos \theta$$
$$= 2\left(\frac{x}{2}\right)\left(\frac{\sqrt{4 - x^2}}{2}\right)$$
$$= \frac{x\sqrt{4 - x^2}}{2}$$

Formula (9) may be solved for $\sin^2 \alpha$ as follows:

$$\cos 2\alpha = 1 - 2 \sin^2 \alpha$$

$$2 \sin^2 \alpha = 1 - \cos 2\alpha$$

$$\sin^2 \alpha = \frac{1 - \cos 2\alpha}{2}$$

Since this result holds for all values of α, we may change the form of α by writing $\alpha = \frac{\theta}{2}$ and substitute into the preceding identity to obtain this *half-angle formula* for the sine.

(12) $\qquad \sin^2 \frac{\theta}{2} = \frac{1 - \cos \theta}{2} \qquad$ or $\qquad \sin \frac{\theta}{2} = \pm\sqrt{\frac{1 - \cos \theta}{2}}$

When using this formula for $\sin \frac{\theta}{2}$, the appropriate sign will depend on the terminal side of $\frac{\theta}{2}$.

Example 4   Evaluate $\sin 15°$ by using a half-angle formula.

Solution   Note that $15° = \frac{30°}{2}$ is in quadrant I. Therefore we use the plus sign in the half-angle formula for $\sin \frac{\theta}{2}$:

$$\sin 15° = \sqrt{\frac{1 - \cos 30°}{2}} = \sqrt{\frac{1 - \frac{\sqrt{3}}{2}}{2}} = \frac{\sqrt{2 - \sqrt{3}}}{2}$$

Formula (10) may be used to produce this half-angle formula for the cosine.

(13) $\qquad \cos^2 \frac{\theta}{2} = \frac{1 + \cos \theta}{2} \qquad$ or $\qquad \cos \frac{\theta}{2} = \pm\sqrt{\frac{1 + \cos \theta}{2}}$

A half-angle formula for the tangent is obtained by dividing (12) by (13).

$$\tan^2 \frac{\theta}{2} = \frac{\sin^2 \frac{\theta}{2}}{\cos^2 \frac{\theta}{2}} = \frac{\frac{1 - \cos \theta}{2}}{\frac{1 + \cos \theta}{2}} = \frac{1 - \cos \theta}{1 + \cos \theta}$$

(14) $\qquad \tan^2 \frac{\theta}{2} = \frac{1 - \cos \theta}{1 + \cos \theta} \qquad$ or $\qquad \tan \frac{\theta}{2} = \pm\sqrt{\frac{1 - \cos \theta}{1 + \cos \theta}}$

To use (13) and (14) for a specific value $\frac{\theta}{2}$, we use the plus or minus sign depending on the location of the terminal side of $\frac{\theta}{2}$.

Another form for $\tan \frac{\theta}{2}$ may be found in Exercise 25.

Here is a summary of the half-angle formulas in terms of an angle θ.

Half-angle formulas

$$\sin \frac{\theta}{2} = \pm\sqrt{\frac{1 - \cos \theta}{2}} \qquad\qquad \sin^2 \frac{\theta}{2} = \frac{1 - \cos \theta}{2}$$

$$\cos \frac{\theta}{2} = \pm\sqrt{\frac{1 + \cos \theta}{2}} \qquad\qquad \cos^2 \frac{\theta}{2} = \frac{1 + \cos \theta}{2}$$

$$\tan \frac{\theta}{2} = \pm\sqrt{\frac{1 - \cos \theta}{1 + \cos \theta}} \qquad\qquad \tan^2 \frac{\theta}{2} = \frac{1 - \cos \theta}{1 + \cos \theta}$$

CAUTION! LEARN TO AVOID MISTAKES LIKE THESE:

WRONG	RIGHT
$\sin 4x = 4 \sin x$	$\sin 4x = \sin 2(2x) = 2 \sin 2x \cos 2x$
$\cos^4 x = \dfrac{1 + \cos^2 2x}{2}$	$\cos^4 x = \left(\dfrac{1 + \cos 2x}{2}\right)^2$

EXERCISES 9.4

Use half-angle formulas to show each of the following.

1. $\cos 75° = \frac{1}{2}\sqrt{2 - \sqrt{3}}$ **2.** $\sin\left(-\frac{\pi}{8}\right) = -\frac{1}{2}\sqrt{2 - \sqrt{2}}$

3. (a) Show that $\tan^2 (22°30') = \dfrac{(2 - \sqrt{2})^2}{2}$.

(b) Use the result in part (a) to show that $\tan (22°30') = \sqrt{2} - 1$.

(c) Use Appendix Table I to approximate $\sqrt{2} - 1$ and compare to $\tan (22°30')$ from Table V.

4. If $\cos \theta = \frac{12}{13}$ and θ is in the first quadrant, use double-angle formulas to find:

(a) $\sin 2\theta$ (b) $\cos 2\theta$ (c) $\tan 2\theta$

5. If $\frac{\pi}{2} < \theta < \pi$ and $\cos \theta = -.36$, find:

(a) $\sin \theta$ (b) $\tan \theta$ (c) $\sin 2\theta$ (d) $\cos 2\theta$

6. Use the information from the triangle to derive $\cos 2\theta = 1 - \frac{2}{9}x^2$.

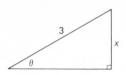

7. Use the information in Exercise 6 to derive $9 \sin 2\theta = x\sqrt{9 - x^2}$.

Verify the indicated identities in Exercises 8 through 17.

8. $\sin 3x = 3 \sin x - 4 \sin^3 x$

9. $\tan 3x = \dfrac{3 \tan x - \tan^3 x}{1 - 3 \tan^2 x}$

10. $\sin 2x = \dfrac{2 \tan x}{\sec^2 x}$

11. $\cot 2x = \dfrac{\cot^2 x - 1}{2 \cot x}$

12. $\csc 2x = \frac{1}{2} \csc x \sec x$

13. $\cos 2x = \cos^4 x - \sin^4 x$

14. $\sin 4x = 8 \sin x \cos^3 x - 4 \sin x \cos x$

15. $\cos 4x = 8 \cos^4 x - 8 \cos^2 x + 1$

16. $\sin^4 x = \frac{1}{8} \cos 4x - \frac{1}{2} \cos 2x + \frac{3}{8}$ $\left(Hint: \sin^4 x = (\sin^2 x)^2 = \left(\dfrac{1 - \cos 2x}{2} \right)^2. \right)$

17. $\cos^4 x = \frac{1}{8} \cos 4x + \frac{1}{2} \cos 2x + \frac{3}{8}$

18. Derive a double-angle formula for the secant.

*19. **(a)** Add the formulas for $\cos(\alpha + \beta)$ and $\cos(\alpha - \beta)$ to derive this *product formula*:

$$\cos \alpha \cos \beta = \tfrac{1}{2}[\cos(\alpha + \beta) + \cos(\alpha - \beta)]$$

(b) Use a similar analysis to prove these product formulas:

$$\sin \alpha \sin \beta = \tfrac{1}{2}[\cos(\alpha - \beta) - \cos(\alpha + \beta)]$$
$$\sin \alpha \cos \beta = \tfrac{1}{2}[\sin(\alpha + \beta) + \sin(\alpha - \beta)]$$
$$\cos \alpha \sin \beta = \tfrac{1}{2}[\sin(\alpha + \beta) - \sin(\alpha - \beta)]$$

*20. Use the results of Exercise 19 to express each of the following as a sum or difference:

(a) $\sin 6x \sin 2x$

(b) $2 \cos x \cos 4x$

(c) $3 \cos 5x \sin(-2x)$

(d) $4 \sin x \cos \dfrac{x}{2}$

*21. Substitute $\alpha = \dfrac{u + v}{2}$ and $\beta = \dfrac{u - v}{2}$ into the formulas in Exercise 19 and derive these *sum formulas*:

$$\cos u + \cos v = 2 \cos \dfrac{u + v}{2} \cos \dfrac{u - v}{2}$$

$$\cos v - \cos u = 2 \sin \dfrac{u + v}{2} \sin \dfrac{u - v}{2}$$

$$\sin u + \sin v = 2 \sin \dfrac{u + v}{2} \cos \dfrac{u - v}{2}$$

$$\sin u - \sin v = 2 \cos \dfrac{u + v}{2} \sin \dfrac{u - v}{2}$$

*22. Use the results of Exercise 21 to express each of the following as a product:

(a) $\sin 4x + \sin 2x$ (b) $\cos 6x - \cos 3x$

(c) $2 \sin 5x - 2 \sin x$ (d) $\dfrac{1}{2} \cos \dfrac{x}{2} + \dfrac{1}{2} \cos \dfrac{5x}{2}$

*23. Prove: $\dfrac{\cos 2x + \cos 8x}{\sin 2x + \sin 8x} = \cot 5x$. (*Hint:* Use Exercise 21.)

*24. Prove: $\sin x + \frac{1}{2} \sin 3x - \frac{1}{2} \sin x = 2 \sin x \cos^2 x$. (*Hint:* Use Exercise 8.)

*25. (a) The sign of $\tan \dfrac{\theta}{2}$ and $\sin \theta$ is the same. Verify this for these cases:

(i) $\dfrac{\pi}{2} < \theta < \pi$; (ii) $\pi < \theta < \dfrac{3\pi}{2}$; (iii) $\dfrac{3\pi}{2} < \theta < 2\pi$.

(b) Show that $\tan^2 \dfrac{\theta}{2} = \dfrac{(1 - \cos \theta)^2}{\sin^2 \theta}$ and deduce that $\tan \dfrac{\theta}{2} = \dfrac{1 - \cos \theta}{\sin \theta}$. (*Hint:* Note that $|\cos \theta| < 1$ and use part (a).)

(c) Show that $\tan \dfrac{\theta}{2} = \dfrac{\sin \theta}{1 + \cos \theta}$ and explain how this result is demonstrated in the figure.

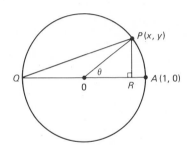

*26. (a) Let θ be an acute angle of a triangle where $\tan \theta = \dfrac{x}{3}$. Show that

$$\ln (\sec \theta + \tan \theta) = \ln \left(\dfrac{\sqrt{x^2 + 9} + x}{3} \right).$$

(b) Let $F(x) = \ln \left(\dfrac{\sqrt{x^2 + 9} + x}{3} \right)$ and evaluate $F(4) - F(0)$.

9.5

Trigonometric Equations

Can you find the points of intersection of the sine and cosine curves? This amounts to finding all values x for which $\sin x = \cos x$. To solve this **trigonometric equation** for x, first divide by $\cos x$.

$$\sin x = \cos x$$

$$\dfrac{\sin x}{\cos x} = 1$$

Note that when $\cos x = 0$, $\sin x \neq 0$ and therefore no roots are lost when we divide by $\cos x$. Using $\dfrac{\sin x}{\cos x} = \tan x$, we now need to solve $\tan x = 1$. For x in the interval $-\dfrac{\pi}{2} < x < \dfrac{\pi}{2}$, $\tan x = 1$ when $x = \dfrac{\pi}{4}$. Then, since the period of the tangent function is π, the answers are $x = \dfrac{\pi}{4} + k\pi$, where k is any integer. In case we confine the answers to the interval $-\dfrac{\pi}{2} < x < \dfrac{\pi}{2}$, there is only the one answer $\dfrac{\pi}{4}$.

Example 1 Solve: $\cos^2 x = \cos x$.

Solution We first solve for x in the interval $0 \le x < 2\pi$. Subtract $\cos x$ from both sides of $\cos^2 x = \cos x$ to obtain

$$\cos^2 x - \cos x = 0$$

Factor out $\cos x$.

$$\cos x(\cos x - 1) = 0$$

Then

$$\cos x = 0 \quad \text{or} \quad \cos x - 1 = 0$$

Hence $\cos x = 0$ for $x = \dfrac{\pi}{2}$ or $\dfrac{3\pi}{2}$ and $\cos x = 1$ for $x = 0$. Since the period of the cosine is 2π, we obtain all the solutions by taking each solution in $[0, 2\pi]$ and adding all multiples of 2π. The final answer may be presented in this form, where k is any integer.

$$x = \begin{cases} \dfrac{\pi}{2} + 2k\pi \\[2mm] \dfrac{3\pi}{2} + 2k\pi \\[2mm] 2k\pi \end{cases} \quad \text{or} \quad x = \begin{cases} 90° + k(360°) \\ 270° + k(360°)° \\ k(360°) \end{cases}$$

Caution: Do *not* begin by dividing $\cos^2 x = \cos x$ by $\cos x$. This step would produce $\cos x = 1$ and we would have lost all the roots of $\cos x = 0$. (This error is comparable to solving the equation $x^2 = x$ by first dividing each side by x.)

Example 2 Solve: $\sin 2x = \sin x$ for $0 \le x < 2\pi$.

Solution Using the double-angle formula for $\sin 2x$ we may write:

$$2 \sin x \cos x = \sin x$$

or

$$2 \sin x \cos x - \sin x = 0$$

Factor the left side.

$$\sin x(2 \cos x - 1) = 0$$

Then

$$\sin x = 0 \quad \text{or} \quad 2 \cos x = 1$$
$$\sin x = 0 \quad \text{or} \quad \cos x = \tfrac{1}{2}$$

Now, $\sin x = 0$ for $x = 0$ and π; $\cos x = \tfrac{1}{2}$ for $x = \dfrac{\pi}{3}$ and $\dfrac{5\pi}{3}$. The solutions

are 0, $\dfrac{\pi}{3}$, π, and $\dfrac{5\pi}{3}$.

Example 3   Solve: $\dfrac{\sec x}{\cos x} - \dfrac{1}{2} \sec x = 0.$

Solution

$$\frac{\sec x}{\cos x} - \frac{1}{2} \sec x = 0$$

$$\sec x\left(\frac{1}{\cos x} - \frac{1}{2}\right) = 0$$

$$\sec x = 0 \quad \text{or} \quad \frac{1}{\cos x} - \frac{1}{2} = 0$$

$$\sec x = 0 \quad \text{or} \quad \frac{1}{\cos x} = \frac{1}{2}$$

$$\sec x = 0 \quad \text{or} \quad \cos x = 2$$

Since $\sec x \geq 1$ or $\sec x \leq -1$ for all x, $\sec x = 0$ has no roots. Also, since $|\cos x| \leq 1$, $\cos x = 2$ has no roots. Thus the given equation has no solutions.

Example 4   Solve: $\cos^2 x + \tfrac{1}{2} \sin x - \tfrac{1}{2} = 0$ for $0 \leq x < 2\pi$.

Solution   Multiply by 2 to clear fractions.

$$2(\cos^2 x + \tfrac{1}{2} \sin x - \tfrac{1}{2}) = 2 \cdot 0$$
$$2 \cos^2 x + \sin x - 1 = 0$$

Convert to an equivalent form involving only $\sin x$ by using $\cos^2 x = 1 - \sin^2 x$.

$$2(1 - \sin^2 x) + \sin x - 1 = 0$$
$$2 \sin^2 x - \sin x - 1 = 0$$

The left side is quadratic in sin x, which is factorable. (Let $u = \sin x$ and factor $2u^2 - u - 1$.)

$$(2 \sin x + 1)(\sin x - 1) = 0$$

$$2 \sin x + 1 = 0 \quad \text{or} \quad \sin x - 1 = 0$$

$$\sin x = -\tfrac{1}{2} \quad \text{or} \quad \sin x = 1$$

Now, $\sin x = -\frac{1}{2}$ for $x = \dfrac{7\pi}{6}$ and $\dfrac{11\pi}{6}$; $\sin x = 1$ for $x = \dfrac{\pi}{2}$. The solutions are $\dfrac{\pi}{2}, \dfrac{7\pi}{6},$ and $\dfrac{11\pi}{6}$.

Example 5   Use Table V to approximate the solutions of $3 \tan x - 7 = 0$ in degree measure, where $0° \le x < 360°$.

Solution   Since $3 \tan x - 7 = 0$, we get $\tan x = \frac{7}{3} = 2.3333$ (four decimal places). We use the closest entry in Table V to obtain $x = 66°50'$. Adding $180°$ produces the second answer, $246°50'$, which is in the third quadrant.

EXERCISES 9.5

Solve each equation in terms of radians.

1. $\cos x = 1$ 2. $\sin x = 1$ 3. $\sin x = \frac{1}{2}$

4. $\cos^2 x = \frac{1}{2}$ 5. $\sec x = -1$ 6. $\csc x = 2$

7. $\sin 2x = 1$ 8. $2 \cos \dfrac{x}{3} = 1$ 9. $\frac{1}{2} \sin^2 x = 1$

10. $\tan x = \sqrt{3}$ 11. $\tan 2x = \sqrt{3}$ 12. $2 \sin (x - 1) = \sqrt{2}$

13. $2 \cos (x + 1) = -2$ 14. $\dfrac{1}{\sec x} = 2$ 15. $\dfrac{1}{\cos x} = 0$

Solve each equation for x in the interval $0 \le x < 2\pi$.

16. $\sin x(\cos x + 1) = 0$ 17. $(\cos x - 1)(2 \sin x + 1) = 0$

18. $\sin x + \cos x = 0$ 19. $\sin x - \sqrt{3} \cos x = 0$

20. $\sec x + \tan x = 0$

21. $\sin x + \cos x = 1$ (*Hint:* Square both sides.)

22. $2 \tan x = \sin x$ 23. $\sin^2 x + \sin x - 2 = 0$

24. $2(\cos^2 x - \sin^2 x) = \sqrt{2}$ 25. $\sin 2x = \cos x$

26. $\sin^2 x + 2 \cos^2 x = 2$ 27. $\sin^2 x + \cos^2 x = 1.5$

28. $\sec^2 x = 2 \tan x$ 29. $2 \cos^2 x + 9 \cos x - 5 = 0$

30. $\cos 2x - \cos x = 0$ 31. $\cos^2 2x = \cos 2x$

32. $2 \cos^2 x - \sqrt{3} \cos x = 0$ **33.** $2 \sin^4 x - 3 \sin^2 x + 1 = 0$
*34. $\sec x + \tan x = 1$ (*Hint:* Square both sides.)
*35. $2 \tan x = 1 - \tan^2 x$ (*Hint:* Convert to double-angle form.)

Use Table V to approximate the solutions of each equation to the nearest multiple of 10′ for x in the interval [0°, 360°).

36. $3 \sin x = 2$ **37.** $7 \sec x - 15 = 0$
38. $12 \cos^2 x + 5 \cos x - 3 = 0$ **39.** $4 \cot^2 x - 12 \cot x + 9 = 0$
40. $\tan^2 x + \tan x - 1 = 0$ (*Hint:* Use the quadratic formula.)
41. $\cos^2 x - \sin^2 x = -\frac{3}{4}$ **42.** $4 \sin x - 5 \cos x = 0$

9.6
Inverse
Circular
Functions

The sine function is not one-to-one. This fact is apparent from its graph, since a horizontal line through a range value y intersects the curve more than once; more than one x corresponds to y.

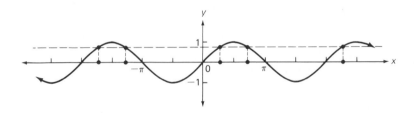

If we restrict the domain to $\left[-\dfrac{\pi}{2}, \dfrac{\pi}{2}\right]$ then $y = \sin x$ is one-to-one because for each range value there corresponds exactly one domain value.

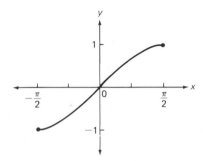

Notice that the range of this restricted function is the same as for the original function: $-1 \le y \le 1$.

Since $y = \sin x$ for $-\frac{\pi}{2} \leq x \leq \frac{\pi}{2}$ is one-to-one we know that there is an inverse function. The graph of the inverse is obtained by reflecting the graph of $y = \sin x$ through the line $y = x$, and the equation of the inverse is obtained by interchanging the variables in $y = \sin x$. We also know that the domain and range of the restricted function $y = \sin x$ become the range and domain of the inverse, respectively.

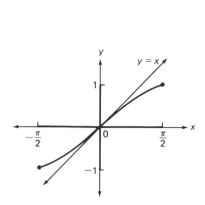

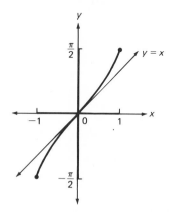

Graph of $y = \sin x$

for $-\frac{\pi}{2} \leq x \leq \frac{\pi}{2}$

and $-1 \leq y \leq 1$

Graph of inverse $x = \sin y$

for $-1 \leq x \leq 1$

and $-\frac{\pi}{2} \leq y \leq \frac{\pi}{2}$

The equation of the inverse, $x = \sin y$, does not express y explicitly as a function of x. To do this we create some new terminology. We say that $x = \sin y$ means

y is the angle whose sine is x.

To shorten this definition we replace "angle whose sine is x" by "arc sin of x," and this phrase is further abbreviated to "arcsin x." Thus $y = \arcsin x$ is, by definition, equivalent to $x = \sin y$. To sum up, here is the basic information about this new function.

The Inverse Sine Function (The arcsin function)

Defining equation: $y = \arcsin x$ (means $x = \sin y$)

Domain: $-1 \leq x \leq 1$

Range: $-\frac{\pi}{2} \leq y \leq \frac{\pi}{2}$ $(-90° \leq y \leq 90°)$

Note: In many books the inverse sine function is written in the form $y = \sin^{-1} x$. The -1 here is not an exponent; this form does *not* represent $\frac{1}{\sin x}$.

Example 1   Evaluate arcsin $\frac{1}{2}$.

Solution   Let $y = \arcsin \frac{1}{2}$. Then y is the angle whose sine is equal to $\frac{1}{2}$. Thus $y = \frac{\pi}{6}$(or 30°). To check this we use the fact that $y = \arcsin x$ is equivalent to $x = \sin y$. Hence

$$\sin y = \sin \frac{\pi}{6} = \frac{1}{2} = x$$

Caution: Even though $\sin \frac{5\pi}{6} = \frac{1}{2}$, we do not have $\arcsin \frac{1}{2} = \frac{5\pi}{6}$ because the range of the arcsin function consists of the numbers in the interval $\left[-\frac{\pi}{2}, \frac{\pi}{2} \right]$.

You will find it helpful in finding values like arcsin $\frac{1}{2}$ to remember that $y = \arcsin x$ is negative for $-1 \leq x < 0$ and positive for $0 < x \leq 1$. You can see this from its graph.

Example 2   Use Table V to approximate arcsin $(-.3960)$.

Solution   The closest entry in the sine column is .4072 radians or 23°20′. Then, since $x = -.3960 < 0$, we find arcsin $(-.3960) = -.4072$ or $-23°20′$.

We will now follow a similar development to produce the inverse cosine function known as *arccos*. First restrict $y = \cos x$ to $0 \leq x \leq \pi$. We can now find the inverse of this *restricted* function.

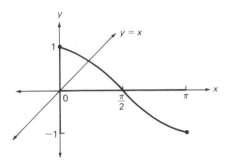

Graph of $y = \cos x$
for $0 \leq x \leq \pi$
and $-1 \leq y \leq 1$

Reflect the curve for $0 \le x \le \pi$ through the line $y = x$ to obtain the graph of the inverse whose equation is $x = \cos y$. We then define $y = \arccos x$ to mean $x = \cos y$ and call it the *inverse cosine function*. (This function is often written in the form $y = \cos^{-1} x$.)

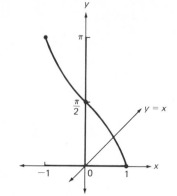

The Inverse Cosine Function

$y = \arccos x$ (equivalent to $x = \cos y$)

Domain: $-1 \le x \le 1$

Range: $0 \le y \le \pi$

Example 3   Find $\arccos\left(-\tfrac{1}{2}\right)$.

Solution   Since $\cos 120° = -\tfrac{1}{2}$ we have $\arccos\left(-\tfrac{1}{2}\right) = 120°$.

Example 4   Find $\sin\left(\arccos \tfrac{1}{2}\right)$.

Solution   Since $\arccos \dfrac{1}{2} = \dfrac{\pi}{2}$ we have

$$\sin\left(\arccos \frac{1}{2}\right) = \sin \frac{\pi}{3} = \frac{\sqrt{3}}{2}$$

Caution: Do not confuse $\sin x(\arccos x)$ with $\sin(\arccos x)$. The first represents the product of the two numbers $\sin x$ and $\arccos x$. The second represents the composite of the two functions arccos (the inner function) and sin (the outer function).

Example 5   Evaluate $\cos\left(\arcsin \tfrac{2}{3}\right)$ without using Table V.

Solution   We know that $y = \arcsin \tfrac{2}{3}$ is an acute angle. (Why?) Now construct a right triangle with y an acute angle and $\sin y = \tfrac{2}{3}$.

From the figure we see that $\cos\left(\arcsin\dfrac{2}{3}\right) = \cos y = \dfrac{\sqrt{5}}{3}.$

The sine and cosine functions are "connected" through a variety of identities. The most fundamental of these identities is $\sin^2\theta + \cos^2\theta = 1$; another is $\cos\left(\dfrac{\pi}{2} - x\right) = \sin x$. We almost expect the inverse sine and cosine functions to be connected by some identity. Such an identity does exist, as is illustrated in the next example.

Example 6   Show that $\arcsin x + \arccos x = \dfrac{\pi}{2}$ for all x in the common domain $-1 \le x \le 1$.

Solution   Let $\arcsin x = y$. Then $x = \sin y$. Now for any y value we have $\cos\left(\dfrac{\pi}{2} - y\right) = \sin y$. Therefore

$$x = \sin y = \cos\left(\frac{\pi}{2} - y\right)$$

But $x = \cos\left(\dfrac{\pi}{2} - y\right)$ is equivalent to $\dfrac{\pi}{2} - y = \arccos x$. Then

$$\frac{\pi}{2} = y + \arccos x$$

and substituting for y gives

$$\frac{\pi}{2} = \arcsin x + \arccos x$$

Note that the identity $\arcsin x + \arccos x = \dfrac{\pi}{2}$ says that the angle whose sine is x and the angle whose cosine is x must be complementary.

TEST YOUR UNDERSTANDING

Evaluate the following when possible.

1. arcsin (-1)

2. arccos $\left(\dfrac{\sqrt{3}}{2}\right)$

3. arcsin 2

4. arccos $(-\tfrac{1}{2})$

5. cos (arcsin 0)

6. cos $\left[\arcsin\left(-\dfrac{\sqrt{3}}{2}\right)\right]$

7. sin (arcsin 1)

8. cos (arccos x)

9. sin (arccos $\tfrac{3}{5}$)

Use Table V to approximate the following.

10. arccos $(-.4436)$

11. arcsin $(.7312)$

12. cos (arcsin .0872)

The tangent function has period π and completes a full cycle on the interval $\left(-\dfrac{\pi}{2}, \dfrac{\pi}{2}\right)$. Thus when we restrict $y = \tan x$ to $-\dfrac{\pi}{2} < x < \dfrac{\pi}{2}$ we have a one-to-one function whose range consists of all real numbers. We will now find the inverse of this restricted function.

The next figure shows the graph of $y = \tan x$ for $-\dfrac{\pi}{2} < x < \dfrac{\pi}{2}$ and with range of all real numbers. Asymptotes: $x = \pm\dfrac{\pi}{2}$.

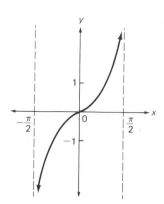

We define the inverse by $y = \arctan x$, which means $x = \tan y$. Its graph is obtained by reflecting the curve in the figure through the line $y = x$. (The inverse tangent function is often written in the form $y = \tan^{-1} x$.)

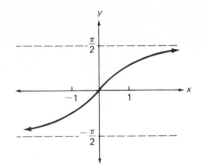

The Inverse Tangent Function

$y = \arctan x$ (equivalent to $x = \tan y$)

Domain: all real numbers

Range: $-\dfrac{\pi}{2} < y < \dfrac{\pi}{2}$

Asymptotes: $y = \dfrac{\pi}{2}, \ y = -\dfrac{\pi}{2}$

Example 7   Evaluate $\arctan \sqrt{3}$.

Solution   Let $y = \arctan \sqrt{3}$. Then y is the angle whose tangent is $\sqrt{3}$. Therefore, since $-\dfrac{\pi}{2} < y < \dfrac{\pi}{2}$, we have $y = \dfrac{\pi}{3}(60°)$.

EXERCISES 9.6

Evaluate. Use Table V only when necessary.

1. $\arcsin 0$ **2.** $\arccos 0$ **3.** $\arcsin(-1)$

4. $\arccos(-1)$ **5.** $\arctan(-1)$ **6.** $\arcsin \dfrac{\sqrt{2}}{2}$

7. $\arccos\left(-\dfrac{\sqrt{2}}{2}\right)$ **8.** $\arctan \dfrac{\sqrt{3}}{3}$ **9.** $\arctan 115$

10. $\arcsin \frac{3}{2}$ **11.** $\arcsin(.5592)$ **12.** $\arccos(-2)$

13. $\arccos(.1475)$ **14.** $\arctan(.6128)$ **15.** $\sin(\arctan \frac{1}{12})$

16. $\tan(\arcsin \frac{1}{2})$ **17.** $\cos[\arcsin(-1)]$ **18.** $\tan(\arccos \frac{4}{5})$

19. $\sin\left(\arcsin \dfrac{\sqrt{3}}{2}\right)$ **20.** $\cos(\arctan 2.35)$ **21.** $\sin(\arccos \frac{12}{13})$

22. $\arcsin\left(\cos \dfrac{\pi}{3}\right)$ **23.** $\arccos(\tan 45°)$ **24.** $\arctan(\tan 30°)$

25. Explain why $\sin(\arcsin x) = x$ for $-1 \le x \le 1$.

26. Explain why $\tan(\arctan) = x$ for all x.

Graph the curves in Exercises 27 through 30.

27. $y = 2 \arcsin x$ **28.** $y = \arcsin(x - 2)$

29. $y = 2 + \arctan x$ **30.** $y = -\arcsin x$

31. Let $f(x) = \arcsin x$ and $g(x) = 3x + 2$ form the composites $f \circ g$ and $g \circ f$.

32. Let $f(x) = \arctan x$ and $g(x) = x^2$ form the composites $f \circ g$ and $g \circ f$.

33. Let $h(x) = \ln (\arccos x)$. Find f and g so that $h = f \circ g$.

34. Let $k(x) = \arctan \sqrt{x^2 - 1}$. Find f, g, and h so that $k = f \circ g \circ h$.

35. If $y = \text{arccot } x$ is the inverse of the cotangent function restricted to $0 < x < \pi$, what are the domain and range of the inverse cotangent? Graph.

36. Prove that $\arctan x + \text{arccot } x = \dfrac{\pi}{2}$. $\left(Hint: \text{Use } \tan \left(\dfrac{\pi}{2} - y\right) = \cot y.\right)$

37. Explain how the graph of $y = \text{arccot } x$ can be obtained from $y = \arctan x$ by using the identity in Exercise 36.

38. If $y = \text{arcsec } x$ is the inverse of $y = \sec x$ restricted to $0 \le x \le \pi, x \ne \dfrac{\pi}{2}$, what are the domain and range of the inverse secant? Graph.

39. If $y = \text{arccsc } x$ is the inverse of $y = \csc x$ restricted to $-\dfrac{\pi}{2} \le x \le \dfrac{\pi}{2}, x \ne 0$, what are the domain and range of the inverse cosecant? Graph.

*40. Find $\sin (\arcsin \tfrac{2}{3} + \arctan \tfrac{4}{3})$. (*Hint:* Use an addition formula.)

*41. Find $\cos (2 \arcsin \tfrac{1}{3})$. (*Hint:* Use a double-angle formula.)

*42. Find $\sin (\tfrac{1}{2} \arccos \tfrac{2}{3})$. (*Hint:* Use a half-angle formula.)

*43. Prove that $\arcsin (-x) = -\arcsin x$.

*44. Prove that $\cos (\arcsin x) = \sqrt{1 - x^2}$.

9.7

Trigonometric Form of Complex Numbers (optional)

Complex numbers were introduced in Section 2.9. We learned that complex numbers may be written in the form $a + bi$, where a and b are real numbers and i is the imaginary unit: $i = \sqrt{-1}$. Also, since $i^2 = -1$, $i^3 = -i$, and $i^4 = 1$, any positive integral power of i can be reduced to one of $\pm 1, \pm i$; for example, $i^{18} = (i^4)^4 i^2 = 1^4 i^2 = i^2 = -1$.

Addition and multiplication of complex numbers are defined by

$$(a + bi) + (c + di) = (a + c) + (b + d)i$$
$$(a + bi)(c + di) = (ac - bd) + (ad + bc)i$$

Complex numbers also correspond to points in a plane in which the horizontal axis is referred to as the real axis and the vertical axis is called the imaginary axis. Thus a complex number $x + iy$ corresponds to the point (x, y). Here are two typical cases; note that z is used to name the complex number $x + iy$.

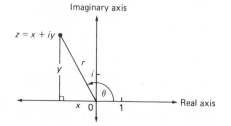
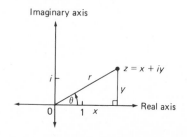

   The point $z = x + iy$ determines a right triangle as shown. This is similar
to the reference triangle used in conjunction with the unit circle. The only difference here is that the hypotenuse does not have to be 1.

   Using r as the distance (positive) from the origin, we have $r^2 = x^2 + y^2$, or
$r = \sqrt{x^2 + y^2}$, r is the **modulus** of $z = x + iy$. We sometimes refer to $x + iy$ as
the **rectangular form** of a complex number. For θ as indicated, we have

$$x = r \cos \theta \quad \text{and} \quad y = r \sin \theta$$

Consequently,

$$z = x + iy = r \cos \theta + i(r \sin \theta)$$

or

$$z = r(\cos \theta + i \sin \theta)$$

which is known as the **trigonometric form** of z. The modulus $r \geq 0$ is a unique
real number. There are many angles θ that can be used, as long as the terminal
sides coincide. Angle θ is also called the **argument** of z, and we will usually select
θ to satisfy $0 \leq \theta < 2\pi$.

Example 1   Convert $\sqrt{3} + i$ and $-1 - i$ into trigonometric form.

Solution   Graph the points and observe that $\sqrt{3} + i$ determines a 30°–60°–90°
triangle and $-1 - i$ determines a 45°–45°–90° triangle.

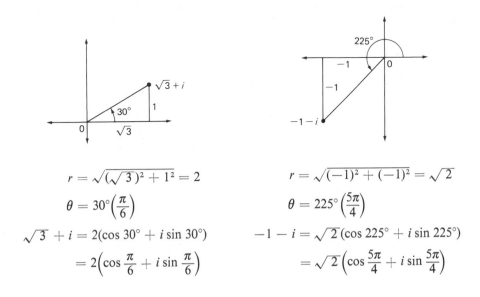

$$r = \sqrt{(\sqrt{3})^2 + 1^2} = 2$$

$$\theta = 30° \left(\frac{\pi}{6}\right)$$

$$\sqrt{3} + i = 2(\cos 30° + i \sin 30°)$$

$$= 2\left(\cos \frac{\pi}{6} + i \sin \frac{\pi}{6}\right)$$

$$r = \sqrt{(-1)^2 + (-1)^2} = \sqrt{2}$$

$$\theta = 225° \left(\frac{5\pi}{4}\right)$$

$$-1 - i = \sqrt{2}(\cos 225° + i \sin 225°)$$

$$= \sqrt{2}\left(\cos \frac{5\pi}{4} + i \sin \frac{5\pi}{4}\right)$$

Example 2 Use Table V to convert $3(\cos 50° + i \sin 50°)$ into the form $x + iy$.

Solution

$$3(\cos 50° + i \sin 50°) = 3(.6428 + .7660i)$$
$$= 1.9284 + 2.2980i$$

Using the trigonometric form, we can derive an interesting result for multiplication of complex numbers. Let

$$z_1 = r_1(\cos \theta_1 + i \sin \theta_1) \quad \text{and} \quad z_2 = r_2(\cos \theta_2 + i \sin \theta_2)$$

and multiply.

$$z_1 z_2 = [r_1(\cos \theta_1 + i \sin \theta_2)][r_2(\cos \theta_2 + i \sin \theta_2)]$$
$$= r_1 r_2[(\cos \theta_1 \cos \theta_2 - \sin \theta_1 \sin \theta_2) + i(\sin \theta_1 \cos \theta_2 + \cos \theta_1 \sin \theta_2)]$$

Now use addition formulas (3) and (1), page 388, to obtain

$$z_1 z_2 = r_1 r_2[\cos (\theta_1 + \theta_2) + i \sin (\theta_1 + \theta_2)]$$

Therefore, *to multiply two complex numbers we multiply their moduli and add their arguments.*

Example 3 Let $z_1 = 2 + 2\sqrt{3}\,i$ and $z_2 = -1 - \sqrt{3}\,i$.
(a) Evaluate $z_1 z_2$ by using the definition of multiplication on page 410.
(b) Evaluate $z_1 z_2$ by using trigonometric forms and verify that the result is the same as in part (a).

Solution
(a) $z_1 z_2 = (2 + 2\sqrt{3}\,i)(-1 - \sqrt{3}\,i)$
$$= (-2 + 6) + (-2\sqrt{3} - 2\sqrt{3})i = 4 - 4\sqrt{3}\,i$$

(b) Converting $z_1 z_2$ to trigonometric form we get $z_1 = 4\left(\cos \dfrac{\pi}{3} + i \sin \dfrac{\pi}{3}\right)$ and $z_2 = 2\left(\cos \dfrac{4\pi}{3} + i \sin \dfrac{4\pi}{3}\right)$. Then

$$z_1 z_2 = \left[4\left(\cos \frac{\pi}{3} + i \sin \frac{4\pi}{3}\right)\right]\left[2\left(\cos \frac{4\pi}{3} + i \sin \frac{4\pi}{3}\right)\right]$$

$$= 4 \cdot 2\left[\cos \left(\frac{\pi}{3} + \frac{4\pi}{3}\right) + i \sin \left(\frac{\pi}{3} + \frac{4\pi}{3}\right)\right]$$

$$= 8\left(\cos \frac{5\pi}{3} + i \sin \frac{5\pi}{3}\right)$$

$$= 8\left[\frac{1}{2} + i\left(-\frac{\sqrt{3}}{2}\right)\right]$$

$$= 4 - 4\sqrt{3}\,i$$

Let $z_1 = -2 + 2i$ and $z_2 = 3\sqrt{3} - 3i$. Convert to trigonometric form; evaluate $z_1 z_2$.

If in the trigonometric form of the product of two complex numbers we use $z_1 = z_2 = z = r(\cos\theta + i\sin\theta)$, then

$$z^2 = r^2(\cos 2\theta + i\sin 2\theta)$$

Now use this result for z^2 to find z^3 as follows:

$$z^3 = z^2 z$$
$$= [r^2(\cos 2\theta + i\sin 2\theta)][r(\cos\theta + i\sin\theta)]$$
$$= r^3(\cos 3\theta + i\sin 3\theta)$$

In this manner it can be shown that $z^n = r^n(\cos n\theta + i\sin n\theta)$ for each positive integer n. Since $z = r(\cos\theta + i\sin\theta)$, we get the following result, known as **De Moivre's theorem**:

$$[r(\cos\theta + i\sin\theta)]^n = r^n(\cos n\theta + i\sin n\theta)$$

Example 4   Use De Moivre's theorem to evaluate $(1 - i)^8$.

Solution   Converting $z = 1 - i$ to trigonometric form, we obtain

$$z = 2^{1/2}\left(\cos\frac{7\pi}{4} + i\sin\frac{7\pi}{4}\right)$$

Now use De Moivre's theorem.

$$z^8 = (2^{1/2})^8\left[\cos\left(8\cdot\frac{7\pi}{4}\right) + i\sin\left(8\cdot\frac{7\pi}{4}\right)\right]$$
$$= 2^4(\cos 14\pi + i\sin 14\pi)$$
$$= 16(\cos 0 + i\sin 0)$$
$$= 16(1 + 0)$$
$$= 16$$

The preceding example shows that $(1 - i)^8 = 16$. This says that $1 - i$ is an 8th root of 16. There are seven more 8th roots of 16. For the integers $n \geq 2$ every complex number has n distinct nth roots. De Moivre's theorem can be used to derive the following formula, which enables us to find such roots.

The nth-root formula

If n is a positive integer and if $z = r(\cos \theta + i \sin \theta)$ is a nonzero complex number, the nth roots of z are given by

$$r^{1/n}\left(\cos \frac{\theta + 2k\pi}{n} + i \sin \frac{\theta + 2k\pi}{n}\right) \qquad \text{for } k = 0, 1, 2, \ldots, n - 1$$

The work that follows will demonstrate how to use this formula. We will not prove it.

Suppose we want to find the three cube roots of $z = 8i$. First convert to trigonometric form. Thus

$$8i = 0 + 8i = 8\left(\cos \frac{\pi}{2} + i \sin \frac{\pi}{2}\right)$$

Now use the nth-root formula with $n = 3, r = 8$, and $\theta = \dfrac{\pi}{2}$. The values $k = 0, 1, 2$ produce the cube roots as follows:

$$k = 0: \quad 8^{1/3}\left(\cos \frac{\frac{\pi}{2} + 0}{3} + i \sin \frac{\frac{\pi}{2} + 0}{3}\right) = 2\left(\cos \frac{\pi}{6} + i \sin \frac{\pi}{6}\right) = \sqrt{3} + i$$

$$k = 1: \quad 8^{1/3}\left(\cos \frac{\frac{\pi}{2} + 2\pi}{3} + i \sin \frac{\frac{\pi}{2} + 2\pi}{3}\right) = 2\left(\cos \frac{5\pi}{6} + i \sin \frac{5\pi}{6}\right) = -\sqrt{3} + i$$

$$k = 2: \quad 8^{1/3}\left(\cos \frac{\frac{\pi}{2} + 4\pi}{3} + i \sin \frac{\frac{\pi}{2} + 4\pi}{3}\right) = 2\left(\cos \frac{3\pi}{2} + i \sin \frac{3\pi}{2}\right) = -2i$$

You can check these results by showing that the cube of each of these roots equals $8i$.

Note that for larger values of k no new roots are produced. You can verify that for $k = 3$ we get $\sqrt{3} + i$; for $k = 4$ we get $-\sqrt{3} + i$; for $k = 5$ we get $-2i$; and so forth.

The nth-root formula shows that each root has modulus $r^{1/n}$ and that the arguments of a pair of successive roots differ by $\dfrac{2\pi}{n}$. Consequently the roots lie on a circle, centered at the origin, with radius $r^{1/n}$, and they are equally spaced on this circle, as can be observed for the cube roots of $8i$.

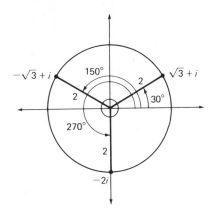

If we let $z = 1$, then the nth-root formula will give the n distinct nth roots of 1. Since $r = 1^{1/n} = 1$, these roots will lie on the unit circle (center at zero) and be equally spaced. Thus, for example, the six 6th roots of unity will determine a regular hexagon. (See Exercises 32 through 37.)

EXERCISES 9.7

Convert each of the following into trigonometric form and plot each point in a coordinate plane.

1. $1 + i$

2. $3 + 3i$

3. $-1 + \sqrt{3}\,i$

4. $-2 + 2\sqrt{3}\,i$

5. $\dfrac{\sqrt{3}}{2} - \dfrac{1}{2}i$

6. 5

7. -10

8. $-4i$

9. $-10 - 10i$

Express each of the following complex numbers in the rectangular form $x + yi$.

10. $2(\cos 30° + i \sin 30°)$

11. $3(\cos 120° + i \sin 120°)$

12. $\dfrac{1}{2}\left(\cos \dfrac{\pi}{2} + i \sin \dfrac{\pi}{2}\right)$

13. $5\left(\cos \dfrac{3\pi}{2} + i \sin \dfrac{3\pi}{2}\right)$

14. $9(\cos 0° + i \sin 0°)$

15. $\sqrt{2}(\cos 60° + i \sin 60°)$

16. $1(\cos \pi + i \sin \pi)$

17. $4(\cos 315° + i \sin 315°)$

18. $3(\cos 65° + i \sin 65°)$ (Use Table V.)

Use De Moivre's theorem to evaluate each of the following and express the answer in the rectangular form $x + yi$.

19. $(-1 - i)^4$

20. $(1 - \sqrt{3}\,i)^8$

21. $\left(\frac{1}{2} + \frac{\sqrt{3}}{2}i\right)^6$

22. $\left(\cos\frac{\pi}{5} + i\sin\frac{\pi}{5}\right)^{15}$

23. $(\cos 6° + i\sin 6°)^{10}$

24. $(-3i)^6$

25. $\left(-\frac{\sqrt{2}}{2} + \frac{\sqrt{2}}{2}i\right)^{10}$

26. $(2 - 2i)^4$

27. $(\cos 45° + i\sin 45°)^8$

For Exercises 28 through 37 write the roots in rectangular form whenever the argument is a special angle. Otherwise leave the roots in trigonometric form.

28. Find the cube roots of 27.

29. Find the 4th roots of 16.

30. Find the 4th roots of $4 - 4\sqrt{3}\,i$.

31. Find the cube roots of $-4\sqrt{2} - 4\sqrt{2}i$.

32. Let $z = 1$ in the nth-root formula and write the general form of the nth root of 1.

33. Find the square roots of 1.

34. Find the cube roots of 1.

35. Find all roots of the equation $z^4 - 1 = 0$.

36. Find all roots of the equation $z^5 - 1 = 0$. Sketch the regluar polygon determined by these roots.

37. Find the 6th roots of unity. Plot the points and sketch the hexagon.

38. Prove that $(\cos\theta + i\sin\theta)^{-1} = \cos\theta - i\sin\theta$. (*Hint:* Multiply the numerator and denominator of $\dfrac{1}{\cos\theta + i\sin\theta}$ by the conjugate of $\cos\theta + i\sin\theta$.)

*39. (a) Let n be a positive integer. Prove that

$$[r(\cos\theta + i\sin\theta)]^{-n} = r^{-n}(\cos n\theta - i\sin n\theta)$$

(*Hint:* Use Exercise 38.)

(b) Explain why $[r(\cos\theta + i\sin\theta)]^n = r^n(\cos n\theta + i\sin n\theta)$ holds for all integers n.

*40. (a) Two complex numbers $a + bi$ and $c + di$ are equal if and only if $a = c$ and $b = d$. Use this criterion of equality in conjunction with De Moivre's theorem to derive the double-angle formulas for the sine and cosine. (*Hint:* Expand the right side of $\cos 2\theta + i\sin 2\theta = (\cos\theta + i\sin\theta)^2$.)

(b) Follow the procedure that is described in part (a) to derive the formulas $\cos 3\theta = 4\cos^3\theta - 3\cos\theta$ and $\sin 3\theta = 3\sin\theta - 4\sin^3\theta$.

41. Polynomials that were not factorable by using real coefficients can now be factored by using complex numbers. Thus $x^2 + 1 = (x + i)(x - i)$. Factor these polynomials:

    (a) $x^2 + 4$ **(b)** $3x^2 + 27$ **(c)** $x^2 + 2$ **(d)** $5x^2 + 15$

42. A quadratic equation $ax^2 + bx + c = 0$ has no real roots when the discriminant $b^2 - 4ac < 0$. In such cases the *complex roots* are given by

$$x = \frac{-b \pm \sqrt{b^2 - 4ac}}{2a} = \frac{b \pm \sqrt{-(4ac - b^2)}}{2a} = \frac{-b \pm i\sqrt{4ac - b^2}}{2a}$$

where $4ac - b^2 > 0$. Solve these equations either by factoring or by using the quadratic formula.

(a) $x^2 + 1 = 0$ (b) $x^2 + 6x + 25 = 0$

(c) $2x^2 + x = -5$ (d) $6x^2 = 7x - 3$

REVIEW EXERCISES FOR CHAPTER 9

The solutions to the following exercises can be found within the text of Chapter 9. Try to answer each question without referring to the text.

Section 9.1

1. Graph the curves $y = \sin x$, $y = \frac{1}{2} \sin x$, and $y = 2 \sin x$ on the same set of axes for $-2\pi \le x \le 2\pi$.

2. Find the period of $y = \sin 2x$ and graph for $0 \le x \le 2\pi$.

3. Find the period and amplitude of $y = 2 \cos \frac{1}{2}x$ and graph for $0 \le x \le 4\pi$.

4. Show that $h(x) = \cos \dfrac{x}{x+1}$ is the composite of two functions such that the inner function is rational.

Section 9.2

5. Graph $y = \tan x$ for $\dfrac{-3\pi}{2} < x < \dfrac{3\pi}{2}$.

6. Prove that $y = \tan x$ is symmetric through the origin.

7. Graph $y = \cot x$ for $-\pi < x < \pi$.

8. Graph $y = -\tan 3x$ for $\dfrac{-\pi}{2} < x < \dfrac{\pi}{2}$.

9. Graph $y = \sec x$ for one full cycle.

10. State the domain and range of $y = \cos x$.

11. State the domain and range of $y = \tan x$.

Section 9.3

12. Use an addition formula and the fact that $\dfrac{7\pi}{12} = \dfrac{\pi}{4} + \dfrac{\pi}{3}$ to evaluate $\sin \dfrac{7\pi}{12}$.

13. Evaluate $\cos 15°$ by using an addition formula.

14. Complete these reduction formulas:

$$\sin (\theta - 2\pi) = \qquad\qquad \sin \left(\frac{\pi}{2} - \theta\right) =$$

$$\cos (\pi + \theta) = \qquad\qquad \cos \left(\frac{\pi}{2} - \theta\right) =$$

$$\tan (\pi - \theta) = \qquad\qquad \cos (\theta + 2\pi) =$$

15. Explain how the graph of $y = \cot x$ can be found by using the formula
$$\cot x = \tan\left(\frac{\pi}{2} - x\right).$$

Section 9.4

16. Complete these double-angle formulas:

$$\sin 2\theta = \qquad\qquad \cos 2\theta =$$

17. If $\sin \theta = \frac{3}{5}$, where $0 < \theta < \frac{\pi}{2}$, find $\sin 2\theta$.

18. Let $\sin \theta = \frac{x}{2}$ for $0 < \theta < \frac{\pi}{2}$. Show that $\sin 2\theta = \frac{x\sqrt{4 - x^2}}{2}$.

19. Complete these half-angle formulas:

$$\cos^2 \frac{\theta}{2} = \qquad\qquad \tan^2 \frac{\theta}{2} =$$

20. Use a half-angle formula to evaluate $\sin 15°$.

Section 9.5

21. Solve $\cos^2 x = \cos x$.
22. Solve $\sin 2x = \cos x$ for $0 \le x < 2\pi$.
23. Solve $\dfrac{\sec x}{\cos x} - \dfrac{1}{2} \sec x = 0$.
24. Solve $\cos^2 x + \frac{1}{2} \sin x - \frac{1}{2} = 0$ for $0 \le x < 2\pi$.

Section 9.6

25. State the domain and range of $y = \arcsin x$ and graph.
26. State the domain and range of $y = \arccos x$ and graph.
27. State the domain and range of $y = \arctan x$ and graph.
28. Evaluate $\arcsin \frac{1}{2}$.
29. Evaluate $\arccos \left(-\frac{1}{2}\right)$.
30. Evaluate $\arctan \sqrt{3}$.
31. Evaluate $\sin \left(\arccos \frac{1}{2}\right)$.
32. Evaluate $\cos \left(\arcsin \frac{2}{3}\right)$.
33. Prove: $\arcsin x + \arccos x = \dfrac{\pi}{2}$.

Section 9.7

34. Convert $\sqrt{3} + i$ into trigonometric form.
35. Using trigonometric forms find the product $(2 + 2\sqrt{3}\,i)(-1 - \sqrt{3}\,i)$.
36. Use De Moivre's theorem to evaluate $(1 - i)^8$.
37. Find all three of the cube roots of $8i$.

1. Find the amplitude and period of $y = 2 \sin 2x$ and graph for x in the interval $0 \leq x \leq 2\pi$.

2. State the domain and range of $y = \sec x$ and graph for x in the interval $-\dfrac{\pi}{2} < x < \dfrac{\pi}{2}$.

3. Use $165° = 135° + 30°$ and an appropriate addition formula to evaluate $\cos 165°$.

4. Use an addition formula to derive $\sin\left(\dfrac{\pi}{4} - \theta\right) = \dfrac{\sqrt{2}}{2}(\cos\theta - \sin\theta)$.

5. Use a half-angle formula to evaluate $\sin\left(-\dfrac{\pi}{12}\right)$.

6. If $\sin\theta = \frac{1}{3}$ and θ is acute, find $\sin 2\theta$.

7. Prove $\csc 4x = \frac{1}{4}(\csc x)(\sec x)(\sec 2x)$.

8. Solve $\sin^2 x - \cos^2 x + \sin x = 0$ for x in the interval $0 \leq x < 2\pi$.

9. (a) State the domain and range of $y = \arctan x$ and graph.

   (b) Evaluate $\arctan\left(-\dfrac{\sqrt{3}}{3}\right)$.

10. Evaluate each of the following:

    (a) $\arcsin\dfrac{\sqrt{2}}{2}$ (b) $\sin\left(\arccos\dfrac{3}{7}\right)$ (c) $\cos\left(2\arcsin\dfrac{1}{2}\right)$

†11. Use De Moivre's theorem to find $(-1 + \sqrt{3}\,i)^5$ and express the answer in the rectangular form $x + yi$.

Answers to the Test Your Understanding Exercises

Page 379

1.

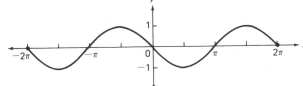

2.

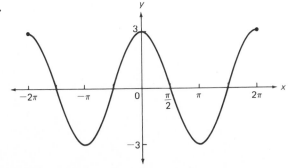

3. 10

4. $\frac{2}{3}$

5. Shift 2 units right.

6. Shift 2 units upward.

Page 390

1. $\frac{1}{4}(\sqrt{6} - \sqrt{2})$; $\frac{1}{4}(\sqrt{6} + \sqrt{2})$

2. Same as number 1.

3. $\frac{1}{4}(\sqrt{6} - \sqrt{2})$; $-\frac{1}{4}(\sqrt{6} + \sqrt{2})$

4. Same as number 2.

5. $\frac{1}{4}(\sqrt{2} - \sqrt{6})$

6. $\frac{1}{4}(\sqrt{6} - \sqrt{2})$

Page 408

1. $-\frac{\pi}{2}$

2. $\frac{\pi}{6}$

3. Undefined.

4. $\frac{2\pi}{3}$

5. 1

6. $\frac{1}{2}$

7. 1

8. x

9. $\frac{4}{5}$

10. 116°20′

11. 47°

12. .9962

Page 413

$$z_1 = 2\sqrt{2}\left(\cos\frac{3\pi}{4} + i\sin\frac{3\pi}{4}\right)$$

$$z_2 = 6\left(\cos\frac{11\pi}{6} + i\sin\frac{11\pi}{6}\right)$$

$$z_1 z_2 = 12\sqrt{2}\left(\cos\frac{31\pi}{12} + i\sin\frac{31\pi}{12}\right)$$

$$= 12\sqrt{2}\left(\cos\frac{7\pi}{12} + i\sin\frac{7\pi}{12}\right)$$

Ten

Sequences and Series

10.1

Sequences A function f, you will recall, is a correspondence between two sets of numbers, the domain and the range, such that for each domain value there corresponds exactly one range value. Ordinarily, f is given by an equation that relates the domain value x with the associated range value $y = f(x)$. For example,

$$f(x) = x^2$$

defines a function whose domain is understood to be the set of all real numbers. A new function would be obtained from the same equation if, for some reason, the domain were restricted, for example, to the set of numbers in the interval $[1, 6]$. The graph of this new function is the part of the parabola shown below.

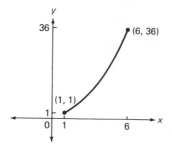

A **sequence** is a special type of function. It is a function whose domain consists of a set of integers, usually a set of consecutive integers beginning with 1. A member of the domain of a sequence is usually represented by a variable such as n, k, i, and so on. Using a letter such as s for the name of a sequence, and letting n stand for the domain values, we use the symbol s_n (read "s-sub-n") to represent the corresponding range values. These range values are often called the **terms** of the sequence, and s_n is the **general**, or **nth, term**.

The equation

$$s_n = n^2$$

for $n = 1, 2, 3, 4, 5, 6$ defines a function that is a sequence. The domain consists of the numbers $1, 2, 3, 4, 5, 6$, and the terms (range values) are

$$s_1 = 1 \qquad s_2 = 4 \qquad s_3 = 9 \qquad s_4 = 16 \qquad s_5 = 25 \qquad s_6 = 36$$

Just as we graphed the function given by $f(x) = x^2$ on the domain $[1, 6]$, we can also graph the sequence given by $s_n = n^2$ for $n = 1, 2, 3, 4, 5, 6$, as shown at the top of the next page.

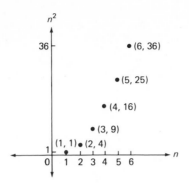

Example 1 Find the range values of the sequence given by $s_n = \dfrac{1}{n}$ for the domain $n = 1, 2, 3, 4, 5$, and graph.

Solution The range values and graph are as follows.

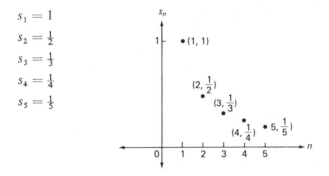

$$s_1 = 1$$
$$s_2 = \tfrac{1}{2}$$
$$s_3 = \tfrac{1}{3}$$
$$s_4 = \tfrac{1}{4}$$
$$s_5 = \tfrac{1}{5}$$

A sequence whose domain is a finite set of integers, as in Example 1, is called a **finite sequence**. A sequence whose domain consists of an infinite number of integers is said to be an **infinite sequence**. The formula $s_n = \dfrac{1}{n}$ produces an infinite sequence when the domain is the set of all positive integers.

In practice, the specific mention of the domain of a sequence is often left out. In that case, the domain is assumed to be the set of all positive integers for which the formula makes sense. That is, just as we would assume that the function

$$f(x) = \frac{1}{x - 1}$$

has a domain of all real numbers other than $x = 1$, we assume that the sequence

$$s_n = \frac{1}{n - 1}$$

has the domain of all consecutive integers beginning with $n = 2$.

Example 2   List the first seven terms of the sequence given by $s_n = \frac{(-1)^k}{k^2}$.

Solution

$$s_1 = \frac{(-1)^1}{1^2} = -1$$

$$s_2 = \frac{(-1)^2}{2^2} = \frac{1}{4}$$

$$s_3 = \frac{(-1)^3}{3^2} = -\frac{1}{9}$$

$$s_4 = \frac{(-1)^4}{4^2} = \frac{1}{16}$$

$$s_5 = \frac{(-1)^5}{5^2} = -\frac{1}{25}$$

$$s_6 = \frac{(-1)^6}{6^2} = \frac{1}{36}$$

$$s_7 = \frac{(-1)^7}{7^2} = -\frac{1}{49}$$

Example 3   Write the first four terms of the sequence $s_n = \frac{1}{(n - 1)(n - 2)}$.

Solution

$$s_3 = \frac{1}{2 \cdot 1} = \frac{1}{2}$$

$$s_4 = \frac{1}{3 \cdot 2} = \frac{1}{6}$$

$$s_5 = \frac{1}{4 \cdot 3} = \frac{1}{12}$$

$$s_6 = \frac{1}{5 \cdot 4} = \frac{1}{20}$$

Explain why the sequence begins with s_3.

Sometimes a sequence is presented by listing its first few terms, and then the general term. For instance:

$$-3, -1, 1, \ldots, 2n - 5, \ldots$$

is a listing of the terms of the sequence given by $s_n = 2n - 5$. To obtain the terms of this sequence, we let $n = 1, 2, 3$, and so on in $s_n = 2n - 5$.

Example 4   Find the tenth term of the sequence $-\dfrac{3}{2}, -4, 6, \ldots, \dfrac{n + 2}{n - 3}, \ldots$

Solution   The first term is obtained by letting $n = 1$. Since 3 is not in the domain of $s_n = \dfrac{n + 2}{n - 3}$, the third term is found by letting $n = 4$. Therefore the tenth term is found by letting $n = 11$, which is $s_{11} = \dfrac{13}{8}$.

A sequence can also be given by a verbal description instead of a formula. If, for example, we ask for the sequence of the odd integers beginning with -3, this is the same sequence given by the formula

$$s_n = 2n - 5$$

where we replace n by the set of positive integers.

Example 5   Write the first four terms of the sequence given by $s_n = \left(1 + \dfrac{1}{n}\right)^n$. Round off to two decimal places when appropriate.

Solution

$$s_1 = (1 + \tfrac{1}{1})^1 = 2$$
$$s_2 = (1 + \tfrac{1}{2})^2 = (\tfrac{3}{2})^2 = \tfrac{9}{4} = 2.25$$
$$s_3 = (1 + \tfrac{1}{3})^3 = (\tfrac{4}{3})^3 = \tfrac{64}{27} = 2.37$$
$$s_4 = (1 + \tfrac{1}{4})^4 = (\tfrac{5}{4})^4 = \tfrac{625}{256} = 2.44$$

The terms of the sequence in Example 5 are getting successively larger. But the increase from term to term is getting smaller. That is, the differences between successive terms are

$$0.25 \qquad 0.12 \qquad 0.07$$

and they are decreasing. If more terms of $s_n = \left(1 + \dfrac{1}{n}\right)^n$ were computed, you would see that while the terms keep on increasing, the amount by which each new term increases keeps getting smaller.

It turns out that no matter how large n is, the value of $\left(1 + \dfrac{1}{n}\right)^n$ is never more than 2.72. In fact, the larger the n that is taken, the closer $\left(1 + \dfrac{1}{n}\right)^n$ gets to the irrational value $e = 2.71828\ldots$. This is the number that was introduced in Chapter 7 in reference to natural logarithms and exponential functions.

EXERCISES 10.1

The domain of each sequence in Exercises 1 through 6 consists of the numbers 1, 2, 3, 4, 5. Write the corresponding range values and graph the sequence.

1. $s_n = 2n - 1$ **2.** $s_n = 10 - n^2$ **3.** $s_n = (-1)^n$

4. $s_k = -\dfrac{6}{k}$ **5.** $s_i = 8(-\tfrac{1}{2})^i$ **6.** $s_i = (\tfrac{1}{2})^{i-3}$

Write the first four terms of the sequence given by the formula in each of Exercises 7 through 30. (Assume the first domain value to be the smallest possible positive integer.)

7. $s_k = (-1)^k k^2$ **8.** $s_j = 3(\tfrac{1}{10})^{j-1}$ **9.** $s_j = 3(\tfrac{1}{10})^j$

10. $s_j = 3(\tfrac{1}{10})^{j+1}$ **11.** $s_j = 3(\tfrac{1}{10})^{2j}$ **12.** $s_n = \dfrac{1}{n - 3}$

13. $s_n = \dfrac{1}{n}$ **14.** $s_n = \dfrac{1}{n + 2}$ **15.** $s_k = 2k - 10$

16. $s_k = 1 + (-1)^k$ **17.** $s_n = -2 + (n - 1)(3)$

18. $s_n = a + (n - 1)(d)$ **19.** $s_i = \dfrac{i}{i + 1}$

20. $s_i = 64^{1/i}$ **21.** $s_n = \left(1 + \dfrac{1}{n}\right)^{n-1}$

22. $s_n = \dfrac{1}{2^n}$ **23.** $s_n = -2(\tfrac{3}{4})^{n-1}$

24. $s_k = ar^{k-1}$ **25.** $s_k = \dfrac{k}{2^k}$

26. $s_n = \dfrac{(-1)^n}{n}$ **27.** $s_k = \dfrac{k}{k + 1} - \dfrac{k + 1}{k}$

28. $s_n = \left(1 + \dfrac{1}{n - 1}\right)^n$ **29.** $s_n = 4$

30. $s_n = \dfrac{1}{n^2 - 2n - 3}$

31. Write the eighth term of $s_n = \dfrac{n}{n^2 - 9}$.

32. Find the hundredth term of $s_k = 1 + 2(k - 1)$.

33. Find the seventh term of $s_k = 3(0.1)^{k-1}$.

34. Write the twentieth term of $s_n = (-1)^{n-1}$.

35. Write the twelfth term of $s_i = i$.

36. Write the twelfth term of $s_i = (i - 1)^2$.

37. Write the twelfth term of $s_i = (i - 1)^3$.

38. Find the eighth term of $s_n = \dfrac{n}{n^2 - 5n + 4}$. (*Note: $n \neq 1$ and $n \neq 4$.*)

39. Write the first four terms of the sequence of even increasing integers beginning with 4.

40. Write the first four terms of the sequence of decreasing odd integers beginning with 3.

41. Write the first three positive multiples of 5 and find the formula for the nth term.

42. Write the first three powers of 5 and find the formula for the nth term.

43. Write the first three powers of -5 and find the formula for the nth term.

*44. Write the first three terms of the sequence of reciprocals of the negative integers and find the formula for the nth term.

*45. The numbers 1, 3, 6, and 10 are called **triangular numbers** because they correspond to the number of dots in this triangular arrays of dots.

Find the next three triangular numbers.

10.2
Sums of Finite Sequences

How long would it take you to add up all the integers from 1 to 1000? Here is a quick way. First list the numbers of the sequence to be added:

$$1, 2, 3, \ldots, 998, 999, 1000$$

Then the sum of the first and last terms is

$$1 + 1000 = 1001$$

The second and the second from last give

$$2 + 999 = 1001$$

The third and the third from last give

$$3 + 998 = 1001$$

And so on. Since there are 500 such pairs to be added, the total is

$$500(1001) = 500,500$$

Do not be upset if you did not see this for yourself; very few people would make such a discovery. (One person who did was a 10-year-old boy by the name of Karl Friedrich Gauss (1777–1855), but he grew up to be one of the greatest mathematicians of all time.) As you will see later in Section 10.3, the solution to this question is a special case of a general situation that calls for the sum of a certain kind of sequence.

For any finite sequence we can add up all its terms and say that we have found the *sum of the sequence*. The indicated sum of a sequence is called a **series**. For example, with the sequence

$$1, 3, 5, 7, 9, 11$$

we can associate the series

$$1 + 3 + 5 + 7 + 9 + 11$$

The sum of the terms in this series can easily be found, by adding, to be 36. As another example, the sequence $s_n = \dfrac{1}{n}$ for $n = 1, 2, 3, 4, 5$ has the sum

$$1 + \frac{1}{2} + \frac{1}{3} + \frac{1}{4} + \frac{1}{5} = \frac{60 + 30 + 20 + 15 + 12}{60} = \frac{137}{60}$$

Example 1   Find the sum of the first seven terms of $s_k = 2k$.

Solution

$$s_1 + s_2 + s_3 + s_4 + s_5 + s_6 + s_7 = 2 + 4 + 6 + 8 + 10 + 12 + 14 = 56$$

There is a very handy notational device available for expressing the sum of a sequence. The Greek letter $\sum$ (capital sigma) is used for this purpose. Referring to Example 1, the sum of the seven terms is expressed by the symbol $\displaystyle\sum_{k=1}^{7} s_k$; that is,

$$\sum_{k=1}^{7} s_k = s_1 + s_2 + s_3 + s_4 + s_5 + s_6 + s_7$$

Just think of the sigma as the *command to add*. What you add are the terms s_k for consecutive values of k, starting with $k = 1$ up to and including $k = 7$. With this symbolism, the question in Example 1 can now be stated by asking for the value of $\sum_{k=1}^{7} s_k$, where $s_k = 2k$.

Example 2   Find $\sum_{n=1}^{5} s_n$, where $s_n = \dfrac{2}{n}$.

Solution

$$\sum_{n=1}^{5} s_n = s_1 + s_2 + s_3 + s_4 + s_5$$

$$= \frac{2}{1} + \frac{2}{2} + \frac{2}{3} + \frac{2}{4} + \frac{2}{5}$$

$$= \frac{120 + 60 + 40 + 30 + 24}{60}$$

$$= \frac{274}{60} = \frac{137}{30}$$

Example 3   Find $\sum_{k=1}^{8} (2k + 1)$.

Solution   It is understood here that we are to find the sum of the first eight terms of the sequence whose general term is $s_k = 2k + 1$.

$$\sum_{k=1}^{8} (2k + 1) = (2 \cdot 1 + 1) + (2 \cdot 2 + 1) + (2 \cdot 3 + 1)$$

$$+ (2 \cdot 4 + 1) + (2 \cdot 5 + 1) + (2 \cdot 6 + 1)$$

$$+ (2 \cdot 7 + 1) + (2 \cdot 8 + 1)$$

$$= 3 + 5 + 7 + 9 + 11 + 13 + 15 + 17$$

$$= 80$$

Example 4   Find $\sum_{i=1}^{5} s_i$, where $s_i = (-1)^i(i + 1)$.

Solution

$$\sum_{i=1}^{5} s_i = s_1 + s_2 + s_3 + s_4 + s_5$$

$$= (-1)^1(1 + 1) + (-1)^2(2 + 1) + (-1)^3(3 + 1)$$

$$+ (-1)^4(4 + 1) + (-1)^5(5 + 1)$$

$$= -2 + 3 - 4 + 5 - 6$$

$$= -4$$

EXERCISES 10.2

Find the sum of the first five terms of the sequence given by the formulas in Exercises 1 through 6.

1. $s_n = 3n$ **2.** $s_k = (-1)^k \frac{1}{k}$ **3.** $s_i = i^2$

4. $s_i = i^3$ **5.** $s_k = \dfrac{3}{10^k}$ **6.** $s_n = -6 + 2(n - 1)$

7. Find $\displaystyle\sum_{n=1}^{8} s_n$, where $s_n = 2^n$.

8. Find $\displaystyle\sum_{n=0}^{8} s_n$, where $s_n = \dfrac{1}{2^n}$. (Note that the sequence being added begins for $n = 0$.)

9. Find $\displaystyle\sum_{k=1}^{20} s_k$, where $s_k = 3$.

Find each of the following sums for $n = 7$.

10. $2 + 4 + \cdots + 2n$ **11.** $2 + 4 + \cdots + 2^n$

12. $-7 + 2 + \cdots + (9n - 16)$ **13.** $3 + \frac{3}{2} + \cdots + 3(\frac{1}{2})^{n-1}$

Compute each of the following sums.

14. $\displaystyle\sum_{k=1}^{6} (5k)$ **15.** $5\left(\displaystyle\sum_{k=1}^{6} k\right)$ **16.** $\displaystyle\sum_{n=1}^{4} (n^2 + n)$

17. $\displaystyle\sum_{n=1}^{4} n^2 + \displaystyle\sum_{n=1}^{4} n$ **18.** $\displaystyle\sum_{i=1}^{8} (i - 2i^2)$ **19.** $\displaystyle\sum_{k=1}^{4} \dfrac{k}{2^k}$

20. $\displaystyle\sum_{k=1}^{7} (-1)^k$ **21.** $\displaystyle\sum_{k=1}^{8} (-1)^k$ **22.** $\displaystyle\sum_{k=3}^{7} (2k - 5)$

23. $\displaystyle\sum_{j=1}^{6} [-3 + (j - 1)5]$ **24.** $\displaystyle\sum_{k=-3}^{3} 10^k$

25. $\displaystyle\sum_{k=-3}^{3} \dfrac{1}{10^k}$ **26.** $\displaystyle\sum_{k=1}^{5} 4(-\frac{1}{2})^{k-1}$

27. $\displaystyle\sum_{i=1}^{4} (-1)^i 3^i$ **28.** $\displaystyle\sum_{n=1}^{3} \left(\dfrac{n+1}{n} - \dfrac{n}{n+1}\right)$

29. $\displaystyle\sum_{n=1}^{3} \dfrac{n+1}{n} - \displaystyle\sum_{n=1}^{3} \dfrac{n}{n+1}$ **30.** $\displaystyle\sum_{k=1}^{8} \dfrac{1 + (-1)^k}{2}$

31. $\displaystyle\sum_{k=1}^{3} (0.1)^{2k}$

***32.** Follow the argument at the beginning of this section, where we found the sum of the first 1000 positive integers, and find a formula for the sum of the first n positive integers.

***33.** (a) Find $\displaystyle\sum_{k=1}^{n} (2k - 1)$ for each of the following values of n: 2, 3, 4, 5, 6.

    (b) On the basis of results in part (a), find a formula for the sum of the first n odd numbers.

*34. Show that $\sum\limits_{k=1}^{9} \ln \dfrac{k+1}{k} = 1.$ $\left(Hint: \ln \dfrac{a}{b} = \ln a - \ln b.\right)$

*35. Prove: $\sum\limits_{k=1}^{n} s_k + \sum\limits_{k=1}^{n} t_k = \sum\limits_{k=1}^{n} (s_k + t_k).$

*36. Prove: $\sum\limits_{k=1}^{n} cs_k = c \sum\limits_{k=1}^{n} s_k.$

*37. Prove: $\sum\limits_{k=1}^{n} (s_k + c) = \left(\sum\limits_{k=1}^{n} s_k\right) + nc,$ c a constant.

10.3 Arithmetic Sequences and Series

Here are the first five terms of the sequence whose general term is $s_k = 7k - 2$:

$$5, 12, 19, 26, 33$$

Do you notice any special pattern? It does not take long to observe that each term is 7 more than the preceding term. This sequence is an example of an *arithmetic sequence*.

Arithmetic sequence

A sequence is said to be *arithmetic* if each term, after the first, is obtained from the preceding term by adding a common value.

Here are the first four terms of three different arithmetic sequences:

$$2, 4, 6, 8, \ldots$$
$$-\tfrac{1}{2}, -1, -\tfrac{3}{2}, -2, \ldots$$
$$11, 2, -7, -16, \ldots$$

For the first sequence, the common value (or difference) that is added to each term to get the next is 2. Thus it is easy to see that 10, 12, and 14 are the next three terms. You might guess that the nth term is $s_n = 2n$.

The second sequence has the common difference $-\tfrac{1}{2}$. This can be found by subtracting the first term from the second, or the second from the third, and so forth. The nth term is $s_n = -\tfrac{1}{2}n$.

The third sequence has -9 as its common difference, but it is not so easy to see what the general term is. Rather than employ a hit-or-miss process in trying to find the general term of this sequence, we will instead consider arithmetic sequences in general, thus making it possible to write the general term of any such sequence.

Let s_n be the nth term of an arithmetic sequence. Denote its first term by the letter a; that is, $s_1 = a$. Also, let d be the common difference. Then the first four terms are

$$s_1 = a$$
$$s_2 = s_1 + d = a + d$$
$$s_3 = s_2 + d = (a + d) + d = a + 2d$$
$$s_4 = s_3 + d = (a + 2d) + d = a + 3d$$

The pattern is clear. Without further computation we see that:

$$s_5 = a + 4d$$
$$s_6 = a + 5d$$

Since the coefficient of d is always 1 less than the number of the term, the nth term must be given as

$$s_n = a + (n - 1)d$$

You can check this result by substituting in the values $n = 1, 2, 3$, and so on to obtain the preceding terms.

This formula says that the nth term of an arithmetic sequence is completely identified by its first term a and its common difference d.

Let us return to the sequence given earlier: $11, 2, -7, -16, \ldots$. It is now easy to find its nth term, with $a = 11$ and $d = -9$:

$$s_n = 11 + (n - 1)(-9)$$
$$= -9n + 20$$

TEST YOUR UNDERSTANDING

Each of the following gives the first few terms of an arithmetic sequence. Find the nth term in each case.

1. $5, 10, 15, \ldots$ **2.** $6, 2, -2, \ldots$

3. $\frac{1}{10}, \frac{1}{5}, \frac{3}{10}, \ldots$ **4.** $-5, -13, -21, \ldots$

5. $1, 2, 3, \ldots$ **6.** $-3, -2, -1, \ldots$

Find the nth term s_n of the arithmetic sequence with the given values for the first term and the common difference.

7. $a = \frac{2}{3}; d = \frac{2}{3}$ **8.** $a = 53; d = -12$

9. $a = 0; d = \frac{1}{5}$ **10.** $a = 2; d = 1$

Adding the terms of a finite sequence may not be much work when the number of terms to be added is small. When many terms are to be added, however, the amount of time and effort needed can be overwhelming. For example, to add the first 10,000 terms of the arithmetic sequence beginning with

$$246, 261, 276, \ldots$$

would call for an enormous effort unless some shortcut could be found. The indicated sum of an arithmetic sequence is called an **arithmetic series**. Fortunately, there is an easy way to find such sums. This method (in disguise) was already used in the question at the start of Section 10.2. Let us look at the general situation. Let S_n denote the sum of the first n terms of the arithmetic sequence given by $s_k = a + (k - 1)d$:

$$S_n = \sum_{k=1}^{n} [a + (k - 1)d]$$
$$= a + [a + d] + [a + 2d] + \cdots + [a + (n - 1)d]$$

Put this sum in reverse order and write the two equalities together as follows:

$$S_n = \quad a \quad + \quad [a + d] \quad + \cdots + [a + (n - 2)d] + [a + (n - 1)d]$$
$$S_n = [a + (n - 1)d] + [a + (n - 2)d] + \cdots + \quad [a + d] \quad + \quad a$$

Now add to get

$$2S_n = [2a + (n - 1)d] + [2a + (n - 1)d] + \cdots$$
$$+ [2a + (n - 1)d] + [2a + (n - 1)d]$$

On the right-hand side of this equation there are n terms of the form $2a + (n - 1)d$. Therefore:

$$2S_n = n[2a + (n - 1)d]$$

Divide by 2 to solve for S_n:

$$S_n = \frac{n}{2}[2a + (n - 1)d]$$

Returning to the sigma notation, we can summarize our results this way:

Sum of an arithmetic series

$$\sum_{k=1}^{n} [a + (k - 1)d] = \frac{n}{2}[2a + (n - 1)d]$$

Example 1 Find S_{20} for the arithmetic sequence whose first term is $a = 3$ and whose common difference is $d = 5$.

Solution Substituting $a = 3$, $d = 5$, and $n = 20$ into the formula for S_n, we have

$$S_n = \tfrac{20}{2}[2(3) + (20 - 1)5]$$
$$= 10(6 + 95)$$
$$= 1010$$

Example 2 Find the sum of the first 10,000 terms of the arithmetic sequence beginning with 246, 261, 276, ...

Solution Since $a = 246$ and $d = 15$,

$$S_{10,000} = \tfrac{10,000}{2}[2(246) + (10,000 - 1)15]$$
$$= 5000(150,477)$$
$$= 752,385,000$$

Example 3 Find the sum of the first n positive integers.

Solution First observe that the problem calls for the sum of the sequence $s_k = k$ for $k = 1, 2, \ldots, n$. This is an arithmetic sequence with $a = 1$ and $d = 1$. Therefore

$$\sum_{k=1}^{n} k = \frac{n}{2}[2 + (n - 1)1] = \frac{n(n + 1)}{2}$$

With the result of Example 3 we are able to check the answer for the sum of the first 1000 positive integers, found at the beginning of Section 10.2, as follows:

$$\sum_{k=1}^{1000} k = \frac{1000(1001)}{2} = 500,500$$

The form $s_k = a + (k - 1)d$ for the general term of an arithmetic sequence easily converts to

$$s_k = dk + (a - d)$$

It is this latter form that is ordinarily used when the general term of a *specific* arithmetic sequence is given. For example, we would usually begin with the form

$$s_k = 3k + 5$$

instead of

$$s_k = 8 + (k - 1)3$$

The important thing to notice in the form $s_k = dk + (a - d)$ is that the common difference is the coefficient of k.

Example 4 Find $\displaystyle\sum_{k=1}^{50} (-6k + 10)$.

Solution First note that $s_k = -6k + 10$ is an arithmetic sequence with $d = -6$ and $a = s_1 = 4$.

$$\sum_{k=1}^{50} (-6k + 10) = \tfrac{50}{2}[2(4) + (50 - 1)(-6)]$$

$$= -7150$$

EXERCISES 10.3

Each of the following gives the first two terms of an arithmetic sequence. Write the next three terms; find the nth term; and find the sum of the first 20 terms.

1. $1, 3, \ldots$　　　　**2.** $2, 4, \ldots$　　　　**3.** $2, -4, \ldots$

4. $1, -3, \ldots$　　　**5.** $\frac{15}{2}, 8, \ldots$　　　**6.** $-\frac{4}{3}, -\frac{11}{3}, \ldots$

7. $\frac{2}{5}, -\frac{1}{5}, \ldots$　　　**8.** $-\frac{1}{2}, \frac{1}{4}, \ldots$　　　**9.** $50, 100, \ldots$

10. $-27, -2, \ldots$　　**11.** $-10, 10, \ldots$　　**12.** $225, 163, \ldots$

Find the indicated sum by using ordinary addition; also find the sum by using the formula for the sum of an arithmetic series.

13. $5 + 10 + 15 + 20 + 25 + 30 + 35 + 40 + 45 + 50 + 55 + 60 + 65$

14. $-33 - 25 - 17 - 9 - 1 + 7 + 15 + 23 + 31 + 39$

15. $\frac{3}{4} + 1 + \frac{5}{4} + \frac{3}{2} + \frac{7}{4} + 2 + \frac{9}{4} + \frac{5}{2} + \frac{11}{4}$

16. $128 + 71 + 14 - 43 - 100 - 157$

In Exercises 17 through 24 find S_{100} for the arithmetic sequence with the given values for a and d.

17. $a = 3; d = 3$　　　　　　**18.** $a = 1; d = 8$

19. $a = -91; d = 21$　　　　**20.** $a = -7; d = -10$

21. $a = \frac{1}{7}; d = 5$ **22.** $a = \frac{2}{5}; d = -4$

23. $a = 725; d = 100$ **24.** $a = 0.1; d = 10$

25. Find S_{28} for the sequence $-8, 8, \ldots, 16n - 24, \ldots$.

26. Graph the sequences given in Exercises 18 and 22 for the domain $n = 1, 2, 3, 4$.

27. Find the sum of the first 50 positive multiples of 12.

28. (a) Find the sum of the first 100 positive even numbers.
    (b) Find the sum of the first n positive even numbers.

29. (a) Find the sum of the first 100 positive odd numbers.
    (b) Find the sum of the first n positive odd numbers.

Evaluate the sums in Exercises 30 through 37.

30. $\displaystyle\sum_{k=1}^{12} [3 + (k - 1)9]$ **31.** $\displaystyle\sum_{k=1}^{9} [-6 + (k - 1)\frac{1}{2}]$

32. $\displaystyle\sum_{k=1}^{20} (4k - 15)$ **33.** $\displaystyle\sum_{k=1}^{30} (10k - 1)$

34. $\displaystyle\sum_{k=1}^{40} (-\frac{1}{3}k + 2)$ **35.** $\displaystyle\sum_{k=1}^{49} (\frac{3}{4}k - \frac{1}{2})$

36. $\displaystyle\sum_{k=1}^{20} 5k$ **37.** $\displaystyle\sum_{k=1}^{n} 5k$

38. Find u such that $7, u, 19$ is an arithmetic sequence.

39. Find u such that $-7, u, \frac{5}{2}$ is an arithmetic sequence.

40. Find the twenty-third term of the arithmetic sequences $6, -4, \ldots$.

41. Find the thirty-fifth term of the arithmetic sequence $-\frac{2}{3}, -\frac{1}{5}, \ldots$.

42. An object is dropped from an airplane and falls 32 feet during the first second. During each successive second it falls 48 feet more than in the preceding second. How many feet does it travel during the first 10 seconds? How far does it fall during the tenth second?

43. Suppose you save $10 one week and that each week thereafter you save 50¢ more than the preceding week. How much will you have saved by the end of one year?

44. A pyramid of blocks has 26 blocks in the bottom row and 2 fewer blocks in each successive row thereafter. How many blocks are there in the pyramid?

*45. Find $\displaystyle\sum_{n=6}^{20} (5n - 3)$.

*46. Find the sum of all the even numbers between 33 and 427.

*47. If $\displaystyle\sum_{k=1}^{30} [a + (k - 1)d] = -5865$ and $\displaystyle\sum_{k=1}^{20} [a + (k - 1)d] = -2610$, find a and d.

*48. Listing the first few terms of a sequence like $2, 4, 6, \ldots$, without stating its general term or describing what kind of sequence it is makes it impossible to predict the next term. To see this, show that both $t_n = 2n$ and $s_n = 2n + (n - 1)(n - 2)(n - 3)$ produce these first three terms but their fourth terms are different.

Suppose a ball is dropped from a height of 4 feet and bounces straight up and down, always bouncing up exactly one-half the distance it just came down. How far will the ball have traveled if you catch it after it reaches the top of the fifth bounce? The following figure will help you to answer this question. For the sake of clarity, the bounces have been separated in the figure.

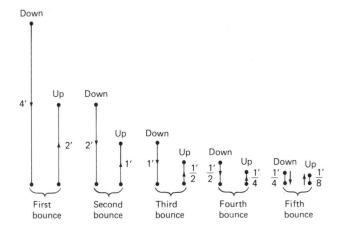

From this diagram we can determine how far the ball has traveled on each bounce. On the first bounce it goes 4 feet down and 2 feet up, for a total of 6 feet; on the second bounce the total distance is $2 + 1 = 3$ feet; and so on. These distances form the following sequence of five terms (one for each bounce):

$$6, 3, \tfrac{3}{2}, \tfrac{3}{4}, \tfrac{3}{8}$$

This sequence has the special property that, after the first term, each successive term can be obtained by multiplying the preceding term by $\tfrac{1}{2}$; that is, the second term, 3, is half the first, 6, and so on. This is an example of a *geometric sequence*. Later we will develop a formula for finding the sum of such a sequence; in the meantime, we can find the total distance the ball has traveled during the five bounces by adding the five terms as follows:

$$6 + 3 + \frac{3}{2} + \frac{3}{4} + \frac{3}{8} = \frac{48 + 24 + 12 + 6 + 3}{8} = 11\frac{5}{8}$$

Geometric sequence

A sequence is said to be *geometric* if each term, after the first, is obtained by multiplying the preceding term by a common value.

437

Here are the first four terms of three different geometric sequences.

$$2, -4, 8, -16, \ldots$$

$$1, \tfrac{1}{3}, \tfrac{1}{9}, \tfrac{1}{27}, \ldots$$

$$5, -5, 5, -5, \ldots$$

By inspection you can determine that the common multipliers for these sequences are -2, $\tfrac{1}{3}$, and -1, respectively. We will find their nth terms by deriving the formula for the nth term of any geometric sequence.

Let s_n be the nth term of a geometric sequence, and let a be its first term. The common multiplier, which is also called the **common ratio**, is denoted by r. Here are the first five terms:

$$s_1 = a$$

$$s_2 = ar$$

$$s_3 = ar^2$$

$$s_4 = ar^3$$

$$s_5 = ar^4$$

Notice that the exponent of r is 1 less than the number of the term. This observation allows us to write the nth term as

$$s_n = ar^{n-1}$$

With this result, the first four terms and the nth terms of the three given geometric sequences are as follows:

$$2, -4, 8, -16, \ldots, 2(-2)^{n-1}$$

$$1, \tfrac{1}{3}, \tfrac{1}{9}, \tfrac{1}{27}, \ldots, 1(\tfrac{1}{3})^{n-1}$$

$$5, -5, 5, -5, \ldots, 5(-1)^{n-1}$$

You can substitute the values $n = 1, 2, 3$, and 4 into these forms and see that the given first four terms are obtained in each case.

Example 1 Find the hundredth term of the geometric sequence having $r = \tfrac{1}{2}$ and $a = \tfrac{1}{2}$.

Solution   The nth term of this sequence is given by

$$s_n = \frac{1}{2}\left(\frac{1}{2}\right)^{n-1}$$

$$= \frac{1}{2}\left(\frac{1}{2^{n-1}}\right)$$

$$= \frac{1}{2^n}$$

Thus:

$$s_{100} = \frac{1}{2^{100}}$$

Example 2   Find the nth term of the geometric sequence beginning with

$$6, 9, \tfrac{27}{2}, \ldots$$

and find the seventh term.

Solution   The common ratio r must satisfy $6r = 9$ (or $9r = \frac{27}{2}$). Then $r = \frac{3}{2}$ and

$$s_n = 6\left(\frac{3}{2}\right)^{n-1}$$

Thus:

$$s_7 = 6\left(\frac{3}{2}\right)^{7-1} = 6\left(\frac{3}{2}\right)^6$$

$$= (3 \cdot 2) \cdot \frac{3^6}{2^6} = \frac{3^7}{2^5} = \frac{2187}{32}$$

Note: The common ratio r can always be found by taking the ratio of the $(n+1)$st term to the nth term. That is,

$$r = \frac{ar^n}{ar^{n-1}}$$

And in this example,

$$r = \tfrac{9}{6} = \tfrac{3}{2}$$

Example 3   Write the kth term of the geometric sequence $s_k = \left(\frac{1}{2}\right)^{2k}$ in the form ar^{k-1} and find the value of a and r.

Solution

$$s_k = (\tfrac{1}{2})^{2k} = [(\tfrac{1}{2})^2]^k = (\tfrac{1}{4})^k$$
$$= \tfrac{1}{4}(\tfrac{1}{4})^{k-1} \qquad \text{(this is now in the form } ar^{n-1})$$

Then $a = \tfrac{1}{4}$ and $r = \tfrac{1}{4}$.

Note: The first term a can *also* be found by simply computing s_1 in the given formula for s_k; the value of r, however, is not so obvious in this original form.

TEST YOUR UNDERSTANDING

Write the first five terms of the geometric sequences with the given general term. Also write the nth term in the form ar^{n-1} and find r.

1. $s_n = (\tfrac{1}{2})^{n-1}$ **2.** $s_n = (\tfrac{1}{2})^{n+1}$

3. $s_n = (-\tfrac{1}{2})^n$ **4.** $s_n = (-\tfrac{1}{3})^{3n}$

Find r and the nth term of the geometric sequence with the given first two terms.

5. $\tfrac{1}{5}, 2$ **6.** $27, -12$

Let us return to the original problem of this section. We found that the total distance the ball traveled was $11\tfrac{5}{8}$ feet. This is the sum of the first five terms of the geometric sequence whose nth term is $6(\tfrac{1}{2})^{n-1}$. Adding these five terms was easy. But what about adding the first 100 terms? There is a formula for the sum of a geometric sequence that will enable us to find such answers efficiently.

The indicated sum of a geometric sequence is called a **geometric series**. Just as with arithmetic series, there is a formula for finding such sums. To discover this formula, let $s_k = ar^{k-1}$ be a geometric sequence and denote the sum of the first n terms by $S_n = \sum_{k=1}^{n} ar^{k-1}$. Then

$$S_n = a + ar + ar^2 + \cdots + ar^{n-2} + ar^{n-1}$$

Multiplying this equation by r gives

$$rS_n = ar + ar^2 + \cdots + ar^{n-1} + ar^n$$

Now consider these two equations:

$$S_n = a + ar + ar^2 + \cdots + ar^{n-2} + ar^{n-1}$$
$$rS_n = \qquad ar + ar^2 + \cdots + ar^{n-2} + ar^{n-1} + ar^n$$

Subtract and factor:

$$S_n - rS_n = a - ar^n$$
$$(1 - r)S_n = a(1 - r^n)$$

Divide by $1 - r$ to solve for S_n:

$$S_n = \frac{a(1 - r^n)}{1 - r}$$

Returning to the sigma notation, we can summarize our results this way:

Sum of a geometric series

$$\sum_{k=1}^{n} ar^{k-1} = \frac{a(1 - r^n)}{1 - r}$$

This formula can be used to verify the earlier result for the bouncing ball:

$$\sum_{k=1}^{5} 6\left(\frac{1}{2}\right)^{k-1} = \frac{6[1 - (\frac{1}{2})^5]}{1 - \frac{1}{2}}$$
$$= \frac{6(1 - \frac{1}{32})}{\frac{1}{2}}$$
$$= \frac{93}{8}$$
$$= 11\frac{5}{8}$$

Example 4 Find the sum of the first 100 terms of the geometric sequence given by $s_k = 6(\frac{1}{2})^{k-1}$ and show that the answer is very close to 12.

Solution

$$S_{1000} = \frac{6\left(1 - \frac{1}{2^{100}}\right)}{1 - \frac{1}{2}}$$
$$= 12\left(1 - \frac{1}{2^{100}}\right)$$

Next observe that the fraction $\frac{1}{2^{100}}$ is so small that $1 - \frac{1}{2^{100}}$ is very nearly equal to 1, and therefore S_{100} is very close to 12.

Example 5   Evaluate $\sum\limits_{k=1}^{8} 3(\frac{1}{10})^{k+1}$.

Solution   We get $3(\frac{1}{10})^{k+1} = \frac{3}{100}(\frac{1}{10})^{k-1}$. Then $a = 0.03$, $r = 0.1$, and

$$S_8 = \frac{0.03[1 - (0.1)^8]}{1 - 0.1}$$

$$= \frac{0.03(1 - 0.00000001)}{0.9}$$

$$= 0.033333333$$

Geometric sequences have many applications, as illustrated in the next example. You will find others in the exercises at the end of this section.

Example 6   Suppose that you save $128 in January and that each month thereafter you only manage to save half of what you saved the previous month. How much do you save in the tenth month, and what is your total savings after 10 months?

Solution   The amounts saved each month form a geometric sequence with $a = 128$ and $r = \frac{1}{2}$. Then $s_n = 128(\frac{1}{2})^{n-1}$ and

$$s_{10} = 128\left(\frac{1}{2}\right)^9 = \frac{2^7}{2^9} = \frac{1}{4} = 0.25$$

This means that you saved 25¢ in the tenth month. Your total savings is given by

$$S_{10} = \frac{128\left(1 - \frac{1}{2^{10}}\right)}{1 - \frac{1}{2}}$$

$$= 256\left(1 - \frac{1}{2^{10}}\right)$$

$$= 256 - \frac{256}{2^{10}}$$

$$= 256 - \frac{2^8}{2^{10}}$$

$$= 255.75$$

EXERCISES 10.4

The first three terms of a geometric sequence are given in Exercises 1 through 12. Write the next three terms and also find the formula for the nth term.

1. $2, 4, 8, \ldots$
2. $2, -4, 8, \ldots$
3. $1, 3, 9, \ldots$
4. $2, -2, 2, \ldots$
5. $-3, 1, -\frac{1}{3}, \ldots$
6. $100, 10, 1, \ldots$
7. $-1, -5, -25, \ldots$
8. $12, -6, 3, \ldots$
9. $-6, -4, -\frac{8}{3}, \ldots$
10. $-64, 16, -4, \ldots$
11. $\frac{1}{1000}, \frac{1}{10}, 10, \ldots$
12. $\frac{27}{8}, \frac{3}{2}, \frac{2}{3}, \ldots$

In Exercises 13 through 15 find the sum of the first six terms of the indicated sequence by using ordinary addition and also by using the formula for the sum of a geometric series.

13. The sequence in Exercise 1.
14. The sequence in Exercise 5.
15. The sequence in Exercise 9.

16. Find the tenth term of the geometric sequence $2, 4, 8, \ldots$.
17. Find the fourteenth term of the geometric sequence $\frac{1}{8}, \frac{1}{4}, \frac{1}{2}, \ldots$.
18. Find the fifteenth term of the geometric sequence $\dfrac{1}{100{,}000}, \dfrac{1}{10{,}000}, \dfrac{1}{1000}, \ldots$.
19. What is the one-hundred-first term of the geometric sequence having $a = 3$ and $r = -1$?
20. For the geometric sequence with $a = 100$ and $r = \frac{1}{10}$, use the formula $s_n = ar^{n-1}$ to find which term is equal to $\dfrac{1}{10^{10}}$.

Find the sums in Exercises 21 through 29.

21. $\displaystyle\sum_{k=1}^{10} 2^{k-1}$
22. $\displaystyle\sum_{j=1}^{10} 2^{j+2}$
23. $\displaystyle\sum_{k=1}^{n} 2^{k-1}$
24. $\displaystyle\sum_{k=1}^{8} 3(\tfrac{1}{10})^{k-1}$
25. $\displaystyle\sum_{k=1}^{5} 3^{k-4}$
26. $\displaystyle\sum_{k=1}^{6} (-3)^{k-2}$
27. $\displaystyle\sum_{j=1}^{5} (\tfrac{2}{3})^{j-2}$
28. $\displaystyle\sum_{k=1}^{8} 16(\tfrac{1}{2})^{k+2}$
29. $\displaystyle\sum_{k=1}^{8} 16(-\tfrac{1}{2})^{k+2}$

30. Find $u > 0$ such that $2, u, 98$ forms a geometric sequence.
31. Find $u < 0$ such that $\frac{1}{7}, u, \frac{25}{63}$ forms a geometric sequence.
32. Suppose that the amount you save in any given month is twice the amount you saved in the previous month. How much will you have saved at the end of one year if you save $1 in January? How much if you saved 25¢ in January?
33. A certain bacterial culture doubles in number every day. If there were 1000 bacteria at the end of the first day, how many will there be after 10 days? How many after n days?
34. A radioactive substance is decaying so that at the end of each month there is only one-third as much as there was at the beginning of the month. If there were 75 grams of the substance at the beginning of the year, how much is left after 6 months?

*35. Suppose that an automobile depreciates 10% in value each year for the first 5 years. What is it worth after 5 years if its original cost was $5280? (*Hint:* Use $a = 5280$ and $n = 6$.)

*36. Compound-interest problems can be explained in terms of geometric sequences. The basic idea is that an investment P will earn i percent interest for the first year. Then the new total, consisting of the original investment P plus the interest Pi, will earn i percent interest for the second year, and so on. In such a situation we say that P earns i percent interest **compounded annually**.

(a) If $1000 is invested in a bank paying 6% interest compounded annually, the amount at the end of the first year is

$$1000 + 1000(0.06) = 1060$$

After 2 years the amount becomes

$$1060 + 1060(0.06) = 1123.60$$

What is the amount after 3 years?

(b) Let A_1 be the amount after 1 year on an investment of P dollars at i percent interest compounded annually. Then

$$A_1 = P + Pi = P(1 + i)$$

After the second year:

$$A_2 = P(1 + i) + P(1 + i)i$$
$$= P(1 + i)(1 + i)$$
$$= P(1 + i)^2$$

Find A_3.

(c) Referring to part (b), find the formula for A_n, the amount after n years, and show that this is the nth terms of a geometric sequence having ratio $r = 1 + i$.

*37. A sum of $800 is invested at 5% interest compounded annually.

(a) What is the amount after n years?

(b) What is the amount after 5 years? (Use common logarithms to get an approximation.)

*38. Use common logarithms to find the amount of money an investment of $1500 earns at the interest rate of 8% compounded annually for 5 years.

10.5

**Infinite
Geometric
Series**

In decimal form the fraction $\frac{3}{4}$ becomes 0.75, which means $\frac{75}{100}$. This can also be written as $\frac{7}{10} + \frac{5}{100}$. What about $\frac{1}{3}$? As a decimal we can write

$$\tfrac{1}{3} = 0.333 \ldots$$

where the dots mean that the 3 repeats endlessly. When we attempt to express this decimal as the sum of fractions, all we can write is

$$\tfrac{3}{10} + \tfrac{3}{100} + \tfrac{3}{1000} + \cdots$$

This raises the interesting question of the addition of an infinite number of numbers. Can any sense be made of this? Certainly if we attempt to add all the even natural numbers by writing

$$2 + 4 + 6 + \cdots + 2n + \cdots$$

it is obvious that no finite answer is possible. But for the preceding situation, we want to say

$$\tfrac{1}{3} = \tfrac{3}{10} + \tfrac{3}{100} + \tfrac{3}{1000} + \cdots$$

because the decimal $0.333\ldots$ is nothing more than an abbreviation for $\tfrac{3}{10} + \tfrac{3}{100} + \tfrac{3}{1000} + \cdots$.

The numbers being added here are the terms of the *infinite geometric sequence* with first term $a = \tfrac{3}{10}$ and common ratio $r = \tfrac{1}{10}$. Thus the kth term is

$$
\begin{aligned}
ar^{k-1} &= \frac{3}{10}\left(\frac{1}{10}\right)^{k-1} = 3\left(\frac{1}{10}\right)\left(\frac{1}{10}\right)^{k-1} \\
&= 3\left(\frac{1}{10}\right)^{k} \\
&= \frac{3}{10^{k}}
\end{aligned}
$$

The sum of the first n terms is found by using the formula

$$S_n = \sum_{k=1}^{n} ar^{k-1} = \frac{a(1 - r^n)}{1 - r}$$

Here are some cases:

$$S_1 = \frac{\dfrac{3}{10}\left(1 - \dfrac{1}{10}\right)}{1 - \dfrac{1}{10}} = \frac{1}{3}\left(1 - \frac{1}{10}\right) = 0.3$$

$$S_2 = \frac{\dfrac{3}{10}\left(1 - \dfrac{1}{10^2}\right)}{1 - \dfrac{1}{10}} = \frac{1}{3}\left(1 - \frac{1}{10^2}\right) = 0.33$$

$$S_3 = \frac{\dfrac{3}{10}\left(1 - \dfrac{1}{10^3}\right)}{1 - \dfrac{1}{10}} = \frac{1}{3}\left(1 - \frac{1}{10^3}\right) = 0.333$$

$$S_{10} = \frac{\frac{3}{10}\left(1 - \frac{1}{10^{10}}\right)}{1 - \frac{1}{10}} = \frac{1}{3}\left(1 - \frac{1}{10^{10}}\right) = 0.3333333333$$

$$S_n = \frac{\frac{3}{10}\left(1 - \frac{1}{10^n}\right)}{1 - \frac{1}{10}} = \frac{1}{3}\left(1 - \frac{1}{10^n}\right) = 0.\underbrace{333\ldots3}_{n \text{ places}}$$

You can see that as more and more terms are added, the closer and closer the answer gets to $\frac{1}{3}$. This can be observed by studying the form for the sum of the first n terms:

$$S_n = \frac{1}{3}\left(1 - \frac{1}{10^n}\right)$$

Here it is clear that the bigger n is, the closer $\frac{1}{10^n}$ is to 0, the closer $1 - \frac{1}{10^n}$ is to 1, and, finally, the closer S_n is to $\frac{1}{3}$. While it is true that S_n is never exactly equal to $\frac{1}{3}$, for very large n the difference between S_n and $\frac{1}{3}$ is negligible. Saying this another way:

By taking n large enough, S_n can be made as close to $\frac{1}{3}$ as we like.

Thus we now say that the sum of all the terms is $\frac{1}{3}$. In symbols:

$$\frac{3}{10} + \frac{3}{10^2} + \frac{3}{10^3} + \cdots + \frac{3}{10^n} + \cdots = \frac{1}{3}$$

The summation symbol, $\sum$, can also be used here after an adjustment in notation. Traditionally, the symbol ∞ has been used to suggest an infinite number of objects. So we use this symbol and make the transition from the sum of a finite number of terms

$$S_n = \sum_{k=1}^{n} \frac{3}{10^k} = \frac{3}{10} + \frac{3}{10^2} + \cdots + \frac{3}{10^n} = \frac{1}{3}\left(1 - \frac{3}{10^n}\right)$$

to the sum of an infinite number of terms

$$S_\infty = \sum_{k=1}^{\infty} \frac{3}{10^k} = \frac{3}{10} + \frac{3}{10^2} + \cdots + \frac{3}{10^n} + \cdots = \frac{1}{3}$$

Infinite geometric series

The *indicated* sum of all the terms of an infinite geometric sequence is called an infinite geometric series.

Not every geometric sequence produces an infinite geometric series that has a finite sum. For instance, the sequence

$$2, 4, 8, \ldots, 2^n, \ldots$$

is geometric, but the corresponding geometric series

$$2 + 4 + 8 + \cdots + 2^n + \cdots$$

does not have a finite sum.

By now you might suspect that the common ratio r determines whether or not an infinite geometric sequence can be added. This turns out to be true. To see how this works, the general case is considered next.

Let

$$a, ar, ar^2, \ldots, ar^{k-1}, \ldots$$

be an infinite geometric sequence. The sum of the first n terms is

$$S_n = \frac{a(1 - r^n)}{1 - r}$$

Rewrite S_n in this form:

$$S_n = \frac{a}{1 - r}(1 - r^n)$$

At this point the importance of r^n becomes clear. If, as n gets larger, r^n gets very large, the infinite geometric series will not have a finite sum. But when r^n gets arbitrarily close to 0 as n gets larger, then S_n gets very close to $\frac{a}{1 - r}$.

The values of r for which r^n gets arbitrarily close to 0 are precisely those values between -1 and 1; that is, $|r| < 1$. For instance, $\frac{3}{5}$, $-\frac{1}{10}$, and 0.09 are values of r for which r^n gets close to 0; and 1.01, -2, and $\frac{3}{2}$ are values for which r^n does not get close to 0.

To sum up, we have the following useful result:

Sum of an infinite geometric series

If $|r| < 1$, then $\sum_{k=1}^{\infty} ar^{k-1} = \frac{a}{1 - r}$. For other values of r the series has no finite sum.

Example 1 Find the sum of the infinite geometric series

$$27 + 3 + \tfrac{1}{3} + \cdots$$

Solution Since $r = \tfrac{3}{27} = \tfrac{1}{9}$ and $a = 27$, the preceding result gives

$$27 + 3 + \frac{1}{3} + \cdots = \frac{27}{1 - \frac{1}{9}} = \frac{27}{\frac{8}{9}}$$
$$= \frac{243}{8}$$

Example 2 Why does the infinite geometric series $\sum\limits_{k=1}^{\infty} 5(\tfrac{4}{3})^{k-1}$ have no finite sum?

Solution The series has no finite sum because the common ratio $r = \tfrac{4}{3}$ is not between -1 and 1.

Example 3 Find $\sum\limits_{k=1}^{\infty} \dfrac{7}{10^{k+1}}$.

Solution Since $\dfrac{7}{10^{k+1}} = 7\left(\dfrac{1}{10^{k+1}}\right) = 7\left(\dfrac{1}{10^2}\right)\left(\dfrac{1}{10^{k-1}}\right) = \dfrac{7}{100}\left(\dfrac{1}{10}\right)^{k-1}$, it follows that $a = \tfrac{7}{100}$ and $r = \tfrac{1}{10}$. Therefore, by the formula for the sum of an infinite geometric series, we have

$$S_\infty = \sum_{k=1}^{\infty} \frac{7}{10^{k+1}} = \frac{\frac{7}{100}}{1 - \frac{1}{10}} = \frac{7}{100 - 10} = \frac{7}{90}$$

TEST YOUR UNDERSTANDING

Find the common ratio r and then find the sum if the given infinite geometric series has one.

1. $10 + 1 + \tfrac{1}{10} + \cdots$ **2.** $\tfrac{1}{64} + \tfrac{1}{16} + \tfrac{1}{4} + \cdots$

3. $36 - 6 + \tfrac{1}{6} - \cdots$ **4.** $-16 - 4 - 1 - \cdots$

5. $\sum\limits_{k=1}^{\infty} (\tfrac{4}{3})^{k-1}$ **6.** $\sum\limits_{k=1}^{\infty} 3(0.01)^k$

7. $\sum\limits_{i=1}^{\infty} (-1)^i 3^i$ **8.** $\sum\limits_{n=1}^{\infty} 100(-\tfrac{9}{10})^{n+1}$

The introduction to this section indicated how the endless repeating decimal fraction $0.333\ldots$ can be regarded as an infinite geometric series. The next example illustrates how such decimal fractions can be transformed into the rational form $\frac{a}{b}$ (the ratio of two integers) by using the formula for the sum of an infinite geometric series.

Example 4 Express the repeating decimal $0.242424\ldots$ in rational form.

Solution First write

$$0.242424\ldots = \frac{24}{100} + \frac{24}{10,000} + \frac{24}{1,000,000} + \cdots$$

$$= \frac{24}{10^2} + \frac{24}{10^4} + \frac{24}{10^6} + \cdots + \frac{24}{10^{2k}} + \cdots$$

$$= \sum_{k=1}^{\infty} \frac{24}{10^{2k}}$$

Next observe that

$$\frac{24}{10^{2k}} = 24\left(\frac{1}{10^{2k}}\right) = 24\left(\frac{1}{10^2}\right)^k$$

$$= 24\left(\frac{1}{100}\right)^k = \frac{24}{100}\left(\frac{1}{100}\right)^{k-1}$$

Then $a = \frac{24}{100}$, $r = \frac{1}{100}$, and

$$0.242424\ldots = \sum_{k=1}^{\infty} \frac{24}{10^{2k}}$$

$$= \frac{\frac{24}{100}}{1 - \frac{1}{100}}$$

$$= \frac{24}{99}$$

Note: You can check this answer by dividing 99 into 24.

CAUTION! LEARN TO AVOID MISTAKES LIKE THESE:

WRONG	RIGHT
$\displaystyle\sum_{k=1}^{\infty}\left(\frac{1}{2}\right)^{n+1} = \frac{1}{1-\frac{1}{2}}$	$\displaystyle\sum_{n=1}^{\infty}\left(\frac{1}{2}\right)^{n+1} = \sum_{n=1}^{\infty}\frac{1}{4}\left(\frac{1}{2}\right)^{n-1} = \frac{\frac{1}{4}}{1-\frac{1}{2}}$

WRONG	RIGHT
$\displaystyle\sum_{n=1}^{\infty}\left(-\frac{1}{3}\right)^{n-1}=\frac{1}{1-\frac{1}{3}}$	$\displaystyle\sum_{n=1}^{\infty}\left(-\frac{1}{3}\right)^{n-1}=\frac{1}{1-\left(-\frac{1}{3}\right)}$
$\displaystyle\sum_{n=1}^{\infty}3(1.02)^{n-1}=\frac{3}{1-1.02}$	$\displaystyle\sum_{n=1}^{\infty}3(1.02)^{n-1}$ is not a finite sum since $r = 1.02 > 1$

EXERCISES 10.5

Find the sum, if it exists, of each infinite geometric series.

1. $2 + 1 + \frac{1}{2} + \cdots$ **2.** $8 + 4 + 2 + \cdots$

3. $25 + 5 + 1 + \cdots$ **4.** $1 + \frac{4}{3} + \frac{16}{9} + \cdots$

5. $1 - \frac{1}{2} + \frac{1}{4} - \cdots$ **6.** $100 - 1 + \frac{1}{100} - \cdots$

7. $1 + 0.1 + 0.01 + \cdots$ **8.** $52 + 0.52 + 0.0052 + \cdots$

9. $-2 - \frac{1}{4} - \frac{1}{32} - \cdots$ **10.** $-729 + 81 - 9 + \cdots$

Decide whether or not the given infinite geometric series has a sum. If it does, find it by using $S_{\infty} = \dfrac{a}{1-r}$.

11. $\displaystyle\sum_{k=1}^{\infty}\left(\tfrac{1}{3}\right)^{k-1}$ **12.** $\displaystyle\sum_{k=1}^{\infty}\left(\tfrac{1}{3}\right)^{k}$ **13.** $\displaystyle\sum_{k=1}^{\infty}\left(\tfrac{1}{3}\right)^{k+1}$

14. $\displaystyle\sum_{n=1}^{\infty}\frac{1}{2^{n+1}}$ **15.** $\displaystyle\sum_{n=1}^{\infty}\frac{1}{2^{n-2}}$ **16.** $\displaystyle\sum_{k=1}^{\infty}\left(\tfrac{1}{10}\right)^{k-1}$

17. $\displaystyle\sum_{k=1}^{\infty}2(0.1)^{k-1}$ **18.** $\displaystyle\sum_{k=1}^{\infty}\left(-\tfrac{1}{2}\right)^{k-1}$ **19.** $\displaystyle\sum_{n=1}^{\infty}\left(\tfrac{3}{2}\right)^{n-1}$

20. $\displaystyle\sum_{n=1}^{\infty}\left(-\tfrac{1}{3}\right)^{n+2}$ **21.** $\displaystyle\sum_{k=1}^{\infty}(0.7)^{k-1}$ **22.** $\displaystyle\sum_{k=1}^{\infty}5(0.7)^{k}$

23. $\displaystyle\sum_{k=1}^{\infty}5(1.01)^{k}$ **24.** $\displaystyle\sum_{k=1}^{\infty}\left(\tfrac{1}{10}\right)^{k-4}$ **25.** $\displaystyle\sum_{k=1}^{\infty}10\left(\tfrac{2}{3}\right)^{k-1}$

26. $\displaystyle\sum_{k=1}^{\infty}(-1)^{k}$ **27.** $\displaystyle\sum_{k=1}^{\infty}(0.45)^{k-1}$ **28.** $\displaystyle\sum_{k=1}^{\infty}(-0.9)^{k+1}$

29. $\displaystyle\sum_{n=1}^{\infty}7\left(-\tfrac{3}{4}\right)^{n-1}$ **30.** $\displaystyle\sum_{k=1}^{\infty}(0.1)^{2k}$ **31.** $\displaystyle\sum_{k=1}^{\infty}\left(-\tfrac{2}{5}\right)^{2k}$

Find a rational form for each of the following repeating decimals in a manner similar to that in Example 4. Check your answers.

32. $0.444\ldots$ **33.** $0.777\ldots$ **34.** $7.777\ldots$

35. $0.131313\ldots$ **36.** $13.131313\ldots$ **37.** $0.0131313\ldots$

38. $0.050505\ldots$ **39.** $0.999\ldots$ **40.** $0.125125125\ldots$

*41. A certain ball always rebounds one-third the distance it falls. If the ball is dropped from a height of 9 feet, how far does it travel before coming to rest? (See the similar situation at the beginning of Section 10.4.)

*42. After it is set in motion, each swing of a particular pendulum is 40% as long as the preceding swing. What is the total distance that the end of the pendulum travels, before coming to rest, if the first swing is 30 inches long?

10.6

Mathematical Induction (optional)

Study these statements:

$$1 = 1^2$$
$$1 + 3 = 2^2$$
$$1 + 3 + 5 = 3^2$$
$$1 + 3 + 5 + 7 = 4^2$$
$$1 + 3 + 5 + 7 + 9 = 5^2$$

Do you see the pattern? The last statement shows that the sum of the first five positive odd integers is 5^2. What about the sum of the first six positive odd integers? The pattern is the same:

$$1 + 3 + 5 + 7 + 9 + 11 = 6^2$$

It would be reasonable to guess that the sum of the first n positive odd integers is n^2. That is,

$$1 + 3 + 5 + \cdots + (2n - 1) = n^2$$

But a guess is not a proof. It is our objective in this section to learn how to prove a statement that involves an infinite number of cases.

Let us refer to the nth statement above by S_n. Thus S_1, S_2, S_3, S_4, S_5, and S_6 are the first six cases of S_n that we know are true.

Does the truth of the first six cases allow us to conclude that S_n is true for all positive integer n? No! We cannot assume that a few special cases guarantee the general case which includes an infinite number of cases. If we allowed "proving by a finite number of cases," then the following example is such a "proof" that all positive even integers are less than 100.

The first positive even integer is 2, and we know that $2 < 100$. The second is 4, and we know that $4 < 100$. The third is 6, and $6 < 100$. Therefore, since $2n < 100$ for a finite number of cases, we conclude that $2n < 100$ for all n.

This false result should convince you that in trying to prove a collection of statements S_n for all positive integers $n = 1, 2, 3, \ldots$, we need to do more than just check it out for a finite number of cases. We need to call on a type of proof known as **mathematical induction**.

Suppose we had a long (endless) row of dominoes each 2 inches long all standing up in a straight row so that the distance between any two of them is $1\frac{1}{2}$ inches. How can you make them all fall down with the least effort?

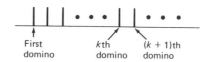

The answer is obvious. Push the first domino down toward the second. Since the first one must fall, and because the space between each pair is less than the length of a domino, they will all (eventually) fall down. The first knocks down the second, the second knocks down the third, and, in general, the kth domino knocks down the $(k + 1)$st. Two things guaranteed this "chain reaction":

1. The first domino fell.

2. If any domino falls, then so will the next.

These two conditions are guidelines in forming the principle of mathematical induction.

The principle of mathematical induction

Let S_n be a statement for each positive integer n. Suppose the following two conditions hold:

1. S_1 is true.

2. If S_k is true, then S_{k+1} is true, where k is any positive integer.

Then S_n is a true statement for all n.

Note that we are not proving this principle; rather, it becomes a basic principle that we accept and use to construct proofs. Condition 1 starts the "chain reaction" and condition 2 keeps it going. It is very important to realize that in condition 2 we are not proving S_k to be true; rather, we must prove this proposition:

If S_k is true, then S_{k+1} is true.

Consequently, a proof by mathematical induction *includes* a proof of the proposition that S_k implies S_{k+1}, a proof within a proof.

Example 1 Prove by mathematical induction that S_n is true for all positive integers n, where S_n is the statement

$$1 + 3 + 5 + \cdots + (2n - 1) = n^2$$

Proof 1. S_1 is true because $1 = 1^2$.
  2. Suppose that S_k is true, where k is a positive integer. Then we have

$$1 + 3 + 5 + \cdots + (2k - 1) = k^2$$

Note: We are *assuming* that S_k is true since we are trying to show that S_k implies S_{k+1}. By assuming S_k to be true, we have not *proved* S_k to be true.
  The next odd number after $2k - 1$ is $2(k + 1) - 1 = 2k + 1$. Let us add this to the preceding equation:

$$
\begin{array}{r}
1 + 3 + 5 + \cdots + (2k - 1) \qquad\qquad = k^2 \\
2k + 1 = 2k + 1 \\
\hline
1 + 3 + 5 + \cdots + (2k - 1) + (2k + 1) = k^2 + 2k + 1
\end{array}
$$

Factor the right side.

$$1 + 3 + 5 + \cdots + (2k + 1) = (k + 1)^2$$

This is the statement S_{k+1}. Therefore we have shown that if S_k is given, then S_{k+1} must follow. This, together with the fact that S_1 is true, allows us to say that S_n is true for all n by the principle of mathematical induction.

Example 2 Prove that $\underbrace{1^2 + 2^2 + 3^2 + \cdots + n^2}_{\text{(sum of first } n \text{ squares)}} = \dfrac{n(n + 1)(2n + 1)}{6}$.

Proof Let S_n be the statement

$$1^2 + 2^2 + 3^2 + \cdots + n^2 = \frac{n(n + 1)(2n + 1)}{6}$$

for any positive integer n.
  1. S_1 is true because $1^2 = \dfrac{1(1 + 1)(2 \cdot 1 + 1)}{6}$.
  2. Suppose that S_k is true for any k. Then we have

$$1^2 + 2^2 + 3^2 + \cdots + k^2 = \frac{k(k + 1)(2k + 1)}{6}$$

Add the next square $(k + 1)^2$ to both sides.

(A) $1^2 + 2^2 + 3^2 + \cdots + k^2 + (k+1)^2 = \dfrac{k(k+1)(2k+1)}{6} + (k+1)^2$

Combine the right side.

$$\frac{k(k+1)(2k+1)}{6} + (k+1)^2 = \frac{k(k+1)(2k+1) + 6(k+1)^2}{6}$$

$$= \frac{2k^3 + 9k^2 + 13k + 6}{6}$$

$$= \frac{(k+1)(k+2)(2k+3)}{6}$$

Substitute back into Equation (A).

$$1^2 + 2^2 + 3^2 + \cdots + (k+1)^2 = \frac{(k+1)(k+2)(2k+3)}{6}$$

To see that this is S_{k+1}, rewrite the right side:

$$1^2 + 2^2 + 3^2 + \cdots + (k+1)^2 = \frac{(k+1)[(k+1)+1][2(k+1)+1]}{6}$$

Now both conditions of the principle of mathematical induction have been satisfied, and it follows that

$$1^2 + 2^2 + 3^2 + \cdots + n^2 = \frac{n(n+1)(2n+1)}{6}$$

is true for all integers $n \geq 1$.

Example 3 Let $t > 0$. Then use mathematical induction to prove that $(1+t)^n > 1 + nt$ for all positive integers $n \geq 2$.

Proof Let S_n be the statement $(1+t)^n > 1 + nt$, where $t > 0$ and n is any integer where $n \geq 2$.
1. For $n = 2$, $(1+t)^2 = 1 + 2t + t^2$. Since $t^2 > 0$, we get

$$1 + 2t + t^2 > 1 + 2t$$

because $(1 + 2t + t^2) - (1 + 2t) = 1 + t^2 > 0$. (Recall that $a > b$ if and only if $a - b > 0$.)
2. Suppose that for $k \geq 2$ we have $(1+t)^k > 1 + kt$. Multiply both sides by $(1+t)$.

$$(1+t)^k(1+t) > (1+kt)(1+t)$$

Then

$$(1 + t)^{k+1} > 1 + (k + 1)t + kt^2$$

But $1 + (k + 1)t + kt^2 > 1 + (k + 1)t$. Therefore, by transitivity of $>$,

$$(1 + t)^{k+1} > 1 + (k + 1)t$$

By (1) and (2) above, the principle of mathematical induction implies that $(1 + t)^n > 1 + nt$ for all integers $n \geq 2$.

EXERCISES 10.6

Use mathematical induction to prove the following statements for all positive integers n.

1. $1 + 2 + 3 + \cdots + n = \dfrac{n(n + 1)}{2}$

2. $2 + 4 + 6 + \cdots + 2n = n(n + 1)$

3. $\displaystyle\sum_{i=1}^{n} 3i = \dfrac{3n(n + 1)}{2}$

4. $1 + 4 + 7 + \cdots + (3n - 2) = \dfrac{n(3n - 1)}{2}$

5. $\dfrac{5}{3} + \dfrac{4}{3} + 1 + \cdots + \left(-\dfrac{1}{3}n + 2\right) = \dfrac{n(11 - n)}{6}$

6. $1 \cdot 2 + 2 \cdot 3 + 3 \cdot 4 + \cdots + n(n + 1) = \dfrac{n(n + 1)(n + 2)}{3}$

7. $\dfrac{1}{1 \cdot 2} + \dfrac{1}{2 \cdot 3} + \dfrac{1}{3 \cdot 4} + \cdots + \dfrac{1}{n(n + 1)} = \dfrac{n}{n + 1}$

8. $3 + 3^2 + 3^3 + \cdots + 3^n = \dfrac{3^{n+1} - 3}{2}$

9. $a^n < 1$, where $0 < a < 1$.

10. Let $0 < a < 1$. Use mathematical induction to prove that $a^n < a$ for all integers $n \geq 2$.

11. Let a and b be real numbers. Then use mathematical induction to prove that $(ab)^n = a^n b^n$ for all positive integers n.

12. Use mathematical induction to prove the generalized distributive property

$$a(b_0 + b_1 + \cdots + b_n) = ab_0 + ab_1 + \cdots + ab_n$$

for all positive integers n, where a and b_i are real numbers. (Assume that parentheses may be inserted into or extracted from an indicated sum of real numbers.)

*13. Give an inductive proof of

$$|a_0 + a_1 + \cdots + a_n| \leq |a_0| + |a_1| + \cdots + |a_n|$$

for all positive integers n, where the a_i are real numbers. (See note for Exercise 12.)

*14. Use induction to prove that if $a_0 a_1 \cdots a_n = 0$, then at least one of the factors is zero, for all positive integers n. (Assume that parentheses may be inserted into or extracted from an indicated product of real numbers.)

*15. Give an inductive proof of

$$1 + \frac{1}{2} + \frac{1}{4} + \cdots + \frac{1}{2^{n-1}} = 2\left(1 - \frac{1}{2^n}\right)$$

for all integers $n \geq 1$.

*16. Use induction to prove that $\sum_{i=1}^{n} r^{i-1} = \frac{1 - r^n}{1 - r}$ for all integers $n \geq 1$.

*17. (a) Prove by induction that

$$\frac{a^n - b^n}{a - b} = a^{n-1} + a^{n-2}b + \cdots + ab^{n-2} + b^{n-1}$$

for all integers $n \geq 2$. *Hint:* Consider

$$\frac{a^{n+1} - b^{n+1}}{a - b} = \frac{a^n a - b^n b}{a - b} = \frac{a^n a - b^n a + b^n a - b^n b}{a - b}$$

(b) How does the result in (a) give the factorization of the difference of two nth powers?

10.7

Permutations and Combinations (optional)

Suppose there are 30 students in your class, it is the first day of the term, and you all decide to get acquainted by shaking hands. Each person shakes hands with every other person. How many handshakes take place? Obviously, this is not an important mathematical question in itself. It suggests, however, that our elementary counting procedures may be too limited. We need to learn some new and more advanced ways to count. In this section, a few new methods will be studied.

Fundamental principle of counting

If a certain task can be completed in m_1 ways, a second task in m_2 ways, a third task in m_3 ways, and so forth, then the total number of ways in which these tasks can be completed together is

$$m_1 \cdot m_2 \cdot m_3 \ldots$$

Suppose you are planning a trip that will consist of visiting cities A, B, and C. How many different trips are possible? The trip begins with a stop at any one of the three cities, the second stop will be at any one of the remaining two cities, and the trip is completed by stopping at the remaining city. According to the fundamental principle there are $3 \cdot 2 \cdot 1 = 6$ trips possible. These six cases can be displayed as follows:

$$ABC \qquad BAC \qquad CAB$$
$$ACB \qquad BCA \qquad CBA$$

The arrangement ABC means that the trip begins with city A, goes to B, and ends at C. The other arrangements have similar interpretations. The order of the elements A, B, C in ABC is crucial here, and we say that each of the six arrangements is a *permutation* of three objects taken three at a time. In general, for n elements (n a positive integer) and r an integer satisfying $0 \leq r \leq n$, we have this definition.

Permutation

A *permutation* of n elements taken r at a time is an arrangement, without repetitions, of r of the n elements.

In symbols, $_nP_r$ is read as *the number of permutations of n elements taken r at a time.*

We could have avoided the use of the fundamental principle in the preceding problem by simply listing all the possible cases. However, the fundamental principle is very useful when the numbers involved are large and such listings become impractical.

Example 1 How many three-letter "words" from the 26 letters of the alphabet are possible? No duplication of letters is permitted.

Solution The first letter may be any one of the 26, the second may be any one of the remaining 25, and the third may be chosen from the remaining 24. Thus the total number of different "words" is $26 \cdot 25 \cdot 24 = 15{,}600$. Since we have taken 3 elements out of 26, without repetitions and in all possible orders, we may say that there are 15,600 permutations; that is, $_{26}P_3 = 26 \cdot 25 \cdot 24 = 15{,}600$.

Note that $_{26}P_3 = 26 \cdot 25 \cdot 24$ has three factors beginning with 26 and with each successive factor decreasing by 1. In general, for $_nP_r$ there will be r factors beginning with n, as shown at the top of the next page.

$$_nP_r = n(n-1)(n-2)(n-3) \cdots [n-(r-1)]$$
$$= n(n-1)(n-2)(n-3) \cdots (n-r+1)$$

A specific application of this formula occurs when $n = r$. In this case we have the permutation of n elements taken n at a time, and the product has n factors.

$$_nP_r = n(n-1)(n-2)(n-3) \cdots 3 \cdot 2 \cdot 1$$

We may abbreviate this formula by using **factorial notation**.

$$_nP_n = n!$$

For example:

$$_3P_3 = 3! = 3 \cdot 2 \cdot 1 = 6$$
$$_4P_4 = 4! = 4 \cdot 3 \cdot 2 \cdot 1 = 24$$
$$_5P_5 = 5! = 5 \cdot 4 \cdot 3 \cdot 2 \cdot 1 = 120$$

For future consistency we find it convenient to define 0! as equal to 1.

Example 2 How many different ways can the four letters of the word MATH be arranged?

Solution Here we have the permutations of four elements taken four at a time. Thus $_4P_4 = 4! = 4 \cdot 3 \cdot 2 \cdot 1 = 24$. This includes such arrangements as MATH, AMTH, TMAH, and HMAT. Can you list all 24 possibilities?

You have seen that $n!$ consists of n factors, beginning with n and successively decreasing to 1. At times it will be useful to display only some of the specific factors in $n!$ as follows.

$$n! = n(n-1)! = n(n-1)(n-2)! = n(n-1)(n-2)(n-3)!$$
$$= n(n-1)(n-2) \cdots 2 \cdot 1$$

In particular,

$$5! = 5 \cdot 4! = 5 \cdot 4 \cdot 3! = 5 \cdot 4 \cdot 3 \cdot 2! = 5 \cdot 4 \cdot 3 \cdot 2 \cdot 1$$

We can use the formula for $_nP_n$ to obtain a different form for $_nP_r$. To do so we write the formula for $_nP_r$ and then multiply numerator and denominator by $(n-r)!$ as follows.

$$_nP_r = \frac{n(n-1)(n-2) \cdots [n-(r-1)]}{1} \cdot \frac{(n-r)!}{(n-r)!}$$

Permutation of n elements taken r at a time

$$_nP_r = \frac{n!}{(n-r)!}$$

Example 3   Evaluate $_7P_4$ by using each of the formulas for $_nP_r$.

Solution

(a) Using $_nP_r = n(n-1)(n-2) \cdots (n-r+1)$ we have

$$_7P_4 = 7 \cdot 6 \cdot 5 \cdot 4 = 840$$

(b) Using $_nP_r = \dfrac{n!}{(n-r)!}$ we have

$$_7P_4 = \frac{7!}{3!} = \frac{7 \cdot 6 \cdot 5 \cdot 4 \cdot \cancel{3!}}{\cancel{3!}} = 840$$

Example 4   A class contains 10 members. They wish to elect officers consisting of a president, vice-president, and secretary-treasurer. How many sets of officers are possible?

Solution   We may think of the three offices to be filled in terms of a first office (president), a second office (vice-president), and a third office (secretary-treasurer). Thus we need to select 3 out of 10 members and arrange them in all possible orders; we need to find the permutations of 10 elements taken 3 at a time.

$$_{10}P_3 = \frac{10!}{7!} = \frac{10 \cdot 9 \cdot 8 \cdot \cancel{7!}}{\cancel{7!}} = 720$$

A permutation may be regarded as an ordered collection of objects. We now consider situations when the order of the objects is not required. In general, letting n be a positive integer and r be any integer satisfying $0 \le r \le n$, we define a *combination* as follows.

Combination

A *combination* is a collection of r distinct elements selected out of n elements without regard to order.

Suppose that the class in Example 4 merely wished to elect a committee of three members. Then a committee consisting of members David, Ellen, and Robert is the same regardless of the order in which they are selected. In other words, using D, E, and R as abbreviations for their names, each of the following permutations is the same combination.

DER   DRE   EDR   ERD   RDE   RED

Each combination of three members in this illustration actually gives rise to $3! = 6$ permutations. To find the number of possible committees we need to find the combinations of 10 elements taken 3 at a time. The *number* of such combination is expressed as $_{10}C_3$ and is read as "the number of combinations of 10 elements taken 3 at a time." Since each of these combinations produces $3! = 6$ permutations, it follows that

$$3!(_{10}C_3) = {}_{10}P_3 \qquad \text{or} \qquad {}_{10}C_3 = \frac{_{10}P_3}{3!}$$

Using $_nP_r = \dfrac{n!}{(n-r)!}$, we obtain these general forms for $_nC_r$.

Combination of n elements taken r at a time

$$_nC_r = \frac{_nP_r}{r!} = \frac{n!}{r!(n-r)!}$$

Frequently the symbol $\dbinom{n}{r}$ is used in place of $_nC_r$. Note that $\dbinom{n}{0} = \dfrac{n!}{0!n!} = 1$; also $\dbinom{n}{n} = \dfrac{n!}{n!0!} = 1$. Can you prove that $\dbinom{n}{r} = \dbinom{n}{n-r}$?

Example 5   Evaluate $\dbinom{10}{2}$.

Solution

$$\binom{10}{2} = \frac{10!}{2!8!} = \frac{10 \cdot 9 \cdot \cancel{8!}}{2 \cdot \cancel{8!}} = 45$$

The question of *order* is the essential ingredient that determines whether a problem involves permutations or combinations. The following examples illustrate this distinction.

Example 6   Using the digits 1 through 9, how many different four-digit numbers can be formed if repetition of digits is not allowed?

Solution   Order is important here; thus 4923 is a different number from 9432. Therefore we need to find the *permutations* of nine elements taken four at a time.

$$_9P_4 = \frac{9!}{5!} = 9 \cdot 8 \cdot 7 \cdot 6 = 3024$$

Example 7   A student has a penny, nickel, dime, quarter, and half-dollar and wishes to leave a tip consisting of exactly two coins. How many different amounts as tips are possible?

Solution   Order is not important here; a tip of 10¢ + 25¢ is the same as one of 25¢ + 10¢. Therefore we need to find the *combinations* of five things taken two at a time.

$$\binom{5}{2} = {_5C_2} = \frac{5!}{2!3!} = \frac{5 \cdot 4}{2} = 10$$

List all 10 possibilities.

Consider this triangular array of numbers of the form $\binom{n}{r}$.

$$\binom{1}{0} \quad \binom{1}{1}$$

$$\binom{2}{0} \quad \binom{2}{1} \quad \binom{2}{2}$$

$$\binom{3}{0} \quad \binom{3}{1} \quad \binom{3}{2} \quad \binom{3}{3}$$

$$\binom{4}{0} \quad \binom{4}{1} \quad \binom{4}{2} \quad \binom{4}{3} \quad \binom{4}{4}$$

$$\binom{5}{0} \quad \binom{5}{1} \quad \binom{5}{2} \quad \binom{5}{3} \quad \binom{5}{4} \quad \binom{5}{5}$$

Evaluating the entries converts the triangular array into this form.

```
        1    1
      1    2    1
    1    3    3    1
  1    4    6    4    1
1    5   10   10    5    1
```

These are the first five rows of **Pascal's triangle**. (See Section 2.8, Exercise 23 through 27.) After the first row the entries in the $(n + 1)$st row are generated by the numbers in the nth row. For example, note the entries circled in the preceding array: $4 + 6 = 10$. This suggests that $\binom{4}{1} + \binom{4}{2} = \binom{5}{2}$. (You can carry out the computations to verify this result.) In general, it can be shown that $\binom{n}{r-1} + \binom{n}{r} = \binom{n+1}{r}$. The proof of this identity would justify our earlier construction of Pascal's triangle.

The nth row of Pascal's triangle is also known to contain the coefficients of the expansion of $(a + b)^n$. Thus the binomial formula can now be written in terms of the binomial coefficients $\binom{n}{r}$ as follows.

$$(a + b)^n =$$

$$\binom{n}{0}a^n + \binom{n}{1}a^{n-1}b + \binom{n}{2}a^{n-2}b^2 + \cdots + \binom{n}{r}a^{n-r}b^r + \cdots + \binom{n}{n-1}ab^{n-1} + \binom{n}{n}b^n$$

Using the sigma notation we have

$$(a + b)^n = \sum_{k=0}^{n} \binom{n}{k}a^{n-k}b^k$$

EXERCISES 10.7

Evaluate.

1. $\dfrac{7!}{6!}$ 2. $\dfrac{12!}{10!}$ 3. $\dfrac{12!}{2!10!}$ 4. $\dfrac{15!}{10!5!}$

5. $_5P_4$ 6. $_5P_5$ 7. $_4P_1$ 8. $_8P_5$

9. $_5C_2$ 10. $_{10}C_1$ 11. $_{10}C_0$ 12. $_4C_3$

13. $\binom{15}{15}$ 14. $\binom{30}{3}$ 15. $\binom{30}{27}$ 16. $\binom{n}{3}$

17. How many three-digit numbers can be formed from the digits 1 through 9
    (a) If repetition of digits is not allowed? (b) If repetition of digits is allowed?

18. In how many different ways can the letters of STUDY be arranged?

19. A class consists of 20 members. In how many different ways can the class select
    (a) A committee of four? (b) A set of four officers?

20. On a test a student must select 8 questions out of a total of 10. In how many different ways can this be done?

21. How many different ways can six people be lined up for a picture?

22. How many straight lines are determined by five points, no three of which are collinear?

23. There are 15 women on a basketball team. In how many different ways can a coach field a team of 5 players?

24. Answer Exercise 23 if two of the players can only play center and the others can play any of the remaining positions.

25. Answer the question stated at the beginning of Section 10.7: How many handshakes take place when each person in a group of 30 shakes hands with every other person?

You may use a calculator to complete the computations in Exercises 26 through 28.

26. A baseball team consists of nine players. How many different batting orders are possible? How many if the pitcher bats last?

27. How many different ways can 5 cards be selected from a deck of 52 cards?

28. (a) Each question in a multiple-choice exam has the four choices indicated by the letters *a*, *b*, *c*, and *d*. If there are eight questions, how many ways can the test be answered? (*Hint:* Use the fundamental counting principle.)
    (b) How many ways are there if no two consecutive answers can have the same answer?

29. When people are seated at a circular table, we consider only their positions relative to each other and are not concerned with the particular seat that a person occupies. How many arrangements are there for seven people to seat themselves around a circular table? (*Hint:* Consider one person's position as fixed.)

30. Review Exercise 29 and conjecture a formula for the number of different circular permutations of *n* distinct objects placed around a circle.

31. Solve for *n*:   (a) $_nP_1 = 10$;   (b) $_nC_2 = 6$.

32. Express as a simplified fraction: (a) $_nP_{n-1}$;   (b) $_nC_{n-1}$.

33. Prove: $\binom{n}{r} = \binom{n}{n-r}$.

*34. Evaluate $_nC_4$ given that $_nP_4 = 1680$.

*35. Solve for *n*: $35\binom{n-3}{3} = 4\binom{n}{3}$.

*36. A class consists of 12 women and 10 men. A committee is to be selected consisting of 3 women and 2 men. How many different committees are possible?

37. Consider this expression: $\binom{n}{0} + \binom{n}{1} + \binom{n}{2} + \cdots + \binom{n}{n-1} + \binom{n}{n}$. Evaluate for (a) $n = 2$;   (b) $n = 3$;   (c) $n = 4$; and (d) $n = 5$.

*38. Use the results of Exercise 37 and conjecture the value of the expression for any positive integer *n*.

*39. How many ways can 5-card hands be selected out of 52 cards so that all 5 cards are in the same suit?

*40. How many ways can 5-card hands be selected out of 52 cards such that 4 of the cards have the same face value? (Four 10's and some fifth card is one such hand.)

*41. How many ways can 5-card hands be selected out of 52 cards so that the 5 cards consist of a pair and three of a kind? (Two kings and three 7's is one such hand.)

*42. Prove: $\binom{n}{r-1} + \binom{n}{r} = \binom{n+1}{r}$.

REVIEW EXERCISES FOR CHAPTER 10

The solutions to the following exercises can be found within the text of Chapter 10. Try to answer each question without referring to the text.

Section 10.1

1. Find the range values of the sequence $s_n = \dfrac{1}{n}$ for the domain $n = 1, 2, 3, 4, 5$, and graph.

2. List the first seven terms of the sequence $s_k = \dfrac{(-1)^k}{k^2}$.

3. Write the first four terms of the sequence $s_n = \dfrac{1}{(n-1)(n-2)}$.

4. Find the tenth term of the sequence $-\dfrac{3}{2}, -4, 6, \ldots, \dfrac{n+2}{n-3}, \ldots$.

Section 10.2

5. Find the sum of the first seven terms of $s_k = 2k$.

6. Find $\displaystyle\sum_{x=1}^{8} (2k + 1)$.

7. Find $\displaystyle\sum_{i=1}^{5} (-1)^i(i + 1)$.

Section 10.3

8. What is the nth term of an arithmetic sequence whose first term is a and whose common difference is d?

9. Write the formula for the sum S_n of the arithmetic sequence $s_k = a + (k - 1)d$.

10. Find $\displaystyle\sum_{k=1}^{50} (-6k + 10)$.

11. Find the sum of the first n positive integers.

Section 10.4

12. What is a geometric sequence?

13. Find the hundredth term of the geometric sequence having $r = \frac{1}{2}$ and $a = \frac{1}{2}$.

14. Find the nth term of the geometric sequence beginning with $6, 9, \frac{27}{2}, \ldots$.

15. Write the kth term of the geometric sequence $s_k = (\frac{1}{2})^{2k}$ in the form ar^{k-1} and find the values of a and r.

16. Write the formula for the sum S_n of a geometric sequence $s_k = ar^{k-1}$.

17. Evaluate $\displaystyle\sum_{k=1}^{5} 6(\frac{1}{2})^{k-1}$.

18. Evaluate $\displaystyle\sum_{k=1}^{8} 3\left(\frac{1}{10}\right)^{k+1}$.

19. Suppose that you save \$128 in January and that each month thereafter you only manage to save half of what you saved the previous month. How much do you save in the tenth month? What is your total savings after 10 months?

Section 10.5

20. For which values of r do we have $\sum\limits_{k=1}^{\infty} ar^{k-1} = \dfrac{a}{1-r}$?

21. Find the sum of the infinite geometric series $27 + 3 + \frac{1}{3} + \cdots$.

22. Evaluate $\sum\limits_{k=1}^{\infty} \dfrac{7}{10^{k+1}}$.

23. Express the repeating decimal $0.242424\ldots$ in rational form (the ratio of two integers).

Section 10.6

24. Prove by mathematical induction:

$$1 + 3 + 5 + \cdots + (2n - 1) = n^2$$

25. Use mathematical induction to prove:

$$1^2 + 2^2 + 3^2 + \cdots + n^2 = \dfrac{n(n+1)(2n+1)}{6}$$

Section 10.7

26. How many three-letter 'words,' with no duplication of letters, can be formed from the 26 letters of the alphabet?

27. A class contains 10 members. They wish to elect officers consisting of a president, vice-president, and secretary-treasurer. How many sets of officers are possible?

28. A student has a penny, nickel, dime, quarter, and half-dollar and wishes to leave a tip consisting of exactly two coins. How many different amounts of tips are possible?

SAMPLE TEST FOR CHAPTER 10

1. (a) Find the first three terms of the sequence given by $s_n = \dfrac{n^2}{6-n}$.

(b) Find the tenth term of the sequence in part (a).

2. An arithmetic sequence has $a = -3$ and $d = \frac{1}{2}$.
(a) Find the forty-ninth term.
(b) What is the sum of the first 20 terms?

3. (a) Write the next three terms of the geometric sequence beginning $-768, 192, -48, \ldots$.
(b) Write the nth term of the sequence in part (a).

4. Use the formula for the sum of a finite geometric sequence to show that

$$\sum_{k=1}^{4} 8(\tfrac{1}{2})^k = \tfrac{15}{2}$$

5. Find $\sum\limits_{j=1}^{101} (4j - 50)$.

Decide whether each of the given infinite geometric series in Exercises 6 and 7 has a sum. Find the sum if it exists; otherwise, give a reason why there is no sum.

6. $\sum\limits_{k=1}^{\infty} 8(\frac{3}{4})^{k+1}$ **7.** $1 + \frac{3}{2} + \frac{9}{4} + \cdots$

8. Change the repeating decimal $0.363636\ldots$ into rational form.

9. Suppose you save \$10 one week and that each week thereafter you save 10¢ more than the week before. How much will you have saved after 1 year?

10. An object is moving along a straight line such that each minute it travels one-third as far as it did during the preceding minute. How far will the object have moved, before coming to rest, if it moves 24 feet during the first minute?

†11. Prove by mathematical induction:

$$5 + 10 + 15 + \cdots + 5n = \frac{5n(n+1)}{2}$$

†12. A basketball squad has 9 members, two of which can only play the center position. If each of the other members can play any position except center, then how many 5 player teams can the coach form?

Answers To The Test Your Understanding Exercises

Page 432

1. $5n$ **4.** $-8n + 3$ **7.** $\frac{2}{3}n$ **9.** $\frac{1}{5}n - \frac{1}{5}$

2. $-4n + 10$ **5.** n **8.** $-12n + 65$ **10.** $n + 1$

3. $\frac{1}{10}n$ **6.** $n - 4$

Page 440

1. $1, \frac{1}{2}, \frac{1}{4}, \frac{1}{8}, \frac{1}{16}; 1(\frac{1}{2})^{n-1}; r = \frac{1}{2}$ **4.** $-\frac{1}{27}, \frac{1}{27^2}, -\frac{1}{27^3}, \frac{1}{27^4}, -\frac{1}{27^5}; -\frac{1}{27}(-\frac{1}{27})^{n-1};$

2. $\frac{1}{4}, \frac{1}{8}, \frac{1}{16}, \frac{1}{32}, \frac{1}{64}; \frac{1}{4}(\frac{1}{2})^{n-1}; r = \frac{1}{2}$ $r = -\frac{1}{27}$

3. $-\frac{1}{2}, \frac{1}{4}, -\frac{1}{8}, \frac{1}{16}, -\frac{1}{32}; -\frac{1}{2}(-\frac{1}{2})^{n-1}; r = -\frac{1}{2}$ **5.** $r = 10; \frac{1}{5}(10)^{n-1}$

6. $r = -\frac{4}{9}; 27(-\frac{4}{9})^{n-1}$

Page 448

1. $r = \frac{1}{10}; S_{\infty} = 11\frac{1}{9}$ **4.** $r = \frac{1}{4}; S_{\infty} = -21\frac{1}{3}$ **7.** $r = -3$; no finite sum.

2. $r = 4$; no finite sum.   **5.** $r = \frac{4}{3}$; no finite sum.   **8.** $r = -\frac{9}{10}; S_{\infty} = \frac{810}{19}$

3. $r = -\frac{1}{6}; S_{\infty} = 30\frac{6}{7}$ **6.** $r = 0.01; S_{\infty} = \frac{1}{33}$

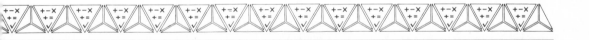

Tables

Table I: Square Roots and Cube Roots

N	$\sqrt{N}$	$\sqrt[3]{N}$	N	$\sqrt{N}$	$\sqrt[3]{N}$	N	$\sqrt{N}$	$\sqrt[3]{N}$	N	$\sqrt{N}$	$\sqrt[3]{N}$
1	1.000	1.000	51	7.141	3.708	101	10.050	4.657	151	12.288	5.325
2	1.414	1.260	52	7.211	3.733	102	10.100	4.672	152	12.329	5.337
3	1.732	1.442	53	7.280	3.756	103	10.149	4.688	153	12.369	5.348
4	2.000	1.587	54	7.348	3.780	104	10.198	4.703	154	12.410	5.360
5	2.236	1.710	55	7.416	3.803	105	10.247	4.718	155	12.450	5.372
6	2.449	1.817	56	7.483	3.826	106	10.296	4.733	156	12.490	5.383
7	2.646	1.913	57	7.550	3.849	107	10.344	4.747	157	12.530	5.395
8	2.828	2.000	58	7.616	3.871	108	10.392	4.762	158	12.570	5.406
9	3.000	2.080	59	7.681	3.893	109	10.440	4.777	159	12.610	5.418
10	3.162	2.154	60	7.746	3.915	110	10.488	4.791	160	12.649	5.429
11	3.317	2.224	61	7.810	3.936	111	10.536	4.806	161	12.689	5.440
12	3.464	2.289	62	7.874	3.958	112	10.583	4.820	162	12.728	5.451
13	3.606	2.351	63	7.937	3.979	113	10.630	4.835	163	12.767	5.463
14	3.742	2.410	64	8.000	4.000	114	10.677	4.849	164	12.806	5.474
15	3.873	2.466	65	8.062	4.021	115	10.724	4.863	165	12.845	5.485
16	4.000	2.520	66	8.124	4.041	116	10.770	4.877	166	12.884	5.496
17	4.123	2.571	67	8.185	4.062	117	10.817	4.891	167	12.923	5.507
18	4.243	2.621	68	8.246	4.082	118	10.863	4.905	168	12.961	5.518
19	4.359	2.668	69	8.307	4.102	119	10.909	4.919	169	13.000	5.529
20	4.472	2.714	70	8.367	4.121	120	10.954	4.932	170	13.038	5.540
21	4.583	2.759	71	8.426	4.141	121	11.000	4.946	171	13.077	5.550
22	4.690	2.802	72	8.485	4.160	122	11.045	4.960	172	13.115	5.561
23	4.796	2.844	73	8.544	4.179	123	11.091	4.973	173	13.153	5.572
24	4.899	2.884	74	8.602	4.198	124	11.136	4.987	174	13.191	5.583
25	5.000	2.924	75	8.660	4.217	125	11.180	5.000	175	13.229	5.593
26	5.099	2.962	76	8.718	4.236	126	11.225	5.013	176	13.267	5.604
27	5.196	3.000	77	8.775	4.254	127	11.269	5.027	177	13.304	5.615
28	5.292	3.037	78	8.832	4.273	128	11.314	5.040	178	13.342	5.625
29	5.385	3.072	79	8.888	4.291	129	11.358	5.053	179	13.379	5.636
30	5.477	3.107	80	8.944	4.309	130	11.402	5.066	180	13.416	5.646
31	5.568	3.141	81	9.000	4.327	131	11.446	5.079	181	13.454	5.657
32	5.657	3.175	82	9.055	4.344	132	11.489	5.092	182	13.491	5.667
33	5.745	3.208	83	9.110	4.362	133	11.533	5.104	183	13.528	5.677
34	5.831	3.240	84	9.165	4.380	134	11.576	5.117	184	13.565	5.688
35	5.916	3.271	85	9.220	4.397	135	11.619	5.130	185	13.601	5.698
36	6.000	3.302	86	9.274	4.414	136	11.662	5.143	186	13.638	5.708
37	6.083	3.332	87	9.327	4.431	137	11.705	5.155	187	13.675	5.718
38	6.164	3.362	88	9.381	4.448	138	11.747	5.168	188	13.711	5.729
39	6.245	3.391	89	9.434	4.465	139	11.790	5.180	189	13.748	5.739
40	6.325	3.420	90	9.487	4.481	140	11.832	5.192	190	13.784	5.749
41	6.403	3.448	91	9.539	4.498	141	11.874	5.205	191	13.820	5.759
42	6.481	3.476	92	9.592	4.514	142	11.916	5.217	192	13.856	5.769
43	6.557	3.503	93	9.644	4.531	143	11.958	5.229	193	13.892	5.779
44	6.633	3.530	94	9.695	4.547	144	12.000	5.241	194	13.928	5.789
45	6.708	3.557	95	9.747	4.563	145	12.042	5.254	195	13.964	5.799
46	6.782	3.583	96	9.798	4.579	146	12.083	5.266	196	14.000	5.809
47	6.856	3.609	97	9.849	4.595	147	12.124	5.278	197	14.036	5.819
48	6.928	3.634	98	9.899	4.610	148	12.166	5.290	198	14.071	5.828
49	7.000	3.659	99	9.950	4.626	149	12.207	5.301	199	14.107	5.838
50	7.071	3.684	100	10.000	4.642	150	12.247	5.313	200	14.142	5.848

Table II: Exponential Functions

x	e^x	e^{-x}	x	e^x	e^{-x}
0.0	1.00	1.000	3.1	22.2	0.045
0.1	1.11	0.905	3.2	24.5	0.041
0.2	1.22	0.819	3.3	27.1	0.037
0.3	1.35	0.741	3.4	30.0	0.033
0.4	1.49	0.670	3.5	33.1	0.030
0.5	1.65	0.607	3.6	36.6	0.027
0.6	1.82	0.549	3.7	40.4	0.025
0.7	2.01	0.497	3.8	44.7	0.022
0.8	2.23	0.449	3.9	49.4	0.020
0.9	2.46	0.407	4.0	54.6	0.018
1.0	2.72	0.368	4.1	60.3	0.017
1.1	3.00	0.333	4.2	66.7	0.015
1.2	3.32	0.301	4.3	73.7	0.014
1.3	3.67	0.273	4.4	81.5	0.012
1.4	4.06	0.247	4.5	90.0	0.011
1.5	4.48	0.223	4.6	99.5	0.010
1.6	4.95	0.202	4.7	110	0.0091
1.7	5.47	0.183	4.8	122	0.0082
1.8	6.05	0.165	4.9	134	0.0074
1.9	6.69	0.150	5.0	148	0.0067
2.0	7.39	0.135	5.5	245	0.0041
2.1	8.17	0.122	6.0	403	0.0025
2.2	9.02	0.111	6.5	665	0.0015
2.3	9.97	0.100	7.0	1097	0.00091
2.4	11.0	0.091	7.5	1808	0.00055
2.5	12.2	0.082	8.0	2981	0.00034
2.6	13.5	0.074	8.5	4915	0.00020
2.7	14.9	0.067	9.0	8103	0.00012
2.8	16.4	0.061	9.5	13360	0.00075
2.9	18.2	0.055	10.0	22026	0.000045
3.0	20.1	0.050			

Table III: Natural Logarithms (Base *e*)

x	$\ln x$	x	$\ln x$	x	$\ln x$
0.0		3.4	1.224	6.8	1.917
0.1	−2.303	3.5	1.253	6.9	1.932
0.2	−1.609	3.6	1.281	7.0	1.946
0.3	−1.204	3.7	1.308	7.1	1.960
0.4	−0.916	3.8	1.335	7.2	1.974
0.5	−0.693	3.9	1.361	7.3	1.988
0.6	−0.511	4.0	1.386	7.4	2.001
0.7	−0.357	4.1	1.411	7.5	2.015
0.8	−0.223	4.2	1.435	7.6	2.028
0.9	−0.105	4.3	1.459	7.7	2.041
1.0	0.000	4.4	1.482	7.8	2.054
1.1	0.095	4.5	1.504	7.9	2.067
1.2	0.182	4.6	1.526	8.0	2.079
1.3	0.262	4.7	1.548	8.1	2.092
1.4	0.336	4.8	1.569	8.2	2.104
1.5	0.405	4.9	1.589	8.3	2.116
1.6	0.470	5.0	1.609	8.4	2.128
1.7	0.531	5.1	1.629	8.5	2.140
1.8	0.588	5.2	1.649	8.6	2.152
1.9	0.642	5.3	1.668	8.7	2.163
2.0	0.693	5.4	1.686	8.8	2.175
2.1	0.742	5.5	1.705	8.9	2.186
2.2	0.788	5.6	1.723	9.0	2.197
2.3	0.833	5.7	1.740	9.1	2.208
2.4	0.875	5.8	1.758	9.2	2.219
2.5	0.916	5.9	1.775	9.3	2.230
2.6	0.956	6.0	1.792	9.4	2.241
2.7	0.993	6.1	1.808	9.5	2.251
2.8	1.030	6.2	1.825	9.6	2.262
2.9	1.065	6.3	1.841	9.7	2.272
3.0	1.099	6.4	1.856	9.8	2.282
3.1	1.131	6.5	1.872	9.9	2.293
3.2	1.163	6.6	1.887	10.0	2.303
3.3	1.194	6.7	1.902		

Table IV: Four-Place Common Logarithms (Base 10)

N	0	1	2	3	4	5	6	7	8	9
1.0	.0000	.0043	.0086	.0128	.0170	.0212	.0253	.0294	.0334	.0374
1.1	.0414	.0453	.0492	.0531	.0569	.0607	.0645	.0682	.0719	.0755
1.2	.0792	.0828	.0864	.0899	.0934	.0969	.1004	.1038	.1072	.1106
1.3	.1139	.1173	.1206	.1239	.1271	.1303	.1335	.1367	.1399	.1430
1.4	.1461	.1492	.1523	.1553	.1584	.1614	.1644	.1673	.1703	.1732
1.5	.1761	.1790	.1818	.1847	.1875	.1903	.1931	.1959	.1987	.2014
1.6	.2041	.2068	.2095	.2122	.2148	.2175	.2201	.2227	.2253	.2279
1.7	.2304	.2330	.2355	.2380	.2405	.2430	.2455	.2480	.2504	.2529
1.8	.2553	.2577	.2601	.2625	.2648	.2672	.2695	.2718	.2742	.2765
1.9	.2788	.2810	.2833	.2856	.2878	.2900	.2923	.2945	.2967	.2989
2.0	.3010	.3032	.3054	.3075	.3096	.3118	.3139	.3160	.3181	.3201
2.1	.3222	.3243	.3263	.3284	.3304	.3324	.3345	.3365	.3385	.3404
2.2	.3424	.3444	.3464	.3483	.3502	.3522	.3541	.3560	.3579	.3598
2.3	.3617	.3636	.3655	.3674	.3692	.3711	.3729	.3747	.3766	.3784
2.4	.3802	.3820	.3838	.3856	.3874	.3892	.3909	.3927	.3945	.3962
2.5	.3979	.3997	.4014	.4031	.4048	.4065	.4082	.4099	.4116	.4133
2.6	.4150	.4166	.4183	.4200	.4216	.4232	.4249	.4265	.4281	.4298
2.7	.4314	.4330	.4346	.4362	.4378	.4393	.4409	.4425	.4440	.4456
2.8	.4472	.4487	.4502	.4518	.4533	.4548	.4564	.4579	.4594	.4609
2.9	.4624	.4639	.4654	.4669	.4683	.4698	.4713	.4728	.4742	.4757
3.0	.4771	.4786	.4800	.4814	.4829	.4843	.4857	.4871	.4886	.4900
3.1	.4914	.4928	.4942	.4955	.4969	.4983	.4997	.5011	.5024	.5038
3.2	.5051	.5065	.5079	.5092	.5105	.5119	.5132	.5145	.5159	.5172
3.3	.5185	.5198	.5211	.5224	.5237	.5250	.5263	.5276	.5289	.5302
3.4	.5315	.5328	.5340	.5353	.5366	.5378	.5391	.5403	.5416	.5428
3.5	.5441	.5453	.5465	.5478	.5490	.5502	.5514	.5527	.5539	.5551
3.6	.5563	.5575	.5587	.5599	.5611	.5623	.5635	.5647	.5658	.5670
3.7	.5682	.5694	.5705	.5717	.5729	.5740	.5752	.5763	.5775	.5786
3.8	.5798	.5809	.5821	.5832	.5843	.5855	.5866	.5877	.5888	.5899
3.9	.5911	.5922	.5933	.5944	.5955	.5966	.5977	.5988	.5999	.6010
4.0	.6021	.6031	.6042	.6053	.6064	.6075	.6085	.6096	.6107	.6117
4.1	.6128	.6138	.6149	.6160	.6170	.6180	.6191	.6201	.6212	.6222
4.2	.6232	.6243	.6253	.6263	.6274	.6284	.6294	.6304	.6314	.6325
4.3	.6335	.6345	.6355	.6365	.6375	.6385	.6395	.6405	.6415	.6425
4.4	.6435	.6444	.6454	.6464	.6474	.6484	.6493	.6503	.6513	.6522
4.5	.6532	.6542	.6551	.6561	.6571	.6580	.6590	.6599	.6609	.6618
4.6	.6628	.6637	.6646	.6656	.6665	.6675	.6684	.6693	.6702	.6712
4.7	.6721	.6730	.6739	.6749	.6758	.6767	.6776	.6785	.6794	.6803
4.8	.6812	.6821	.6830	.6839	.6848	.6857	.6866	.6875	.6884	.6893
4.9	.6902	.6911	.6920	.6928	.6937	.6946	.6955	.6964	.6972	.6981
5.0	.6990	.6998	.7007	.7016	.7024	.7033	.7042	.7050	.7059	.7067
5.1	.7076	.7084	.7093	.7101	.7110	.7118	.7126	.7135	.7143	.7152
5.2	.7160	.7168	.7177	.7185	.7193	.7202	.7210	.7218	.7226	.7235
5.3	.7243	.7251	.7259	.7267	.7275	.7284	.7292	.7300	.7308	.7316
5.4	.7324	.7332	.7340	.7348	.7356	.7364	.7372	.7380	.7388	.7396
N	0	1	2	3	4	5	6	7	8	9

Table IV: Continued

N	0	1	2	3	4	5	6	7	8	9
5.5	.7404	.7412	.7419	.7427	.7435	.7443	.7451	.7459	.7466	.7474
5.6	.7482	.7490	.7497	.7505	.7513	.7520	.7528	.7536	.7543	.7551
5.7	.7559	.7566	.7574	.7582	.7589	.7597	.7604	.7612	.7619	.7627
5.8	.7634	.7642	.7649	.7657	.7664	.7672	.7679	.7686	.7694	.7701
5.9	.7709	.7716	.7723	.7731	.7738	.7745	.7752	.7760	.7767	.7774
6.0	.7782	.7789	.7796	.7803	.7810	.7818	.7825	.7832	.7839	.7846
6.1	.7853	.7860	.7868	.7875	.7882	.7889	.7896	.7903	.7910	.7917
6.2	.7924	.7931	.7938	.7945	.7952	.7959	.7966	.7973	.7980	.7987
6.3	.7993	.8000	.8007	.8014	.8021	.8028	.8035	.8041	.8048	.8055
6.4	.8062	.8069	.8075	.8082	.8089	.8096	.8102	.8109	.8116	.8122
6.5	.8129	.8136	.8142	.8149	.8156	.8162	.8169	.8176	.8182	.8189
6.6	.8195	.8202	.8209	.8215	.8222	.8228	.8235	.8241	.8248	.8254
6.7	.8261	.8267	.8274	.8280	.8287	.8293	.8299	.8306	.8312	.8319
6.8	.8325	.8331	.8338	.8344	.8351	.8357	.8363	.8370	.8376	.8382
6.9	.8388	.8395	.8401	.8407	.8414	.8420	.8426	.8432	.8439	.8445
7.0	.8451	.8457	.8463	.8470	.8476	.8482	.8488	.8494	.8500	.8506
7.1	.8513	.8519	.8525	.8531	.8537	.8543	.8549	.8555	.8561	.8567
7.2	.8573	.8579	.8585	.8591	.8597	.8603	.8609	.8615	.8621	.8627
7.3	.8633	.8639	.8645	.8651	.8657	.8663	.8669	.8675	.8681	.8686
7.4	.8692	.8698	.8704	.8710	.8716	.8722	.8727	.8733	.8739	.8745
7.5	.8751	.8756	.8762	.8768	.8774	.8779	.8785	.8791	.8797	.8802
7.6	.8808	.8814	.8820	.8825	.8831	.8837	.8842	.8848	.8854	.8859
7.7	.8865	.8871	.8876	.8882	.8887	.8893	.8899	.8904	.8910	.8915
7.8	.8921	.8927	.8932	.8938	.8943	.8949	.8954	.8960	.8965	.8971
7.9	.8976	.8982	.8987	.8993	.8998	.9004	.9009	.9015	.9020	.9025
8.0	.9031	.9036	.9042	.9047	.9053	.9058	.9063	.9069	.9074	.9079
8.1	.9085	.9090	.9096	.9101	.9106	.9112	.9117	.9122	.9128	.9133
8.2	.9138	.9143	.9149	.9154	.9159	.9165	.9170	.9175	.9180	.9186
8.3	.9191	.9196	.9201	.9206	.9212	.9217	.9222	.9227	.9232	.9238
8.4	.9243	.9248	.9253	.9258	.9263	.9269	.9274	.9279	.9284	.9289
8.5	.9294	.9299	.9304	.9309	.9315	.9320	.9325	.9330	.9335	.9340
8.6	.9345	.9350	.9355	.9360	.9365	.9370	.9375	.9380	.9385	.9390
8.7	.9395	.9400	.9405	.9410	.9415	.9420	.9425	.9430	.9435	.9440
8.8	.9445	.9450	.9455	.9460	.9465	.9469	.9474	.9479	.9484	.9489
8.9	.9494	.9499	.9504	.9509	.9513	.9518	.9523	.9528	.9533	.9538
9.0	.9542	.9547	.9552	.9557	.9562	.9566	.9571	.9576	.9581	.9586
9.1	.9590	.9595	.9600	.9605	.9609	.9614	.9619	.9624	.9628	.9633
9.2	.9638	.9643	.9647	.9652	.9657	.9661	.9666	.9671	.9675	.9680
9.3	.9685	.9689	.9694	.9699	.9703	.9708	.9713	.9717	.9722	.9727
9.4	.9731	.9736	.9741	.9745	.9750	.9754	.9759	.9763	.9768	.9773
9.5	.9777	.9782	.9786	.9791	.9795	.9800	.9805	.9809	.9814	.9818
9.6	.9823	.9827	.9832	.9836	.9841	.9845	.9850	.9854	.9859	.9863
9.7	.9868	.9872	.9877	.9881	.9886	.9890	.9894	.9899	.9903	.9908
9.8	.9912	.9917	.9921	.9926	.9930	.9934	.9939	.9943	.9948	.9952
9.9	.9956	.9961	.9965	.9969	.9974	.9978	.9983	.9987	.9991	.9996
N	0	1	2	3	4	5	6	7	8	9

Table V: Trigonometric Functions

Degrees	Radians	sin	cos	tan	cot	sec	csc		
0° 00′	.0000	.0000	1.0000	.0000	——	1.000	——	1.5708	90° 00′
10	029	029	000	029	343.8	000	343.8	679	50
20	058	058	000	058	171.9	000	171.9	650	40
30	.0087	.0087	1.0000	.0087	114.6	1.000	114.6	1.5621	30
40	116	116	.9999	116	85.94	000	85.95	592	20
50	145	145	999	145	68.75	000	68.76	563	10
1° 00′	.0175	.0175	.9998	.0175	57.29	1.000	57.30	1.5533	89° 00′
10	204	204	998	204	49.10	000	49.11	504	50
20	233	233	997	233	42.96	000	42.98	475	40
30	.0262	.0622	.9997	.0262	38.19	1.000	38.20	1.5446	30
40	291	291	996	291	34.37	000	34.38	417	20
50	320	320	995	320	31.24	001	31.26	388	10
2° 00′	.0349	.0349	.9994	.0349	28.64	1.001	28.65	1.5359	88° 00′
10	378	378	993	378	26.43	001	26.45	330	50
20	407	407	992	407	24.54	001	24.56	301	40
30	.0436	.0436	.9990	.0437	22.90	1.001	22.93	1.5272	30
40	465	465	989	466	21.47	001	21.49	243	20
50	495	494	988	495	20.21	001	20.23	213	10
3° 00′	.0524	.0523	.9986	.0524	19.08	1.001	19.11	1.5184	87° 00′
10	553	552	985	553	18.07	002	18.10	155	50
20	582	581	983	582	17.17	002	17.20	126	40
30	.0611	.0610	.9981	.0612	16.35	1.002	16.38	1.5097	30
40	640	640	980	641	15.60	002	15.64	068	20
50	669	669	978	670	14.92	002	14.96	039	10
4° 00′	.0698	.0698	.9976	.0699	14.30	1.002	14.34	1.5010	86° 00′
10	727	727	974	729	13.73	003	13.76	981	50
20	756	756	971	758	13.20	003	13.23	952	40
30	.0785	.0785	.9969	.0787	12.71	1.003	12.75	1.4923	30
40	814	814	967	816	12.25	003	12.29	893	20
50	844	843	964	846	11.83	004	11.87	864	10
5° 00′	.0873	.0872	.9962	.0875	11.43	1.004	11.47	1.4835	85° 00′
10	902	901	959	904	11.06	004	11.10	806	50
20	931	929	957	934	10.71	004	10.76	777	40
30	.0960	.0958	.9954	.0963	10.39	1.005	10.43	1.4748	30
40	989	987	951	992	10.08	005	10.13	719	20
50	.1018	.1016	948	.1022	9.788	005	9.839	690	10
6° 00′	.1047	.1045	.9945	.1051	9.514	1.006	9.567	1.4661	84° 00′
10	076	074	942	080	9.255	006	9.309	632	50
20	105	103	939	110	9.010	006	9.065	603	40
30	.1134	.1132	.9936	.1139	8.777	1.006	8.834	1.4573	30
40	164	161	932	169	8.556	007	8.614	544	20
50	193	190	929	198	8.345	007	8.405	515	10
7° 00′	.1222	.1219	.9925	.1228	3.144	1.008	8.206	1.4486	83° 00′
10	251	248	922	257	7.953	008	8.016	457	50
20	280	276	918	287	7.770	008	7.834	428	40
30	.1309	.1305	.9914	.1317	7.596	1.009	7.661	1.4399	30
40	338	334	911	346	7.429	009	7.496	370	20
50	367	363	907	376	7.269	009	7.337	341	10
8° 00′	.1396	.1392	.9903	.1405	7.115	1.010	7.185	1.4312	82° 00′
10	425	421	899	435	6.968	010	7.040	283	50
20	454	449	894	465	6.827	011	6.900	254	40
30	.1484	.1478	.9890	.1495	6.691	1.011	6.765	1.4224	30
40	513	507	886	524	6.561	012	6.636	195	20
50	542	536	881	554	6.435	012	6.512	166	10
9° 00′	.1571	.1564	.9877	.1584	6.314	1.012	6.392	1.4137	81° 00′
		cos	sin	cot	tan	csc	sec	Radians	Degrees

Table V: Continued

Degrees	Radians	sin	cos	tan	cot	sec	csc		
9 00′	.1571	.1564	.9877	.1584	6.314	1.012	6.392	1.4137	81° 00′
10	600	593	872	614	197	013	277	108	50
20	629	622	868	644	084	013	166	079	40
30	.1658	.1650	.9863	.1673	5.976	1.014	6.059	1.4050	30
40	687	679	858	703	871	014	5.955	1.4021	20
50	716	708	853	733	769	015	855	992	10
10 00′	.1745	.1736	.9848	.1763	5.671	1.015	5.759	1.3963	80° 00′
10	774	765	843	793	576	016	665	934	50
20	804	794	838	823	485	016	575	904	40
30	.1833	.1822	.9833	.1853	5.396	1.017	5.487	1.3875	30
40	862	851	827	883	309	018	403	846	20
50	891	880	822	914	226	018	320	817	10
11 00′	.1920	.1908	.9816	.1944	5.145	1.019	5.241	1.3788	79° 00′
10	949	937	811	974	066	019	164	759	50
20	978	965	805	.2004	4.989	020	089	730	40
30	.2007	.1994	.9799	.2035	4.915	1.020	5.016	1.3701	30
40	036	.2022	793	065	843	021	4.945	672	20
50	065	051	787	095	773	022	876	643	10
12 00′	.2094	.2079	.9781	.2126	4.705	1.022	4.810	1.3614	78° 00′
10	123	108	775	156	638	023	745	584	50
20	153	136	769	186	574	024	682	555	40
30	.2182	.2164	.9763	.2217	4.511	1.024	4.620	1.3526	30
40	211	193	757	247	449	025	560	497	20
50	240	221	750	278	390	026	502	468	10
13 00′	.2269	.2250	.9744	.2309	4.331	1.026	4.445	1.3439	77° 00′
10	298	278	737	339	275	027	390	410	50
20	327	306	730	370	219	028	336	381	40
30	.2356	.2334	.9724	.2401	4.165	1.028	4.284	1.3352	30
40	385	363	717	432	113	029	232	323	20
50	414	391	710	462	061	030	182	294	10
14 00′	.2443	.2419	.9703	.2493	4.011	1.031	4.134	1.3265	76° 00′
10	473	447	696	524	3.962	031	086	235	50
20	502	476	689	555	914	032	039	206	40
30	.2531	.2504	.9681	.2586	3.867	1.033	3.994	1.3177	30
40	560	532	674	617	821	034	950	148	20
50	589	560	667	648	776	034	906	119	10
15 00′	.2618	.2588	.9659	.2679	3.732	1.035	3.864	1.3090	75° 00′
10	647	616	652	711	689	036	822	061	50
20	676	644	644	742	647	037	782	032	40
30	.2705	.2672	.9636	.2773	3.606	1.038	3.742	1.3003	30
40	734	700	628	805	566	039	703	974	20
50	763	728	621	836	526	039	665	945	10
16 00′	.2793	.2756	.9613	.2867	3.487	1.040	3.628	1.2915	74° 00′
10	822	784	605	899	450	041	592	886	50
20	851	812	596	931	412	042	556	857	40
30	.2880	.2840	.9588	.2962	3.376	1.043	3.521	1.2828	30
40	909	868	580	994	340	044	487	799	20
50	938	896	572	.3026	305	045	453	770	10
17 00′	.2967	.2924	.9563	.3057	3.271	1.046	3.420	1.2741	73° 00′
10	996	952	555	089	237	047	388	712	50
20	.3025	979	546	121	204	048	356	683	40
30	.3054	.3007	.9537	.3153	3.172	1.049	3.326	1.2654	30
40	083	035	528	185	140	049	295	625	20
50	113	062	520	217	108	050	265	595	10
18 00′	.3142	.3090	.9511	.3249	3.078	1.051	3.236	1.2566	72° 00′
		cos	sin	cot	tan	csc	sec	Radians	Degrees

Table V: Continued

Degrees	Radians	sin	cos	tan	cot	sec	csc		
18° 00′	.3142	.3090	.9511	.3249	3.078	1.051	3.236	1.2566	**72° 00′**
10	171	118	502	281	047	052	207	537	50
20	200	145	492	314	018	053	179	508	40
30	.3229	.3173	.9483	.3346	2.989	1.054	3.152	1.2479	30
40	258	201	474	378	960	056	124	450	20
50	287	228	465	411	932	057	098	421	10
19° 00′	.3316	.3256	.9455	.3443	2.904	1.058	3.072	1.2392	**71° 00′**
10	345	283	446	476	877	059	046	363	50
20	374	311	436	508	850	060	021	334	40
30	.3403	.3338	.9426	.3541	2.824	1.061	2.996	1.2305	30
40	432	365	417	574	798	062	971	275	20
50	462	393	407	607	773	063	947	246	10
20° 00′	.3491	.3420	.9397	.3640	2.747	1.064	2.924	1.2217	**70° 00′**
10	520	449	387	673	723	065	901	188	50
20	549	475	377	706	699	066	878	159	40
30	.3578	.3502	.9367	.3739	2.675	1.068	2.855	1.2130	30
40	607	529	356	772	651	069	833	101	20
50	636	557	346	805	628	070	812	072	10
21 °00′	.3665	.3584	.9336	.3839	2.605	1.071	2.790	1.2043	**69° 00′**
10	694	611	325	872	583	072	769	1.2014	50
20	723	638	315	906	560	074	749	985	40
30	.3752	.3665	.9304	.3939	2.539	1.075	2.729	1.1956	30
40	782	692	293	973	517	076	709	926	20
50	811	719	283	.4006	496	077	689	897	10
22° 00′	.3840	.3746	.9272	.4040	2.475	1.079	2.669	1.1868	**68° 00′**
10	869	773	261	074	455	080	650	839	50
20	898	800	250	108	434	081	632	810	40
30	.3927	.3827	.9239	.4142	2.414	1.082	2.613	1.1781	30
40	956	854	228	176	394	084	595	752	20
50	985	881	216	210	375	085	577	723	10
23° 00′	.4014	.3907	.9205	.4245	2.356	1.086	2.559	1.1694	**67° 00′**
10	043	934	194	279	337	088	542	665	50
20	072	961	182	314	318	089	525	636	40
30	.4102	.3987	.9171	.4348	2.300	1.090	2.508	1.1606	30
40	131	.4014	159	383	282	092	491	577	20
50	160	041	147	417	264	093	475	548	10
24° 00′	.4189	.4067	.9135	.4452	2.246	1.095	2.459	1.1519	**66° 00′**
10	218	094	124	487	229	096	443	490	50
20	247	120	112	522	211	097	427	461	40
30	.4276	.4147	.9100	.4557	2.194	1.099	2.411	1.1432	30
40	305	173	088	592	177	100	396	403	20
50	334	200	075	628	161	102	381	374	10
25° 00′	.4363	.4226	.9063	.4663	2.145	1.103	2.366	1.1345	**65° 00′**
10	392	253	051	699	128	105	352	316	50
20	422	279	038	734	112	106	337	286	40
30	.4451	.4305	.9026	.4770	2.097	1.108	2.323	1.1257	30
40	480	331	013	806	081	109	309	228	20
50	509	358	001	841	066	111	295	199	10
26° 00′	.4538	.4384	.8988	.4877	2.050	1.113	2.281	1.1170	**64° 00′**
10	567	410	975	913	035	114	268	141	50
20	596	436	962	950	020	116	254	112	40
30	.4625	.4462	.8949	.4986	2.006	1.117	2.241	1.1083	30
40	654	488	936	.5022	1.991	119	228	054	20
50	683	514	923	059	977	121	215	1.1025	10
27° 00′	.4712	.4540	.8910	.5095	1.963	1.122	2.203	1.0996	**63° 00′**
		cos	sin	cot	tan	csc	sec	Radians	Degrees

Table V: Continued

Degrees	Radians	sin	cos	tan	cot	sec	csc		
27° 00′	.4712	.4540	.8910	.5095	1.963	1.122	2.203	1.0996	**63° 00′**
10	741	566	897	132	949	124	190	966	50
20	771	592	884	169	935	126	178	937	40
30	.4800	.4617	.8870	.5206	1.921	1.127	2.166	1.0908	30
40	829	643	857	243	907	129	154	879	20
50	858	669	843	280	894	131	142	850	10
28° 00′	.4887	.4695	.8829	.5317	1.881	1.133	2.130	1.0821	**62° 00′**
10	916	720	816	354	868	134	118	792	50
20	945	746	802	392	855	136	107	763	40
30	.4974	.4772	.8788	.5430	1.842	1.138	2.096	1.0734	30
40	.5003	797	774	467	829	140	085	705	20
50	032	823	760	505	816	142	074	676	10
29° 00′	.5061	.4848	.8746	.5543	1.804	1.143	2.063	1.0647	**61° 00′**
10	091	874	732	581	792	145	052	617	50
20	120	899	718	619	780	147	041	588	40
30	.5149	.4924	.8704	.5658	1.767	1.149	2.031	1.0559	30
40	178	950	689	696	756	151	020	530	20
50	207	975	675	735	744	153	010	501	10
30° 00′	.5236	.5000	.8660	.5774	1.732	1.155	2.000	1.0472	**60° 00′**
10	265	025	646	812	720	157	1.990	443	50
20	294	050	631	851	709	159	980	414	40
30	.5323	.5075	.8616	.5890	1.698	1.161	1.970	1.0385	30
40	352	100	601	930	686	163	961	356	20
50	381	125	587	969	675	165	951	327	10
31° 00′	.5411	.5150	.8572	.6009	1.664	1.167	1.942	1.0297	**59° 00′**
10	440	175	557	048	653	169	932	268	50
20	469	200	542	088	643	171	923	239	40
30	.5498	.5225	.8526	.6128	1.632	1.173	1.914	1.0210	30
40	527	250	511	168	621	175	905	181	20
50	556	275	496	208	611	177	896	152	10
32° 00′	.5585	.5299	.8480	.6249	1.600	1.179	1.887	1.0123	**58° 00′**
10	614	324	465	289	590	181	878	094	50
20	643	348	450	330	580	184	870	065	40
30	.5672	.5373	.8434	.6371	.1570	1.186	1.861	1.0036	30
40	701	398	418	412	560	188	853	1.0007	20
50	730	422	403	453	550	190	844	977	10
33° 00′	.5760	.5446	.8387	.6494	1.540	1.192	1.836	.9948	**57° 00′**
10	789	471	371	536	530	195	828	919	50
20	818	495	355	577	520	197	820	890	40
30	.5847	.5519	.8339	.6619	1.511	1.199	1.812	.9861	30
40	876	544	323	661	501	202	804	832	20
50	905	568	307	703	1.492	204	796	803	10
34° 00′	.5934	.5592	.8290	.6745	1.483	1.206	1.788	.9774	**56° 00′**
10	963	616	274	787	473	209	781	745	50
20	992	640	258	830	464	211	773	716	40
30	.6021	.5664	.8241	.6873	1.455	1.213	1.766	.9687	30
40	050	688	225	916	446	216	758	657	20
50	080	712	208	959	437	218	751	628	10
35° 00′	.6109	.5736	.8192	.7002	1.428	1.221	1.743	.9599	**55° 00′**
10	138	760	175	046	419	223	736	570	50
20	167	783	158	089	411	226	729	541	40
30	.6196	.5807	.8141	.7133	1.402	1.228	1.722	.9512	30
40	225	831	124	177	393	231	715	483	20
50	254	854	107	221	385	233	708	454	10
36° 00′	.6283	.5878	.8090	.7265	1.376	1.236	1.701	.9425	**54° 00′**
		cos	sin	cot	tan	csc	sec	Radians	Degrees

Table V: Continued

Degrees	Radians	sin	cos	tan	cot	sec	csc		
36° 00′	.6283	.5878	.8090	.7265	1.376	1.236	1.701	.9425	**54° 00′**
10	312	901	073	310	368	239	695	396	50
20	341	925	056	355	360	241	688	367	40
30	.6370	.5948	.8039	.7400	1.351	1.244	1.681	.9338	30
40	400	972	021	445	343	247	675	308	20
50	429	995	004	490	335	249	668	279	10
37° 00′	.6458	.6018	.7986	.7536	1.327	1.252	1.662	.9250	**53° 00′**
10	487	041	969	581	319	255	655	221	50
20	516	065	951	627	311	258	649	192	40
30	.6545	.6088	.7934	.7673	1.303	1.260	1.643	.9163	30
40	574	111	916	720	295	263	636	134	20
50	603	134	898	766	288	266	630	105	10
38° 00′	.5632	.6157	.7880	.7813	1.280	1.269	1.624	.9076	**52° 00′**
10	661	180	862	860	272	272	618	047	50
20	690	202	844	907	265	275	612	.9018	40
30	.6720	.6225	.7826	.7954	1.257	1.278	1.606	.8988	30
40	749	248	808	.8002	250	281	601	959	20
50	778	271	790	050	242	284	595	930	10
39° 00′	.6807	.6293	.7771	.8098	1.235	1.287	1.589	.8901	**51° 00′**
10	836	316	753	146	228	290	583	872	50
20	865	338	735	195	220	293	578	843	40
30	.6894	.6361	.7716	.8243	1.213	1.296	1.572	.8814	30
40	923	383	698	292	206	299	567	785	20
50	952	406	679	342	199	302	561	756	10
40° 00′	.6981	.6428	.7660	.8391	1.192	1.305	1.556	.8727	**50° 00′**
10	.7010	450	642	441	185	309	550	698	50
20	039	472	623	491	178	312	545	668	40
30	.7069	.6494	.7604	.8541	1.171	1.315	1.540	.8639	30
40	098	517	585	591	164	318	535	610	20
50	127	539	566	642	157	322	529	581	10
41° 00′	.7156	.6561	.7547	.8693	1.150	1.325	1.524	.8552	**49° 00′**
10	185	583	528	744	144	328	519	523	50
20	214	604	509	796	137	332	514	494	40
30	.7243	.6626	.7490	.8847	1.130	1.335	1.509	.8465	30
40	272	648	470	899	124	339	504	436	20
50	301	670	451	952	117	342	499	407	10
42° 00′	.7330	.6691	.7431	.9004	1.111	1.346	1.494	.8378	**48° 00′**
10	359	713	412	057	104	349	490	348	50
20	389	734	392	110	098	353	485	319	40
30	.7418	.6756	.7373	.9163	1.091	1.356	1.480	.8290	30
40	447	777	353	217	085	360	476	261	20
50	476	799	333	271	079	364	471	232	10
43° 00′	.7505	.6820	.7314	.9325	1.072	1.367	1.466	.8203	**47° 00′**
10	534	841	294	380	066	371	462	174	50
20	563	862	274	435	060	375	457	145	40
30	.7592	.6884	.7254	.9490	1.054	1.379	1.453	.8116	30
40	621	905	234	545	048	382	448	087	20
50	650	926	214	601	042	386	444	058	10
44° 00′	.7679	.6947	.7193	.9657	1.036	1.390	. 1.440	.8029	**46° 00′**
10	709	967	173	713	030	394	435	999	50
20	738	988	153	770	024	398	431	970	40
30	.7767	.7009	.7133	.9827	1.018	1.402	1.427	.7941	30
40	796	030	112	884	012	406	423	912	20
50	825	050	092	942	006	410	418	883	10
45° 00′	.7854	.7071	.7071	1.000	1.000	1.414	1.414	.7854	**45° 00′**
		cos	sin	cot	tan	csc	sec	Radians	Degrees

Answers to Odd-Numbered Exercises

Chapter 1: REAL NUMBERS, EQUATIONS, INEQUALITIES

1.1 THE REAL NUMBER LINE (*Page 4*)

1. True.

3. False.

5. False.

7. True.

9. True.

	Counting Numbers	Integers	Rational Numbers	Irrational Numbers	Real Numbers
11.		✓	✓		✓
13.				✓	✓
15.	✓	✓	✓		✓
17.			✓		✓
19.				✓	✓

21.
$$\leftarrow\!\!\!\!\underset{-3\ -2\ -1\ \ 0\ \ 1\ \ 2}{\bullet\ \ \bullet\ \ \bullet\ \ \bullet\ \ \bullet\ \ \bullet}\!\!\!\!\rightarrow$$

23.
$$\leftarrow\!\!\!\!\underset{-1\ -\frac{2}{3}\ -\frac{1}{3}\ \ 0\ \ \frac{1}{3}\ \ \frac{2}{3}\ \ 1\ \ \frac{4}{3}}{\bullet\ \ \bullet\ \ \bullet\ \ \bullet\ \ \bullet\ \ \bullet\ \ \bullet\ \ \bullet}\!\!\!\!\rightarrow$$

25.
$$\leftarrow\!\!\!\!\underset{-\frac{5}{4}\ -1\quad -\frac{1}{4}\ \ 0\quad \frac{3}{4}\ \ 1\quad \frac{7}{4}}{\bullet\ \ \bullet\ \ \ \ \bullet\ \ \bullet\ \ \ \ \bullet\ \ \bullet\ \ \ \ \bullet}\!\!\!\!\rightarrow$$

*27.

*29. π; circumference $= 2\pi(\frac{1}{2}) = \pi$.

1.2 BASIC PROPERTIES OF THE REAL NUMBERS (*Page 10*)

1. Commutative for $\times$.

3. Associative for $+$.

5. Distributive.

7. Commutative for $+$.

9. Associative for $\times$.

11. Additive inverse.

13. Property of zero.

15. Distributive.

17. $8 \div 2 \neq 2 \div 8$

19. $2(12 \div 3) \neq (2 \times 12) \div (2 \times 3)$

21. (i) Distributive; (ii) distributive; (iii) associative for $+$; (iv) commutative for $\times$; (v) commutative for $+$; (vi) distributive; (vii) commutative for $+$.

*23. Let $\frac{0}{0} = x$, where x is some number. Then the definition of division gives $0 \cdot x = 0$. Since any number x will work, the answer to $\frac{0}{0}$ is not unique; therefore $\frac{0}{0}$ is undefined.

1. 7
3. 14
5. −5
7. 9
9. −4
11. −9
13. 5
15. $-\frac{11}{2}$
17. $w = 7; l = 21.$
19. 36, 38.
21. Base = 5; side = 8.
23. 14 ones, 7 fives, 12 tens.
25. $3\frac{1}{2}$ hours
27. Ben is 10 and Bob is 30.
29. 36

1.3 INTRODUCTION TO EQUATIONS (*Page 14*)

*31. 5, 6, 7, 8, 9
*33. Regardless of original dimensions, if they satisfy the stated conditions the new rectangle will always have length = twice the width. If original dimensions are w and $2w − 1$, the new dimensions are $w + 5$ and $2w + 10$; hence $2w + 10 = 2(w + 5)$ for all w.

35. If $a = b$, then the multiplication property of equality gives $a \times \frac{1}{c} = b \times \frac{1}{c}$ for $c \neq 0$. Now using the definition of division we have $a \div c = b \div c$.

37. $\dfrac{p - 2l}{2}$

39. $\dfrac{N - u}{10}$

1. <
3. <
5. <
7. =
9. (a) $x < 1$; (b) $x > -\frac{2}{3}$; (c) $x \geq 5$;
   (d) $-10 < x \leq 7$.
11. True.
13. False; $-10 < 5$ and $-4 < 6$, but $(-10)(-4) = 40 > 30$.

1.4 PROPERTIES OF ORDER (*Page 18*)

15. True.
17. True.
19. (a) $b − a$ is positive; (b) $(b − a)c$;
    (c) $bc − ac$; (d) $bc − ac$; (e) $ac < bc$.

*21. We are given that $1 < a$. Since $0 < 1$ the transitive property gives $0 < a$. Now use Rule 3 to get $a \cdot 1 < a \cdot a$, or $a < a^2$.

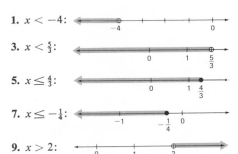

1. $x < -4$:
3. $x < \frac{5}{3}$:
5. $x \leq \frac{4}{3}$:
7. $x \leq -\frac{1}{4}$:
9. $x > 2$:
11. $x < 0$:

1.5 CONDITIONAL INEQUALITIES (*Page 22*)

13. (a) Positive; (b) positive.
15. (a) Positive; (b) positive; (c) negative;
    (d) negative.
17. $-1 < x < 1$
19. $-3 < x < -2$
21. $-5 \leq x \leq \frac{1}{2}$
23. $x < -\frac{1}{3}$ or $x > \frac{1}{2}$
25. $x < -1$ or $x > 0$
27. $x < -5$ or $x > 2$
*29. $x < 2$ or $x > 5$

1.6 ABSOLUTE VALUE *(Page 27)*

1. False.

3. False.

5. True.

7. True.

9. False.

11. $x = \pm 1$

13. $x = \frac{4}{3}$

15. $x = 1; x = 5.$

17. $x > 0$

19. $-1 < x < 1$:

21. $-5 < x < 1$:

23. $-3 \le x \le 2$:

25. $x < -1$ or $x > 3$:

27. $x \le -2$ or $x \ge 1$:

29. All x:

*31. *Case 1:* Suppose $x \ge 0$. By definition of absolute value, $|x| = x$, which implies $x \le |x|$. Also, since $|x| \ge 0$, we get $-|x| \le 0$. Now we have these three statements: $-|x| \le 0, 0 \le x, x \le |x|$, which imply $-|x| \le x \le |x|$.

Case 2: Suppose $x < 0$. By definition of absolute value,
$|x| = -x$. Then $-|x| = x$, which implies $-|x| \le x$. Also, since $|x| \ge 0$, we have $-|x| \le x < 0 \le |x|$ and this gives $-|x| \le x \le |x|$.

*33. (a) Since $x < 0$ and $y > 0$, we find that the product xy is negative. Thus
$$|xy| = -(xy) \quad \text{(definition of absolute value)}$$
$$= (-x)(y)$$
$$= |x||y| \quad (|x| = -x \text{ and } |y| = y)$$
(b) $|x| = -x$ and $|y| = -y$ since each of x and y is negative. Then
$$|x||y| = (-x)(-y)$$
$$= xy$$
$$= |xy| \quad \text{(since } xy > 0)$$

*35. $|x^2| = x^2$ since $x^2 \ge 0$ for any x. Also,
$|x^2| = |x \cdot x| = |x||x| = |x|^2$.

SAMPLE TEST FOR CHAPTER 1 *(Page 29)*

1.

	Natural	Integer	Rational	Irrational	Negative
$-\frac{6}{3}$		✓	✓		✓
0.231			✓		
$\sqrt{12}$				✓	
$\sqrt{\frac{1}{4}}$			✓		
1979	✓	✓	✓		

2. Construct a perpendicular of unit length at the point with coordinate 3. The hypotenuse of the resulting right triangle has length $\sqrt{3^2 + 1^2} = \sqrt{10}$. Now use a compass with this hypotenuse as radius and the origin as center to locate $\sqrt{10}$ on the number line.

3. (a) False; (b) true; (c) false; (d) true; (e) false; (f) false.

4. (i)—(b); (ii)—(j); (iii)—(f); (iv)—(e); (v)—(i); (vi)—k; (vii)—(c).

5. (a) $x = 16$; (b) $x = -11$.

6. Width $= 7$ inches; length $= 19$ inches.

7. (a) $x = \frac{1}{4}, x = \frac{5}{4}$; (b) all $x < -2$.

8. $x < \frac{2}{9}$:

9. $x \le -1$ or $x \ge 2$:

10. $-5 < x < \frac{1}{2}$

Chapter 2: FUNDAMENTALS OF ALGEBRA

2.1 INTEGRAL EXPONENTS (*Page 39*)

1. False; 5^{12}

3. False; 3^{6}

5. False; $2 \cdot 3^{4}$

7. False; 1

9. True

11. False; 1

13. True

15. True

17. True

19. 100,000

21. 10

23. $\frac{15}{4}$

25. 25

27. $\frac{2}{5}$

29. 1

31. 2

33. 10

35. 1

37. -216

39. $\frac{64}{27}$

41. $\frac{1}{x^6}$

43. x^6

45. $-108x^{12}y^8$

47. 1

49. $\frac{4a^6}{b^8}$

51. $(x - 2y)^4$

53. $\frac{y^{10}}{x^{12}}$

55. $\frac{1}{(xy)^5}$

57. $\frac{ab^2}{a + b^2}$

59. $\frac{a^2}{b^3}$

61. (a) Rule 5; (b) preserving Rule 5 resulted in dividing by zero, which is not possible.

2.2 RADICALS AND RATIONAL EXPONENTS (*Page 45*)

1. $11^{1/2}$

3. $9^{1/4}$

5. $6^{2/3}$

7. $(-\frac{1}{5})^{3/5}$

9. $\sqrt{3}$

11. $\sqrt[3]{-9}$

13. $\frac{1}{\sqrt{2}}$

15. $\frac{1}{\sqrt[4]{\frac{3}{4}}}$

17. True

19. True

21. False; -2

23. False; 1.2

25. True

27. 5

29. $\frac{1}{9}$

31. $\frac{1}{16}$

33. 25

35. $\frac{9}{4}$

37. 9

39. 50

41. $\frac{1}{32}$

43. 5

45. 13

47. $\frac{5}{6}$

49. $-\frac{3}{10}$

51. 0.0001

53. $\frac{4a^2}{b^6}$

55. $\frac{1}{a^3b^6}$

57. 1

59. $16a^4b^6$

61. $\frac{a^2}{b}$

63. $\frac{1}{x}$

2.3 SIMPLIFYING RADICALS (*Page 50*)

1. $10\sqrt{2}$

3. $10\sqrt{2}$

5. $6\sqrt[3]{2}$

7. $7\sqrt{2}$

9. $14\sqrt{2}$

11. $3\sqrt[3]{7x}$

13. $|x|\sqrt{2}$

15. $5\sqrt[4]{2}$

17. $5\sqrt{10}$

19. $\frac{33\sqrt{2}}{2}$

21. $5\sqrt{6} - 3\sqrt{2}$

23. $7\sqrt{2x}$

25. $12|x|$

27. $|x|\sqrt{y} + 12|x|\sqrt{2y}$

29. $7a\sqrt{5a}$

31. $4\sqrt{5}$

33. $4x\sqrt{2}$

35. $2\sqrt{2} + 2\sqrt{3}$

37. $\frac{\sqrt{2}}{6}$

39. $\frac{8\sqrt{3}}{|x|}$

41. $4\sqrt[3]{4}$

43. $\sqrt{3} - \sqrt{2}$

45. $\frac{20(3 + \sqrt{2})}{7}$

47. $|xy| = \sqrt{(xy)^2} = \sqrt{x^2y^2} = \sqrt{x^2}\sqrt{y^2} = |x||y|$

2.4 FUNDAMENTAL OPERATIONS WITH POLYNOMIALS (*Page 55*)

1. $6x + 3$

3. $5x + 2$

5. $x^3 - 6x^2 + 6x - 20$

7. $x^3 + 2x^2 + x$

9. $7y + 8$

11. $-4x^2 - 5x + 3$

13. $-20x^4 + 4x^3 + 2x^2$

15. $x^2 + 2x + 1$

17. $4x^2 + 26x - 14$

19. $20x^2 - 42x + 18$

21. $x^2 + \frac{1}{4}x - \frac{1}{8}$

23. $-6x^2 - 3x + 18$

25. $-6x^2 + 3x + 18$

27. $\frac{4}{9}x^2 + 8x + 36$

29. $-63 + 55x - 12x^2$

31. $a^2x^2 - b^2$

33. $x^2 - 0.01$

35. $\frac{1}{25}x^2 - \frac{1}{10}x + \frac{1}{16}$

37. $x - 100$

39. $x - 2$

41. 1

43. $x^4 - 2x^3 + 2x^2 - 31x - 36$

45. $8x^4 + 12x^3 + 6x^2 - 18x - 27$

47. $x^3 - 8$

49. $x^5 - 32$

51. $x^{4n} - x^{2n} - 2$

53. $3x - 6x^2 + 3x^3$

*55. $-3x^3 + 7x^2 + 6x$

57. $4a$

59. $-4x^3 - 3x^2 - 2x - 2$

61. $x^6 - 3x^5 + 6x^4 - 7x^3 + 6x^2 - 3x + 1$

2.5 FACTORING POLYNOMIALS (*Page 62*)

1. $5(x - 1)$

3. $2x(8x^3 + 4x^2 + 2x + 1)$

5. $2ab(2a - 3)$

7. $2x(y + 2x + 4x^3)$

9. $5(a - b)(2x + y)$

11. $(9 + x)(9 - x)$

13. $(2x + 3)(2x - 3)$

15. $(a + 11b)(a - 11b)$

17. $(x + 4)(x^2 - 4x + 16)$

19. $(5x - 4)(25x^2 + 20x + 16)$

21. $(2x + 7y)(4x^2 - 14xy + 49y^2)$

23. $(\sqrt{3} + 2x)(\sqrt{3} - 2x)$

25. $(\sqrt{x} + 6)(\sqrt{x} - 6)$

27. $(2\sqrt{2} + \sqrt{3x})(2\sqrt{2} - \sqrt{3x})$

29. $(\sqrt[3]{7} + a)[(\sqrt[3]{7})^2 - a\sqrt[3]{7} + a^2]$

31. $(3\sqrt[3]{x} + 1)(9\sqrt[3]{x^2} - 3\sqrt[3]{x} + 1)$

33. $(\sqrt[3]{3x} - \sqrt[3]{4})(\sqrt[3]{9x^2} + \sqrt[3]{12x} + \sqrt[3]{16})$

35. $(x - 1)(x + y)$

37. $(x - 1)(y - 1)$

39. $(2 - y^2)(1 + x)$

41. $7(x + h)(x^2 - hx + h^2)$

43. $xy(x + y)(x - y)$

45. $(x^2 + y^2)(x + y)(x - y)$

47. $(a^4 + b^4)(a^2 + b^2)(a + b)(a - b)$

49. $(a - 2)(a^4 + 2a^3 + 4a^2 + 8a + 16)$

51. $(1 - h)(1 + h + h^2 + h^3 + h^4 + h^5 + h^6)$

53. $3(a + 1)(x + 2)(x^2 - 2x + 4)$

55. $(x - y)(a + b)(a^2 - ab + b^2)$

57. $(x^2 + 4y^2)(x + 2y)(x - 2y)^2$

*59. (a) $(x + 2)(x^4 - 2x^3 + 4x^2 - 8x + 16)$;
(b) $x(2x + y)(64x^6 - 32x^5y + 16x^4y^2 - 8x^3y^3 + 4x^2y^4 - 2xy^5 + y^6)$.

2.6 FACTORING TRINOMIALS (*Page 67*)

1. $(x + 2)^2$

3. $(a - 7)^2$

5. $(1 + b)^2$

7. $9(x - y)^2$

9. $(x + 2)(x + 3)$

11. $(x + 3)(x + 17)$

13. $(5a - 1)(4a - 1)$

15. $(3x + 1)^2$

17. $(5a - 1)^2$

19. $(3x + 2)(x + 6)$

21. $(7x + 1)(2x + 5)$

23. $(2a + 5)^2$

25. $(b + 3)(b + 15)$

27. $2(2x - 1)(2x - 3)$

29. $(4a - 3)(3a - 4)$

31. $2(6b + 1)(b - 3)$

33. $3(a + 1)^2$

35. $2(3x - 2)^2$

37. Not factorable.

39. $s(7r - 3)^2$

41. $10(5a + 1)(a - 9)$

43. $5(1 + y)(3 - 2y)$

45. $x^2(2a - b)^2$

47. $8(2x^2 - 3x - 1)$

49. $3(x + 2)(8x - 3)$

*51. $(x + 1)^2(x - 1)^2$

*53. $3(x^2 + 1)^2$

*55. $(a - b)^2(a^2 + ab + b^2)^2$

2.7 FUNDAMENTAL OPERATIONS WITH RATIONAL EXPRESSIONS (*Page 75*)

1. False; $\frac{1}{21}$

3. False; $\frac{ax}{2} - \frac{5b}{6}$

5. True

7. $\frac{2x}{3z}$

9. $-x$

11. $\frac{n+1}{n^2+1}$

13. $\frac{3(x-1)}{2(x+1)}$

15. $\frac{2x+3}{2x-3}$

17. $\frac{a-4b}{a^2-4ab+16b^2}$

19. $\frac{2y}{x}$

21. $\frac{2}{a^2}$

23. $\frac{x}{y}$

25. $\frac{-3a-8b}{6}$

27. $\frac{x^2+1}{3(x+1)}$

29. 1

31. $\frac{2-xy}{x}$

33. $\frac{5y^2-y}{y^2-1}$

35. $\frac{3x}{x+1}$

37. $\frac{8-x}{x^2-4}$

39. $\frac{3(x+9)}{(x-3)(x+3)^2}$

41. $x-3$

43. $\frac{2(n^2+4)}{n+2}$

45. $\frac{1}{2(x+2)}$

47. $-\frac{1}{4x}$

49. $-\frac{1}{4(4+h)}$

*51. $-\frac{1}{ab}$

*53. $\frac{y^2-x^2}{x^3y^3}$

*55. (a) $\dfrac{\dfrac{AD}{B}+C}{D} = \dfrac{AD+BC}{BD} = \dfrac{AD}{BD} + \dfrac{BC}{BD} = \dfrac{A}{B} + \dfrac{C}{D}$

(b) $\left[\dfrac{\left(\dfrac{AB}{D}+C\right)D}{F} + E\right] \cdot F =$

$\left[\dfrac{AB+CD}{F} + E\right] \cdot F = AB + CD + EF$

2.8 THE BINOMIAL EXPANSION (*Page 79*)

1. $x^5 + 5x^4 + 10x^3 + 10x^2 + 5x + 1$

3. $x^7 + 7x^6 + 21x^5 + 35x^4 + 35x^3 + 21x^2 + 7x + 1$

5. $a^4 - 4a^3b + 6a^2b^2 - 4ab^3 + b^4$

7. $243x^5 - 405x^4y + 270x^3y^2 - 90x^2y^3 + 15xy^4 - y^5$

9. $a^{10} + 5a^8 + 10a^6 + 10a^4 + 5a^2 + 1$

11. $1 - 10h + 45h^2 - 120h^3 + 210h^4 - 252h^5 + 210h^6 - 120h^7 + 45h^8 - 10h^9 + h^{10}$

13. $3 + 3h + h^2$

15. $3c^2 + 3ch + h^2$

17. $2(5x^4 + 10x^3h + 10x^2h^2 + 5xh^3 + h^4)$

19. $(1+1)^{10} = 1 + 10 + 45 + 120 + 210 + 252 + 210 + 120 + 45 + 10 + 1 = 1024$

21. $c^{20} + 20c^{19}h + 190c^{18}h^2 + 1140c^{17}h^3 + 4845c^{16}h^4 + \cdots + 4845c^4h^{16} + 1140c^3h^{17} + 190c^2h^{18} + 20ch^{19} + h^{20}$

23. The nth row of the triangle contains the coefficients in the expansion of $(a+b)^n$.

25.
```
   1   7   21   35    35   21   7  1
 1   8   28   56    70   56   28  8  1
1   9  36   84  126  126  84  36  9  1
1  10  45  120 210  252 210 120 45 10 1
```

27. $x^{10} - 10x^9h + 45x^8h^2 - 120x^7h^3 + 210x^6h^4 - 252x^5h^5 + 210x^4h^6 - 120x^3h^7 + 45x^2h^8 - 10xh^9 + h^{10}$

2.9 COMPLEX NUMBERS (*Page 88*)

1. True.

3. True.

5. False.

7. $5 + 2i$

9. $-5 + 0i$

11. $-3i$

13. $-2i$

15. $3i$

17. $12i$

19. $\frac{3}{4}i$

21. -27

23. $-\sqrt{6}$

25. $15i$

27. $12i$

29. $5i\sqrt{2}$

31. $(3 - \sqrt{3})i$

33. $10 + 7i$

35. $5 - 3i$

37. $10 + 2i$

39. $-10 + 6i$

41. $13i$

43. $23 + 14i$

45. $5 - 3i$

47. $\frac{13}{5} + \frac{1}{5}i$

49. $\frac{4}{5} - \frac{3}{5}i$

51. $\frac{3}{13} - \frac{2}{13}i$

53.

$\times$	i	i^2	i^3	i^4
i	-1	$-i$	1	i
i^2	$-i$	1	i	-1
i^3	1	i	-1	$-i$
i^4	i	-1	$-i$	1

55. $[(3 + i)(3 - i)](4 + 3i) = (9 - i^2)(4 + 3i) =$
$10(4 + 3i) = 40 + 30i$;
$(3 + i)[(3 - i)(4 + 3i)] = (3 + i)(15 + 5i) =$
$40 + 30i$.

57. (a) True; (b) true.

59. $\dfrac{a + bi}{c + di} = \dfrac{(a + bi)(c - di)}{(c + di)(c - di)} =$
$\dfrac{(ac + bd) + (bc - ad)i}{c^2 + d^2} = \dfrac{ac + bd}{c^2 + d^2} + \dfrac{bc - ad}{c^2 + d^2}i$

61.

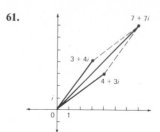

63.

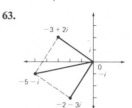

65.

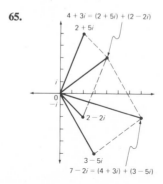

67.

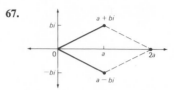

*69. $\sqrt{ab} = \sqrt{-[(-a)b]} = i\sqrt{(-a)b}$
$\qquad$ (since $(-a)b > 0$)
$\qquad = i\sqrt{-a}\,\sqrt{b}$
$\qquad$ (since $-a > 0$ and $b \geq 0$)
$\qquad = \sqrt{-(-a)}\,\sqrt{b}$
$\qquad$ (since $-a > 0$)
$\qquad = \sqrt{a}\,\sqrt{b}$

1. (a) False; (b) true; (c) false; (d) false; (e) true; (f) false.
2. (b)
3. (a)
4. (b)
5. (a) $7cd - 6a^2$; (b) $11x^2 - 14x - 21$.
6. (a) $(4 - 3b)(16 + 12b + 9b^2)$;
   (b) $3x(x^2 + 4)(x + 2)(x - 2)$.
7. (a) $(x - 2)(x^4 + 2x^3 + 4x^2 + 8x + 16)$;
   (b) $(2x + y^2)(x - 3y)$.

SAMPLE TEST FOR CHAPTER 2 (*Page 92*)

8. x
9. $-\dfrac{7 + x}{49x^2}$
10. $\dfrac{3x^2 - 18x + 25}{(x - 3)(x + 3)(2x - 5)}$
†11. $a^{20} - 40a^{19} + 760a^{18} - 9120a^{17} + \cdots$
†12. $43 + 23i$

Chapter 3: LINEAR FUNCTIONS

3.1 INTRODUCTION TO THE FUNCTION CONCEPT (*Page 101*)

1. Function: all reals.
3. Function: all $x > 0$.
5. Not a function.
7. Function: all $x \neq -1$.
9. Not a function.
11. Function: $x < -2$ or $x > 2$.
13. True.
15. False; -3.
17. True.
19. False: $-x^2 + 16$.
21. True.
23. (a) -3; (b) -1; (c) 0.
25. (a) 1; (b) 0; (c) $\frac{1}{4}$.
27. (a) -2; (b) -1; (c) $-\frac{7}{8}$.

29. (a) 2; (b) 0; (c) $\frac{5}{16}$.
31. (a) $-\frac{1}{2}$; (b) -1; (c) -2.
33. (a) -1; (b) does not exist; (c) $\sqrt[3]{2}$.
35. (a) 81; (b) 5; (c) $\frac{25}{36}$; (d) $\frac{1}{36}$.
37. $3h(2) = 24 \neq 48 = h(6)$
39. $x + 3$
41. $-\dfrac{1}{3x}$
43. 2
45. 1
47. $-4 - h$
*49. $-\dfrac{4 + h}{4(2 + h)^2}$

3.2 THE RECTANGULAR COORDINATE SYSTEM (*Page 107*)

1.

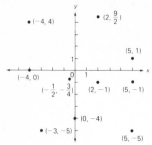

15.

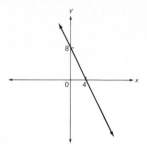

3. $x = -3$ for all y

5. $y = 2$ for $0 \leq x \leq 7$

7. $y = -x$ for $-2 \leq x \leq 1$

17.

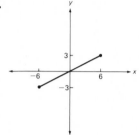

9.

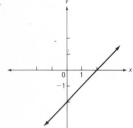

11.

19.

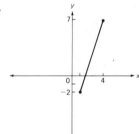

13.

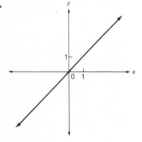

21.

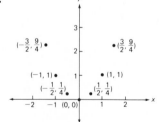

23.

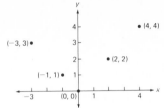

31.

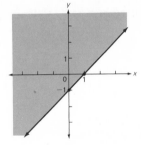

25.

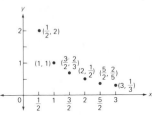

27.

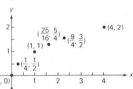

***33.**

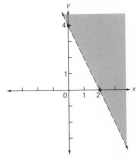

29.

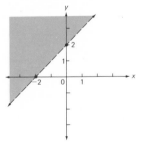

3.3   LINEAR FUNCTIONS   (*Page 115*)

1. $\frac{1}{4}$

3. $\frac{1}{9}$

5. 0

7. $-\frac{17}{28}$

9. $\sqrt{3} - 1$

11.

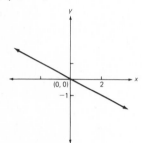

13.

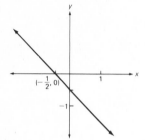

15.

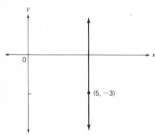

17.

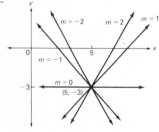

19.

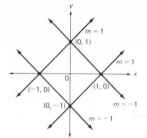

21. Slope $AD = \frac{1}{2}$ = Slope BC
  Slope $AB = -1$ = Slope DC

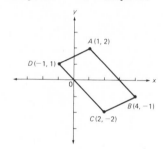

23. $y - 5 = 0$

25. $y - 1 = \frac{1}{2}(x - 2)$

27. $y = 5x$

29. $y + 3 = -\frac{3}{4}(x + 6)$

31. $y + \sqrt{2} = 10(x - \sqrt{2})$

33. $y = -\frac{2}{3}x + 2; m = -\frac{2}{3}, b = 2.$

35. $y = \frac{3}{2}x - 3; m = \frac{3}{2}, b = -3.$

37. $y = -\frac{2}{5}x + 2; m = -\frac{2}{5}, b = 2.$

39. $y = \frac{4}{3}x - \frac{7}{3}; m = \frac{4}{3}, b = -\frac{7}{3}.$

41. $y = \frac{3}{2}x + 3$

43. $y = -2x - 15$

45. $y = 27$

47.

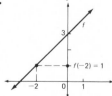

49.

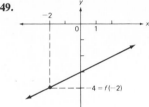

51. (a)

(b) $x = 5$; **(c)** (x, y) is on a vertical line if and only if x is a constant for all values y. Thus, the equation of a vertical line must be of the form $x = c$, where c is a constant.

53. (a)

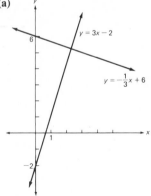

(b) All real numbers; **(c)** one.

55.

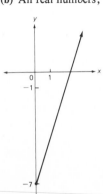

57.

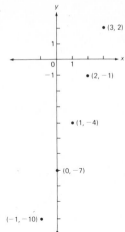

*59. (a) If $c = 0$ then $(0, 0)$ fits the equation and the line would then pass through the origin.

(b) $\dfrac{c}{b} = y$-intercept; $\dfrac{c}{a} = x$-intercept.

(c) $ax + by = c$; $\dfrac{ax}{c} + \dfrac{by}{c} = 1$;

$$\dfrac{x}{\left(\dfrac{c}{a}\right)} + \dfrac{y}{\left(\dfrac{c}{b}\right)} = 1; \dfrac{x}{q} - \dfrac{y}{p} = 1.$$

(d) $\dfrac{x}{\frac{3}{2}} + \dfrac{y}{-5} = 1$ or $10x - 3y = 15$

(e) $m = \dfrac{0 - (-5)}{\frac{3}{2} - (0)} = \dfrac{5}{\frac{3}{2}} = \dfrac{10}{3}$; $y = \dfrac{10}{3}x - 5.$

3.4 SOME SPECIAL FUNCTIONS (*Page 121*)

1. Domain: all reals
Range: $y \geq 0$

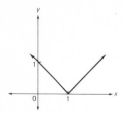

3. Domain: all reals
Range: $y \geq 0$

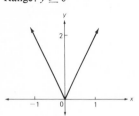

5. Domain: all reals
Range: $y \geq 0$

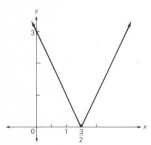

7. Domain: $x \geq -1$
Range: $y \leq 3$

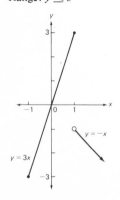

9. Domain: $-2 < x \leq 3$
Range: $-2 < y \leq 4$

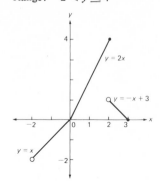

11.

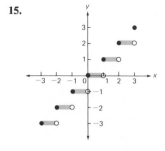

13.

15.

*17.

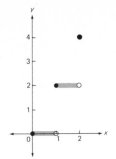

*19.

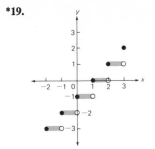

3.5 SYSTEMS OF LINEAR EQUATIONS *(Page 127)*

25. $\left(\frac{17}{18}, \frac{7}{18}\right)$

27. $\left(\frac{7}{2}, 4\right)$

29. $(1, -2, 3)$

31. $(2, -1, -1)$

33. $\left(\dfrac{ce - bf}{ae - bd}, \dfrac{af - cd}{ae - bd}\right)$

*35. $(-20, -20)$

37. 40 field goals; 16 foul shots.

39. \$1700 room and board; \$3600 tuition.

*41. Speed of plane = 410 miles per hour; wind velocity = 10 miles per hour.

*43. 180 quarts of 44¢ oil; 220 quarts of 64¢ oil.

45. \$2800 at 8%; \$3200 at $7\frac{1}{2}$%.

*47. 6 miles by car; 72 miles by train.

1.

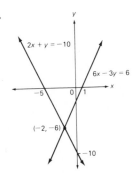

3.

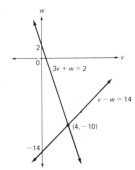

5. $(-4, -1)$

7. $(-26, -62)$

9. $(-16, -38)$

11. $\left(\frac{21}{88}, -\frac{9}{22}\right)$

13. $(5, 1)$

15. $(8, 5)$

17. $(3, -11)$

19. $(1, 0)$

21. $\left(1, \frac{1}{2}\right)$

23. $(1, 0)$

3.6 CLASSIFICATION OF LINEAR SYSTEMS (*Page 133*)

1. Consistent.

3. Inconsistent.

5. Consistent.

7. Dependent.

9. Inconsistent.

11. $y - 7 = -\frac{1}{3}(x - 4)$

13. $y - 1 = -\frac{1}{2}(x + 5)$

15.

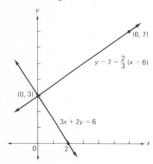

17. Parallel.

19. (a) The slopes of the lines are negative
reciprocals of one another.

(b)

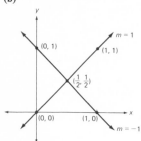

21. Diagonal QS has slope $-\frac{17}{7}$ and has equation
$y = -\frac{17}{7}x - 1$. Diagonal PR has slope $\frac{7}{17}$ and
has equation $y = \frac{7}{17}x + \frac{152}{17}$. Since $-\frac{17}{7} = -\frac{1}{\frac{7}{17}}$,
it follows that $QS \perp PR$.

*23. Horizontal lines have slope zero, which does not
have a reciprocal. Also, vertical lines have no
slope, and therefore a slope comparison cannot
be made.

25. (a) (i)1, (ii) -6;   (b) $\dfrac{x_1 + x_2}{2}$.

*27. $t = \frac{49}{2}$

29.

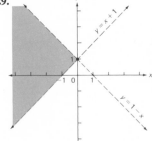

*31.

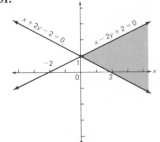

3.7 MATRICES AND DETERMINANTS (*Page 144*)

1. $(-4, -1)$

3. $(0, 9)$

5. $(\frac{1}{8}, \frac{1}{4})$

7. $(1, 0)$

9. $(5, -2)$

11. $(-\frac{2}{3}, 2)$

13. $(5, 2, -1)$

15. 17

17. 94

19. -60

21. 0

23. 35

25. -45

27. 4

29. $(-1, 2)$

31. $(3, 2)$

33. $(1, -1)$

35. $(1, 2, 3)$

37. $(0, -2, 3)$

39. 6

*41. (a) $\begin{vmatrix} a_1 & b_1 \\ a_2 & b_2 \end{vmatrix} = a_1b_2 - a_2b_1 = a_1b_2 - b_1a_2$

$$= \begin{vmatrix} a_1 & a_2 \\ b_1 & b_2 \end{vmatrix}$$

(b) $\begin{vmatrix} a_1 & b_1 & c_1 \\ a_2 & b_2 & c_2 \\ a_3 & b_3 & c_3 \end{vmatrix} = a_1b_2c_3 + a_2b_3c_1 + a_3b_1c_2$

$$- a_1b_3c_2 - a_2b_1c_3 - a_3b_2c_1$$

$$= \begin{vmatrix} a_1 & a_2 & a_3 \\ b_1 & b_2 & b_3 \\ c_1 & c_2 & c_3 \end{vmatrix}$$

*43. (a) $a_2 = ka_1, b_2 = kb_1, c_2 = kc_1,$ so

$$\begin{vmatrix} a_1 & b_1 & c_1 \\ a_2 & b_2 & c_2 \\ a_3 & b_3 & c_3 \end{vmatrix} = \begin{vmatrix} a_1 & b_1 & c_1 \\ ka_1 & kb_1 & kc_1 \\ a_3 & b_3 & c_3 \end{vmatrix}$$

$$= ka_1b_1c_3 + ka_1b_3c_1 + ka_3b_1c_1 - ka_1b_3c_1$$
$$- ka_1b_1c_3 - ka_3b_1c_1 = 0$$

(Similarly for the other rows.)

(b) Let $b_1 = ka_1, b_2 = ka_2, b_3 = ka_3$. Then

$$\begin{vmatrix} a_1 & b_1 & c_1 \\ a_2 & b_2 & c_3 \\ a_3 & b_3 & c_3 \end{vmatrix} = \begin{vmatrix} a_1 & ka_1 & c_1 \\ a_2 & ka_2 & c_2 \\ a_3 & ka_3 & c_3 \end{vmatrix}$$

$$= \begin{vmatrix} a_1 & a_2 & a_3 \\ ka_1 & ka_2 & ka_3 \\ c_1 & c_2 & c_3 \end{vmatrix}$$

(by Exercise 41)

$$= 0 \quad \text{(by part (a))}$$

(Similarly for the other columns.)

1. (a) False; **(b)** false; **(c)** true; **(d)** false.

2.

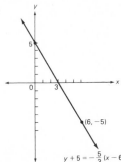

$y + 5 = -\frac{5}{3}(x - 6)$

3. (a) $y = 4x + 12$; **(b)** $y = -\frac{1}{3}x - \frac{16}{3}$.

4. Slope $PQ = 4$, slope $PR = -\frac{1}{3}$, and these numbers are not negative reciprocals of one another.

5. (a) $-4 \le x \le -1$; **(b)** $-4 \le y \le 8$; **(c)** $y = f(x) = 4x + 12$.

*45. $\begin{vmatrix} 27 & 3 \\ 105 & -75 \end{vmatrix} = 3\begin{vmatrix} 9 & 1 \\ 105 & -75 \end{vmatrix} = 45\begin{vmatrix} 9 & 1 \\ 7 & -5 \end{vmatrix}$

$$= -2340$$

$$\begin{vmatrix} 27 & 3 \\ 105 & -75 \end{vmatrix} = 3\begin{vmatrix} 9 & 1 \\ 105 & -75 \end{vmatrix} = 9\begin{vmatrix} 3 & 1 \\ 35 & -75 \end{vmatrix}$$

$$= 45\begin{vmatrix} 3 & 1 \\ 7 & -15 \end{vmatrix}$$

$$= -2340$$

*47. $\begin{vmatrix} a_1 + kb_1 & b_1 \\ a_2 + kb_2 & b_2 \end{vmatrix} = \begin{vmatrix} a_1 & b_1 \\ a_2 & b_2 \end{vmatrix} + \begin{vmatrix} kb_1 & b_1 \\ kb_2 & b_2 \end{vmatrix}$

(by Exercise 46)

$$= \begin{vmatrix} a_1 & b_1 \\ a_2 & b_2 \end{vmatrix} + 0$$

(by Exercise 42)

$$= \begin{vmatrix} a_1 & b_1 \\ a_2 & b_2 \end{vmatrix}$$

SAMPLE TEST FOR CHAPTER 3 *(Page 148)*

6. (a) $y = \frac{5}{2}x + 3$; **(b)** $(0, 3)$.

7. Domain: all real numbers
Range: all $y \ge 0$

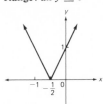

8. Dependent.

9. Consistent; $(3, -5)$.

10. Inconsistent.

†11. $(-1, 2, 4)$

Chapter 4: QUADRATIC FUNCTIONS

4.1   GRAPHING QUADRATIC FUNCTIONS   (*Page 156*)

1.

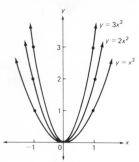

3.

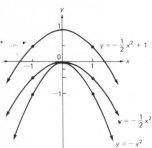

5.

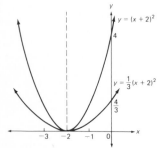

7.

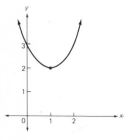

9.

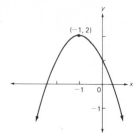

11.

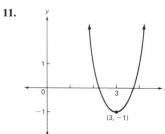

13.

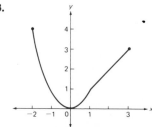

15. 1. (a) $y \geq 0$;   (b) $y \geq 0$;   (c) $y \geq 0$.
   3. (a) $y \leq 0$;   (b) $y \leq 0$;   (c) $y \leq 1$.
   5. (a) $y \geq 0$;   (b) $y \geq 0$.
   7. $y \geq 2$;   **9.** $y \leq 2$;   **11.** $y \geq -1$.

17. Not a function because for $x > 0$ there are two corresponding y-values.

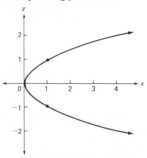

*19. $a = -1, c = 4$.

1. $(x + 1)^2 - 6$
3. $-(x + 3)^2 + 11$
5. $-(x - \frac{3}{2})^2 - \frac{7}{4}$
7. $(x + \frac{5}{2})^2 - \frac{33}{4}$
9. $2(x + 1)^2 - 5$
11. $3(x + 1)^2 - 8$
13. $(x - \frac{1}{4})^2 + \frac{15}{16}$
15. $\frac{3}{4}(x - \frac{2}{3})^2 - \frac{2}{3}$
17. $-5(x + \frac{1}{5})^2 + 1$
*19. $(x - \frac{1}{2})^2 - 3$
*21. $h = -\dfrac{b}{2a}; k = \dfrac{4ac - b^2}{4a}.$

23.

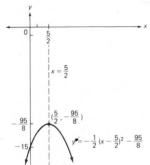

25.

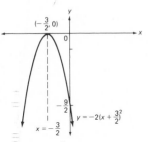

27.

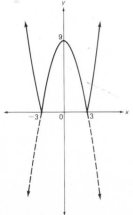

4.2 COMPLETING THE SQUARE (*Page 161*)

29.

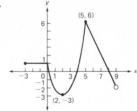

31. $a > 0; k = 0.$
33. $a > 0; k = 2.$
35. $a < 0; k > 0.$
*37. $y = -2x^2 + 3x + 7$

4.3 MAXIMUM AND MINIMUM (*Page 165*)

1.

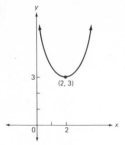

3.

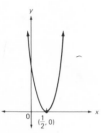

5. Maximum $= 7$ at $x = 5$.

7. Minimum $= 36$ at $x = 2$.

9. Minimum $= 0$ at $x = \frac{7}{2}$.

11. Maximum $= \frac{1}{9}$ at $x = -\frac{1}{3}$.

13. 4

15. $\frac{n}{2}, \frac{n}{2}$.

17. $w = 10$ feet; $l = 10$ feet.

19. 16 feet.

*21. 15 centimeters each.

4.4 THE QUADRATIC FORMULA (*Page 172*)

1. $x = -2$; $x = 5$.

3. $x = 3$

5. $x = -2$; $x = -\frac{1}{3}$.

7. $x = 1 - \sqrt{5}$; $x = 1 + \sqrt{5}$.

9. $x = \dfrac{3 - \sqrt{17}}{4}$; $x = \dfrac{3 + \sqrt{17}}{4}$.

11. 3; $-\frac{1}{2}$.

13. None.

15. $\frac{3}{2}$; 2.

17. None.

19. 0; (b).

21. 36; (c).

23. 13; (d).

25. 41; (d).

27. -7; (a).

29. Once; (a) $(\frac{1}{3}, 0)$; (b) $(0, 1)$; (c) $(\frac{1}{3}, 0)$.

31. $-1 < x < 3$

33. $x \leq -5$ or $x \geq 2$

35. No solution.

37.

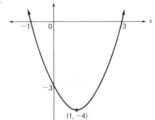

39.

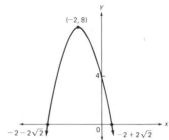

41.

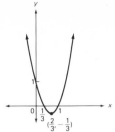

45. -6 or 6

47. $c > -4$

49. $a < -\frac{1}{4}$

*51. (a) $-\dfrac{b}{a}$; (b) $\dfrac{c}{a}$.

53. $x = -3$; $x = -2$; $x = 2$.

43.

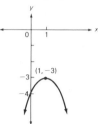

4.5 CIRCLES (*Page 179*)

1.

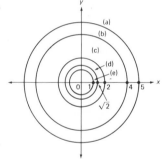

17.

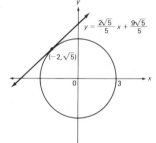

3. $(x - 2)^2 + (y - 5)^2 = 1$; $C(2, 5)$, $r = 1$.

5. $(x - 4)^2 + (y - 0)^2 = 2$; $C(4, 0)$, $r = \sqrt{2}$.

7. $(x - 10)^2 + (y + 10)^2 = 100$; $C(10, -10)$, $r = 10$.

9. $(x + \frac{3}{4})^2 + (y - 1)^2 = 9$; $C(-\frac{3}{4}, 1)$, $r = 3$.

11. $(x - 2)^2 + y^2 = 4$

13. $(x + 3)^2 + (y - 3)^2 = 7$

15.

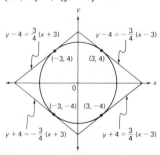

19.

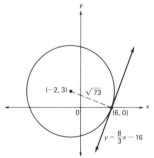

*21. $y = -\frac{1}{2}x + 10$

*23. $5x + 12y = 26$

4.6 SOLVING NONLINEAR SYSTEMS (*Page 183*)

1.

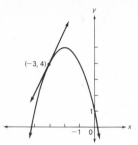

3.

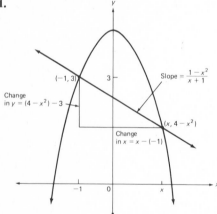

5.

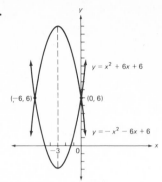

7. No solutions.

9. No solutions.

11. $(-1, -1); (-2, -2)$.

13. $(0, 0); (2, 0)$.

*15. $(3, -3); (3 + \sqrt{3}, -2); (3 - \sqrt{3}, -2)$.

4.7 DIFFERENCE QUOTIENTS AND TANGENTS (*Page 188*)

1.

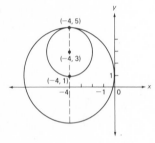

3.

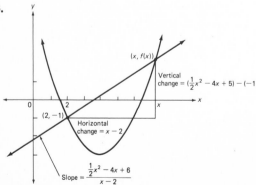

5. (a) $\dfrac{g(x) - g(-2)}{x - (-2)} = x + 4$

x	-4	$-3\frac{1}{2}$	-3	$-2\frac{1}{2}$	$-2\frac{1}{10}$	$-2\frac{1}{100}$
$x + 4$	0	$\frac{1}{2}$	1	$1\frac{1}{2}$	$1\frac{9}{10}$	$1\frac{99}{100}$

(b)

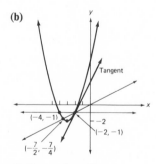

(c) $2; y = 2x + 3$.

7. $x + 2; y = 4x - 8$.

9. $-(3 + x); y = -6x + 9$.

11. $3x - 2; y = x - 2$.

13. $-\frac{1}{3}(2x + 9); y = -7x + 24$.

15. $2c$

17. $6c - 5$

*19. $h = x - c$ implies $x = c + h$. Substitute into

$\dfrac{f(x) - f(c)}{x - c}$ to get

$$\dfrac{f(x) - f(c)}{x - c} = \dfrac{f(c + h) - f(c)}{(c + h) - c}$$

$$= \dfrac{f(c + h) - f(c)}{h}$$

*21. $-6 - h;\ -6$.

*23. $-11 - 2h;\ -11$.

*25. $4a + b$

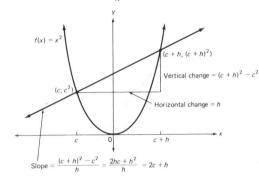

$f(x) = x^2$

$(c + h, (c + h)^2)$

Vertical change $= (c + h)^2 - c^2$

(c, c^2)

Horizontal change $= h$

Slope $= \dfrac{(c + h)^2 - c^2}{h} = \dfrac{2hc + h^2}{h} = 2c + h$

4.8   THE ELLIPSE AND THE HYPERBOLA   (*Page 199*)

1. Ellipse; center $(0, 0)$, vertices $(\pm 5, 0)$, foci $(\pm 3, 0)$.

3. Hyperbola; center $(0, 0)$, vertices $(\pm 6, 0)$, foci $(\pm\sqrt{61}, 0)$, asymptotes $y = \pm\frac{5}{6}x$.

5. Ellipse; center $(0, 0)$, vertices $(0, \pm 5)$, foci $(0, \pm\sqrt{21})$.

7. Ellipse; center $(0, 1)$, vertices $(0, -1)$ and $(0, 3)$, foci $(0, 1 - \sqrt{3})$ and $(0, 1 + \sqrt{3})$.

9. Ellipse:

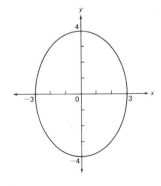

11. Hyperbola:

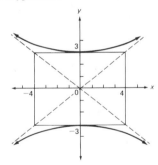

13. Ellipse:

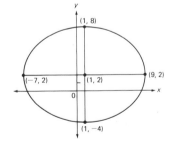

15. Hyperbola:

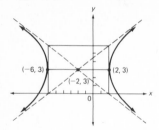

17. $(x - 1)^2 + (y + 2)^2 = 4$; circle; center $(1, -2)$; $r = 2$.

19. $\dfrac{(x + 1)^2}{4} + \dfrac{y^2}{1} = 1$; ellipse; center $(-1, 0)$;
    vertices $(-3, 0)$ and $(1, 0)$; foci $(-1 - \sqrt{3}, 0)$
    and $(-1 + \sqrt{3}, 0)$.

*21. $\dfrac{(x + 1)^2}{16} - \dfrac{(y - 3)^2}{9} = 1$; hyperbola; center
    $(-1, 3)$; vertices $(-5, 3)$ and $(3, 3)$; foci $(-6, 3)$
    and $(4, 3)$.

23. $\dfrac{x^2}{25} + \dfrac{y^2}{9} = 1$

25. $\dfrac{x^2}{16} + \dfrac{y^2}{25} = 1$

27. $\dfrac{(x - 2)^2}{16} + \dfrac{(y + 3)^2}{36} = 1$

29. $\dfrac{x^2}{16} - \dfrac{y^2}{20} = 1$

31. $\dfrac{(y - 3)^2}{9} - \dfrac{(x + 2)^2}{7} = 1$

*33. $\dfrac{x^2}{16} - \dfrac{y^2}{4} = 1$

35. (a) $\sqrt{x^2 + (y - p)^2}$; (b) $y + p$.
    (c) $\sqrt{x^2 + (y - p)^2} = y + p$
    $$x^2 + (y - p)^2 = (y + p)^2$$
    $$x^2 = 4py$$

37. $x^2 = -12y$

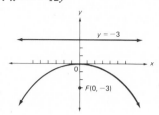

39. $(0, -1)$; $y = 1$.

*41. $(x - 2)^2 = 8(y + 5)$; $y = \frac{1}{8}x^2 - \frac{1}{2}x - \frac{9}{2}$.

SAMPLE TEST FOR CHAPTER 4 (*Page 203*)

1.

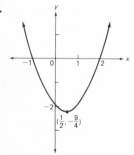

2. (a) $y = -5(x - 2)^2 + 19$; (b) $(2, 19)$;
   (c) $x = 2$; (d) domain: all real numbers;
   range: all $y \le 19$.

3. (a) $x = -\frac{1}{3}, x = 3$; (b) $2 - \sqrt{11}, 2 + \sqrt{11}$.

4. 5; two irrational numbers.

5. 169; two rational numbers.

6. Maximum $= 20$ at $x = -6$.

7. (a)

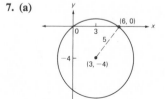

   (b) $y = -\frac{3}{4}x + \frac{9}{2}$

8. $C(-\frac{1}{2}, 7)$; $r = 5$.

9. $(0, 2)$; $(4, 2)$; $(2 - \sqrt{3}, 1)$; $(2 + \sqrt{3}, 1)$.

10. 6 feet.

†11. (a) $x + 4$; (b) $9x - 25$.

†12. $\dfrac{(x + 2)^2}{1} - \dfrac{(y - 3)^2}{4} = 1$. A hyperbola with
    center at $(-2, 3)$, and vertices at $(-3, 3)$, and
    $(-1, 3)$.

Chapter 5: POLYNOMIAL AND RATIONAL FUNCTIONS

5.1   HINTS FOR GRAPHING   (*Page 210*)

9. $y = (x - 3)^4 + 2$

11. $y = |x + \frac{3}{4}|$

13.

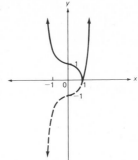

1.

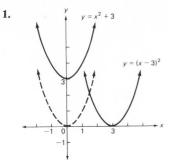

3.

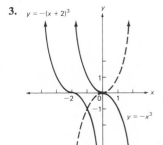

5.

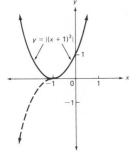

7.

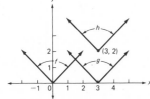

*15.

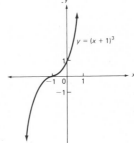

*17.

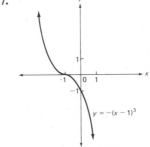

*19. (a) $x^2 + 2x + 4$;   (b) 12.

5.2   GRAPHING SOME SPECIAL RATIONAL FUNCTIONS   *(Page 214)*

1.

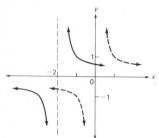

7. Asymptotes: $x = -4$, $y = -2$.

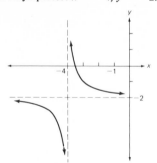

3.

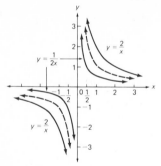

9. Asymptotes: $x = 2$, $y = 0$.

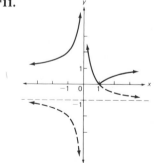

5.

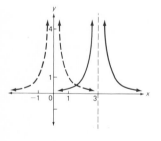

*11.

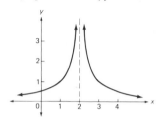

5.3   SYNTHETIC DIVISION   *(Page 220)*

1. $x^2 + x - 2$; $r = 0$.

3. $2x^2 - x - 2$; $r = 9$.

5. $x^2 + 7x + 7$; $r = 22$.

7. $x^3 - 5x^2 + 17x - 36$; $r = 73$.

9. $2x^3 + 2x^2 - x + 3$; $r = 1$.

11. $x^2 + 3x + 9$; $r = 0$.

13. $x^2 - 3x + 9$; $r = 0$.

15. $x^3 + 2x^2 + 4x + 8$; $r = 0$.

17. $x^3 - 2x^2 + 4x - 8$; $r = 32$.

19. $x^3 + \frac{1}{2}x^2 + \frac{5}{6}x + \frac{7}{12}$; $r = \frac{47}{60}$.

*21. $x + 2$; $r = 0$.

5.4   THE REMAINDER AND FACTOR THEOREMS   *(Page 223)*

1. 8

3. 59

5. 0

7. 1

9. 67

11. -4

For Exercises 13 through 23 use synthetic division to obtain $p(c) = 0$, showing $x - c$ to be a factor of $p(x)$. The remaining factors of $p(x)$ are obtained by factoring the quotient obtained in the synthetic division.

13. $(x + 1)(x + 2)(x + 3)$
15. $(x - 2)(x + 3)(x + 4)$
17. $-(x + 2)(x - 3)(x + 1)$
19. $(x - 5)(3x + 4)(2x - 1)$

21. $(x + 2)^2(x + 1)(x - 1)$
23. $x^2(x + 3)^2(x + 1)(x - 1)$
*25. -36
*27. 2

5.5 THE RATIONAL ROOT THEOREM *(Page 229)*

1. $2; 3; -7.$
3. $-\frac{1}{2}; 4.$
5. -1
7. $-3; -\sqrt{2}; \sqrt{2}.$
9. $-2; 3.$
11. $-\sqrt{3}; \sqrt{3}; 2; 3.$
13. $-3; 3; -2 - \sqrt{2}; -2 + \sqrt{2}.$
15. $-\frac{1}{3}; 2; \frac{5}{2}.$

17. $(x - 2)(x - 4)(x + 3)$
19. $(x + 2)(x + 4)(x - 6)$
21. $(-2, -27); (2, 1); (3, 8).$
*23. $(-3, -161); (5, 335); (\frac{1}{3}, \frac{253}{27}).$
*25. By the rational root theorem, the only possible rational roots are:
$$\pm 1, \pm\tfrac{1}{2}, \pm 2, \pm 4, \pm 8$$
Using synthetic division we see that none of these are roots.

5.6 POLYNOMIAL AND RATIONAL FUNCTIONS *(Page 235)*

1. $f(x) < 0$ on both $(-\infty, 1)$ and $(2, 3)$; $f(x) > 0$ on both $(1, 2)$ and $(3, \infty)$.
3. $f(x) < 0$ on $(-\infty, -4)$ and $(\frac{1}{3}, 2)$; $f(x) > 0$ on $(-4, 0), (0, \frac{1}{3})$, and $(2, \infty)$.
5. $x < 1$ or $x > 4$
7. $x < -1$ or $\frac{1}{3} < x < 10$
9. $f(x) < 0$ on $(-\infty, -4)$ and on $(-2, 0)$; $f(x) > 0$ on $(-4, -2)$ and on $(0, \infty)$.
11. $f(x) < 0$ on $(-\infty, -1)$ and on $(0, \frac{7}{2})$; $f(x) > 0$ on $(-1, 0)$ and on $(\frac{7}{2}, \infty)$.
13. Same as Exercise 1.
15. $f(x) < 0$ on both $(-\infty, -1)$ and $(0, 1)$; $f(x) > 0$ on both $(-1, 0)$ and $(1, \infty)$.
17. $f(x) < 0$ on both $(-\infty, -1)$ and $(\frac{1}{2}, 3)$; $f(x) > 0$ on both $(-1, \frac{1}{2})$ and $(3, \infty)$.
19. 20
21. 7
23. 4
25. $\frac{17}{4}$
27. 2
29. $-\frac{5}{2}; 2.$

31. $-\frac{3}{2}; 0; 5.$
33. No solution.
*35. $x > 1$

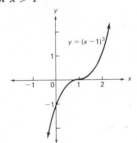

*37.

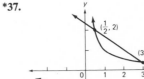

5.7 DECOMPOSING RATIONAL FUNCTIONS *(Page 242)*

1. $\dfrac{1}{x + 1} + \dfrac{1}{x - 1}$
3. $\dfrac{2}{x - 3} - \dfrac{1}{x + 2}$
5. $\dfrac{6}{x - 4} + \dfrac{3}{x - 2} - \dfrac{4}{x + 1}$

7. $\dfrac{3}{x - 2} + \dfrac{3}{(x - 2)^2}$
9. $\dfrac{6}{5x + 2} - \dfrac{3}{3x - 4}$
11. $\dfrac{1}{x - 2} + \dfrac{3}{(x - 2)^2} - \dfrac{2}{(x - 2)^3}$

13. $x + \dfrac{1}{x-1} - \dfrac{1}{x+1}$

15. $3x^2 + 1 + \dfrac{1}{2x-1} - \dfrac{3}{(2x-1)^2}$

17. $10x - 5 + \dfrac{6}{x-3} + \dfrac{14}{x+2}$

*19. $x^2 - 1 - \dfrac{1}{x-1} + \dfrac{2}{(x-1)^2} + \dfrac{6}{(x-1)^3}$

SAMPLE TEST FOR CHAPTER 5 (*Page 244*)

1. No asymptotes.

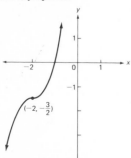

2. Horizontal asymptote: $y = 2$
Vertical asymptote: $x = 0$

3. $-3 \,\big|\, 2 + 5 + 0 - \;\,1 - 21 + \;\,7$
         $\underline{\quad\;\; - 6 + 3 - \;\,9 + 30 - 27}$
         $\;\;\; 2 - 1 + 3 - 10 + \;\,9 \,\big|\, {-20}$

quotient: $2x^4 - x^3 + 3x^2 - 10x + 9$
remainder: -20

4. (a) When $p(x)$ is divided by $x - \frac{1}{3}$ the remainder is $\frac{2}{3}$. Then, by the remainder theorem, we have $p(\frac{1}{3}) = \frac{2}{3}$.

(b) Since $p(\frac{1}{3}) \neq 0$, the factor theorem says that $x - \frac{1}{3}$ is not a factor of $p(x)$.

5. $2 \,\big|\, 1 - 4 + 7 - 12 + 12$
         $\underline{\quad\;\; + 2 - 4 + \;\,6 - 12}$
         $\;\; 1 - 2 + 3 - \;\,6 + \big|\, 0 = r$

Since $r = 0$, $x - 2$ is a factor of $p(x)$, and we get
$$\begin{aligned} p(x) &= (x - 2)(x^3 - 2x^2 + 3x - 6) \\ &= (x - 2)[x^2(x - 2) + 3(x - 2)] \\ &= (x - 2)(x^2 + 3)(x - 2) \\ &= (x - 2)^2(x^2 + 3) \end{aligned}$$

6. $(x + 3)^2(x^2 - x + 1)$

7. $-3; \frac{1}{2}; 1 - \sqrt{3}; 1 + \sqrt{3}.$

8. $f(x) < 0$ on both $(-\infty, -3)$ and $(0, 2)$;
$f(x) > 0$ on both $(-3, 0)$ and $(2, \infty)$.

9. $-\frac{3}{2}; 2.$

10. $(2, 8); (3, 27); (-5, -125).$

†11. $\dfrac{2}{x+5} - \dfrac{1}{x-5}$

Chapter 6: RADICAL FUNCTIONS

6.1 GRAPHING SOME SPECIAL RADICAL FUNCTIONS (*Page 251*)

1. Domain: $x \geq -2$

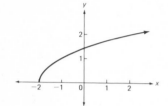

3. Domain: $x \geq 3$

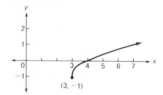

5. Domain: $x \leq 0$

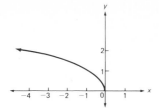

7. Domain: all real x

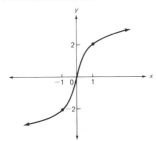

9. Domain: all real x

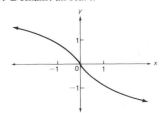

11. Domain: $x > 0$
Asymptotes: $x = 0, y = -1$.

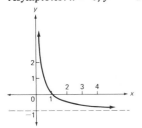

13. (a) $f(-x) = \dfrac{1}{\sqrt[3]{-x}} = \dfrac{1}{-\sqrt[3]{x}} = -\dfrac{1}{\sqrt[3]{x}} = -f(x)$

(b) All $x \neq 0$.

(c)

x	$\frac{1}{27}$	$\frac{1}{8}$	1	8
y	3	2	1	$\frac{1}{2}$

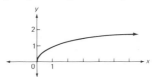

(d) $x = 0; y = 0$.

*15. $y = \sqrt[4]{x}$ is equivalent to $y^4 = x$ for $x \geq 0$.

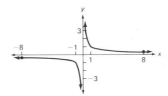

*17.

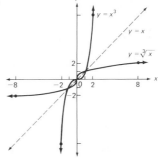

6.2 RADICAL EQUATIONS (*Page 257*)

1. 17

3. 10

5. $0; \frac{5}{2}$.

7. 2

9. 5

11. No solutions.

13. 9

15. 10

17. 4

19. No solutions.

21. $\frac{1}{3}$

23. $\frac{27}{8}$

25.

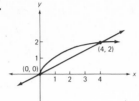

27.

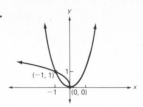

*29. $A = 4\pi r^2$; 16π.

*31. $(9, 3)$

6.3 DETERMINING THE SIGNS OF RADICAL FUNCTIONS (*Page 261*)

1. (a) $x \geq 2$; (b) $f(x) > 0$ for all $x > 2$; (c) 2.

3. (a) All real x; (b) $f(x) > 0$ for all $x \neq 2$;
(c) 2.

5. (a) $-3 \leq x \leq 3$, $x \neq 0$; (b) $f(x) < 0$ on
$(-3, 0)$; $f(x) > 0$ on $(0, 3)$; (c) $-3, 3$.

7. (a) $x > 0$; (b) $f(x) < 0$ on $(0, 1)$; $f(x) > 0$ on
$(1, \infty)$; (c) 1.

9. (a) All real x; (b) $f(x) < 0$ on $(-4, 2)$;
$f(x) > 0$ on $(2, \infty)$ and on $(-\infty, -4)$;
(c) $-4, 2$.

11. (a) $x > 0$; (b) $f(x) < 0$ on $(0, 3)$; $f(x) > 0$ on
$(3, \infty)$; (c) 3.

*13. (a) All $x \neq 0$; (b) $f(x) < 0$ on $(0, 4)$; $f(x) > 0$
on $(-\infty, 0)$ and $(4, \infty)$; (c) 4.

6.4 COMBINING FUNCTIONS (*Page 266*)

1. (a) $-1, 5, 4$; (b) $5x - 1$, all reals; (c) 4.

3. (a) $-2, \frac{7}{2}, -7$; (b) $6x^2 - 5x - 6$, all reals;
(c) -7.

5. (a) $2, 1$; (b) $6x + 1$, all reals; (c) 1.

7. (a) $x^2 + \sqrt{x}$; $x \geq 0$; (b) $x^{3/2}$; $x > 0$;
(c) x; $x \geq 0$.

9. (a) $x^3 - 1 + \dfrac{1}{x}$; all $x \neq 0$; (b) $x^4 - x$;

all $x \neq 0$; (c) $\dfrac{1}{x^3} - 1$; all $x \neq 0$.

11. (a) $x^2 + 6x + 8 + \sqrt{x - 2}$; $x \geq 2$;

(b) $\dfrac{x^2 + 6x + 8}{\sqrt{x - 2}}$; $x > 2$;

(c) $x + 6 + 6\sqrt{x - 2}$; $x \geq 2$.

13. (a) $\dfrac{1}{2\sqrt[3]{x} - 1}$; (b) $\dfrac{2}{\sqrt[3]{x}} - 1$; (c) $\dfrac{1}{\sqrt[3]{2x - 1}}$.

15. x

*17. $g(x) = \sqrt[3]{x}$

6.5 DECOMPOSITION OF COMPOSITE FUNCTIONS (*Page 270*)

(Other answers are possible for Exercises 1–17.)

1. $g(x) = 3x + 1$; $f(x) = x^2$.

3. $g(x) = 1 - 4x$; $f(x) = \sqrt{x}$.

5. $g(x) = \dfrac{x + 1}{x - 1}$; $f(x) = x^2$.

7. $g(x) = 3x^2 - 1$; $f(x) = x^{-3}$.

9. $g(x) = \dfrac{x}{x - 1}$; $f(x) = \sqrt{x}$.

11. $g(x) = (x^2 - x - 1)^3$; $f(x) = \sqrt{x}$.

13. $f(x) = 2x + 1$; $g(x) = x^{1/2}$; $h(x) = x^3$.

15. $f(x) = \dfrac{x}{x + 1}$; $g(x) = x^5$; $h(x) = x^{1/2}$.

17. $f(x) = x^2 - 9$; $g(x) = x^2$; $h(x) = x^{1/3}$.

6.6 CROSS SECTIONS OF SOLIDS OF REVOLUTION (*Page 275*)

1. (a)

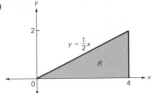

(b)

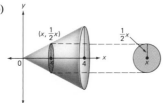

(c) $A(x) = \pi(\frac{1}{2}x)^2 = \frac{1}{4}\pi x^2$

3. $A(x) = \pi(3^2) - \pi(3 - \frac{1}{2}x)^2 = \pi(3x - \frac{1}{4}x^2)$

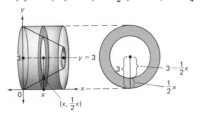

1. Domain: All reals

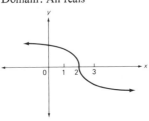

2. Domain: $x > 0$
Asymptotes: $y = 2, x = 0$

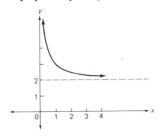

5. $A(x) = \pi(2^2) - \pi(\frac{1}{2}x^2)^2 = \pi(4 - \frac{1}{4}x^4)$

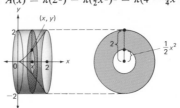

7. $A(x) = \pi(3 - \frac{1}{2}x^2)^2 - \pi(1)^2$
$\quad = \pi(8 - 3x^2 + \frac{1}{4}x^4)$

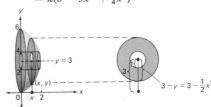

9. $A(x) = \pi x^{2/3}$

11. $A(x) = \pi(x^2 - x^4)$

13. $A(x) = \pi(x^4 - 5x^2 + 4x)$

15. $A(x) = \pi\left(x^2 - 5x + \frac{25}{4} - \frac{1}{x^2}\right)$

SAMPLE TEST FOR CHAPTER 6 (*Page 278*)

3. $\frac{2}{9}$

4. $-\frac{64}{27}; \frac{1}{8}$.

5. $f(x) < 0$ on $(2, 4)$; $f(x) > 0$ on $(-\infty, 2)$ and $(4, \infty)$.

6. $f(x) < 0$ on $(0, 4)$; $f(x) > 0$ on $(4, \infty)$.

7. $(f - g)(x) = \dfrac{1}{x^2 - 1} - \sqrt{x + 2}$; all $x \geq -2$ and $x \neq \pm 1$. $\dfrac{f}{g}(x) = \dfrac{1}{(x^2 - 1)\sqrt{x + 2}}$; all $x > -2$ and $x \neq \pm 1$.

8. $(f \circ g)(x) = \dfrac{1}{1 - x}$; all $x > 0$ and $x \neq 1$.

$(g \circ f)(x) = \dfrac{1}{\sqrt{1 - x^2}}$; $-1 < x < 1$.

(Other answers are possible for Exercises 9 and 10.)

9. $g(x) = x - 2; f(x) = x^{2/3}$.

10. $h(x) = 2x - 1; g(x) = x^{3/2}; f(x) = \dfrac{1}{x}$.

†11. $A(x) = \pi(1 + \sqrt{x - 2})^2 - \pi$
$\quad = \pi(2\sqrt{x - 2} + x - 2)$

Chapter 7: EXPONENTIAL AND LOGARITHMIC FUNCTIONS

7.1 INVERSE FUNCTIONS *(Page 286)*

1. $(f \circ g)(x) = \frac{1}{3}(3x + 9) - 3 = x$
$(g \circ f)(x) = 3(\frac{1}{3}x - 3) + 9 = x$

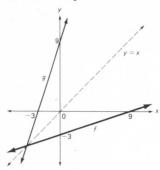

3. $(f \circ g)(x) = (\sqrt[3]{x} - 1 + 1)^3 = x$
$(g \circ f)(x) = \sqrt[3]{(x + 1)^3} - 1 = x$

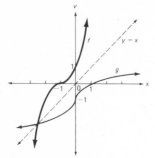

5. $(f \circ g)(x) = \dfrac{1}{\dfrac{1}{x} + 1 - 1} = x$

$(g \circ f)(x) = \dfrac{1}{\dfrac{1}{x - 1}} + 1 = x$

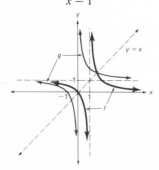

7. $g(x) = \sqrt[3]{x} + 5$

9. $g(x) = \frac{3}{2}x + \frac{3}{2}$

11. $g(x) = \sqrt[5]{x} + 1$

13. $g(x) = x^{5/3}$

15. $g(x) = 2 + \dfrac{2}{x}$

17. $g(x) = \dfrac{1}{\sqrt[5]{x}}$

19. $g(x) = \sqrt{x} - 1$

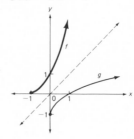

21. $g(x) = \dfrac{1}{x^2}$

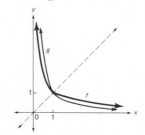

23. $g(x) = \sqrt{x + 4}$

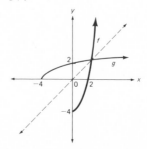

*25. $y = mx + k$, where $m = -1$.

1.

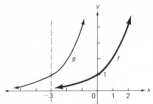

3.

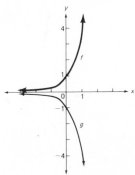

5.

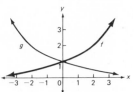

7.

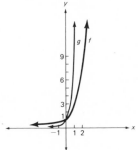

7.2 EXPONENTIAL FUNCTIONS *(Page 292)*

9.

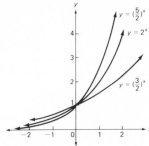

11. 6

13. ± 3

15. 1

17. $-2; 1.$

19. -5

21. $\frac{1}{2}$

23. $\frac{3}{2}$

25. $-\frac{1}{2}$

27. $\frac{2}{3}$

*29. $y = -3$

*31.

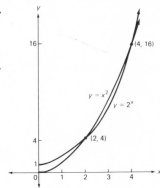

7.3 LOGARITHMIC FUNCTIONS (*Page 296*)

1. $g(x) = \log_4 x$

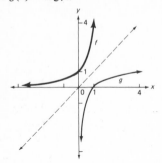

3. $g(x) = \log_{1/3} x$

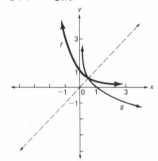

5. Shift 2 units left; $x > -2$; $x = -2$.

7. Shift 2 units upward; $x > 0$; $x = 0$.

9. $\log_2 256 = 8$

11. $\log_{1/3} 3 = -1$

13. $\log_{17} 1 = 0$

15. $10^{-4} = 0.0001$

17. $(\sqrt{2})^2 = 2$

19. $12^{-3} = \frac{1}{1728}$

21. 4

23. -3

25. $\frac{1}{216}$

27. 5

29. 4

31. $\frac{1}{3}$

33. $\frac{2}{3}$

35. 9

37. -2

39. $-\frac{1}{3}$

41. 9

*43. -1

*45. $g(x) = 3^x - 3$
$(f \circ g)(x) = \log_3 (3^x - 3 + 3) = \log_3 3^x = x$
$(g \circ f)(x) = 3^{\log_3 (x+3)} - 3 = (x + 3) - 3 = x$

7.4 THE LAWS OF LOGARITHMS (*Page 302*)

1. $\log_b 3 + \log_b x - \log_b (x + 1)$

3. $\frac{1}{2} \log_b (x^2 - 1) - \log_b x =$
$\frac{1}{2} \log_b (x + 1) + \frac{1}{2} \log_b (x - 1) - \log_b x$

5. $-2 \log_b x$

7. $\log_b \dfrac{x + 1}{x + 2}$

9. $\log_b \sqrt{\dfrac{x^2 - 1}{x^2 + 1}}$

11. $\log_b \dfrac{x^3}{2(x + 5)}$

13. $\log_b 27 + \log_b 3 = \log_b 81$ (Law 1)
$\log_b 243 - \log_b 3 = \log_b 81$ (Law 2)

15. $-2 \log_b \frac{4}{9} = \log_b (\frac{4}{9})^{-2}$ (Law 3)
$= \log_b \frac{81}{16}$

17. 20

19. $\frac{1}{20}$

21. 17

23. 8

25. 2

27. 7

29. 1.01

31. 5

*33. Let $r = \log_b M$ and $s = \log_b N$. Then $b^r = M$
and $b^s = N$. Divide:
$$\frac{M}{N} = \frac{b^r}{b^s} = b^{r-s}$$
Convert to log form and substitute:
$$\log_b \frac{M}{N} = r - s = \log_b M - \log_b N$$

*35. -1; 0.

*37. 8

*39. $B^{\log_B N} = N$
$\log_b B^{\log_B N} = \log_b N$ (take log base b)
$(\log_B N)(\log_b B) = \log_b N$ (Law 3)
$\log_B N = \dfrac{\log_b N}{\log_b B}$ (divide by $\log_b B$)

1.

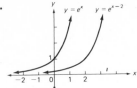

3.

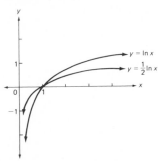

5.

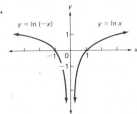

7. Since $f(x) = 1 + \ln x$, shift 1 unit upward.

9. Since $f(x) = \frac{1}{2} \ln x$, multiply the ordinates by $\frac{1}{2}$.

11. Since $f(x) = \ln (x - 1)$, shift 1 unit to the right.

13. $x > -2$

15. $x > \frac{1}{2}$

17. $\ln 5 + \ln x - \ln(x^2 - 4) =$
$\ln 5 + \ln x - \ln(x + 2) - \ln(x - 2)$

19. $\ln(x - 1) + 2 \ln(x + 3) - \frac{1}{2} \ln (x^2 + 2)$

21. $\frac{3}{2} \ln x + \frac{1}{2} \ln (x + 1)$

23. $\ln \sqrt{x} (x^2 + 5)$

7.5 THE BASE e (*Page 308*)

25. $\ln (x^2 - 1)^3$

27. $-100 \ln 27$

29. $\frac{1}{3}$

31. 0

33. No roots.

35. 8

37. 5

(Other answers are possible for Exercises 39–45.)

39. Let $g(x) = -x^2 + x$ and $f(x) = e^x$. Then
$(f \circ g)(x) = f(g(x)) = f(-x^2 + x) = e^{-x^2+x} = h(x)$.

41. Let $g(x) = \dfrac{x}{x + 1}$ and $f(x) = \ln x$. Then
$(f \circ g)(x) = f(g(x)) = f\left(\dfrac{x}{x+1}\right) = \ln \dfrac{x}{x + 1} = h(x)$.

43. Let $h(x) = 3x - 1, g(x) = x^2$, and $f(x) = e^x$.
Then
$$(f \circ g \circ h)(x) = f(g(h(x))) = f(g(3x - 1))$$
$$= f((3x - 1)^2)$$
$$= e^{(3x-1)^2} = F(x)$$

45. Let $h(x) = e^x + 1, g(x) = \sqrt{x}$, and $f(x) = \ln x$.
Then
$$(f \circ g \circ h)(x) = f(g(h(x))) = f(g(e^x + 1))$$
$$= f(\sqrt{e^x + 1})$$
$$= \ln \sqrt{e^x + 1} = F(x)$$

47. $f(x) > 0$ on the interval $(-\infty, \frac{1}{2})$; $f(x) < 0$ on
the interval $(\frac{1}{2}, \infty)$.

49. $f(x) > 0$ on the interval $\left(\dfrac{1}{e}, \infty\right)$; $f(x) < 0$ on
the interval $\left(0, \dfrac{1}{e}\right)$.

*51. $(e^x + e^{-x})^2 - (e^x - e^{-x})^2 =$
$(e^{2x} + 2 + e^{-2x}) - (e^{2x} - 2 + e^{-2x}) = 4$

*53. $x = \ln (y + \sqrt{y^2 + 1})$

7.6 EXPONENTIAL GROWTH AND DECAY (*Page 312*)

1. 2010

3. 27

5. $\frac{1}{2} \ln 100$

7. $\frac{1}{4} \ln \frac{1}{3}$

9. 667,000

11. 1.83 days

13. (a) $\frac{1}{3} \ln \frac{4}{5}$; (b) 6.4 grams; (c) 15.5 years.

15. 93.2 seconds

*17. 4240 years

*19. 114,651 years

7.7 COMMON LOGARITHMS (*Page 318*)

1. 43,000,000
3. 2770
5. 0.000125
7. 1.22
9. 0.00887
11. 6.58
13. $1.209 per gallon
15. 2.27 years

17. $5950

*19. For $1 \leq x < 10$ we get $\log 1 \leq \log x < \log 10$ because $f(x) = \log x$ is an increasing function. Substituting $0 = \log 1$ and $1 = \log 10$ into the preceding inequality gives $0 \leq \log x < 1$.

*21. (a) 2.825; (b) 9.273; (c) 1.378; (d) 5.489; (e) 4.495; (f) 7.497; (g) 1.021; (h) 9.508; (i) 2.263.

SAMPLE TEST FOR CHAPTER 7 (*Page 322*)

1. (i) a; (ii) c; (iii) e; (iv) h; (v) d.
2. (a) Each range value corresponds to exactly one domain value.
   (b) $g(x) = (x + 1)^3$; $(f \circ g)(x) = f(g(x)) = f((x + 1)^3) = \sqrt[3]{(x + 1)^3} - 1 = x$.
3. (a) $\frac{1}{2}$; (b) 9.
4. (a) $\frac{2}{3}$; (b) -2.
5. (a) All real values x; $y = -4$; (b) all $x > -4$; $x = -4$.
6. 5
7.

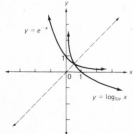

8. $3 \ln x - \ln (x + 1) - \frac{1}{2} \ln (x^2 + 2)$
9. $\dfrac{\ln 5}{2 \ln 4}$
10. $25 \ln 2$
†11. 1,330,000

Chapter 8: TRIGONOMETRY

8.1 TRIGONOMETRIC RATIOS (*Page 331*)

1. $\frac{3}{5}$
3. $\frac{4}{3}$
5. $\frac{5}{4}$
7. $\frac{4}{5}$

9. $\frac{5}{3}$
11. $\frac{3}{4}$

(Alternative forms of some of the following answers can be obtained by rationalizing denominators.)

	sin A	cos A	tan A	cot A	sec A	csc A
13.	$\dfrac{2}{\sqrt{13}}$	$\dfrac{3}{\sqrt{13}}$	$\dfrac{2}{3}$	$\dfrac{3}{2}$	$\dfrac{\sqrt{13}}{3}$	$\dfrac{\sqrt{13}}{2}$
15.	$\dfrac{4}{\sqrt{41}}$	$\dfrac{5}{\sqrt{41}}$	$\dfrac{4}{5}$	$\dfrac{5}{4}$	$\dfrac{\sqrt{41}}{5}$	$\dfrac{\sqrt{41}}{4}$
17.	$\dfrac{1}{\sqrt{5}}$	$\dfrac{2}{\sqrt{5}}$	$\dfrac{1}{2}$	2	$\dfrac{\sqrt{5}}{2}$	$\sqrt{5}$
19.	$\dfrac{3}{4}$	$\dfrac{\sqrt{7}}{4}$	$\dfrac{3}{\sqrt{7}}$	$\dfrac{\sqrt{7}}{3}$	$\dfrac{4}{\sqrt{7}}$	$\dfrac{4}{3}$
21.	$\dfrac{\sqrt{21}}{5}$	$\dfrac{2}{5}$	$\dfrac{\sqrt{21}}{2}$	$\dfrac{2}{\sqrt{21}}$	$\dfrac{5}{2}$	$\dfrac{5}{\sqrt{21}}$
	cos B	sin B	cot B	tan B	csc B	sec B

23. 1

25. $\dfrac{x^2}{z^2}$

27. 1

29. 1

31. $\sin A = \dfrac{3}{4}$; $\cos A = \dfrac{\sqrt{7}}{4}$; $\tan A = \dfrac{3}{\sqrt{7}}$;

$\cot A = \dfrac{\sqrt{7}}{3}$; $\sec A = \dfrac{4}{\sqrt{7}}$; $\csc A = \dfrac{4}{3}$.

33. $(\sin A)(\cos B) = (\sqrt{1-x^2})(x) = x\sqrt{1-x^2}$

35. $\dfrac{4\sin^2 A}{\cos A} = \dfrac{4\left(\dfrac{x}{4}\right)^2}{\dfrac{\sqrt{16-x^2}}{4}} = \dfrac{x^2}{\sqrt{16-x^2}}$

*37. $(\sin A)(\tan A) = \left(\dfrac{x}{2}\right)\left(\dfrac{x}{\sqrt{4-x^2}}\right) = \dfrac{x^2}{2\sqrt{4-x^2}}$

1. $\dfrac{\pi}{4}$

3. $\dfrac{\pi}{2}$

5. $\dfrac{3\pi}{2}$

7. $\dfrac{5\pi}{6}$

9. $\dfrac{5\pi}{4}$

11. $\dfrac{7\pi}{6}$

13. $\dfrac{11\pi}{6}$

15. $\dfrac{5\pi}{12}$

17. 180°

19. 360°

21. 100°

8.2 ANGLE MEASURE (*Page 335*)

23. 120°

25. 225°

27. 300°

29. 50°

31. 12π square centimeters

33. 54π square centimeters

35. 126π square centimeters

*37. 8 square inches

8.3 SOME SPECIAL ANGLES (*Page 341*)

1.

3.

5.

7. 14.4

9. 21.2

11. 34.6

13. 15.0

15. 14.1

17. 17.3

19. 77.9

21. $\tan 35° = .7$; $\cot 35° = 1.4$.

23. $AB = 24$; $\sin 35° = .58$; $\cos 35° = .83$.

8.4 BASIC IDENTITIES (*Page 347*)

The following are illustrative proofs. A variety of proofs for each exercise is possible.

1. $(\sin \theta)(\cot \theta)(\sec \theta) = (\sin \theta)\left(\dfrac{\cos \theta}{\sin \theta}\right)\left(\dfrac{1}{\cos \theta}\right) = 1$

3. $\cos^2 \theta (\tan^2 \theta + 1) = \cos^2 \theta \left(\dfrac{\sin^2 \theta}{\cos^2 \theta} + 1\right) =$
$\sin^2 \theta + \cos^2 \theta = 1$

5. $\tan \theta + \cot \theta = \dfrac{\sin \theta}{\cos \theta} + \dfrac{\cos \theta}{\sin \theta} =$
$\dfrac{\sin^2 \theta + \cos^2 \theta}{\sin \theta \cos \theta} = \dfrac{1}{\sin \theta \cos \theta}$

7. $(\sec \theta + \tan \theta)(1 - \sin \theta) =$
$\sec \theta + \tan \theta - \sec \theta \sin \theta - \tan \theta \sin \theta =$
$\sec \theta + \tan \theta - \dfrac{\sin \theta}{\cos \theta} - \dfrac{\sin^2 \theta}{\cos \theta} =$
$\sec \theta - \dfrac{\sin^2 \theta}{\cos \theta} = \dfrac{1}{\cos \theta} - \dfrac{\sin^2 \theta}{\cos \theta} = \dfrac{1 - \sin^2 \theta}{\cos \theta} =$
$\dfrac{\cos^2 \theta}{\cos \theta} = \cos \theta$

9. $\sec \theta - \cos \theta = \dfrac{1}{\cos \theta} - \cos \theta = \dfrac{1 - \cos^2 \theta}{\cos \theta} =$
$\dfrac{\sin^2 \theta}{\cos \theta} = \sin \theta \tan \theta$

11. $\dfrac{\cot \theta - 1}{1 - \tan \theta} = \dfrac{\dfrac{\cos \theta}{\sin \theta} - 1}{1 - \dfrac{\sin \theta}{\cos \theta}} = \dfrac{\dfrac{\cos \theta - \sin \theta}{\sin \theta}}{\dfrac{\cos \theta - \sin \theta}{\cos \theta}} =$
$\dfrac{\dfrac{1}{\sin \theta}}{\dfrac{1}{\cos \theta}} = \dfrac{\csc \theta}{\sec \theta}$

13. $\dfrac{1 + \tan^2 \theta}{1 + \cot^2 \theta} = \dfrac{\sec^2 \theta}{\csc^2 \theta} = \dfrac{\dfrac{1}{\cos^2 \theta}}{\dfrac{1}{\sin^2 \theta}} = \dfrac{\sin^2 \theta}{\cos^2 \theta} =$
$\tan^2 \theta = \sec^2 \theta - 1$

15. $(\csc \theta - \cot \theta)^2 = \left(\dfrac{1}{\sin \theta} - \dfrac{\cos \theta}{\sin \theta}\right)^2 =$
$\dfrac{(1 - \cos \theta)^2}{\sin^2 \theta} = \dfrac{(1 - \cos \theta)^2}{1 - \cos^2 \theta} =$
$\dfrac{(1 - \cos \theta)^2}{(1 - \cos \theta)(1 + \cos \theta)} = \dfrac{1 - \cos \theta}{1 + \cos \theta}$

17. $(\sin^2 \theta + \cos^2 \theta)^5 = 1^5 = 1$

19. $\dfrac{\tan^2 \theta + 1}{\tan^2 \theta} = 1 + \dfrac{1}{\tan^2 \theta} = 1 + \cot^2 \theta = \csc^2 \theta$

*21. $\dfrac{\tan \theta}{\sec \theta - 1} = \dfrac{\tan \theta (\sec \theta + 1)}{(\sec \theta - 1)(\sec \theta + 1)} =$
$\dfrac{\tan \theta (\sec \theta + 1)}{\sec^2 \theta - 1} = \dfrac{\tan \theta (\sec \theta + 1)}{\tan^2 \theta} =$
$\dfrac{\sec \theta + 1}{\tan \theta}$

23. $\dfrac{\cot \dfrac{\pi}{3} - 1}{1 - \tan \dfrac{\pi}{3}} = \dfrac{\dfrac{1}{\sqrt{3}} - 1}{1 - \sqrt{3}} \neq \dfrac{\dfrac{2}{\sqrt{3}}}{\dfrac{2}{\sqrt{3}}} = \dfrac{\sec \dfrac{\pi}{3}}{\csc \dfrac{\pi}{3}}$

1. $\sin \theta = .5783$; $\cos \theta = .8158$; $\tan \theta = .7089$;
$\cot \theta = 1.411$; $\sec \theta = 1.226$; $\csc \theta = 1.729$.

3. .6691

5. .2126

7. 57.30

9. .0465

11. 1.006

13. 1.060

15. $81°00' = 1.4137$ radians

17. $46°10' = .8058$ radians

19. $6°20' = .1105$ radians

21. 12.7

23. 18.2

25. 71°30'

27. 21°10'

29. 274.7 feet

31. 1925 feet

*33. 12.3

*35. 1028 feet using cotangent ratio; 1029 feet using
tangent ratio.

1. $\sin \dfrac{2\pi}{3} = \dfrac{\sqrt{3}}{2}$; $\quad \cot \dfrac{2\pi}{3} = -\dfrac{\sqrt{3}}{3}$

$\cos \dfrac{2\pi}{3} = -\dfrac{1}{2}$; $\quad \sec \dfrac{2\pi}{3} = -2$

$\tan \dfrac{2\pi}{3} = -\sqrt{3}$; $\quad \csc \dfrac{2\pi}{3} = \dfrac{2\sqrt{3}}{3}$

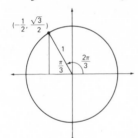

3. $\sin \left(-\dfrac{7\pi}{4} \right) = \dfrac{\sqrt{2}}{2}$; $\quad \cot \left(-\dfrac{7\pi}{4} \right) = 1$

$\cos \left(-\dfrac{7\pi}{4} \right) = \dfrac{\sqrt{2}}{2}$; $\quad \sec \left(-\dfrac{7\pi}{4} \right) = \sqrt{2}$

$\tan \left(-\dfrac{7\pi}{4} \right) = 1$; $\quad \csc \left(-\dfrac{7\pi}{4} \right) = \sqrt{2}$

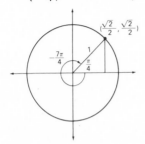

5. Ratios are the same as in Exercise 3.

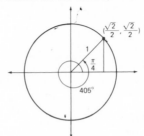

7. $(0, -1)$; $\sin \theta = -1$; $\cos \theta = 0$; $\tan \theta$ is
undefined; $\cot \theta = 0$; $\sec \theta$ is undefined;
$\csc \theta = -1$.

9. $(0, 1)$; $\sin \theta = 1$; $\cos \theta = 0$; $\tan \theta$ is undefined;
$\cot \theta = 0$; $\sec \theta$ is undefined; $\csc \theta = 1$.

11. $\theta = -\dfrac{3\pi}{2}$

13. .8391

15. $-.9877$

17. Undefined.

19. 0

21. Undefined.

23. 0

25. .0872

27. 7.115

29. .3007

31. $-\dfrac{\sqrt{3}}{3}$

33. -1.015

35. $-.9346$

37. (a) $\left(\dfrac{2}{3}\right)^2 + \left(\dfrac{\sqrt{5}}{3}\right)^2 = \dfrac{4}{9} + \dfrac{5}{9} = \dfrac{9}{9} = 1$

(b)

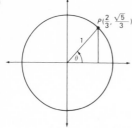

(c) $\sin\theta = \dfrac{\sqrt{5}}{3}$; $\cos\theta = \dfrac{2}{3}$; $\tan\theta = \dfrac{\sqrt{5}}{2}$;

$\cot\theta = \dfrac{2\sqrt{5}}{5}$; $\sec\theta = \dfrac{3}{2}$; $\csc\theta = \dfrac{3\sqrt{5}}{5}$.

*39. $y = -\dfrac{\sqrt{13}}{4} = -.9014$; $\theta = -64°20'$.

*41. $-\dfrac{\sqrt{5}}{5}$

*43. $AB = \dfrac{AB}{1} = \dfrac{AB}{OA} = \dfrac{y}{x} = \tan\theta$; OB.

45. $\dfrac{\pi}{2}$

47. 0; π.

49. $\dfrac{\pi}{6}$; $\dfrac{5\pi}{6}$.

51. $\dfrac{3\pi}{4}$; $\dfrac{5\pi}{4}$.

8.7 THE LAW OF SINES AND THE LAW OF COSINES *(Page 368)*

1. $a = \sqrt{93}$

3. $a = c = 18\sqrt{3}$

5. $a = 6.1$; $\angle B = 26°20'$.

7. $\angle C = 90°$

9. $c = 11.2$; $\angle B = 24°$.

11. $\angle B = 100°$; $a = 5.1$; $c = 10.0$.

13. $\angle C = 67°40,$; $a = 5.8$; $c = 6.0$.

15. $\angle B = 86°40,$; $c = 9.7$.

17. No solution.

19. $\angle C = 27°30'$

21. 151 meters

23. 1.6 miles

25. None when $a \leq b$; one when $a > b$.

SAMPLE TEST FOR CHAPTER 8 *(Page 371)*

1. (a) $\frac{15}{8}$; **(b)** $\frac{8}{17}$; **(c)** $\frac{17}{8}$.

2. $\sin A = \dfrac{\sqrt{5}}{3}$; $\cos A = \dfrac{2}{3}$; $\tan A = \dfrac{\sqrt{5}}{2}$.

3. $AB = \sqrt{9 + x^2}$; $3(\sin^2 A)(\sec A) =$

$3\left(\dfrac{x^2}{x^2 + 9}\right)\left(\dfrac{\sqrt{x^2 + 9}}{3}\right) = \dfrac{x^2}{\sqrt{x^2 + 9}}$

4. (a) $\dfrac{\sqrt{3}}{2}$; **(b)** 0; **(c)** $-\dfrac{\sqrt{2}}{2}$.

5. (a) $-\dfrac{\sqrt{3}}{3}$; **(b)** -1; **(c)** 2.

6. $20\sqrt{2}$

7. 373 feet

8. $\cot^2\theta - \cos^2\theta = \dfrac{\cos^2\theta}{\sin^2\theta} - \cos^2\theta =$

$\cos^2\theta\left(\dfrac{1}{\sin^2\theta} - 1\right) = \cos^2\theta\left(\dfrac{1 - \sin^2\theta}{\sin^2\theta}\right) =$

$\cos^2\theta\left(\dfrac{\cos^2\theta}{\sin^2\theta}\right) = \cos^2\theta \cot^2\theta$

9. $(\sec\theta - \tan\theta)^2 = \left(\dfrac{1}{\cos\theta} - \dfrac{\sin\theta}{\cos\theta}\right)^2 =$

$\dfrac{(1 - \sin\theta)^2}{\cos^2\theta} = \dfrac{(1 - \sin\theta)^2}{1 - \sin^2\theta} =$

$\dfrac{(1 - \sin\theta)^2}{(1 - \sin\theta)(1 + \sin\theta)} = \dfrac{1 - \sin\theta}{1 + \sin\theta}$

10. (a) $4\sqrt{7}, \dfrac{\pi}{3}$; **(b)** $\sin B = \dfrac{8\sin 60°}{4\sqrt{7}} = \dfrac{\sqrt{21}}{7}$.

Chapter 9: THE CIRCULAR FUNCTIONS

9.1 GRAPHING THE SINE AND COSINE FUNCTIONS (*Page 380*)

1.

x	$-\pi$	$-\dfrac{5\pi}{6}$	$-\dfrac{2\pi}{3}$	$-\dfrac{\pi}{2}$	$-\dfrac{\pi}{3}$	$-\dfrac{\pi}{6}$	0
$y = \sin x$	0	$-\dfrac{1}{2}$	$-\dfrac{\sqrt{3}}{2}$	-1	$-\dfrac{\sqrt{3}}{2}$	$-\dfrac{1}{2}$	0

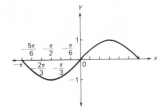

3.

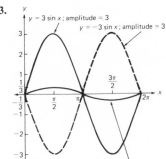

5. Amplitude $= 1$; period $= \pi$.

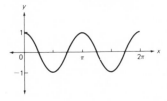

7. Amplitude $= \dfrac{3}{2}$; period $= \dfrac{\pi}{2}$.

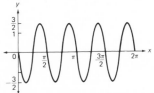

9. Amplitude $= 1$; period $= 4\pi$.

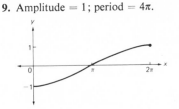

11. $p = 8\pi$

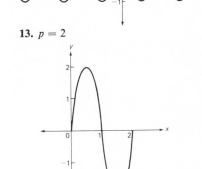

13. $p = 2$

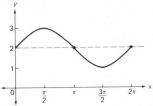

15. $a = 1$; $b = 3$.

17. Amplitude $= 1$.

19. Shift $y = \sin 2x$ by $\dfrac{1}{2}$ unit to the right.

*21. **(i)** Graph $y = \sin bx$ having period $\dfrac{2\pi}{b}$ and amplitude 1.

(ii) Multiply the ordinates of $y = \sin bx$ by a to get the graph of $y = a \sin bx$.

(iii) Shift $y = a \sin bx$ by $-\dfrac{c}{b}$ units to the right to obtain $y = a \sin b\left[x - \left(-\dfrac{c}{b}\right)\right] = a \sin b\left(x + \dfrac{c}{b}\right) = a \sin(bx + c)$.

(iv) Shift the curve in part (iii) by d units upward.

23. $(f \circ g)(x) = 2 \cos x + 5; (g \circ f)(x) = \cos(2x + 5)$.

25. $g(x) = 5x^2; f(x) = \cos x$.

27. $h(x) = 1 - 2x; g(x) = \sqrt[3]{x}; f(x) = \cos x$.

9.2 GRAPHING THE OTHER TRIGONOMETRIC FUNCTIONS (*Page 387*)

1.

x	-1.4	-1.3	$-\dfrac{\pi}{3}$	$-\dfrac{\pi}{4}$	$-\dfrac{\pi}{6}$	0
$y = \tan x$	-5.8	-3.6	$-\sqrt{3}$	-1	$-\dfrac{\sqrt{3}}{3}$	0

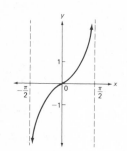

3. $p = \dfrac{\pi}{3}$

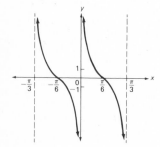

5. $p = 2\pi$

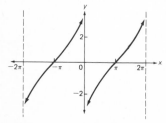

7.

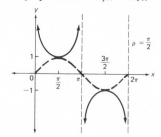

9. Asymptotes: $x = 0, x = \pi, x = 2\pi$.

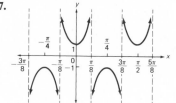

11. $p = 6\pi$

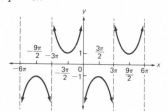

13.

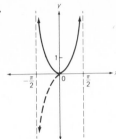

15. $(f \circ g)(x) = \tan^2 x; (g \circ f)(x) = \tan x^2.$

17. $g(x) = \cot x; f(x) = x^3.$

*19. $0 < QP = \sin \theta < \overset{\frown}{AP} = \theta; 0 < \dfrac{\sin \theta}{\theta} < 1.$

9.3 THE ADDITION FORMULAS *(Page 393)*

1. $\sin 75° = \sin (45° + 30°) =$
$\sin 45° \cos 30° + \cos 45° \sin 30° = \frac{1}{4}(\sqrt{6} + \sqrt{2})$
$\cos 75° = \cos (45° + 30°) =$
$\cos 45° \cos 30° - \sin 45° \sin 30° = \frac{1}{4}(\sqrt{6} - \sqrt{2})$
$\tan 75° = \tan (45° + 30°) =$
$\dfrac{\tan 45° + \tan 30°}{1 - \tan 45° \tan 30°} = \dfrac{\sqrt{3} + 1}{\sqrt{3} - 1}$

3. $\sin \dfrac{\pi}{12} = \sin \left(\dfrac{\pi}{3} - \dfrac{\pi}{4} \right) =$

$\sin \dfrac{\pi}{3} \cos \dfrac{\pi}{4} - \cos \dfrac{\pi}{3} \sin \dfrac{\pi}{4} = \dfrac{1}{4}(\sqrt{6} - \sqrt{2})$

$\cos \dfrac{\pi}{12} = \cos \left(\dfrac{\pi}{3} - \dfrac{\pi}{4} \right) =$

$\cos \dfrac{\pi}{3} \cos \dfrac{\pi}{4} + \sin \dfrac{\pi}{3} \sin \dfrac{\pi}{4} = \dfrac{1}{4}(\sqrt{2} + \sqrt{6})$

$\tan \dfrac{\pi}{12} = \tan \left(\dfrac{\pi}{3} - \dfrac{\pi}{4} \right) = \dfrac{\tan \dfrac{\pi}{3} - \tan \dfrac{\pi}{4}}{1 + \tan \dfrac{\pi}{3} \tan \dfrac{\pi}{4}} =$

$\dfrac{\sqrt{3} - 1}{1 + \sqrt{3}}$

5. $\frac{1}{2} \sin \theta + \dfrac{\sqrt{3}}{2} \cos \theta$

7. $\cos \left(\dfrac{\pi}{2} - x \right) = \sin x;$

$\cos \left[\dfrac{\pi}{2} - \left(\dfrac{\pi}{2} - \theta \right) \right] = \sin \left(\dfrac{\pi}{2} - \theta \right);$

$\cos \theta = \sin \left(\dfrac{\pi}{2} - \theta \right).$

9. $\sin(\alpha - \beta) = \sin [\alpha + (-\beta)] =$
$\sin \alpha \cos (-\beta) + \cos \alpha \sin (-\beta) =$
$\sin \alpha \cos \beta - \cos \alpha \sin \beta$

11. $\sin (\pi - \theta) = \sin \pi \cos \theta - \cos \pi \sin \theta =$
$-(-1) \sin \theta = \sin \theta$

13. $\tan (\pi - \theta) = \dfrac{\tan \pi - \tan \theta}{1 + \tan \pi \tan \theta} = \dfrac{0 - \tan \theta}{1 + 0} =$
$-\tan \theta$

15. $\sec \left(\dfrac{\pi}{2} - \theta \right) = \dfrac{1}{\cos \left(\dfrac{\pi}{2} - \theta \right)} =$

$\dfrac{1}{\cos \dfrac{\pi}{2} \cos \theta + \sin \dfrac{\pi}{2} \sin \theta} = \dfrac{1}{0 + \sin \theta} = \csc \theta$

17. $\cot (\alpha + \beta) = \dfrac{\cos (\alpha + \beta)}{\sin (\alpha + \beta)} =$

$\dfrac{\cos \alpha \cos \beta - \sin \alpha \sin \beta}{\sin \alpha \cos \beta + \cos \alpha \sin \beta} =$

$\dfrac{\dfrac{\cos \alpha \cos \beta}{\sin \alpha \sin \beta} - \dfrac{\sin \alpha \sin \beta}{\sin \alpha \sin \beta}}{\dfrac{\sin \alpha \cos \beta}{\sin \alpha \sin \beta} + \dfrac{\cos \alpha \sin \beta}{\sin \alpha \sin \beta}} = \dfrac{\cot \alpha \cot \beta - 1}{\cot \beta + \cot \alpha}$

19. Since $\sin \theta = -\cos \left(\theta + \dfrac{\pi}{2} \right)$, shift $y = \cos \theta$

by $\dfrac{\pi}{2}$ units to the left and reflect through the

θ-axis.

*21. $\dfrac{f(x + h) - f(x)}{h} = \dfrac{\sin (x + h) - \sin x}{h} =$

$\dfrac{\sin x \cos h + \cos x \sin h - \sin x}{h} =$

$\dfrac{\sin x (\cos h - 1)}{h} + \dfrac{\cos x \sin h}{h} =$

$\left(\dfrac{\cos h - 1}{h} \right) \sin x + \left(\dfrac{\sin h}{h} \right) \cos x$

9.4 THE DOUBLE- AND HALF-ANGLE FORMULAS (*Page 397*)

1. $\cos 75° = \cos \dfrac{150°}{2} = \sqrt{\dfrac{1 + \cos 150°}{2}} =$

$\sqrt{\dfrac{1 - \dfrac{\sqrt{3}}{2}}{2}} = \dfrac{1}{2}\sqrt{2 - \sqrt{3}}$

3. (a) $\tan^2 (22°30') = \tan^2 \dfrac{45°}{2} = \dfrac{1 - \cos 45°}{1 + \cos 45°} =$

$\dfrac{1 - \dfrac{\sqrt{2}}{2}}{1 + \dfrac{\sqrt{2}}{2}} = \dfrac{2 - \sqrt{2}}{2 + \sqrt{2}} =$

$\dfrac{2 - \sqrt{2}}{2 + \sqrt{2}} \cdot \dfrac{2 - \sqrt{2}}{2 - \sqrt{2}} = \dfrac{(2 - \sqrt{2})^2}{2}$

(b) $\tan 22°30' = \sqrt{\dfrac{(2 - \sqrt{2})^2}{\sqrt{2}}} = \dfrac{2 - \sqrt{2}}{\sqrt{2}} =$

$\dfrac{(2 - \sqrt{2})\sqrt{2}}{2} = \dfrac{2\sqrt{2} - 2}{2} =$

$\dfrac{2(\sqrt{2} - 1)}{2} = \sqrt{2} - 1$

(c) $\sqrt{2} - 1 = 1.414 - 1 = .414$ (by Table I);
$\tan 22°30' = .4142$ (by Table V).

5. (a) .93; (b) -2.58; (c) $-.67$; (d) $-.74$.

7. $9 \sin 2\theta = 18 \sin \theta \cos \theta = 18\left(\dfrac{x}{3}\right)\left(\dfrac{\sqrt{9 - x^2}}{3}\right) =$

$2x\sqrt{9 - x^2}$

9. $\tan 3x = \tan (2x + x) = \dfrac{\tan 2x + \tan x}{1 - \tan 2x \tan x} =$

$\dfrac{\dfrac{2 \tan x}{1 - \tan^2 x} + \tan x}{1 - \dfrac{2 \tan x}{1 - \tan^2 x} \cdot \tan x} =$

$\dfrac{2 \tan x + \tan x - \tan^3 x}{1 - \tan^2 x - 2 \tan^2 x} = \dfrac{3 \tan x - \tan^3 x}{1 - 3 \tan^2 x}$

11. $\cot 2x = \dfrac{1}{\tan 2x} = \dfrac{1}{\dfrac{2 \tan x}{1 - \tan^2 x}} = \dfrac{1 - \tan^2 x}{2 \tan x} =$

$\dfrac{\dfrac{1}{\tan^2 x} - \dfrac{\tan^2 x}{\tan^2 x}}{\dfrac{2 \tan x}{\tan^2 x}} = \dfrac{\cot^2 x - 1}{2 \cot x}$

13. $\cos^4 x - \sin^4 x =$
$(\cos^2 x - \sin^2 x)(\cos^2 x + \sin^2 x) =$
$(\cos^2 x - \sin^2 x) = \cos 2x$

15. $\cos 4x = \cos 2(2x) = 2 \cos^2 2x - 1 =$
$2(2 \cos^2 x - 1)^2 - 1 =$
$2(4 \cos^4 x - 4 \cos^2 x + 1) - 1 =$
$8 \cos^4 x - 8 \cos^2 x + 1$

17. $\cos^4 x = (\cos^2 x)^2 = \left(\dfrac{1 + \cos 2x}{2}\right)^2 =$

$\dfrac{1}{4}(1 + 2 \cos 2x + \cos^2 2x) =$

$\dfrac{1}{4}\left(1 + 2 \cos 2x + \dfrac{1 + \cos 4x}{2}\right) =$

$\dfrac{1}{8} \cos 4x + \dfrac{1}{2} \cos 2x + \dfrac{3}{8}$

*19. (a) $\cos \alpha \cos \beta - \sin \alpha \sin \beta = \cos (\alpha + \beta)$
$\cos \alpha \cos \beta + \sin \alpha \sin \beta = \cos (\alpha - \beta)$
$\overline{}$
$2 \cos \alpha \cos \beta = \cos (\alpha + \beta) + \cos (\alpha - \beta)$
$\cos \alpha \cos \beta = \frac{1}{2}[\cos (\alpha + \beta) + \cos (\alpha - \beta)]$

(b) $\cos \alpha \cos \beta + \sin \alpha \sin \beta = \cos (\alpha - \beta)$
$\cos \alpha \cos \beta - \sin \alpha \sin \beta = \cos (\alpha + \beta)$
$\overline{}$
$2 \sin \alpha \sin \beta = \cos (\alpha - \beta) - \cos (\alpha + \beta)$
$\sin \alpha \sin \beta = \frac{1}{2}[\cos (\alpha - \beta) - \cos (\alpha + \beta)]$

$\sin \alpha \cos \beta + \cos \alpha \sin \beta = \sin (\alpha + \beta)$
$\sin \alpha \cos \beta - \cos \alpha \sin \beta = \sin (\alpha - \beta)$
$\overline{}$
$2 \sin \alpha \cos \beta = \sin (\alpha + \beta) + \sin (\alpha - \beta)$
$\sin \alpha \cos \beta = \frac{1}{2}[\sin (\alpha + \beta) + \sin (\alpha - \beta)]$

$\sin \alpha \cos \beta + \cos \alpha \sin \beta = \sin (\alpha + \beta)$
$\sin \alpha \cos \beta - \cos \alpha \sin \beta = \sin (\alpha - \beta)$
$\overline{}$
$2 \cos \alpha \sin \beta = \sin (\alpha + \beta) - \sin (\alpha - \beta)$
$\cos \alpha \sin \beta = \frac{1}{2}[\sin (\alpha + \beta) - \sin (\alpha - \beta)]$

*21. $2 \cos \dfrac{u + v}{2} \cos \dfrac{u - v}{2} =$

$\cos \left(\dfrac{u + v}{2} - \dfrac{u - v}{2}\right) + \cos \left(\dfrac{u + v}{2} + \dfrac{u - v}{2}\right) =$

$\cos v + \cos u$
Similarly for the other formulas.

*23. $\dfrac{\cos 2x + \cos 8x}{\sin 2x + \sin 8x} = \dfrac{2 \cos \dfrac{2x + 8x}{2} \cos \dfrac{2x - 8x}{2}}{2 \sin \dfrac{2x + 8x}{2} \cos \dfrac{2x - 8x}{2}} =$

$\dfrac{\cos 5x}{\sin 5x} = \cot 5x$

*25. (a) (i) $\dfrac{\pi}{2} < \theta < \pi$ implies $\dfrac{\pi}{4} < \dfrac{\theta}{2} < \dfrac{\pi}{2}$. Then

$\dfrac{\theta}{2}$ is in quadrant I, so $\tan \dfrac{\theta}{2}$ and

$\sin \theta$ are both positive.

(ii) $\pi < \theta < \dfrac{3\pi}{2}$ implies $\dfrac{\pi}{2} < \dfrac{\theta}{2} < \dfrac{3\pi}{4}$. Then

$\dfrac{\theta}{2}$ is in quadrant II, so $\tan \dfrac{\theta}{2}$

and $\sin \theta$ are both negative.

(iii) Similar to part (ii).

(b) $\tan^2 \dfrac{\theta}{2} = \dfrac{1 - \cos\theta}{1 + \cos\theta} = \dfrac{(1 - \cos\theta)^2}{1 - \cos^2\theta} =$

$\dfrac{(1 - \cos\theta)^2}{\sin^2\theta}$

$\tan \dfrac{\theta}{2} = \pm\sqrt{\dfrac{(1 - \cos\theta)^2}{\sin^2\theta}} =$

$\pm\dfrac{\sqrt{(1 - \cos\theta)^2}}{\sqrt{\sin^2\theta}} = \pm\dfrac{1 - \cos\theta}{\sin\theta}$

since $1 - \cos\theta > 0$ and the signs in $\sqrt{\sin^2\theta} = \pm\sin\theta$ are absorbed into the $\pm$ signs preceding the fraction. Finally,

$\tan \dfrac{\theta}{2} = \dfrac{1 - \cos\theta}{\sin\theta}$ because $1 - \cos\theta > 0$

and $\sin\theta$ and $\tan \dfrac{\theta}{2}$ have the same sign.

(c) $\tan \dfrac{\theta}{2} = \dfrac{1 - \cos\theta}{\sin\theta} = \dfrac{1 - \cos^2\theta}{\sin\theta(1 + \cos\theta)} =$

$\dfrac{\sin\theta}{1 + \cos\theta}$. Since $\angle PQR = \dfrac{1}{2}\widehat{AP} = \dfrac{\theta}{2}$,

$\tan \angle PQR = \tan \dfrac{\theta}{2} = \dfrac{RP}{QR} = \dfrac{RP}{1 + OR} =$

$\dfrac{\sin\theta}{1 + \cos\theta}$.

9.5 TRIGONOMETRIC EQUATIONS (*Page 402*)

In the following answers k represents any integer.

1. $2k\pi$

3. $\dfrac{\pi}{6} + 2k\pi$; $\dfrac{5\pi}{6} + 2k\pi$.

5. $\pi + 2k\pi$

7. $\dfrac{\pi}{4} + k\pi$

9. No solutions.

11. $\dfrac{\pi}{6} + \dfrac{k\pi}{2}$; $\dfrac{2\pi}{3} + \dfrac{k\pi}{2}$.

13. $\pi + 2k\pi - 1$

15. No solutions.

17. $0; \dfrac{7\pi}{6}; \dfrac{11\pi}{6}$.

19. $\dfrac{\pi}{3}; \dfrac{4\pi}{3}$.

21. $0; \dfrac{\pi}{2}$.

23. $\dfrac{\pi}{2}$

25. $\dfrac{\pi}{6}; \dfrac{5\pi}{6}; \dfrac{\pi}{2}; \dfrac{3\pi}{2}$.

27. No solutions.

29. $\dfrac{\pi}{3}; \dfrac{5\pi}{3}$.

31. $0; \dfrac{\pi}{4}; \dfrac{3\pi}{4}$.

33. $\dfrac{\pi}{4}; \dfrac{\pi}{2}; \dfrac{3\pi}{4}; \dfrac{5\pi}{4}; \dfrac{3\pi}{2}; \dfrac{7\pi}{4}$.

*35. $\dfrac{\pi}{8}; \dfrac{5\pi}{8}$.

37. $62°10'; 297°50'$.

39. $33°40'; 213°40'$.

41. $69°20'; 290°40'; 110°40'; 249°20'$.

9.6 INVERSE CIRCULAR FUNCTIONS (*Page 409*)

1. 0

3. $-\dfrac{\pi}{2}$

5. $-\dfrac{\pi}{4}$

7. $\dfrac{3\pi}{4}$

9. 1.5621 (89°30')

11. .5934 (34°)

13. 1.4224 (81°30')

15. $\dfrac{\sqrt{145}}{145}$

17. 0

19. $\dfrac{\sqrt{3}}{2}$

21. $\frac{5}{13}$

23. 0

25. Sin and arcsin are inverse functions.

27.

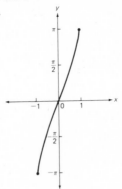

29.

31. $(f \circ g)(x) = \arcsin(3x + 2)$;
$(g \circ f)(x) = 3 \arcsin x + 2$.

33. $g(x) = \arccos x$; $f(x) = \ln x$.

35. $y = \text{arccot } x$
Domain: all reals
Range: $0 < y < \pi$

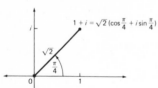

37. Since $\text{arccot } x = \dfrac{\pi}{2} - \arctan x$,

reflect $y = \arctan x$ through the x-axis

and shift $\dfrac{\pi}{2}$ units upward.

39. $y = \text{arccsc } x$
Domain: all reals
Range: $-1 \leq y \leq 1$, $y \neq 0$

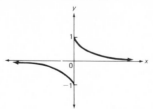

*41. $\frac{7}{9}$

*43. Let $y = \arcsin x$. Then $x = \sin y$;
$-x = -\sin y = \sin(-y)$; $-y = \arcsin(-x)$;
$-\arcsin x = \arcsin(-x)$.

9.7 TRIGONOMETRIC FORM OF COMPLEX NUMBERS (*Page 415*)

1.

$1 + i = \sqrt{2}\left(\cos\frac{\pi}{4} + i\sin\frac{\pi}{4}\right)$

3.

$-1 + \sqrt{3}\,i = 2\left(\cos\frac{2\pi}{3} + i\sin\frac{2\pi}{3}\right)$

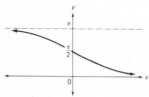

5.

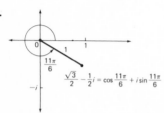

$\frac{\sqrt{3}}{2} - \frac{1}{2}i = \cos\frac{11\pi}{6} + i\sin\frac{11\pi}{6}$

7.

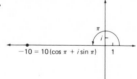

$-10 = 10(\cos\pi + i\sin\pi)$

9.

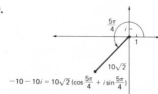

$-10 - 10i = 10\sqrt{2}\left(\cos\frac{5\pi}{4} + i\sin\frac{5\pi}{4}\right)$

11. $-\dfrac{3}{2} + \dfrac{3\sqrt{3}}{2}i$

13. $-5i$

15. $\dfrac{\sqrt{2}}{2} + \dfrac{\sqrt{6}}{2}i$

17. $2\sqrt{2} - 2\sqrt{2}\,i$

19. -4

21. 1

23. $\dfrac{1}{2} + \dfrac{\sqrt{3}}{2}i$

25. $-i$

27. 1

29. $2; -2; 2i; -2i.$

31. $2\left(\cos\dfrac{5\pi}{12} + i\sin\dfrac{5\pi}{12}\right); 2\left(\cos\dfrac{13\pi}{12} + i\sin\dfrac{13\pi}{12}\right);$
$2\left(\cos\dfrac{21\pi}{12} + i\sin\dfrac{21\pi}{12}\right).$

33. $-1; 1.$

35. $-1; 1; -i, i.$

37.

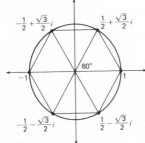

1. Amplitude $= 2$;
period $= \pi$.

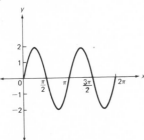

2. Domain: all $x \ne \dfrac{\pi}{2} + k\pi$
Range: all $y \ge 1$ or $y \le -1$

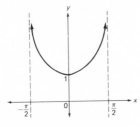

*39. (a) $[r(\cos\theta + i\sin\theta)]^{-n} = r^{-n}(\cos\theta + i\sin\theta)^{-n}$
$= r^{-n}[(\cos\theta + i\sin\theta)^{-1}]^n$
$= r^{-n}[(\cos\theta - i\sin\theta)]^n$ (by Exercise 38)
$= r^{-n}[\cos(-\theta) + i\sin(-\theta)]^n$
$= r^{-n}[\cos(-n\theta) + i\sin(-n\theta)]$ (since $n > 0$)
$= r^{-n}(\cos n\theta - i\sin n\theta)$

(b) Assuming the rule $z^0 = 1$, the formula holds
for $n = 0$. The formula is given in Section 9.7
for positive integers n. If in part (a) we let
$m = -n$, where m is a negative integer, then
$[r(\cos\theta + i\sin\theta)]^m = [r(\cos\theta + i\sin\theta)]^{-n} =$
$r^{-n}(\cos n\theta - i\sin n\theta) =$
$r^{-n}[\cos(-n\theta) + i\sin(-n\theta)] =$
$r^m[\cos(m\theta) + i\sin(m\theta)]$
which shows that the formula holds for
negative integers.

41. (a) $(x + 2i)(x - 2i);$ **(b)** $3(x + 3i)(x - 3i);$
(c) $(x + \sqrt{2}\,i)(x - \sqrt{2}\,i);$
(d) $5(x + \sqrt{3}\,i)(x - \sqrt{3}\,i).$

SAMPLE TEST FOR CHAPTER 9   (*Page 418*)

3. $-\tfrac{1}{4}(\sqrt{6} + \sqrt{2})$

4. $\sin\left(\dfrac{\pi}{4} - \theta\right) = \sin\dfrac{\pi}{4}\cos\theta - \cos\dfrac{\pi}{4}\sin\theta =$
$\dfrac{\sqrt{2}}{2}\cos\theta - \dfrac{\sqrt{2}}{2}\sin\theta = \dfrac{\sqrt{2}}{2}(\cos\theta - \sin\theta)$

5. $-\tfrac{1}{2}\sqrt{2 - \sqrt{3}}$

6. $\dfrac{2\sqrt{8}}{9} = \dfrac{4\sqrt{2}}{9}$

7. $\csc 4x = \dfrac{1}{\sin 4x} = \dfrac{1}{2\sin 2x\cos 2x} =$
$\dfrac{1}{4\sin x\cos x\cos 2x} = \dfrac{1}{4}\csc x\sec x\sec 2x$

8. $\dfrac{\pi}{6}; \dfrac{5\pi}{6}; \dfrac{3\pi}{2}.$

9. (a) Domain: all reals

Range: $-\dfrac{\pi}{2} < y < \dfrac{\pi}{2}$

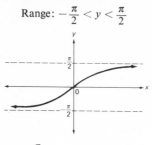

(b) $-\dfrac{\pi}{6}$

10. (a) $\dfrac{\pi}{4}$; (b) $\dfrac{2}{7}\sqrt{10}$; (c) $\dfrac{1}{2}$.

†**11.** $-16 - 16\sqrt{3}\,i$

Chapter 10: SEQUENCES AND SERIES

10.1 SEQUENCES (*Page 426*)

1. $s_1 = 1; s_2 = 3; s_3 = 5; s_4 = 7; s_5 = 9.$

3. $s_1 = -1; s_2 = 1; s_3 = -1; s_4 = 1; s_5 = -1.$

5. $s_1 = -4; s_2 = 2; s_3 = -1; s_4 = \tfrac{1}{2}; s_5 = -\tfrac{1}{4}.$

7. $-1; 4; -9; 16.$

9. $\dfrac{3}{10}; \dfrac{3}{100}; \dfrac{3}{1000}; \dfrac{3}{10,000}.$

11. $\dfrac{3}{100}; \dfrac{3}{10,000}; \dfrac{3}{1,000,000}; \dfrac{3}{100,000,000}.$

13. $1; \tfrac{1}{2}; \tfrac{1}{3}; \tfrac{1}{4}.$

15. $-8; -6; -4; -2.$

17. $-2; 1; 4; 7.$

19. $\tfrac{1}{2}; \tfrac{2}{3}; \tfrac{3}{4}; \tfrac{4}{5}.$

21. $1; \tfrac{3}{2}; \tfrac{16}{9}; \tfrac{125}{64}.$

23. $-2; -\tfrac{3}{2}; -\tfrac{9}{8}; -\tfrac{27}{32}.$

25. $\tfrac{1}{2}; \tfrac{1}{2}; \tfrac{3}{8}; \tfrac{1}{4}.$

27. $-\tfrac{3}{2}; -\tfrac{5}{6}; -\tfrac{7}{12}; -\tfrac{9}{20}.$

29. $4; 4; 4; 4.$

31. $\tfrac{1}{8}$

33. 0.000003

35. 12

37. 1331

39. $4; 6; 8; 10.$

41. $5, 10, 15; s_n = 5n.$

43. $-5, 25, -125; s_n = (-5)^n.$

*45. $15; 21; 28.$

10.2 SUMS OF FINITE SEQUENCES (*Page 430*)

1. 45

3. 55

5. 0.33333

7. 510

9. 60

11. 254

13. $\frac{381}{64}$

15. 105

17. 40

19. $\frac{13}{8}$

21. 0

23. 57

25. 1111.111

27. 60

29. $\frac{35}{12}$

31. 0.010101

*33. (a) 4, 9, 16, 25, 36; **(b)** n^2.

*35. $\displaystyle\sum_{k=1}^{n} s_k + \sum_{k=1}^{n} t_k =$
$(s_1 + s_2 + \cdots + s_n) + (t_1 + t_2 + \cdots + t_n) =$
$(s_1 + t_1) + (s_2 + t_2) + \cdots + (s_n + t_n) =$
$\displaystyle\sum_{k=1}^{n} (s_k + t_k)$

*37. $\displaystyle\sum_{k=1}^{n} (s_k + c) =$
$(s_1 + c) + (s_2 + c) + \cdots + (s_n + c) =$
$(s_1 + s_2 + \cdots + s_n) + (c + c + \cdots + c) =$
$\left(\displaystyle\sum_{k=1}^{n} s_k\right) + nc$

10.3 ARITHMETIC SEQUENCES AND SERIES (*Page 435*)

1. 5, 7, 9; $2n - 1$; 400.

3. $-10, -16, -22$; $-6n + 8$; $-1100.$

5. $\frac{17}{2}, 9, \frac{19}{2}$; $\frac{1}{2}n + 7$; 245.

7. $-\frac{4}{3}, -\frac{7}{3}, -2$; $-\frac{3}{3}n + 1$; $-106.$

9. 150, 200, 250; $50n$; 10,500.

11. 30, 50, 70; $20n - 30$; 3600.

13. 455

15. $\frac{63}{4}$

17. 15,150.

19. 94,850

21. $\dfrac{173{,}350}{7}$

23. 567,500

25. 5824

27. 15,300

29. (a) 10,000; **(b)** n^2.

31. -36

33. 4620

35. $\frac{3577}{4}$

37. $\frac{5}{2}n(n + 1)$

39. $-\frac{9}{4}$

41. $\frac{228}{15} = \frac{76}{5}$

43. \$1183

*45. 930

*47. $a = -7; d = -13.$

10.4 GEOMETRIC SEQUENCES AND SERIES (*Page 443*)

1. 16, 32, 64; 2^n.

3. 27, 81, 243; 3^{n-1}.

5. $\frac{1}{9}, -\frac{1}{27}, \frac{1}{81}$; $-3(-\frac{1}{3})^{n-1}$.

7. $-125, -625, -3125$; -5^{n-1}.

9. $-\frac{16}{9}, -\frac{32}{27}, -\frac{64}{81}$; $-6(\frac{2}{3})^{n-1}$.

11. 1000, 100,000, 10,000,000; $\frac{1}{1000}(100)^{n-1}$.

13. 126

15. $-\frac{1330}{81}$

17. 1024

19. 3

21. 1023

23. $2^n - 1$

25. $\frac{121}{27}$

27. $\frac{211}{54}$

29. $-\frac{85}{64}$

31. $-\frac{5}{21}$

33. 512,000; $1000(2^{n-1})$.

*35. \$3117.79

*37. (a) $\$800(1.05)^n$; **(b)** \$1020.

10.5 INFINITE GEOMETRIC SERIES (*Page 450*)

1. 4

3. $\frac{125}{4}$

5. $\frac{2}{3}$

7. $\frac{10}{9}$

9. $-\frac{16}{7}$

11. $\frac{3}{2}$

13. $\frac{1}{6}$

15. 4

17. $\frac{20}{9}$

19. No finite sum.

21. $\frac{10}{3}$

23. No finite sum.

25. 30

27. $\frac{20}{11}$

29. 4

31. $\frac{4}{21}$

33. $\frac{7}{9}$

35. $\frac{13}{99}$

37. $\frac{13}{990}$

39. 1

*41. 18 feet

10.6 MATHEMATICAL INDUCTION (*Page 455*)

For these exercises S_n represents the given statement where n is an integer ≥ 1 ($n \geq 2$ when appropriate). The second part of each proof begins with the hypothesis S_k, where k is an arbitrary positive integer.

1. Since $1 = \frac{1(1+1)}{2}$, S_1 is true. Assume S_k and add $k + 1$ to obtain
$$1 + 2 + 3 + \cdots + k + (k+1) =$$
$$\frac{k(k+1)}{2} + (k+1) = \frac{k^2 + 3k + 2}{2} =$$
$$\frac{(k+1)(k+2)}{2} = \frac{(k+1)[(k+1)+1]}{2}$$
Therefore S_{k+1} holds. Since S_1 is true and S_k implies S_{k+1}, the principle of mathematical induction makes S_n true for all integers $n \geq 1$. *Note:* The preceding sentence is an appropriate final statement for the remaining proofs. For the sake of brevity, however, it will not be repeated.

3. Since $\sum_{i=1}^{1} 3i = 3 = \frac{3(1+1)}{2}$, S_1 is true. Assume S_k and add $3(k+1)$ to obtain the following.
$$\sum_{i=1}^{k+1} 3i = \left(\sum_{i=1}^{k} 3i \right) + 3(k+1) =$$
$$\frac{3k(k+1)}{2} + 3(k+1)$$
$$\sum_{i=1}^{k+1} 3i = \frac{3(k^2 + 3k + 2)}{2} =$$
$$\frac{3(k+1)(k+2)}{2} = \frac{3(k+1)[(k+1)+1]}{2}$$
Therefore S_{k+1} holds.

5. Since $\frac{5}{3} = \frac{1(11-1)}{6}$, S_1 is true. Assume S_k and add $-\frac{1}{3}(k+1) + 2$ to obtain
$$\frac{5}{3} + \frac{4}{3} + 1 + \cdots + \left(-\frac{1}{3}k + 2 \right) +$$
$$\left[-\frac{1}{3}(k+1) + 2 \right] =$$
$$\frac{k(11-k)}{6} + \left[-\frac{1}{3}(k+1) + 2 \right] =$$
$$\frac{10 + 9k - k^2}{6} = \frac{(k+1)(10-k)}{6} =$$
$$\frac{(k+1)[11 - (k+1)]}{6}$$
Therefore S_{k+1} holds.

7. Since $\frac{1}{1 \cdot 2} = \frac{1}{1+1}$, S_1 is true. Assume S_k and add $\frac{1}{(k+1)[(k+1)+1]}$ to obtain
$$\frac{1}{1 \cdot 2} + \frac{1}{2 \cdot 3} + \cdots + \frac{1}{k(k+1)} +$$
$$\frac{1}{(k+1)(k+2)} = \frac{k}{k+1} + \frac{1}{(k+1)(k+2)} =$$
$$\frac{k^2 + 2k + 1}{(k+1)(k+2)} = \frac{(k+1)^2}{(k+1)(k+2)} = \frac{k+1}{k+2} =$$
$$\frac{k+1}{(k+1)+1}$$
Therefore S_{k+1} holds.

9. Since $a^1 < 1$, S_1 is true. Assume S_k and multiply by a to obtain $a \cdot a^k < a \cdot 1$ or $a^{k+1} < a$. But $a < 1$; hence $a^{k+1} < 1$ and therefore S_{k+1} holds.

11. S_1 is true since $(ab)^1 = a^1 b^1$. Assume that $(ab)^k = a^k b^k$ and multiply by ab to obtain
$$(ab)^k (ab) = (a^k b^k) ab$$
$$(ab)^{k+1} = (a^k a)(b^k b)$$
$$(ab)^{k+1} = a^{k+1} b^{k+1}$$
Therefore S_{k+1} holds.

*13. S_1 is true since $|a_0 + a_1| \leq |a_0| + |a_1|$. Assume S_k. Then
$$|a_0 + a_1 + \cdots + a_k + a_{k+1}| =$$
$$|(a_0 + a_1 + \cdots + a_k) + a_{k+1}| \leq$$
$$|a_0 + a_1 + \cdots + a_k| + |a_{k+1}| \leq \quad \text{(by } S_1\text{)}$$
$$(|a_0| + |a_1| + \cdots + |a_k|) + |a_{k+1}| = \quad \text{(by } S_k\text{)}$$
$$|a_0| + |a_1| + \cdots + |a_{k+1}|$$
Therefore S_{k+1} holds.

*15. Since $1 = 2\left(1 - \frac{1}{2^1}\right)$, S_1 is true. Assume S_k

and add $\frac{1}{2^k}$ to obtain

$$1 + \frac{1}{2} + \frac{1}{4} + \cdots + \frac{1}{2^{k-1}} + \frac{1}{2^k} =$$
$$2\left(1 - \frac{1}{2^k}\right) + \frac{1}{2^k} = 2\left(1 - \frac{1}{2^k}\right) + \frac{2}{2^{k+1}} =$$
$$2\left(1 - \frac{1}{2^k} + \frac{1}{2^{k+1}}\right) = 2\left(1 + \frac{-2 + 1}{2^{k+1}}\right) =$$
$$2\left(1 - \frac{1}{2^{k+1}}\right)$$
Therefore S_{k+1} holds.

*17. (a) Since $\dfrac{a^2 - b^2}{a - b} = a + b = a^{2-1} + b^{2-1}$, S_2 is true. Assume S_k. Then

$$\frac{a^{k+1} - b^{k+1}}{a - b} = \frac{a^k a - b^k b}{a - b} =$$
$$\frac{a^k a - b^k a + b^k a - b^k b}{a - b} =$$
$$\frac{a(a^k - b^k) + b^k(a - b)}{a - b} = \frac{a(a^k - b^k)}{a - b} + b^k =$$
$$a[a^{k-1} + a^{k-2} b + \cdots + ab^{k-2} + b^{k-1}] + b^k$$
$$\text{(by } S_k\text{)}$$
$$= a^k + a^{k-1} b + \cdots + a^2 b^{k-1} + ab^{k-1} + b^k$$
Therefore S_{k+1} holds.

(b) Since $\dfrac{a^n - b^n}{a - b} = a^{n-1} + a^{n-2} b + \cdots +$ $ab^{n-2} + b^{n-1}$, multiplying by $a - b$ gives

$$a^n - b^n =$$
$$(a - b)(a^{n-1} + a^{n-2} b + \cdots + ab^{n-2} + b^{n-1})$$

10.7 PERMUTATIONS AND COMBINATIONS *(Page 462)*

1. 7

3. 66

5. 120

7. 4

9. 10

11. 1

13. 1

15. 4060

17. (a) 504; **(b)** 729.

19. (a) 4845; **(b)** 116,280.

21. 720

23. 3003

25. 435

27. 2,598,960

29. 720

31. (a) 10; **(b)** 4.

33. $\dbinom{n}{r} = \dfrac{n!}{r!(n-r)!} = \dfrac{n!}{(n-r)! \, r!} = \dbinom{n}{n-r}$

*35. 7

37. (a) 2^2; **(b)** 2^3; **(c)** 2^4; **(d)** 2^5.

*39. $4\dbinom{13}{5} = 5148$

*41. $13\dbinom{4}{2} 12 \dbinom{4}{3} = 3744$

SAMPLE TEST FOR CHAPTER 10   (*Page 465*)

1. (a) $\frac{1}{5}, 1, 3$;   (b) $-\frac{121}{5}$ (note that 6 is not in the domain).

2. (a) 21;   (b) 35.

3. (a) $12, -3, \frac{3}{4}$;   (b) $-768(-\frac{1}{4})^{n-1}$.

4. $\displaystyle\sum_{k=1}^{4} 8\left(\frac{1}{2}\right)^k = \frac{4\left(1 - \frac{1}{2^4}\right)}{1 - \frac{1}{2}} = \frac{15}{2}$

5. 15,554

6. 18

7. No finite sum since $r = \frac{3}{2} > 1$.

8. $\frac{4}{11}$

9. $652.60

10. 36 feet

†11. For $n = 1$, we have $5 = \dfrac{5 \cdot 1(1 + 1)}{2}$. If

$5 + 10 + \cdots + 5k = \dfrac{5k(k + 1)}{2}$, then

$5 + 10 + \cdots + 5k + 5(k + 1) =$

$\dfrac{5k(k + 1)}{2} + 5(k + 1) =$

$\dfrac{5k(k + 1) + 10(k + 1)}{2} = \dfrac{5(k + 1)(k + 2)}{2}$

Thus the statement holds for the $k + 1$ case.

†12. 70

Index

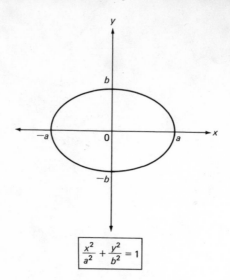

$$\frac{x^2}{a^2} + \frac{y^2}{b^2} = 1$$

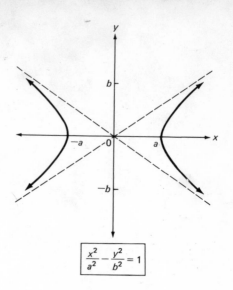

$$\frac{x^2}{a^2} - \frac{y^2}{b^2} = 1$$

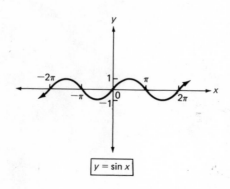

$$y = \sin x$$

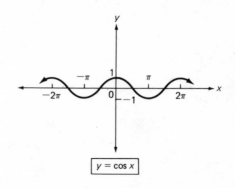

$$y = \cos x$$

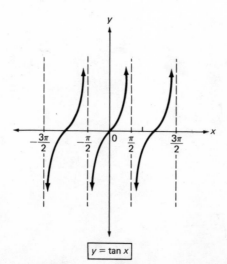

$$y = \tan x$$

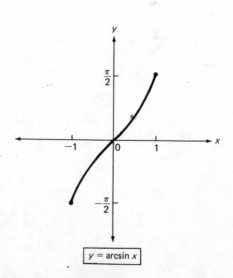

$$y = \arcsin x$$